G Families of Probability Distributions
Theory and Practices

Editors:

Mir Masoom Ali
George and Frances Ball Distinguished Professor Emeritus of Statistics
Ball State University, Muncie, Indiana, USA

Irfan Ali
Department of Statistics & Operations Research
Aligarh Muslim University, Aligarh, India

Haitham M. Yousof
Department of Statistics, Mathematics and Insurance
Faculty of Commerce, Benha University, Egypt

Mohamed Ibrahim
Department of Applied, Mathematical and Actuarial Statistics
Faculty of Commerce, Damietta University, Damietta, Egypt

CRC Press
Taylor & Francis Group
Boca Raton London New York

CRC Press is an imprint of the
Taylor & Francis Group, an **informa** business
A SCIENCE PUBLISHERS BOOK

First edition published 2023
by CRC Press
6000 Broken Sound Parkway NW, Suite 300, Boca Raton, FL 33487-2742

and by CRC Press
4 Park Square, Milton Park, Abingdon, Oxon, OX14 4RN

Library of Congress Cataloging-in-Publication Data (applied for)

ISBN: 978-1-032-14065-0 (hbk)
ISBN: 978-1-032-14068-1 (pbk)
ISBN: 978-1-003-23219-3 (ebk)

DOI: 10.1201/9781003232193

Typeset in Times New Roman
by Radiant Productions

Preface

In many practical fields, including engineering, medicine, and finance, among others, right or left skewness, bi-modality, or multi-modality are characteristics of data sets that can be modelled using statistical distributions. Because of their straightforward shapes and identifiability characteristics, well-known distributions, such as normal, Weibull, gamma, and Lindley, are frequently utilised. However, during the past ten years, much research has concentrated on the more flexible and complicated Generalized or simply G families of continuous distributions to improve their modelling capabilities by including one or more shape parameters.

This book attempts to compile some new results using such distributions that are valuable in theory and application. It is motivated by adding one or more parameters to a distribution function makes it more versatile and more flexible in analysing data. The book also examines the characteristics of a few novel G families and how they might be used for statistical inference. Results are collected that could be added to those already available.

The primary goal of our book is to compile recent advances made by diverse authors in the field of G families of their contributions to these new distributions into an edited book. This book will help present and future scholars studying the G family of probability distributions to generate additional new univariate continuous G families of probability distributions; derive valuable mathematical properties, including entropies, order statistics, quantile spread ordering, ordinary and incomplete moments, moments generating functions, residual life and reversed residual life functions, among others and apply the Farlie Gumbel Morgenstern copula, the modified Farlie Gumbel Morgenstern copula, the Clayton copula, the Renyi entropy copula and the Ali-Mikhail-Haq copula for deriving bivariate and multivariate expansions of the new and existing G families.

This book stands out because it includes a lot of new G families, each with its characteristics and applications to diverse real datasets and simulation studies utilising various estimation methods. In the field of statistical modelling, the book deals with analysing and studying actual data that differ in nature and shape.

Acknowledgement

The editors wish to express their deep gratitude to those who provided input, support, suggestions, and comments and assisted in the editing and proofreading of this book. This book would not have been possible without the valuable scholarly contributions of authors across the globe who contributed chapters to this book and hence the editors sincerely acknowledge the authors' valuable contributions to this book. The editors acknowledge the endless support and motivation from their respective families and friends. The editors are very grateful to members who served in the review process of the book. The editors are very much thankful to the entire CRC Press, Taylor & Francis Group publisher team for keeping faith in the successful completion of the project and showing the right direction to complete this project. Finally, the editors use this opportunity to thank all the readers and expect that this book will continue to inspire and guide them in their future endeavours.

The Editors

Contents

Chapter 1

A New Compound G Family of Distributions

Properties, Copulas, Characterizations, Real Data Applications with Different Methods of Estimation

M Masoom Ali,[1,*] *Nadeem Shafique Butt,*[2] *GG Hamedani,*[3] *Saralees Nadarajah,*[4] *Haitham M Yousof*[5] and *Mohamed Ibrahim*[6]

1. Introduction

The statistical literature contains many new G families of continuous distributions which have been generated either by merging (compounding) common G families of continuous distributions or by adding one or more parameter to the G family. These novel G families have been employed for modeling real-life datasets in many applied studies such as insurance, engineering, econometrics, biology, medicine, statistical forecasting, and environmental sciences. Refer to Gupta et al. (1998) for the exponentiated-G family, Marshall and Olkin (1997) for the Marshall-Olkin-G family, Eugene et al. (2002) for beta generalized-G family, Yousof et al. (2015) for the transmuted exponentiated generalized-G family, Nofal et al. (2017) for the generalized transmuted-G family, Rezaei et al. (2017) for the Topp Leone generated family, Merovci et al. (2017) for the exponentiated transmuted-G family, Brito et al. (2017) for the Topp-Leone odd log-logistic-Gfamily, Yousof et al. (2017a) for Burr type XG family, Aryal and Yousof (2017) for exponentiated generalized-G Poisson family, Hamedani et al. (2017) for type I general exponential class of distributions, Cordeiro et al. (2018) for Burr type XII G family, Korkmaz et al. (2018a) for the exponential Lindley odd log-logistic-G family, Korkmaz et al. (2018b) for the Marshall-Olkin generalized-G Poisson family, Yousof et al. (2018) for Burr-Hatke family of distributions, Hamedani et al. (2018) for the extended G family, Hamedani et al. (2019) for the type II general exponential G family, Nascimento et al. (2019) for the odd Nadarajah-Haghighi family of distributions, Yousof et al. (2020) for the Weibull G Poisson family, Karamikabir et al. (2020) for the Weibull

[1] Department of Mathematical Sciences, Ball State University, Muncie, IN 47306 USA.
[2] Department of Family and Community Medicine, King Abdul Aziz University, Jeddah, Kingdom of Saudi Arabia.
[3] Department of Mathematical and Statistical Sciences, Marquette University, USA.
[4] Department of Mathematics, University of Manchester, Manchester M13 9PL, UK.
[5] Department of Statistics, Mathematics and Insurance, Faculty of Commerce, Benha University, Benha, Egypt.
[6] Department of Applied, Mathematical and Actuarial Statistics, Faculty of Commerce, Damietta University, Damietta, Egypt.
Emails: nshafique@kau.edu.sa; gholamhoss.hamedani@marquette.edu; saralees.nadarajah@manchester.ac.uk; haitham.yousof@fcom.bu.edu.eg; mohamed_ibrahim@du.edu.eg
* Corresponding author: mali@bsu.edu

Topp-Leone generated family, Merovci et al. (2020) for the Poisson Topp Leone G family, Korkmaz et al. (2020) for the Hjorth's IDB generator of distributions, Alizadeh et al. (2020a) for flexible Weibull generated family of distributions, Alizadeh et al. (2020b) for the transmuted odd log-logistic-G family, Altun et al. (2021) for the Gudermannian generated family of distributions and El-Morshedy et al. (2021) for the Poisson generalized exponential G family among others. For more new G families see Hamedani et al. (2021) and Hamedani et al. (2022).

The cumulative distribution function (CDF) and the probability density function (PDF) of the Topp Leone generated G (TLG-G) family (Rezaei et al. (2017)) of distributions are specified by

$$H_{\alpha,\beta,\underline{W}}(x) = \{G_{\underline{W}}(x)^{\beta}\,[2 - G_{\underline{W}}(x)^{\beta}]\}^{\alpha}, \tag{1}$$

and

$$h_{\alpha,\beta,\underline{W}}(x) = 2\alpha\beta g_{\underline{W}}(x)\, G_{\underline{W}}(x)^{\alpha\beta-1}\,[1 - G_{\underline{W}}(x)^{\beta}]\,[2 - G_{\underline{W}}(x)^{\beta}]^{\alpha-1}, \tag{2}$$

respectively, where $\underline{\boldsymbol{W}}$ refers to the parameter vectors of the base-line model. For $\beta = 1$, the TLG-G family reduces to the Topp Leone G (TL-G) family. Suppose $Z_1, Z_2, \ldots, Z_n$ be independent identically distributed random variables with a common CDF of the TLG-G family and N be a random variable with

$$P(N = n; a) = \frac{a^n}{e^a - 1} \times \frac{1}{n!}, n \in \mathbb{N} | a > 0,$$

and define $M_N = \max\{Z_1, Z_2, \ldots, Z_n\}$, then,

$$F(x) = \sum_{n=0}^{\infty} \Pr(M_N \le x \mid N = n) \times \Pr(N = n). \tag{3}$$

Using equations (2) and (3), we can write,

$$F_{\underline{\mathbf{P}}}(x) = \mathcal{C}(a)[1 - exp(-a\{G_{\underline{W}}(x)^{\beta}\,[2 - G_{\underline{W}}(x)^{\beta}]\}^{\alpha})],\ a \in \mathbb{R} - \{0\},\ \alpha > 0, \beta > 0, \tag{4}$$

where $\mathcal{C}(a) = \dfrac{1}{1 - exp(-a)}$ and $\underline{\mathbf{P}} = (a, \alpha, \beta, \underline{\boldsymbol{W}})$. Equation (4) is called the Poisson Topp Leone generated-G (PTLG-G) family of distributions. The new CDF in (4) can be used for presenting a new discrete G family for modeling the count data (refer to Aboraya et al. (2020), Chesneau et al. (2021), Ibrahim et al. (2021) and Yousof et al. (2021) for more details). The corresponding PDFof the PTLG-G family can be expressed as

$$f_{\underline{\mathbf{P}}}(x) = 2a\alpha\beta\, \mathcal{C}(a) \frac{g_{\underline{W}}(x) G_{\underline{W}}(x)^{\alpha\beta-1}[1 - G_{\underline{W}}(x)^{\beta}][2 - G_{\underline{W}}(x)^{\beta}]^{\alpha-1}}{exp(a\{G_{\underline{W}}(x)^{\beta}[2 - G_{\underline{W}}(x)^{\beta}]\}^{\alpha})}, a \in \mathbb{R} - \{0\}, \alpha > 0, \beta > 0. \tag{5}$$

For $\beta = 1$, the PTLG-G family reduces to Poisson Topp Leone G (PTL-G) family (Merovci et al. (2020)). For $\alpha = 1$, the PTLG-G family reduces to quasi–Poisson Topp Leone generated-G (QPTLG-G) family. For $\beta = \alpha = 1$, the PTLG-G family reduces to quasi–Poisson Topp Leone G (QPTL-G) family.

In this paper, after studying the main statistical properties and presenting some bivariate type extensions, we briefly considered and then described different estimation methods, namely, the maximum likelihood estimation (MLE) method, Cramér-von-Mises estimation (CVM) method, ordinary least square estimation (OLS) method, weighted least square estimation (WLSE) method, Anderson Darling estimation (ADE) method, right tail Anderson Darling estimation (RTADE) method, left tail Anderson Darling estimation (LTADE) method. These methods are used in the estimation process of the unknown parameters. Monte Carlo simulation experiments are performed to compare the performances of the proposed estimation methods for both small and large samples. The new PTLG-G family may be useful in modelling:

I. The real-life datasets with "monotonically increasing hazard rate" as illustrated in Section 6 (Figures 2 and 3 (top left plots)).

II. The real-life datasets do not have extreme values, as shown in Section 6 (Figures 2 and 3 (bottom right plots) and (bottom left plots)).

III. The real-life datasets for which nonparametric Kernel density estimations are left-skewed bimodal and right-skewed bimodal are as given in Section 6 (Figures 2 and 3 (top right plots)).

The PTLG-G family proved adequately superior to many other well-known G families, as illustrated below:

I. In modelling the failure times of aircraft windshield items, the PTLG-G family is better than the odd log-logistic-G family, the generalized mixture-G family, the transmuted Topp-Leone-G family, the Gamma-G family, the Burr-Hatke-G family, the McDonald-G family, the exponentiated-G family, the Kumaraswamy-G family, and the proportional reversed hazard rate-G family under the consistent-information criteria, Akaike information criteria, Hannan-Quinn information criteria and Bayesian information criteria.

II. In modelling the service of aircraft windshield items, the PTLG-G family is better than the odd log-logistic-G family, the generalized mixture-G family, the transmuted Topp-Leone-G family, the Gamma-G family, the Burr-Hatke-G family, the McDonald-G family, the exponentiated-G family, the Kumaraswamy-G family, and the proportional reversed hazard rate-G family under the consistent-information criteria, Akaike information criteria, Hannan-Quinn information criteria and Bayesian information criteria.

2. Copula

For modelling of the bivariate real data sets, we shall derive some new bivariate PTLG-G (Bv-PTLG-G) type distributions using "Farlie-Gumbel-Morgenstern copula" (FGMC) copula (see Morgenstern (1956), Farlie (1960), Gumbel (1960) and Gumbel (1961)), Johnson and Kotz (1975 and 1977)), modified FGMC (see Balakrishnan and Lai (2009)), "Clayton copula" (see Nelsen (2007)), "Renyi's entropy copula (REC) (Pougaza and Djafari (2010))" and "Ali-Mikhail-Haq copula (AMHC)" (see Ali et al. (1978)). The multivariate PTLG-G (Mv-PTLG-G) type can be easily derived based on the Clayton copula. However, future works may be allocated to study these new models (see also Shehata and Yousof (2021a,b) and Shehata et al. (2021)).

2.1 BvPTLG-G type via Clayton copula

Let $X_1 \sim \text{PTLG}-\text{G}(\underline{\mathbf{P}}_1)$ and $X_2 \sim \text{PTLG}-\text{G}(\underline{\mathbf{P}}_2)$. Depending on the continuous marginals $\bar{u} = 1 - u$ and $\bar{m} = 1 - m$, the Clayton copula can be considered as

$$C_{\text{K}}(\bar{u}, \bar{m}) = \left[\max\left(\bar{u}^{-\text{K}} + \bar{m}^{-\text{K}} - 1\right); 0\right]^{-\frac{1}{\text{K}}}, \text{K} \in [-1, \infty) - \{0\}, \bar{u} \in (0,1) \text{ and } \bar{m} \in (0,1)$$

Let $\bar{u} = 1 - F_{\underline{\mathbf{P}}_1}(x_1)\,|_{\underline{\mathbf{P}}_1}$, $\bar{m} = 1 - F_{\underline{\mathbf{P}}_2}(x_2)\,|_{\underline{\mathbf{P}}_2}$ and

$$F_{\underline{\mathbf{P}}_i}(x_i)\,|_{i=1,2} = \frac{1}{1 - e^{-a_i}}\left\{1 - exp\left(-a_i\{G_{\underline{W}}(x)^{\beta_i}[2 - G_{\underline{W}}(x)^{\beta_i}]\}^{\alpha_i}\right)\right\}.$$

Then, the BvPTLG-G type distribution can be obtained from $C_{\text{K}}(\bar{u}, \bar{m})$. A straightforward multivariate extension via Clayton copula can be derived.

2.2 BvPTLG-G type via REC

The REC can be derived using the continuous marginal functions $u = 1 - \bar{u} = F_{\underline{\mathbf{P}}_1}(x_1) \in (0,1)$ and $m = 1 - \bar{m} = F_{\underline{\mathbf{P}}_1}(x_2) \in (0,1)$ as follows,

$$F(x_1, x_2) = C(F_{\underline{\mathbf{P}}_1}(x_1), F_{\underline{\mathbf{P}}_1}(x_2)) = x_2 u + x_1 m - x_1 x_2.$$

2.3 BvPTLG-G type via FGMC

Considering the FGMC, the joint CDF can be written as $C_K(u, m) = um + umK\overline{um}$, where the continuous marginal functions are $u \in (0,1)$, $m \in (0,1)$ and $K \in [-1,1]$. Setting $\overline{u} = \overline{u}_{\underline{P}_1}|_{\underline{P}_1>0}$, and $\overline{m} = \overline{m}_{\underline{P}_2}|_{\underline{P}_2>0}$, we then have $F(x_1, x_2) = um(1 + K\overline{um})$. Then, the joint PDF can be expressed as $C_K(u, m) = 1 + Ku^*m^*$, where $u^* = 1 - 2u$ and $m^* = 1 - 2m$ or $f_K(x_1, x_2) = f_{\underline{P}_1}(x_1) f_{\underline{P}_2}(x_2) c(F_{\underline{P}_1}(x_1)\, F_{\underline{P}_2}(x_2))$, where the two functions $f_K(x_1, x_2)$ and $c_K(u, m)$ are PDFs corresponding to the joint CDFs $F_K(x_1, x_2)$ and $c_K(u, m)$.

2.4 BvPTLG-G type via modified FGMC

The modified formula of the modified FGMC can be expressed as $c_K(u, m) = K\mathbf{O}(u)^*\, \boldsymbol{T}(m)^* + um$, with $\mathbf{O}(u)^* = u\mathbf{O}(u)$ and $\boldsymbol{T}(m)^* = m\boldsymbol{T}(m)$ where $\mathbf{O}(u) \in (0,1)$ and $\boldsymbol{T}(m) \in (0,1)$ are two continuous functions where $\mathbf{O}(u = 0) = \mathbf{O}(u = 1) = \boldsymbol{T}(m = 0) = \boldsymbol{T}(m = 1) = 0$. The following four types can be derived and considered:

Type I: The new bivariate version via modified FGMC Type I can be written as

$$C_K(u, m) = K\mathbf{O}(u)^*\, \boldsymbol{T}(m)^* + um.$$

Type II: Consider $\mathbf{A}(u; K_1)$ and $\mathbf{B}(m; K_2)$ which satisfy the above conditions where

$$\mathbf{A}(u; K_1)|_{(K_1>0)} = u^{K_1}\, (1 - u)^{1-K_1}$$

and

$$\mathbf{B}(m; K_2)|_{(K_2>0)} = m^{K_2}\, (1 - m)^{1-K_2},$$

Then, the corresponding bivariate version (modified FGMC Type II) can be derived as

$$C_{q,K_1,K_2}\,(u, m) = um + qum\mathbf{A}(u; K_1)\mathbf{B}(m; K_2).$$

Type III: Let $\widetilde{\mathbf{A}(u)} = u[log(1 + \overline{u})]|_{(\overline{u}=1-u)}$ and $\widetilde{\mathbf{B}(m)} = m[log(1 + \overline{m})]|_{(\overline{m}=1-m)}$. Then, the associated CDF of the BvPTLG-G-FGM (modified FGMC type III) is

$$C_K(u, m) = um + umK\widetilde{\mathbf{A}(u)}\,\widetilde{\mathbf{B}(m)}\,.$$

Type IV: Using the quantile concept, the CDF of the BvPTLG-G-FGM (modified FGMC Type IV) model can be obtained as

$$C(u, m) = uF^{-1}(u) + mF^{-1}\,(m) - F^{-1}(u)\, F^{-1}(m)$$

where $F^{-1}(u) = Q(u)$ and $F^{-1}(m) = Q(m)$.

2.5 BvPTLG-G type via AMHC

Under the "stronger Lipschitz condition", the joint CDF of the Archimedean Ali-Mikhail-Haq copula can be written as $C_K(v,m) = \dfrac{1}{1-K\overline{vm}}vm\,|_{K\in(-1,1)}$ the corresponding joint PDF of the Archimedean Ali-Mikhail-Haq copula can be expressed as $C_K(v,m) = \dfrac{1}{[1-K\overline{vm}]^2}\left(1-K+2K\dfrac{vm}{1-K\overline{vm}}\right)|_{K\in(-1,1)}$, and then for any $\overline{v} = 1 - F_{\underline{P}_1}(x_1) = |_{[\overline{v}=(1-v)\in(0,1)]}$ and $\overline{v} = 1 - F_{\underline{P}_2}(x_2) = |_{[\overline{m}=(1-m)\in(0,1)]}$ we have

$$C_K(x_1, x_2) = \frac{1}{1-K[1-F_{\underline{P}_1}(x_1)][1-F_{\underline{P}2}(x2)]}[F_{\underline{P}_1}(x_1)F_{\underline{P}2}(x2)]\,|_{K\in(-1,1)},$$

and

$$c_{K}(x_1,x_2)=\frac{1}{\{1-K[1-F_{\underline{P}_1}(x_1)][1-F_{\underline{P}2}(x_2)]\}^2}$$
$$\times\left(1-K+2K\left\{\frac{F_{\underline{P}_1}(x_1)F_{\underline{P}2}(x_2)}{1-K[1-F_{\underline{P}_1}(x_1)][1-F_{\underline{P}2}(x_2)]}\right\}\right)|_{K\in(-1,1)}.$$

3. Mathematical properties

3.1 Linear representation

First, expanding the quantity $A_{\underline{P}}(x)$ where,

$$A_{\underline{P}}(x)=exp\left(-a\{G_{\underline{W}}(x)^{\beta}[2-G_{\underline{W}}(x)^{\beta}]\}^{\alpha}\right),$$

which then leads to,

$$A_{\underline{P}}(x)=\sum_{l=0}^{+\infty}\frac{(-a)^l}{l!}\{G_{\underline{W}}(x)^{\beta}[2-G_{\underline{W}}(x)^{\beta}]\}^{\alpha l}.$$

Substituting the expansion of $A_{a,\alpha,\beta,\underline{W}}(x)$ in (5), we have,

$$f_{\underline{P}}(x)=\sum_{l=0}^{+\infty}\frac{a^{l+1}\alpha 2^{\alpha(l+1)}(-1)^l\,\mathcal{C}(a)}{l!}\left[1-\frac{1}{2}G_{\underline{W}}(x)^{\beta}\right]^{\alpha(l+1)-1}g_{\underline{W}}(x)G_{\underline{W}}(x)^{\beta(\alpha+l)-1}[1-G_{\underline{W}}(x)^{\beta}]. \tag{6}$$

Consider the power series,

$$\left(1-\frac{\xi_1}{\xi_2}\right)^{\xi_3}=\sum_{d=0}^{\infty}(-1)^d\binom{\xi_3}{d}\left(\frac{\xi_1}{\xi_2}\right)^d, \tag{7}$$

which holds for $\left|\frac{\xi_1}{\xi_2}\right|<1$ and $\xi_3>0$ real non-integer. Using (7), the PTLG-G class in (6) can be written as,

$$f_{\underline{P}}(x)=\sum_{l,d=0}^{+\infty}\frac{\alpha^{l+1}\alpha 2^{\alpha(l+1)-d}(-1)^{l+d}}{l!/\,\mathcal{C}(a)}\binom{\alpha(l+1)-1}{d}\begin{bmatrix}g_{\underline{w}}(x)G_{\underline{w}}(x)^{\beta(\alpha+\alpha l+d)-1}\\-g_{\underline{w}}(x)G_{\underline{w}}(x)^{\beta(\alpha+\alpha l+d+1)-1}\end{bmatrix},$$

which can be summarized as,

$$f_{\underline{P}}(x)=\sum_{l,d=0}^{+\infty}\left\{C_{l,d}\pi_{\beta(\alpha+\alpha l+d)}(x;\underline{W})-C^*_{l,d}\pi_{\beta(\alpha+\alpha l+d+1)}(x;\underline{W})\right\}, \tag{8}$$

where,

$$C_{l,d}=\frac{1}{\beta(\alpha+\alpha l+d)}h_{l,d},\;C^*_{l,d}=\frac{1}{\beta(\alpha+\alpha l+d+1)}h_{l,d},$$
$$h_{l,d}=\frac{\alpha a^{l+1}(-1)^{l+d}2^{\alpha(l+1)-d}}{l!/\,\mathcal{C}(a)}\binom{\alpha(l+1)-1}{d}$$

and $\pi_{\Delta}(x)=\Delta g(x)G(x)^{\Delta-1}$. Equation (8) reveals that the density of the PTLG-G family can then be expressed as a linear representation of exp-G PDFs. Also, the CDF of the PTLG-G family can be expressed as a mixture of exp-G CDFs. By integrating (8), we get,

$$F_{\underline{P}}(x)=\sum_{l,d=0}^{-\infty}\{C_{l,d}\prod\nolimits_{\beta(\alpha+\alpha l+d)}(x;\underline{W})-C^*_{l,d}\prod\nolimits_{\beta(\alpha+\alpha l+d+1)}(x;\underline{W})\},$$

where $\mathbf{\Pi}_{\Delta}(x)$ is the CDF of the exp-G family with power parameter $\mathbf{\Delta}$.

3.2 *Moments*

The rth ordinary moment of X where X follows PTLG-G family with parameters (α, α, β, W) is given by $\mu'_r = \mathbb{E}(X^r) = \int_{-\infty}^{\infty} x^r f_{\underline{P}}(x)dx$. Then we obtain,

$$\mu'_{r,X} = \sum_{l,d=0}^{-\infty} \{C_{l,d}\mathbb{E}(Y^r_{\beta(\alpha+\alpha l+d)}) - C^*_{l,d}\mathbb{E}(Y^r_{\beta(\alpha+\alpha l+d)})\}. \tag{10}$$

where Y_ξ has a density of the exp-G model with power parameter ξ. The expected value $\mathbb{E}(X)$ can be derived from (10) when $r = 1$. The integrations in $\mathbb{E}(Y^r_{\beta(\alpha+\alpha l+d)})$ and $(Y^r_{\beta(\alpha+\alpha l+d+1)})$ can be performed numerically for most parent distributions (see Table 2). The nth central moment of X, variance ($V(X)$), skewness ($S(X)$), kurtosis (K(X)) and dispersion index (DI(X)) measures can be derived using well-known relationships. The sth incomplete moment, say $\mathbf{I}_{s,X}(t)$, of X can be expressed from (9) as $\mathbf{I}_{s,X}(t) = \int_{-\infty}^{t} x^s f(x)dx$. Then,

$$\mathbf{I}_{s,X}(t) = \sum_{l,d=0}^{+\infty} \begin{bmatrix} C_{l,d}\int_{-\infty}^{t} x^s \pi_{\beta(\alpha+\alpha l+d)}(x;\underline{W})dx \\ C^*_{l,d}\int_{-\infty}^{t} x^s \pi_{\beta(\alpha+\alpha l+d+1)}(x;\underline{W})dx \end{bmatrix}. \tag{11}$$

The mean deviation about the mean and mean deviation about the median of X are given by $A_{1,X} = (|X - \mu'_1|) = 2\mu'_1\ F(\mu'_1) - 2\mathbf{I}_1\ (\mu'_{1,X})$ and $A_{2,X} = \mathbb{E}(|X - M|) = \mu'_1 - 2I_{1,X}\ (M)$, respectively, where $\mu'_{1,X} = E(X)$, $M = Median\ (X) = Q\left(\frac{1}{2}\right)$ is the median, and $F(\mu'_{1,X})$ is obtained from (4) and $\mathbf{I}_{1,X}(t)$ is the first incomplete moment given by (11) with $s = 1$ as

$$\mathbf{I}_{1,X}(t) = \sum_{l,d=0}^{+\infty} \left\{C_{l,d}\mathbf{J}_{\beta(\alpha+\alpha l+d)}(x;\underline{W})dx - C^*_{l,d}\mathbf{J}_{\beta(\alpha+\alpha l+d+1)}(x;\underline{W})dx\right\},$$

where $\mathbf{J}_\Delta(x) = \int_{-\infty}^{t} x\pi_\Delta(x)dx$ is the first incomplete moment of the exp-G distribution. The moment generating function $M_X(t) = \mathbb{E}(e^{tX})$ of X can be derived as

$$M_X(t) = \sum_{l,d=0}^{+\infty} \left\{C_{l,d}M_{\beta(\alpha+\alpha l+d)}(t;\underline{W}) - C^*_{l,d}M_{\beta(\alpha+\alpha l+d+1)}(t;\underline{W})\right\},$$

where $M_\Delta(t)$ is the moment generating function of Y_Δ.

3.3 *Moment of the residual life*

The nth moment of the residual life is given by $V_{n,X}(t) = \mathbb{E}[(X-t)^n \mid X > t, n \in \mathbb{N}]$. Then, the nth moment of the residual life of X can be given as $V_{n,X}(t) = \frac{1}{1 - F_{\underline{P}}(t)} \int_t^{\infty} (x-t)^n f_{\underline{P}}(t)dx$. Therefore, using (8) we have,

$$V_{n,X}(t) = \frac{1}{1 - F_{\underline{P}}(t)} \sum_{r=0}^{n} \binom{n}{r} (-t)^{n-r} \sum_{l,d=0}^{+\infty} \begin{bmatrix} C_{l,d}\int_t^{\infty} x^r \pi_{\beta(\alpha+\alpha l+d)}(x;\underline{W})dx \\ C^*_{l,d}\int_t^{\infty} x^r \pi_{\beta(\alpha+\alpha l+d+1)}(x;\underline{W})dx \end{bmatrix}.$$

The life expectation can then be defined by $V_{1,X}(t) = \mathbb{E}[(X-t)|X > t, n = 1]$ which represents the expected additional life length for a unit that is alive at age t. The MRL of X can be obtained by setting $n = 1$ in the last equation.

3.4 Moment of the reversed residual life

The nth moment of the reversed residual life is given by $\boldsymbol{U}_{n,X}(t) = \mathbb{E}\ [(t - X)^n \mid X \leq t, t > 0, n \in \mathbb{N}]$. The nth moment of the reversed residual life of X can be given as $\boldsymbol{U}_{n,X}(t) = \frac{1}{F_{\underline{\mathbf{P}}}(t)} \int_0^t (t-x)^n f_{a,\alpha,\beta,\underline{\boldsymbol{W}}}(t)dx$. Then, the nth moment of the reversed residual life of X becomes,

$$\boldsymbol{U}_{n,X}(t) = \frac{1}{F_{\underline{\mathbf{P}}}(t)} \sum_{r=0}^{n} (-1)^r \binom{n}{r} t^{n-r} \sum_{l,d=0}^{+\infty} \begin{bmatrix} C_{l,d} \int_0^t x^r \pi_{\beta(\alpha+\alpha l+d)}(x;\underline{\boldsymbol{W}})dx \\ -C^*_{l,d} \int_0^t x^r \pi_{\beta(\alpha+\alpha l+d+1)}(x;\underline{\boldsymbol{W}})dx \end{bmatrix}.$$

The mean inactivity time (MIT) is given by $\boldsymbol{U}_{1,X}(t) = \mathbb{E}[(t - X) \mid X \leq t, t > 0, n = 1]$ and it refers to the waiting time elapsed since the failure of an item on the condition that this failure had occurred in $(0,t)$.

3.5 Probability weighted moments

The $(s, r)^{th}$ PWM of X following the PTLG-G family, says $\mathbb{R}_{s,r}$, is formally defined by $\mathbb{R}_{s,r} = E\{X^s F(X)^r\}$. Using equations (4) and (5), we can write

$$f_{\underline{\mathbf{P}}}(x)F_{\underline{\mathbf{P}}}(x)^r = \sum_{l,d=0}^{+\infty} \begin{Bmatrix} \mathrm{K}_{l,d}(r)\pi_{\beta(\alpha+\alpha l+d)}(x;\underline{\boldsymbol{W}}) \\ -\mathrm{K}^*_{l,d}(r)\pi_{\beta(\alpha+\alpha l+d+1)}(x;\underline{\boldsymbol{W}}) \end{Bmatrix}$$

where,

$$\mathrm{K}_{l,d}(r) = \frac{1}{\beta(\alpha+\alpha l+d)} \mathcal{P}^{(r)}_{l,d},\ K^*_{l,d}(r) \frac{1}{\beta(\alpha+\alpha l+d+1)} \mathcal{P}^{(r)}_{l,d},$$

$$\mathcal{P}^{(r)}_{l,d} = \sum_{p=0}^{+\infty} \frac{\alpha a^{1+l}(1+p)^l(-1)^{p+l+d}2^{\alpha(l+1)-d}}{l!} \mathcal{C}(a) \binom{r}{p} \binom{\alpha(l+1)-1}{d},$$

and $\pi_{\Delta}(x)$ is defined above. Then, the $(s,r)^{th}$ PWM of X can be expressed as,

$$\mathbb{R}_{s,r} = \sum_{l,d=0}^{+\infty} \{K_{l,d}(r)\mathbb{E}(Y^s_{\beta(\alpha+\alpha l+d)}) - K^*_{l,d}(r)\mathbb{E}(Y^s_{\beta(\alpha+\alpha l+d+1)})\}dx.$$

3.6 Order statistics

Let $X_1,\ldots, X_n$ be a random sample from the PTLG-G family of distributions and let $X_{1:n}, X_{2:n},\ldots, X_{n:n}$ be the corresponding order statistics. The PDF of the ith order statistic, say $X_{i:n}$, can be written as

$$f_{\mathrm{i}:n}(x) = \frac{1}{B(\mathrm{i}, n-\mathrm{i}+1)} f(x) \sum_{j=0}^{n-\mathrm{i}} (-1)^j \binom{n-\mathrm{i}}{j} F^{j+\mathrm{i}-1}(x), \tag{12}$$

where, $B(\cdot,\cdot)$ is the beta function. Substituting (4) and (5) in equation (12) and using a power series expansion, we get,

$$f_{\underline{\mathbf{P}}}(x)F_{\underline{\mathbf{P}}}(x)^{j+\mathrm{i}-1} = \sum_{l,d=0}^{+\infty} \begin{Bmatrix} \mathrm{K}_{l,d}(j+\mathrm{i}-1)\pi_{\beta(\alpha+\alpha l+d)}(x;\underline{\boldsymbol{W}}) \\ -\mathrm{K}^*_{l,d}(j+\mathrm{i}-1)\pi_{\beta(\alpha+\alpha l+d+1)}(x;\underline{\boldsymbol{W}}) \end{Bmatrix},$$

where,

$$\mathrm{K}_{l,d}(j+\mathrm{i}-1) = \frac{1}{\beta(\alpha+\alpha l+d)} q^{(j+\mathrm{i}-1)}_{l,d},\ \mathrm{K}^*_{l,d}(j+\mathrm{i}-1) = \frac{1}{\beta(\alpha+\alpha l+d+1)} q^{(j+\mathrm{i}-1)}_{l,d},$$

$$q^{(r)}_{l,d} = \sum_{p=0}^{+\infty} \frac{\alpha a^{1+l}(1+p)^l(-1)^{p+l+d}2^{\alpha(l+1)-d}}{l!} \mathcal{C}(a) \binom{r}{p} \binom{\alpha(l+1)-1}{d},$$

Then, the PDF of $X_{i:n}$ can be written as,

$$f_{i:n}(x)=\sum_{j=0}^{n-i}\frac{(-1)^j}{B(i,n-i+1)}\binom{n-i}{j}\sum_{l,d=0}^{+\infty}\left\{\begin{array}{l}K_{l,d}(j+i-1)\pi_{\beta(\alpha+\alpha l+d)}(x;\underline{W})\\-K^*_{l,d}(j+i-1)\pi_{\beta(\alpha+\alpha l+d+1)}(x;\underline{W})\end{array}\right\}.$$

Then, the density function of the PTLG-G order statistics is a mixture of exp-G PDFs. Based on the last result, we note that the main properties of $X_{i:n}$ follow from those properties of $Y_{\beta(\alpha+l+d)}$ and $Y_{\beta(\alpha+l+d)+1}$ which are the exp-G PDFs with bower parameters $\beta(\alpha+l+d)$ and $\beta(\alpha+l+d)+1$ respectively. For example, the sth moment of $X_{i:n}$ can be expressed as,

$$\mathbb{E}(X^s_{i:n})=\sum_{j=0}^{n-i}\frac{(-1)^j}{B(i,n-i+1)}\binom{n-i}{j}\sum_{l,d=0}^{+\infty}\left\{\begin{array}{l}K_{l,d}(j+i-1)\mathbb{E}\left(Y^s_{\beta(\alpha+\alpha l+d)}\right)\\-K^*_{l,d}(j+i-1)\mathbb{E}\left(Y^s_{\beta(\alpha+\alpha l+d)}\right)\end{array}\right\}. \quad (13)$$

4. Characterizations

In this section, we present certain characterizations of the PTLG-G distribution in the following cases: (i) based on two truncated moments, (ii) in terms of the reverse hazard function. We present our characterizations (i) and (ii) in two subsections.

4.1 Characterizations based on two truncated moments

This subsection deals with the characterizations of PTLG-G distribution in terms of a simple relationship between two truncated moments. For the first characterization, we use a theorem of Glänzel (1987) see Theorem 4.1.1 below as stated in Glänzel (1987).

Theorem 4.1.1. Let (Ω, F, P) be a given probability space and let $H=[d,e]$ be an interval for some $d<e$ ($d=-\infty$, $e=\infty$ might as well be allowed). Let $X:\Omega\rightarrow H$ be a continuous random variable with the cumulative distribution function $F_{\underline{P}}$ and let Q_1 and Q_2 be two real functions defined on H such that,

$$E[Q_1(X)\mid X\geq x]=E[Q_2(X)\mid X\geq x]\,\xi(x),\; x\in H,$$

is defined with some real function ξ. Assume that $Q^1(X)$, $Q_2(X)\in C^1(H)$, $C\in C^2(H)$ and $F_{\underline{P}}$ is a twice continuously differentiable and strictly monotone function on the set H. Finally, assume that the equation $\xi(X)[Q_1(X)]=Q_2(X)$ has no real solution in the interior of H. Then $F_{\underline{P}}$ is uniquely determined by the functions $Q_1(X)$, $Q_2(X)$ and ξ, particularly,

$$F_{\underline{P}}(x)=\int_a^x C\left|\frac{\xi'(u)}{\xi(u)Q_1(u)-Q_2(u)}\right|exp(-s(u))du,$$

where the function $s(.)$ is a solution of the differential equation $s'(.)=\dfrac{\xi' Q_1(.)}{\xi(.)[Q_1(.)]-Q_2(.)}$ and C is the normalization constant, such that $\int_H dF_{\underline{P}}(x)=1$.

Remark 4.1.1. The goal in the above theorem is to have $\xi(x)$ as simple as possible.

Proposition 4.1.1. Let $X:\Omega\rightarrow\mathbb{R}$ be a continuous random variable and let,

$$Q_2(x)=\frac{exp(a\{G_{\underline{W}}(x)^\beta[2-G_{\underline{W}}(x)^\beta]\}^\alpha)}{[1-G_{\underline{W}}(x)^\beta][2-G_{\underline{W}}(x)^\beta]^{\alpha-1}},$$

and $Q_2(x)=Q_1(x)\,G_{\underline{W}}(x)^{\alpha\beta}$ for $x\in\mathbb{R}$. The random variable X has PDF (5) if and only if the function ξ defined in Theorem 4.1.1 has the form,

Proof. Let X be a random variable with PDF (5), then

$$\left(1 - F_{\underline{P}}(x)\right)E[Q_1(X) \mid X \geq x] = 2a[1 - G_{\underline{W}}(x)^{\alpha\beta}],\ x \in \mathbb{R},$$

and

$$\left(1 - F_{\underline{P}}(x)\right)E[Q_2(X) \mid X \geq x] = a[1 - G_{\underline{W}}(x)^{\alpha\beta}],\ x \in \mathbb{R},$$

and finally

$$\xi(x)\, Q_1(x) - Q_2(x) = \frac{1}{2} Q_1(x)[1 - G_{\underline{W}}(x)^{\alpha\beta}] > 0,\ x \in \mathbb{R}.$$

Conversely, if ξ is given as above, then,

$$s'(x) = \frac{\xi'(x)Q_1(x)}{\xi(x)Q_1(x) - Q_2(x)} = \frac{\alpha\beta g_{\underline{W}}(x)G_{\underline{W}}(x)^{\alpha\beta-1}}{1 - G_{\underline{W}}(x)^{\alpha\beta}},\ x \in \mathbb{R},$$

and hence $s(x) = -\log[1 - G_{\underline{W}}(x)^{\alpha\beta}],\ x \in \mathbb{R}$. Now, according to Theorem 4.1.1, X has PDF (5).

Corollary 4.1.1. Let $X : \Omega \rightarrow \mathbb{R}$ be a continuous random variable and let $Q_1(x)$ be as in Proposition 4.1.1. Then X has PDF (5) if and only if there exist functions $Q_2(x)$ and ξ defined in Theorem 4.1.1 satisfying the first-order differential equation,

$$\frac{\xi'(x)Q_1(x)}{\xi(x)Q_1(x) - Q_2(x)} = \frac{\alpha\beta g_{\underline{W}}(x)G_{\underline{W}}(x)^{\alpha\beta-1}}{1 - G_{\underline{W}}(x)^{\alpha\beta}},\ x \in \mathbb{R}.$$

Corollary 4.1.2. The general solution of the above differential equation is,

where is a constant. A set of functions satisfying the above differential equation is given in Proposition 4.1.1 with It should, however, be mentioned that there are other triplets satisfying the conditions of Theorem 4.1.1.

4.2 Characterization based on reverse hazard function

The reverse hazard function, $r_{F_{\underline{P}}}$, of a twice differentiable distribution function, $F_{\underline{P}}$, is defined as,

$$rF_{\underline{P}}(x) = \frac{f(x)}{F(x)},\ x \in \text{support of } F_{\underline{P}}.$$

This subsection is devoted to a characterization of the PTLG-G distribution in terms of the reverse hazard function.

Proposition 4.2.1. Let $X : \Omega \rightarrow \mathbb{R}$ be a continuous random variable. Then X has PDF (5) if and only if its reverse hazard function $r_{F_{\underline{P}}}(x)$ satisfies the following first-order differential equation,

$$r'_{F_{\underline{P}}}(x) - \frac{(\alpha\beta - 1)g_{\underline{W}}(x)}{G_{\underline{W}}(x)} r_{F_{\underline{P}}}(x)$$

$$= 2a\alpha\beta G_{\underline{W}}(x)^{\alpha\beta-1} \frac{d}{dx}\left\{\frac{g_{\underline{W}}(x)[1 - G_{\underline{W}}(x)^{\beta}][2 - G_{\underline{W}}(x)^{\beta}]^{\alpha-1}}{exp\left\{a\{G_{\underline{W}}(x)^{\beta}[2 - G_{\underline{W}}(x)^{\beta}]\}\right\} - 1}\right\},\ x \in \mathbb{R},$$

with boundary condition $lim_{x\rightarrow\infty} r_{F_{\underline{P}}}(x) = 0$.

Proof. If X has PDF (5), then clearly the above differential equation holds. Now, if the differential equation holds, then,

$$\frac{d}{dx}\left\{r_{F_{\underline{P}}}(x)G_{\underline{W}}(x)^{1-\alpha\beta}\right\} = 2a\alpha\beta \frac{d}{dx}\left[\frac{g_{\underline{W}}(x)[1 - G_{\underline{W}}(x)^{\beta}][2 - G_{\underline{W}}(x)^{\beta}]^{\alpha-1}}{exp\left(a\{G_{\underline{W}}(x)^{\beta}[2 - G_{\underline{W}}(x)^{\beta}]\}^{\alpha}\right) - 1}\right],$$

or

$$r_{F_{\underline{P}}}(x)=2a\alpha\beta\frac{g_{\underline{W}}(x)G_{\underline{W}}(x)^{\alpha\beta-1}[1-G_{\underline{W}}(x)^{\beta}][2-G_{\underline{W}}(x)^{\beta}]^{\alpha-1}}{exp\left(a\{G_{\underline{W}}(x)^{\beta}[2-G_{\underline{W}}(x)^{\beta}]\}^{\alpha}\right)-1},x\in\mathbb{R},$$

which is the hazard function of the PTLG-G distribution.

5. Studying a special model

In this section, a new special PTLG-G model based on the Lomax distribution called the PTLG-G Lomax (PTLGL) distribution is considered. The following contributions will be considered: Some plots for the PDFs of the PTLGL distribution for some selected parameter values and some plots for the HRFs of the PTLGL distribution for some selected parameter values are sketched (see Figure 1). Two theorems related to the ordinary and incomplete moments of the exponentiated Lomax (exp-L) distribution are presented. Theorem 5.1 and Theorem 5.2 are employed for deriving relevant mathematical properties of the PTLGL distribution (see Table 1). Numerical results for the variance, mean, kurtosis, skewness, and the PTLGL distribution are listed in Table 2. Figure 1 gives some PDF and HRF plots of the PTLGL model with $c = 1$. Based on Figure 1 (right plot), the PDF of the PTLGL can be " asymmetric right-skewed" and "symmetric" with many useful shapes.Based on Figure 1 (left plot), the HRF of the PTLGL can be "constant","decreasing", "upside-down", "increasing" and " increasing-constant".

Below, we present two theorems related to the exp-L distribution. The two theorems are employed in deriving the mathematical properties in Table 1.

Theorem 5.1. Let $Y_{\Delta+1}$ be a random variable having the exponentiated Lomax (exp-L) distribution with power parameter $\mathbf{\Delta}+1$. Then, the CDF of the exp-L model can be expressed as,

$$G_{\Delta+1,\lambda,c}(x)=\left[1-\left(\frac{1}{c}x+1\right)^{-\lambda}\right]^{\Delta+1}$$

Then, the rth ordinary moment of $Y_{\Delta+1}$ is given by,

$$\mathbb{E}(X^{r})=\sum_{m=0}^{r}(\Delta+1)c^{r}(-1)^{m}\binom{r}{m}B\left(\Delta+1,\frac{m-r}{\lambda}+1\right)|\,\lambda>r,$$

where, $B(\varsigma_1,\varsigma_2)=\int_0^1 u^{\varsigma_1-1}(1-u)^{\varsigma_2-1}du$ is the complete beta function.

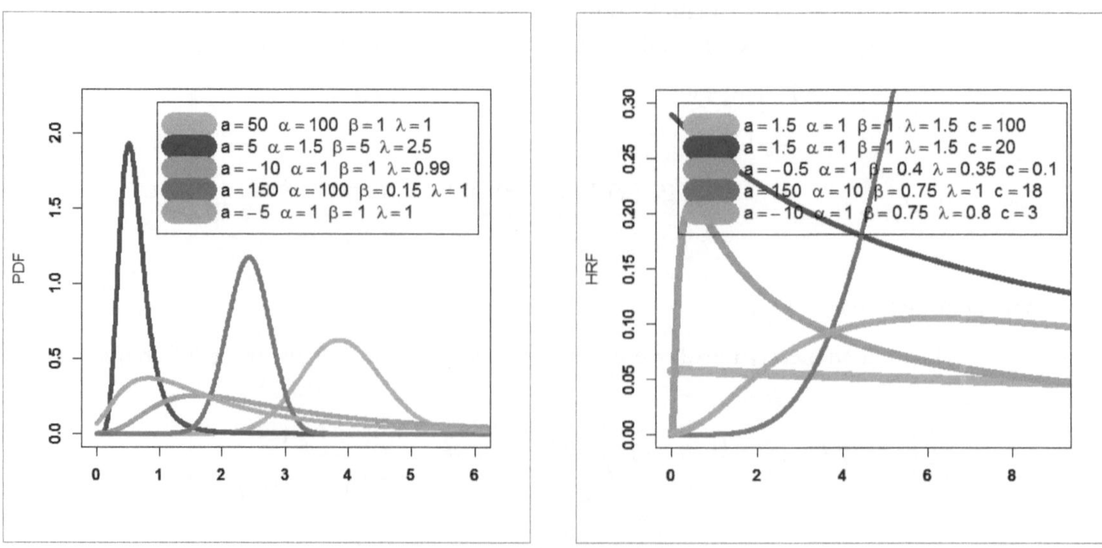

Figure 1: PDFs and HRFs for the PTLGL model.

Table 1: Theoretical results of the PTLGL model.

Property	Result	Support
$\mathbb{E}(X^r)$	$\sum_{l,d=0}^{+\infty}\sum_{m=0}^{r} c^r(-1)^m\binom{r}{m}\left\{\begin{array}{c} C_{l,d}[\beta(\alpha+\alpha l+d)]B\left([\beta(\alpha+\alpha l+d)],\frac{m-r}{\lambda}+1\right) \\ -C^*_{l,d}[\beta(\alpha+\alpha l+d+1)]B\left([\beta(\alpha+\alpha l+d+1)],\frac{m-r}{\lambda}+1\right)\end{array}\right\}$	$\lambda > r$
$M_x(t)$	$\sum_{l,d=0}^{+\infty}\sum_{m=0}^{r} \frac{t^r}{r!}c^r(-1)^m\binom{r}{m}$ $\times\left\{\begin{array}{c} C_{l,d}[\beta(\alpha+\alpha l+d)]B\left([\beta(\alpha+\alpha l+d)],\frac{m-r}{\lambda}+1\right) \\ -C^*_{l,d}[\beta(\alpha+\alpha l+d+1)]B\left([\beta(\alpha+\alpha l+d+1)],\frac{m-r}{\lambda}+1\right)\end{array}\right\}.$	$\lambda > r$
$\mathbf{I}_{s,X}(t)$	$\sum_{l,d=0}^{+\infty}\sum_{m=0}^{s} c^s(-1)^m\binom{s}{m}$ $\times\left\{\begin{array}{c} C_{l,d}[\beta(\alpha+\alpha l+d)]B_t\left([\beta(\alpha+\alpha l+d)],\frac{m-s}{\lambda}+1\right) \\ -C^*_{l,d}[\beta(\alpha+\alpha l+d+1)]B_t\left([\beta(\alpha+\alpha l+d+1)],\frac{m-s}{\lambda}+1\right)\end{array}\right\}.$	$\lambda > s$
$\mathbf{I}_{s,X}(t)$	$\sum_{l,d=0}^{+\infty}\sum_{m=0}^{1} c(-1)^m\binom{1}{m}\left\{\begin{array}{c} C_{l,d}[\beta(\alpha+\alpha l+d)]B_t\left([\beta(\alpha+\alpha l+d)],\frac{m-r}{\lambda}+1\right) \\ -C^*_{l,d}[\beta(\alpha+\alpha l+d+1)]B_t\left([\beta(\alpha+\alpha l+d+1)],\frac{m-r}{\lambda}+1\right)\end{array}\right\}$	$\lambda > 1$
$V_{n,X}(t)$	$\frac{1}{1-F_{\underline{V}}(t)}\sum_{l,d=0}^{+\infty}\sum_{m=0}^{n} c^n(-1)^m\binom{n}{m}$ $\times\left(\begin{array}{c} C_{l,d,v}(V,n)[\beta(\alpha+\alpha l+d)]\left\{\begin{array}{c} B\left([\beta(\alpha+\alpha l+d)],\frac{m-n}{\lambda}+1\right) \\ -B_t\left([\beta(\alpha+\alpha l+d)],\frac{m-n}{\lambda}+1\right)\end{array}\right\} \\ -C^*_{l,d,v}(V,n)[\beta(\alpha+\alpha l+d+1)]\left\{\begin{array}{c} B\left([\beta(\alpha+\alpha l+d+1)],\frac{m-n}{\lambda}+1\right) \\ -B_t\left([\beta(\alpha+\alpha l+d+1)],\frac{m-n}{\lambda}+1\right)\end{array}\right\}\end{array}\right),$ where, $C_{l,d,v}(V,n) = C_{l,d}\sum_{h=0}^{n}\binom{n}{h}(-t)^{n-h},$ and $C^*_{l,d,v}(V,n) = C^*_{l,d}\sum_{h=0}^{n}\binom{n}{h}(-t)^{n-h}$	$t > 0,$ $n \in N,$ $\lambda > n$

Table 1 contd. ...

...Table 1 contd.

Property	Result	Support
$V_{1,X}(t)$	$$\frac{1}{1-F_{\underline{V}}(t)}\sum_{l,d=0}^{+\infty}\sum_{m=0}^{n}c(-1)^{m}\binom{1}{m}$$ $$\times\left(\begin{array}{l} C_{l,d,v}(V,1)[\beta(\alpha+\alpha l+d)]\left\{\begin{array}{l} B\left([\beta(\alpha+\alpha l+d)],\frac{m-1}{\lambda}+1\right)\\ -B_t\left([\beta(\alpha+\alpha l+d)],\frac{m-1}{\lambda}+1\right)\end{array}\right\}\\ -C^{*}_{l,d,v}(V,1)[\beta(\alpha+\alpha l+d+1)]\left\{\begin{array}{l} B\left([\beta(\alpha+\alpha l+d+1)],\frac{m-1}{\lambda}+1\right)\\ -B_t\left([\beta(\alpha+\alpha l+d+1)],\frac{m-1}{\lambda}+1\right)\end{array}\right\}\end{array}\right),$$ where, $$C_{l,d,v}(V,1)=C_{l,d}\sum\nolimits_{h=0}^{1}\binom{1}{h}(-t)^{1-h},$$ and $$C^{*}_{l,d,v}(V,1)=C^{*}_{l,d}\sum\nolimits_{h=0}^{1}\binom{1}{h}(-t)^{n-h}.$$	$t>0$, $n=1$, $\lambda>1$
$U_{n,X}(t)$	$$\frac{1}{F_{\underline{V}}(t)}\sum_{l,d=0}^{+\infty}\sum_{m=0}^{n}c^{n}(-1)^{m}\binom{n}{m}$$ $$\times\left(\begin{array}{l} C_{l,d,v}(U,n)[\beta(\alpha+\alpha l+d)]B_t\left([\beta(\alpha+\alpha l+d)],\frac{m-n}{\lambda}+1\right)\\ C^{*}_{l,d,v}(U,n)[\beta(\alpha+\alpha l+d+1)]B_t\left([\beta(\alpha+\alpha l+d+1)],\frac{m-1}{\lambda}+1\right)\end{array}\right),$$ where, $$C_{l,d,v}(U,n)=C_{l,d}\sum\nolimits_{h=0}^{n}(-1)^{h}\binom{n}{h}t^{n-h},$$ and $$C^{*}_{l,d,v}(U,n)=C^{*}_{l,d}\sum\nolimits_{h=0}^{n}(-1)^{h}\binom{n}{r}t^{n-h}.$$	$t>0$, $n\in N$, $\lambda>n$
$U_{1,X}(t)$	$$\frac{1}{F_{\underline{V}}(t)}\sum_{l,d=0}^{+\infty}\sum_{m=0}^{n}c(-1)^{h}\binom{1}{m}$$ $$\times\left(\begin{array}{l} C_{l,d,v}(U,1)[\beta(\alpha+\alpha l+d)]B_t\left([\beta(\alpha+\alpha l+d)],\frac{m-n}{\lambda}+1\right)\\ C^{*}_{l,d,v}(U,1)[\beta(\alpha+\alpha l+d+1)]B_t\left([\beta(\alpha+\alpha l+d+1)],\frac{m-1}{\lambda}+1\right)\end{array}\right),$$ where, $$C_{l,d,v}(U,1)=C_{l,d}\sum\nolimits_{h=0}^{1}(-1)^{h}\binom{1}{r}t^{1-h},$$ and $$C^{*}_{l,d,v}(U,1)=C^{*}_{l,d}\sum\nolimits_{h=0}^{1}(-1)^{h}\binom{1}{r}t^{1-h}.$$	$t>0$, $n=1$, $\lambda>n$
$\mathbb{R}_{s,r}$	$$\sum_{l,d=0}^{+\infty}\sum_{m=0}^{S}c^{S}(-1)m\binom{s}{m}$$ $$\times\left(\begin{array}{l} \mathrm{K}_{l,d}(r)[\beta(\alpha+\alpha l+d)]B\left([\beta(\alpha+\alpha l+d)],\frac{m-S}{\lambda}+1\right)\\ -\mathrm{K}^{*}_{l,d}(r)[\beta(\alpha+\alpha l+d+1)]B\left([\beta(\alpha+\alpha l+d+1)],\frac{m-S}{\lambda}+1\right)\end{array}\right).$$	$\lambda>s$

Table 1 contd. ...

...Table 1 contd.

Property	Result	Support
$\mathbb{E}(X_{i:n}^{s})$	$\sum_{j=0}^{n-i}\frac{(-1)^{j}}{B(i,n-i+1)}\binom{n-i}{j}\sum_{l,d=0}^{+\infty}\sum_{m=0}^{S}c^{s}(-1)^{m}\binom{s}{m} \times\left(\begin{array}{l}K_{l,d}(j+i-1)[\beta(\alpha+\alpha l+d)]B\left([\beta(\alpha+\alpha l+d)],\frac{m-s}{\lambda}+1\right)\\ -K_{l,d}^{*}(j+i-1)[\beta(\alpha+\alpha l+d+1)]B\left([\beta(\alpha+\alpha l+d+1)],\frac{m-s}{\lambda}+1\right)\end{array}\right).$	$\lambda > s$

Theorem 5.2. Let $Y_{\Delta+1}$ be a random variable having the exp-L distribution with power parameter $\mathbf{\Delta} + 1$. Then, the rth incomplete moment of $Y_{\Delta+1}$ is given by,

$$\mathbf{I}_{r,Y}(t)=\sum_{m=0}^{r}(\Delta+1)c^{r}(-1)^{m}\binom{r}{m}B_{t}\left(\Delta+1,\frac{m-r}{\lambda}+1\right)\mid \lambda>r,$$

where, $B_t(\varsigma_1, \varsigma_2) = \int_0^1 u^{\varsigma_1-1}(1-u)^{\varsigma_2-1}du$ is the incomplete beta function.

Table 2 below gives a numerical analysis for the mean ($\mathbb{E}(X)$), variance (V(X)), skewness (S(X)), kurtosis (K(X)) and the dispersion (DI(X)) for PTLGL distribution. Based on the results listed in Table 2, it is noted that $\mathbb{E}(X)$ decreases as α increases; $\mathbb{E}(X)$ increases as α increases; $\mathbb{E}(X)$ decreases as β increases; S(X) ∈ (2.316822, 8105.145); K(X) ranging from 2.247183 to and DI(X) is always more than one, which means that the PTLGL model will be suitable for the "over-dispersed" data sets.

Table 2: $\mathbb{E}(X)$, V(X), (S(X)) and kurtosis K for PTLGL model.

λ,c	*a*	*α*	*β*	$\mathbb{E}$(X)	V	S	K	DI
2,2	–200	1	3	13.82128	46.424300	5.545125	356.415	3.358899
2,2	–100			11.25278	32.974520	5.520506	362.5764	2.930343
2,2	-50			9.081657	23.473030	5.48318	367.0563	2.584664
2,2	5	1	2	1.67×10^{-5}	0.4034740	5.011508	466.8554	24147.98
2,2		2		3.34×10^{-5}	1.0325210	2.903812	143.5774	30898.23
2,2		3		5.01×10^{-5}	1.6932800	2.316822	80.60721	33780.93
2,2	1.25	1.5	0.3	8.35×10^{-6}	0.1275966	13.06459	2338.006	15277.52
2,2			0.4	1.49×10^{-5}	0.2174070	10.31513	1430.786	14643.04
2,2			0.5	2.32×10^{-5}	0.3264422	8.658873	990.9689	14072.31
0.5,0.25	0.1	1.5	1.5	8.538794	79956.260	177.0491	41749.03	9363.883
0.5,0.25	1			5.518941	48957.670	226.2184	68170.23	8870.845
0.5,0.25	10			0.1845575	38.186930	8105.145	87450930	206.9108
0.5,0.25	–100	1	0.5	57.78778	622701.80	63.46806	5361.263	10775.66
0.5,0.25		10		427.9407	60500080	20.47917	555.0093	14137.49
0.5,0.25		10000		26955.06	924499115	0.736274	2.247183	34297.79
0.5,0.25	–100	10000	1	6382.283	464195758	3.228576	11.87127	72731.94
0.5,0.25			1.5	362.8912	315368350	15.63219	248.3138	86904.37
0.5,0.25			2	6.777539	619562.20	116.9011	13746.34	91414.04

6. Estimation methods

This section briefly describes and considers different classical estimation methods: the MLE method, CVM method, OLS method, WLSE method, ADE method, RTADE method, and LTADE method. All these methods are discussed in the statistical literature with more details. In this work, we may ignore some of its derivation details for avoiding repetition.

6.1 The MLE method

Let $x_1,\ldots, x_m$ be a random sample from the PTLG-G distribution. For determining the maximum likelihood estimates (MLEs), first we derive the log-likelihood function,

$$\ell(\underline{\mathbf{P}}) = m\log 2 - m(1-e^{-a}) + m\log a + m\log\alpha + m\log\beta$$
$$+\sum_{i=1}^{m}\log g_{\underline{W}}(x_{[i,m]}) + (\alpha\beta-1)\sum_{i=1}^{m}\log G_{\underline{W}}(x_{[i,m]})\sum_{i=1}^{m}\log[2-G_{\underline{W}}(x_{[i,m]})^{\beta}]$$
$$+\sum_{i=1}^{m}\log[1-G_{\underline{W}}(x_{[i,m]})^{\beta}] + (\alpha-1) - a\left\{G_{\underline{W}}(x_i)^{\beta}[2-G_{\underline{W}}(x_{[i,m]})^{\beta}]\right\}\alpha.$$

The components of the score vector are,

$$U_a = \frac{\partial}{\partial a}\ell(\underline{\mathbf{P}}), U_\alpha = \frac{\partial}{\partial \alpha}\ell(\underline{\mathbf{P}}),\ U_\beta = \frac{\partial}{\partial \beta}\ell(\underline{\mathbf{P}}) \text{ and } U_{\underline{W}} = \frac{\partial}{\partial_{\underline{W}}}\ell(\underline{\mathbf{P}}).$$

Setting the nonlinear system of equations $U_a = U_a = U_\beta = 0$ and $U_{\underline{W}} = 0$ and solving them simultaneously yield the MLEs. To solve these equations, it is usually more convenient to use nonlinear optimization methods such as the quasi-Newton algorithm to numerically maximize $\ell(\underline{\mathbf{P}})$.

6.2 The CVME method

The CVME of a, α, β and $\underline{\boldsymbol{W}}$ are obtained via minimizing the following expression with respect to a, α, β and $\underline{\boldsymbol{W}}$ respectively, where,

$$CVME(\underline{\mathbf{P}}) = \frac{1}{12}m^{-1} + \sum_{i=1}^{m}[F_{a,b,\theta}(x_{[i,m]}) - c_{(i,m)}]^2,$$

and $c_{(i,m)} = [(2i-1)/2m]$ and

$$CVME(\underline{\mathbf{P}}) = \sum_{i=1}^{m}\left(C(a)\left[1-exp\left(-a\left\{G_{\underline{W}}(x_{[i,m]})^{\beta}[2-G_{\underline{W}}(x_{[i,m]})^{\beta}]\right\}^{\alpha}\right)\right]-c_{(i,m)}\right)^2.$$

The CVME of a, α, β and $\underline{\boldsymbol{W}}$ are obtained by solving the following non-linear equations:

$$0=\sum_{i=1}^{m}\left[C(a)\left[1-exp\left(-a\left\{G_{\underline{W}}(x_{[i,m]})^{\beta}\left[2-G_{\underline{W}}(x_{[i,m]})^{\beta}\right]\right\}^{\alpha}\right)\right]-c_{(i,m)}\right]\nabla_{(a)}(x_{[i,m]};\underline{\mathbf{P}}),$$

$$0=\sum_{i=1}^{m}\left[C(a)\left[1-exp\left(-a\left\{G_{\underline{W}}(x_{[i,m]})^{\beta}\left[2-G_{\underline{W}}(x_{[i,m]})^{\beta}\right]\right\}^{\alpha}\right)\right]-c_{(i,m)}\right]\nabla_{(\alpha)}(x_{[i,m]};\underline{\mathbf{P}}),$$

$$0=\sum_{i=1}^{m}\left[C(a)\left[1-exp\left(-a\left\{G_{\underline{W}}(x_{[i,m]})^{\beta}\left[2-G_{\underline{W}}(x_{[i,m]})^{\beta}\right]\right\}^{\alpha}\right)\right]-c_{(i,m)}\right]\nabla_{(\beta)}(x_{[i,m]};\underline{\mathbf{P}}),$$

and

$$0=\sum_{i=1}^{m}\left[C(a)\left[1-exp\left(-a\left\{G_{\underline{W}}(x_{[i,m]})^{\beta}\left[2-G_{\underline{W}}(x_{[i,m]})^{\beta}\right]\right\}^{\alpha}\right)\right]-c_{(i,m)}\right]\nabla_{(\underline{W})}(x_{[i,m]};\underline{\mathbf{P}}),$$

where,

$$\nabla_{(a)}(x_{[i,m]};a,b,\theta)=\frac{\partial F_{\underline{\mathbf{P}}}(x_{[i,m]})}{\partial a},\nabla_{(\alpha)}(x_{[i,m]};a,b,\theta)=\frac{\partial F_{\underline{\mathbf{P}}}(x_{[i,m]})}{\partial \alpha},$$

$$\nabla_{(\beta)}(x_{[i,m]};a,b,\theta)=\frac{\partial F_{\underline{\mathbf{P}}}(x_{[i,m]})}{\partial \beta}\text{ and }\nabla_{(\beta)}(x_{[i,m]};a,b,\theta)=\frac{\partial F_{\underline{\mathbf{P}}}(x_{[i,m]})}{\partial_{\underline{W}}}.$$

6.3 The OLSE method

Let $F_{\underline{\mathbf{P}}}(x_{[i,m]})$ denotes the CDF of the PTLG-G family and let $x_1 < x_2 < \cdots < x_m$ be the m ordered RS. The OLSEs are obtained upon minimizing

$$OLSE(\underline{\mathbf{P}})=\sum_{i=1}^{m}[F_{\underline{\mathbf{P}}}(x_{[i,m]})-b_{(i,m)}]^2.$$

Then, we have,

$$OLSE(\underline{\mathbf{P}})=\sum_{i=1}^{m}\left[C(a)\left[1-exp\left(-a\left\{G_{\underline{W}}(x_{[i,m]})^{\beta}\left[2-G_{\underline{W}}(x_{[i,m]})^{\beta}\right]\right\}^{\alpha}\right)\right]-b_{(i,m)}\right]^2,$$

where, $b_{(i,m)}=\frac{i}{m+1}$. The LSEs are obtained via solving the following non-linear equations

$$0=\sum_{i=1}^{m}\left[C(a)\left[1-exp\left(-a\left\{G_{\underline{W}}(x_{[i,m]})^{\beta}\left[2-G_{\underline{W}}(x_{[i,m]})^{\beta}\right]\right\}^{\alpha}\right)\right]-b_{(i,m)}\right]\nabla_{(a)}(x_{[i,m]};\underline{\mathbf{P}}),$$

$$0=\sum_{i=1}^{m}\left[C(a)\left[1-exp\left(-a\left\{G_{\underline{W}}(x_{[i,m]})^{\beta}\left[2-G_{\underline{W}}(x_{[i,m]})^{\beta}\right]\right\}^{\alpha}\right)\right]-b_{(i,m)}\right]\nabla_{(\alpha)}(x_{[i,m]};\underline{\mathbf{P}}),$$

$$0=\sum_{i=1}^{m}\left[C(a)\left[1-exp\left(-a\left\{G_{\underline{W}}(x_{[i,m]})^{\beta}\left[2-G_{\underline{W}}(x_{[i,m]})^{\beta}\right]\right\}^{\alpha}\right)\right]-b_{(i,m)}\right]\nabla_{(\beta)}(x_{[i,m]};\underline{\mathbf{P}}),$$

and

$$0=\sum_{i=1}^{m}\left[C(a)\left[1-exp\left(-a\left\{G_{\underline{W}}(x_{[i,m]})^{\beta}\left[2-G_{\underline{W}}(x_{[i,m]})^{\beta}\right]\right\}^{\alpha}\right)\right]-b_{(i,m)}\right]\nabla_{(\underline{W})}(x_{[i,m]};\underline{\mathbf{P}}).$$

6.4 The WLSE method

The WLSE are obtained by minimizing the function **WLSE** $(a, \alpha, \beta, \underline{\mathbf{W}})$ with respect to a, α, β and $\underline{\mathbf{W}}$ where,

$$WLSE(\underline{\mathbf{P}})=\sum_{i=1}^{m}\omega_{(i,m)}[F_{\underline{\mathbf{P}}}(x_{[i,m]})-b_{(i,m)}]^2,$$

and

$$\omega_{(i,m)}=[(1+m)^2(2+m)]/[i(1+m-i)].$$

The WLSEs are obtained by solving

$$0=\sum_{\mathrm{i}=1}^{m}\left[C(a)\left[1-exp\left(-a\left\{G_{\underline{W}}(x_{[\mathrm{i},m]})^{\beta}\left[2-G_{\underline{W}}(x_{[\mathrm{i},m]})^{\beta}\right]\right\}^{\alpha}\right)\right]-b_{(\mathrm{i},m)}\right]\omega_{(\mathrm{i},m)}\nabla_{(a)}(x_{[\mathrm{i},m]};\underline{\mathbf{P}}),$$

$$0=\sum_{\mathrm{i}=1}^{m}\omega_{(\mathrm{i},m)}\left[C(a)\left[1-exp\left(-a\left\{G_{\underline{W}}(x_{[\mathrm{i},m]})^{\beta}\left[2-G_{\underline{W}}(x_{[\mathrm{i},m]})^{\beta}\right]\right\}^{\alpha}\right)\right]-b_{(\mathrm{i},m)}\right]\nabla_{(\alpha)}(x_{[\mathrm{i},m]};\underline{\mathbf{P}}),$$

$$0=\sum_{\mathrm{i}=1}^{m}\omega_{(\mathrm{i},m)}\left[C(a)\left[1-exp\left(-a\left\{G_{\underline{W}}(x_{[\mathrm{i},m]})^{\beta}\left[2-G_{\underline{W}}(x_{[\mathrm{i},m]})^{\beta}\right]\right\}^{\alpha}\right)\right]-b_{(\mathrm{i},m)}\right]\nabla_{(\beta)}(x_{[\mathrm{i},m]};\underline{\mathbf{P}}).$$

and

$$0=\sum_{\mathrm{i}=1}^{m}\omega_{(\mathrm{i},m)}\left[C(a)\left[1-exp\left(-a\left\{G_{\underline{W}}(x_{[\mathrm{i},m]})^{\beta}\left[2-G_{\underline{W}}(x_{[\mathrm{i},m]})^{\beta}\right]\right\}^{\alpha}\right)\right]-b_{(\mathrm{i},m)}\right]\nabla_{(\underline{W})}(x_{[\mathrm{i},m]};\underline{\mathbf{P}}).$$

6.5 The ADE method

The ADE of a, α, β and $\underline{\boldsymbol{W}}$ are obtained by minimizing the function,

$$ADE_{(x_{[\mathrm{i},m]},x_{[-\mathrm{i}+1+m:m]})}(\underline{\mathbf{P}})=-m-m^{-1}\sum_{\mathrm{i}=1}^{m}(2\mathrm{i}-1)\left\{\begin{array}{c}log\,F_{\underline{\mathbf{P}}}(x_{[\mathrm{i},m]})\\+log[1-F_{\underline{\mathbf{P}}}(x_{[1+m-\mathrm{i}:m]})]\end{array}\right\}.$$

The parameter estimates of a, α, β and $\underline{\boldsymbol{W}}$ follow by solving the following nonlinear equations:

$$0=\partial\left[ADE_{(x_{[\mathrm{i},m]},x_{[1+m-\mathrm{i}:m]})}(\underline{\mathbf{P}})\right]/\partial a,0=\partial\left[ADE_{(x_{[\mathrm{i},m]},x_{[1+m-\mathrm{i}:m]})}(\underline{\mathbf{P}})\right]/\partial\alpha,$$

$$0=\partial\left[ADE_{(x_{[\mathrm{i},m]},x_{[1+m-\mathrm{i}:m]})}(\underline{\mathbf{P}})\right]/\partial\beta\,\text{and}\,0=\partial\left[ADE_{(x_{[\mathrm{i},m]},x_{[1+m-\mathrm{i}:m]})}(\underline{\mathbf{P}})\right]/\partial\underline{\boldsymbol{W}}.$$

6.6 The RTADE method

The RTADE of a, α, β and $\underline{\boldsymbol{W}}$ are obtained by minimizing

$$\mathrm{RTADE}_{(x_{[\mathrm{i},m]},x_{[-\mathrm{i}+1+m:m]})}(\underline{\mathbf{P}})=\frac{1}{2}m-2\sum_{\mathrm{i}=1}^{m}F_{\underline{\mathbf{P}}}(x_{[\mathrm{i},m]})-\frac{1}{m}\sum_{\mathrm{i}=1}^{m}(2\mathrm{i}-1)\left\{log[1-F_{\underline{\mathbf{P}}}(x_{[1+m-\mathrm{i}:m]})]\right\}.$$

The estimates of a, b and θ are obtained by solving the following nonlinear equations:

$$0=\partial[\mathrm{RTAD}_{(x_{[\mathrm{i},m]},x_{[1+m-\mathrm{i}:m]})}(\underline{\mathbf{P}})]/\partial a,\ 0=\partial\,[\mathrm{RTADE}_{(x_{[\mathrm{i},m]},x_{[1+m-\mathrm{i}:m]})}(\underline{\mathbf{P}})]/\partial\alpha,$$

$$0=\partial[\mathrm{RTAD}_{(x_{[\mathrm{i},m]},x_{[1+m-\mathrm{i}:m]})}(\underline{\mathbf{P}})]/\partial\beta,\ 0=\partial\,[\mathrm{RTADE}_{(x_{[\mathrm{i},m]},x_{[1+m-\mathrm{i}:m]})}(\underline{\mathbf{P}})]/\partial\underline{\boldsymbol{W}}.$$

6.7 The LTADE method

The LTADE of a, α, β and $\underline{\boldsymbol{W}}$ are obtained by minimizing,

$$\mathrm{L\,TADE}_{(x_{[\mathrm{i},m]})}(\underline{\mathbf{P}})=-\frac{3}{2}m+2\sum_{\mathrm{i}=1}^{m}F_{\underline{\mathbf{P}}}(x_{[\mathrm{i},m]})-\frac{1}{m}\sum_{\mathrm{i}=1}^{m}(2\mathrm{i}-1)\,log\,F_{\underline{\mathbf{P}}}(x_{[\mathrm{i},m]}).$$

The parameter estimates of δ, θ and β are obtained by solving the following nonlinear equations:

$$0=\partial[\mathrm{LTAD}_{(x_{[\mathrm{i},m]})}(\underline{\mathbf{P}})]/\partial a,\ 0=\partial\,[\mathrm{LTADE}_{(x_{[\mathrm{i},m]})}(\underline{\mathbf{P}})]/\partial\alpha,$$

$$0=\partial[\mathrm{LTAD}_{(x_{[\mathrm{i},m]}}(\underline{\mathbf{P}})]/\partial\beta\ \text{and}\ 0=\partial\,[\mathrm{LTADE}_{(x_{[\mathrm{i},m]})}(\underline{\mathbf{P}})]/\partial\underline{\boldsymbol{W}}.$$

7. Comparing methods

7.1 Simulations for competitive estimation methods

A numerical simulation is performed to compare the classical estimation methods. The simulation study is based on $N = 1000$ generated data sets from the PTLGL model defined in Section 5 where $m = 50$, 100, 150 and 300 and

Blend ↓ Initial →	a_0	α_0	β_0	λ_0	c_0
I	−2	1.5	1.5	0.6	0.1
II	1.2	1.2	1.2	1.2	0.3
III	−1.2	2	2.5	2	0.5

The estimates are compared in terms of their biases, the root mean-standard error (RMSE). The mean of the absolute difference between the theoretical and the estimates (D-abs) and the maximum absolute difference between the true parameters and estimates (D-max) are also reported.

Tables 3, 4 and 5 give the simulation results. From Tables 3, 4, and 5 we note that the RMSE tends to zero when m increases, which implies the incidence of consistency property.

Table 3: Simulation results for blend I.

Methods	m	RMSE					D	
		$\hat{a}$	$\hat{\alpha}$	$\hat{\beta}$	$\hat{\lambda}$	$\hat{c}$	abs	max
MLE	20	0.781899	0.211464	0.074903	0.005958	0.001024	0.029454	0.043999
OLS		0.780846	0.2189683	0.0758624	0.006782	0.000931	0.014568	0.022046
WLS		0.791896	0.2115054	0.077497	0.0060914	0.0010457	0.0094253	0.014446
CVM		0.793498	0.222415	0.076071	0.006433	0.000982	0.017160	0.026129
ADE		0.752596	0.200083	0.072198	0.005903	0.000960	0.011598	0.017838
RTADE		0.967502	0.296462	0.090337	0.005451	0.001212	0.034909	0.051486
LTADE		0.760520	0.187759	0.071923	0.008726	0.000951	0.004153	0.007710
MLE	50	0.280263	0.066134	0.026733	0.002284	0.000350	0.006209	0.009787
OLS		0.312155	0.083412	0.030663	0.002686	0.000363	0.009815	0.014591
WLS		0.311915	0.074718	0.029695	0.002439	0.000389	0.006040	0.009152
CVM		0.314216	0.083993	0.030698	0.002522	0.000380	0.010983	0.016405
ADE		0.298349	0.075533	0.028905	0.002365	0.000368	0.009224	0.013740
RTADE		0.381704	0.107703	0.036506	0.002297	0.000457	0.018436	0.027157
LTADE		0.289757	0.070427	0.028190	0.002968	0.000363	0.003952	0.006431
MLE	100	0.141951	0.032639	0.013284	0.001076	0.000170	0.007603	0.011291
OLS		0.149553	0.039150	0.014721	0.001285	0.000172	0.008413	0.012366
WLS		0.151366	0.034563	0.014151	0.001157	0.000186	0.007877	0.011635
CVM		0.150093	0.039309	0.014733	0.001204	0.000179	0.009013	0.013288
ADE		0.142256	0.035504	0.013810	0.001117	0.000173	0.008373	0.012307
RTADE		0.175811	0.047481	0.016824	0.001082	0.000203	0.012165	0.017849
LTADE		0.137238	0.033120	0.013493	0.001384	0.000172	0.006271	0.009304
MLE	300	0.047571	0.010863	0.004528	0.000349	0.000057	0.001991	0.002952
OLS		0.052247	0.013298	0.005113	0.000448	0.000058	0.000357	0.000618
WLS		0.054718	0.012101	0.005025	0.000386	0.000065	0.003253	0.004853
CVM		0.052296	0.013311	0.005113	0.000420	0.000061	0.000558	0.000935
ADE		0.050503	0.012346	0.004868	0.000390	0.000059	0.000526	0.000845
RTADE		0.058356	0.015174	0.005544	0.000362	0.000065	0.002464	0.003640
LTADE		0.050990	0.012053	0.004992	0.000509	0.00062	0.000920	0.001376

Table 4: Simulation results for blend II.

Methods	m	RMSE					D	
		$\hat{a}$	$\hat{\alpha}$	$\hat{\beta}$	$\hat{\lambda}$	$\hat{c}$	abs	max
MLE	20	0.703433	0.068196	0.035148	0.072391	0.007238	0.024048	0.036846
OLS		0.716239	0.081305	0.038636	0.084582	0.007433	0.014173	0.021749
WLS		0.695036	0.068648	0.034998	0.075706	0.007510	0.012948	0.019316
CVM		0.726111	0.083086	0.038959	0.085188	0.007353	0.015584	0.024444
ADE		0.687264	0.069222	0.034893	0.075277	0.007139	0.012287	0.018917
RTADE		0.705086	0.119748	0.050572	0.070665	0.008090	0.040076	0.058915
LTADE		0.868825	0.062239	0.032851	0.110364	0.007663	0.010083	0.016389
MLE	50	0.265814	0.022575	0.012358	0.026278	0.002578	0.003940	0.007280
OLS		0.259876	0.027913	0.014016	0.028914	0.002666	0.013770	0.020128
WLS		0.267629	0.023352	0.012672	0.025851	0.002749	0.015659	0.022755
CVM		0.261113	0.028243	0.014089	0.028973	0.002654	0.014404	0.021179
ADE		0.251497	0.024705	0.012865	0.026217	0.002570	0.013592	0.019834
RTADE		0.258310	0.036706	0.017213	0.025333	0.002760	0.023339	0.033973
LTADE		0.290383	0.022483	0.012165	0.033664	0.002808	0.006031	0.009264
MLE	100	0.135498	0.011094	0.006240	0.012999	0.001362	0.008541	0.012652
OLS		0.137583	0.014554	0.007398	0.015174	0.001384	0.008315	0.012131
WLS		0.143940	0.011688	0.006489	0.013513	0.001398	0.006783	0.009986
CVM		0.137908	0.014640	0.007417	0.015191	0.001381	0.008616	0.012629
ADE		0.133280	0.012828	0.006766	0.013889	0.001329	0.007574	0.011049
RTADE		0.134403	0.018140	0.008721	0.013271	0.001385	0.012224	0.017800
LTADE		0.155833	0.011748	0.006458	0.017970	0.001486	0.003948	0.005962
MLE	300	0.042737	0.003521	0.001979	0.004077	0.000419	0.003977	0.005838
OLS		0.042560	0.004327	0.002236	0.004686	0.000422	0.002902	0.004226
WLS		0.049362	0.003822	0.002163	0.004573	0.000477	0.003545	0.005295
CVM		0.042597	0.004338	0.002238	0.004688	0.000421	0.003008	0.004402
ADE		0.042011	0.003905	0.002085	0.004385	0.000413	0.003127	0.004552
RTADE		0.043507	0.005513	0.002719	0.004330	0.000438	0.003099	0.004543
LTADE		0.046642	0.003513	0.001945	0.005292	0.000449	0.003427	0.004983

7.2 Applications for comparing the competitive estimation methods

Two applications to the real data set are considered for comparing the estimation methods. The 1st data set called the "aircraft windshield" represents the data on failure times of 84 aircraft windshield. The 2nd Data set also called the "aircraft windshield" represents the data on service times of 63 aircraft windshields. The two real data were reported by Murthy et al. (2004). The required computations are carried out using the MATHCAD software. In order to compare the estimation methods, we consider the Cramér-von Mises (CVM) and the Anderson-Darling (AD) statistics. These two statistics are widely used to determine how closely a specific CDF fits the empirical distribution of a given data set. Table 6 and Table 7 give the estimates and the test statistics for all the estimation methods. From Table 6 we conclude that the MLE method is the best method with CVM = 0.04630 and AD = 0.47186. From Table 7 we conclude that the MLE method is the best method with CVM = 0.05717 and AD = 0.35320. However, all other methods performed well.

Table 5: Simulation results for blend III.

Methods	*m*	RMSE					D	
		$\hat{a}$	$\hat{\alpha}$	$\hat{\beta}$	$\hat{\lambda}$	$\hat{c}$	abs	max
MLE	20	0.7651606	0.4001974	0.2165702	0.0411279	0.0059717	0.0245247	0.0359504
OLS		0.6591711	0.3669806	0.1997503	0.0414978	0.0053998	0.0091343	0.0136777
WLS		0.693372	0.375594	0.206029	0.038757	0.005734	0.032803	0.047369
CVM		0.734169	0.454807	0.227511	0.040798	0.005851	0.045657	0.066121
ADE		0.696793	0.384389	0.208877	0.038572	0.005690	0.039343	0.056894
RTADE		0.828938	0.564812	0.263819	0.037734	0.006190	0.062648	0.090633
LTADE		0.711550	0.349402	0.202762	0.048661	0.006159	0.024155	0.035170
MLE	50	0.2524062	0.1089245	0.0686767	0.0143282	0.0020404	0.0092676	0.0135787
OLS		0.2799867	0.1502873	0.0853749	0.0169848	0.0023274	0.0161194	0.0233995
WLS		0.293754	0.126963	0.079370	0.016905	0.002382	0.015562	0.022531
CVM		0.289337	0.153758	0.087473	0.017725	0.002392	0.019822	0.028732
ADE		0.280983	0.135956	0.081320	0.016778	0.002317	0.017172	0.024871
RTADE		0.331773	0.193546	0.103037	0.016625	0.002544	0.026703	0.038675
LTADE		0.284320	0.124703	0.078610	0.020253	0.002475	0.011298	0.016442
MLE	100	0.0898733	0.037102	0.0243643	0.0052244	0.0007408	0.0030559	0.0044817
OLS		0.0846877	0.042357	0.0253857	0.0053323	0.0007115	0.0049563	0.0071751
WLS		0.130984	0.052145	0.033951	0.007536	0.001039	0.005642	0.008177
CVM		0.127780	0.063662	0.037961	0.008066	0.001060	0.004655	0.006776
ADE		0.123282	0.056301	0.034942	0.007551	0.001011	0.006121	0.008852
RTADE		0.142757	0.076450	0.043247	0.007547	0.001096	0.001831	0.002794
LTADE		0.124947	0.052689	0.034212	0.008998	0.001090	0.008960	0.012943
MLE	300	0.0420026	0.017779	0.0115895	0.0024205	0.0003438	0.0031073	0.0045127
OLS		0.0438612	0.0217394	0.0131186	0.0027502	0.0003667	0.0002438	0.0004378
WLS		0.049366	0.019694	0.012902	0.002673	0.000390	0.003894	0.005668
CVM		0.045063	0.022351	0.013460	0.002806	0.000376	0.002516	0.003655
ADE		0.043535	0.020087	0.012473	0.002607	0.000358	0.001940	0.002817
RTADE		0.048705	0.025865	0.014805	0.002523	0.000374	0.003293	0.004774
LTADE		0.044840	0.019134	0.012446	0.003177	0.000395	0.001065	0.001571

Table 6: Comparing estimation methods via an application.

Methods	**Estimates**					**Test statistics**	
	$\hat{a}$	$\hat{\alpha}$	$\hat{\beta}$	$\hat{\lambda}$	$\hat{c}$	C*	A*
MLE	–4.91315	0.06419	20.1558	178.8356	221.2843	**0.04630**	**0.47186**
OLS	–8.09745	0.16768	4.24998	20.18322	30.99778	0.06171	0.63323
WLS	–3.82957	0.14885	15.51752	42.96923	46.92708	0.04931	0.50496
CVM	–7.85575	0.19925	3.94773	18.35556	27.74697	0.06207	0.63341
ADE	–5.14699	0.14422	8.48739	28.97846	39.05812	0.04792	0.49513
RTADE	–1.07411	0.24895	16.68032	40.08727	49.21435	0.10472	0.98303
LTADE	–5.06196	0.20184	5.75072	334.9566	508.5706	0.04701	0.48306

Table 7: Comparing estimation methods via an application.

Methods	Estimates					Test statistics	
	$\hat{a}$	$\hat{\alpha}$	$\hat{\beta}$	$\hat{\lambda}$	$\hat{c}$	C*	A*
MLE	–2.78092	0.08558	12.07814	106.42604	151.53397	**0.05717**	**0.35320**
OLS	–3.10846	0.09195	10.59231	30.10565	43.20463	0.06025	0.37151
WLS	–3.50419	0.06986	11.09497	80.91018	119.79611	0.05783	0.35863
CVM	–3.04688	0.09150	11.23096	44.68476	63.03266	0.05760	0.35573
ADE	–3.26377	0.09084	9.70303	53.84903	80.11256	0.05775	0.35723
RTADE	–2.98547	0.08683	13.69396	35.68186	46.07807	0.05772	0.35591
LTADE	–3.35132	0.08810	8.58356	83.39752	137.66545	0.05916	0.36597

8. Comparing competitive models

Two real-life data applications to illustrate the importance and flexibility of the family are presented under the Lamax model The fits of the PTLGL are compared with other Lomax extensions shown in Table 8. The 1st real-life data set (aircraft windshield consists of 84 aircraft windshield items) represents the data of failure times of 84 aircraft windshields. The 2nd real-life Data set (aircraft windshield consists of 63 aircraft windshield items) represents the data of service times of 63 aircraft windshields. The two data sets are considered based on matching/fitting their properties and the plots of the PDF in Figure 1 (right plot). By examining Figure 1 (the right plot) we see that the PDF of the PTLGL model can be "symmetric" and "asymmetric right-skewed" with different shapes. On the other hand, by exploring the two real-life data sets, we noted that two densities are asymmetric densities (see Figure 2 (top right plot) and Figure 3 (top right plot)). Moreover, the theoretical HRF of the PTLG-G family, including the "asymmetric monotonically increasing HRF" shape and the HRF of the two real data sets are "asymmetric monotonically increasing" (see Figure 1 (left plot), Figure 2 (top left plot) and Figure 2 (top left plot)). The two real data were reported by Murthy et al. (2004). The "nonparametric Kernel density estimation (KDE)" tool is employed for exploring the initial PDF shape. The "normality" is also checked by the p "Quantile-Quantile" (Q-Q) plot. The initial

Table 8: The competitive models.

N.	Model	Abbreviation	Author
1	Special generalized mixture-L	SGML	Chesneau and Yousof (2021)
2	Odd log-logistic-L	OLLL	Elgohari and Yousof (2020)
3	Reduced OLL-L	ROLLL	Elgohari and Yousof (2020)
4	Reduced Burr-Hatke-L	RBHL	Yousof et al. (2018c)
5	Transmuted Topp-Leone-L	TTLL	Yousof et al. (2017b)
6	Reduced TTL-L	RTTLL	Yousof et al. (2017b)
7	Gamma-L	GamL	Cordeiro et al. (2015)
8	Kumaraswamy-L	KumL	Lemonte and Cordeiro (2013)
9	McDonald-L	McL	Lemonte et al. (2013)
10	Beta-L	BL	Lemonte et al. (2013)
11	Exponentiated-L	exp-L	Gupta et al. (1998)
12	L	L	Lomax (1954)
13	Proportional reversed hazard rate-L	PRHRL	New

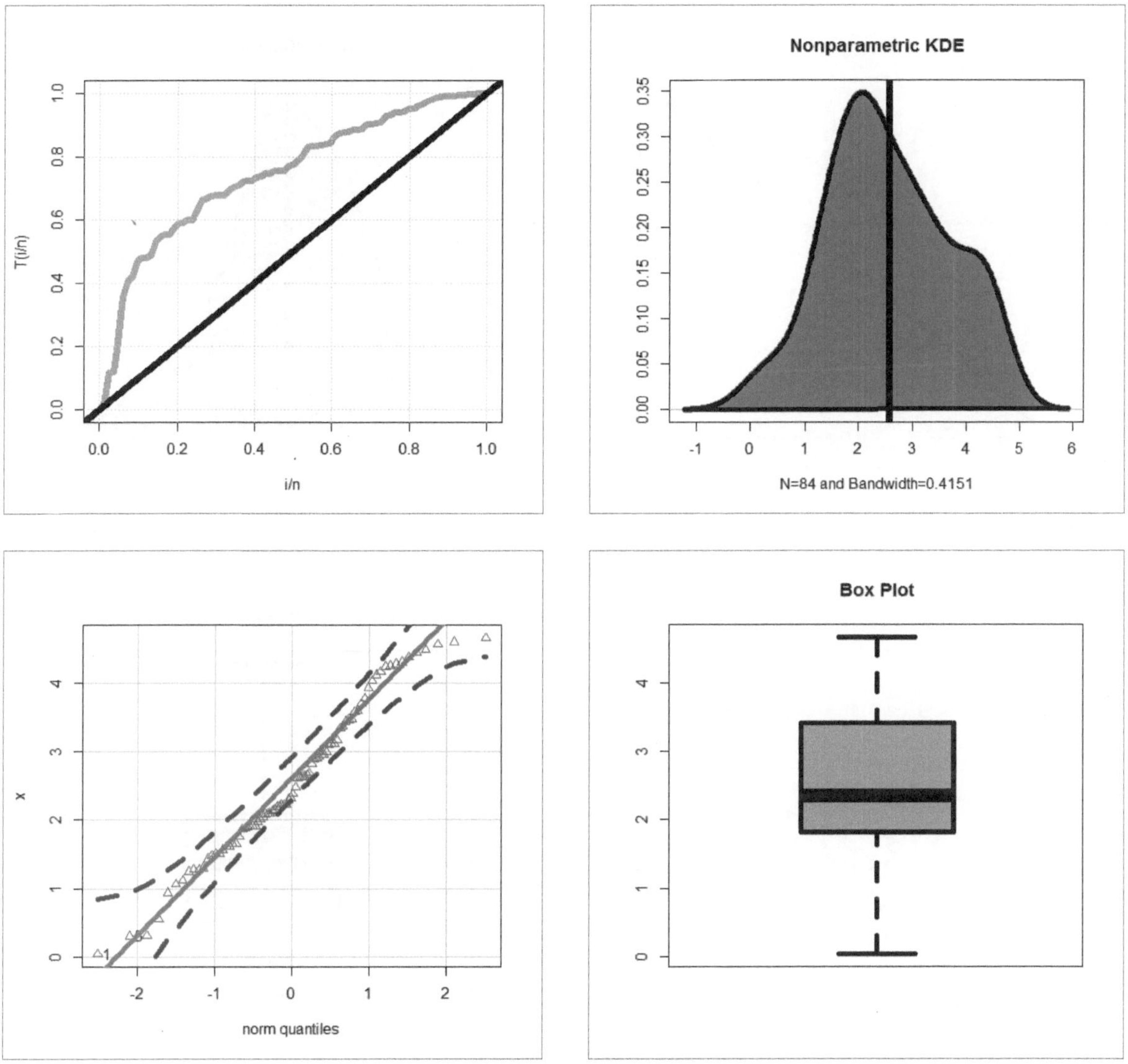

Figure 2: TTT, NKDE, Q-Q and box plot for the 1st data.

HRF shapes are explored via the "total time in test (TTT)" plot. The "box plot" explores the extreme values. Based on Figures 2 and 3 (top left plots), it is shown that the HRFs are "monotonically increasing HRFs" for the two data sets. Based on Figures 2 and 3 (top right plots), it is noted that the PDFs are asymmetric functions for the two data sets. Based on Figures 2 and 3 (bottom left plots), it is noted that "normality" exists. Based on Figures 2 and 3 (bottom right plots) we observe that no extremes are spotted. The following goodness-of-fit (G-O-F) test statistics are used for comparing competitive models: the "Akaike information" (AICr); the "Consistent-AIC" (CAICr); the "Bayesian-IC" (BICr) and the "Hannan-Quinn-IC" (HQICr). Tables 9 and 11 give the MLEs and the corresponding standard errors (SEs) for the two real-life datasets. Tables 10 and 12 list the four G-O-F statistics for the two real-life data sets. Figures 4 and 5 give the Probability-Probability

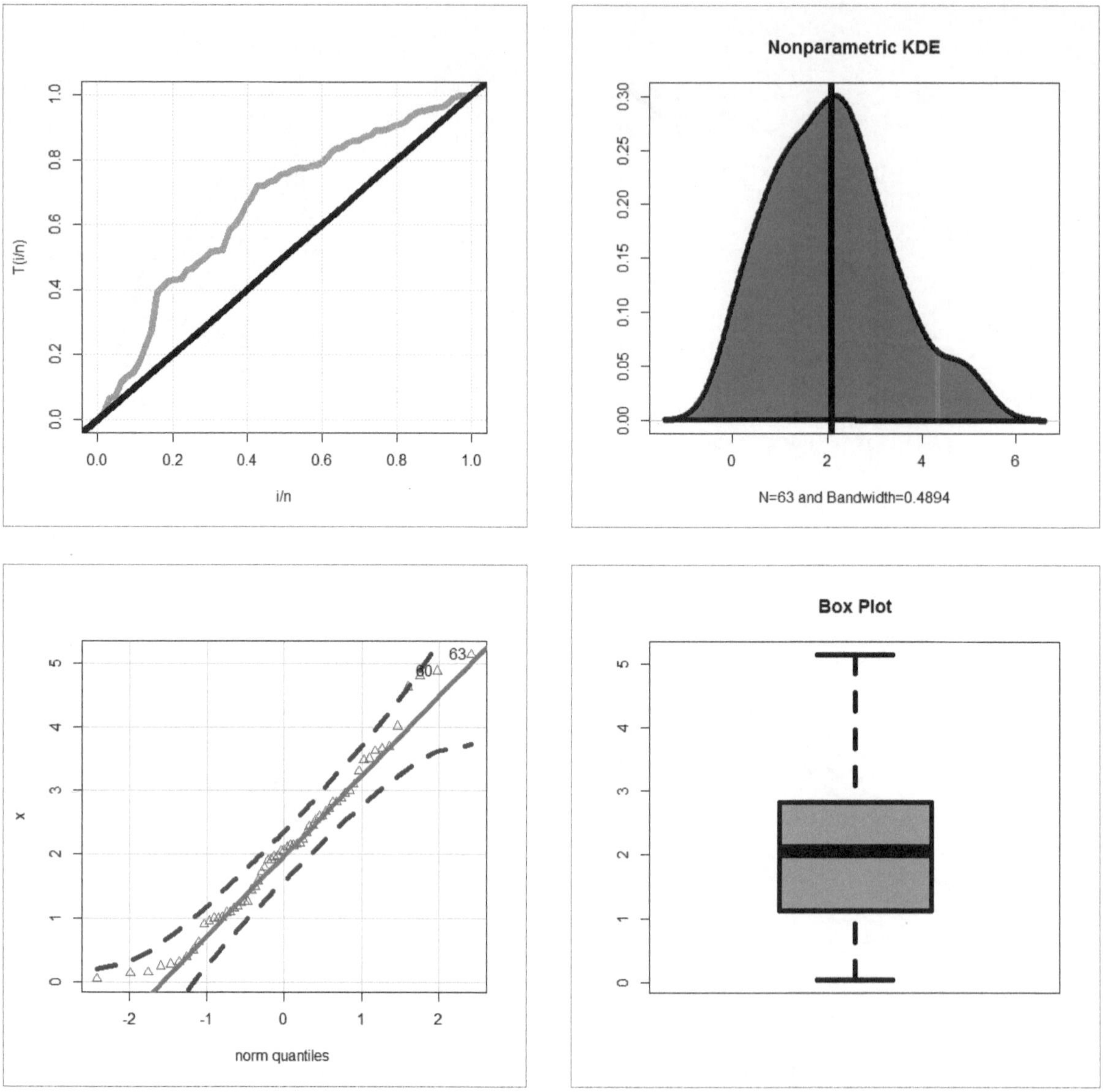

Figure 3: TTT, NKDE, Q-Q and box plot for the 2nd data set.

(P-P) , Kaplan-Meier Survival (KMS), estimated PDF(E-PDF), estimated CDF (E-CDF) and estimated HRF (E-HRF) plots for the two data sets, respectively. Based on Tables 5 and 7, it is noted that the PTLGL model gives the lowest values for all G-O-F statistics with AICr = 269.8712, CAICr = 270.6404, BICr = 282.0253 and HQICr = 274.7570 for the 1st data set, and AICr = 208.582, CAICr = 209.6347, BICr = 218.2977 and HQICr = 212.7966 for the 2nd data set among all fitted competitive models. So, it could be selected as the best model under these G-O-F criteria.

Table 9: MLEs and SEs for the **1st** data set.

Model	**Estimates**				
PTLGL (a,α,β,λ,c)	**−5.4829**	**0.2484**	**4.6215**	**17.199**	**25.023**
	(2.0125)	**(0.6335)**	**(11.269)**	**(20.515)**	**(41.598)**
KL (a,b,λ,c)	2.6150	100.276	5.27710	78.6774	
	(0.3822)	(120.49)	(9.8116)	(186.01)	
TTLL (a,b,λ,c)	−0.8075	2.47663	(15608)	(38628)	
	(0.1396)	(0.5418)	(1602.4)	(123.94)	
BL (a,b,λ,c)	3.60360	33.6387	4.83070	118.837	
	(0.6187)	(63.715)	(9.2382)	(428.93)	
PRHRL (b,λ,c)	3.73×10^{6}	4.71×10^{-1}	4.5×10^{6}		
	1.01×10^{6}	(0.00001)	37.1468		
SGML (b,λ,c)	-1.04×10^{-1}	9.83×10^{6}	1.18×10^{7}		
	(0.1223)	(4843.3)	(501.04)		
RTTLL (b,β,λ)	−0.84732	5.52057	1.15678		
	(0.1001)	(1.1848)	(0.0959)		
OLLL (b,λ,c)	2.32636	$7.17\times e^{5}$	2.3×10^{6}		
	(2.14×10^{-1})	$(1.19\times e^{4})$	(2.6×10^{1})		
exp-L (b,λ,c)	3.62610	20074.5	26257.7		
	(0.6236)	(2041.8)	(99.743)		
GamL (b λ,c)	3.58760	52001.4	37029.7		
	(0.5133)	(7955.0)	(81.16)		
ROLLL (b,λ)	3.89056	0.57316			
	(0.3652)	(0.0195)			
RBHL (λ,c)	1080175	5136722			
	(983309)	(232313)			
L (λ,c)	51425.44	1317902			
	(5933.52)	(296.120)			

Table 10: G-O-F statistics for the 1st data set.

Model	**AICr**	**BICr**	**CAICr**	**HQICr**
PTLGL	**269.8712**	**282.0253**	**270.6404**	**274.757**
OLLL	274.847	282.139	275.147	277.779
TTLL	279.140	288.863	279.646	283.049
GamL	282.808	290.136	283.105	285.756
BL	285.435	295.206	285.935	289.365
exp-L	288.799	296.127	289.096	291.747
ROLLL	289.690	294.552	289.839	291.645
SGML	292.175	299.467	292.475	295.106
RTTLL	313.962	321.254	314.262	316.893
PRHRL	331.754	339.046	332.054	334.686
L	333.977	338.862	334.123	335.942
RBHL	341.208	346.070	341.356	343.162

Table 11: MLEs and SEs for the 2nd data set.

Model	Estimates				
PTLGL (a,α,β,λ,c)	**–2.9426** **(1.4316)**	**0.0689** **(0.0575)**	**15.3547** **(12.667)**	**17.7850** **(18.172)**	**22.6267** **(24.430)**
BL ($a,b\ \lambda,c$)	1.9218 (0.318)	31.2594 (316.84)	4.9684 (50.528)	169.572 (339.21)	
KL ($a,b,\ \lambda,c$)	1.6691 (0.257)	60.5673 (86.013)	2.56490 (4.7589)	65.0640 (177.59)	
TTLL ($a,b,\ \lambda,c$)	(-0.607) (0.2137)	1.78578 (0.4152)	2123.39 (163.92)	4822.79 (200.01)	
RTTLL ($b,\beta,\ \lambda$)	–0.6715 (0.18746)	2.74496 (0.6696)	1.01238 (0.1141)		
PRHRL ($b,\ \lambda,c$)	1.59×10^{6} 2.01×10^{3}	3.93×10^{-1} 0.001×10^{-1}	1.30×10^{6} 0.95×10^{6}		
SGML ($b\ \lambda,c$)	-1.04×10^{-1} (4.1×10^{-10})	6.45×10^{6} (3.21×10^{6})	6.33×10^{6} (3.8573)		
GamL ($b,\ \lambda,c$)	1.9073 (0.3213)	35842.433 (6945.074)	39197.57 (151.653)		
OLLL ($b,\ \lambda,c$)	1.66419 (1.8×10^{-1})	6.340×10^{5} (1.68×10^{4})	2.01×10^{6} 7.22×10^{6}		
exp-L ($b,\ \lambda,c$)	1.9145 (0.348)	22971.15 (3209.53)	32882.0 (162.22)		
RBHL (λ,c)	14055522 (422.01)	53203423 (28.5232)			
ROLLL ($b,\ \lambda$)	2.37233 (0.2683)	0.69109 (0.0449)			
L (λ,c)	99269.8 (11864)	207019.4 (301.237)			

Table 12: G-O-F statistics for the 2nd data set.

Model	AICr	BICr	CAICr	HQICr
PTLGL	**208.582**	**218.2977**	**209.6347**	**212.7966**
KL	209.735	218.308	210.425	213.107
TTLL	212.900	221.472	213.589	216.271
GamL	211.666	218.096	212.073	214.195
SGML	211.788	218.218	212.195	214.317
BL	213.922	222.495	214.612	217.294
exp-L	213.099	219.529	213.506	215.628
OLLL	215.808	222.238	216.215	218.337
PRHRL	224.597	231.027	225.004	227.126
L	222.598	226.884	222.798	224.283
ROLLL	225.457	229.744	225.657	227.143
RTTLL	230.371	236.800	230.778	232.900
RBHL	229.201	233.487	229.401	230.887

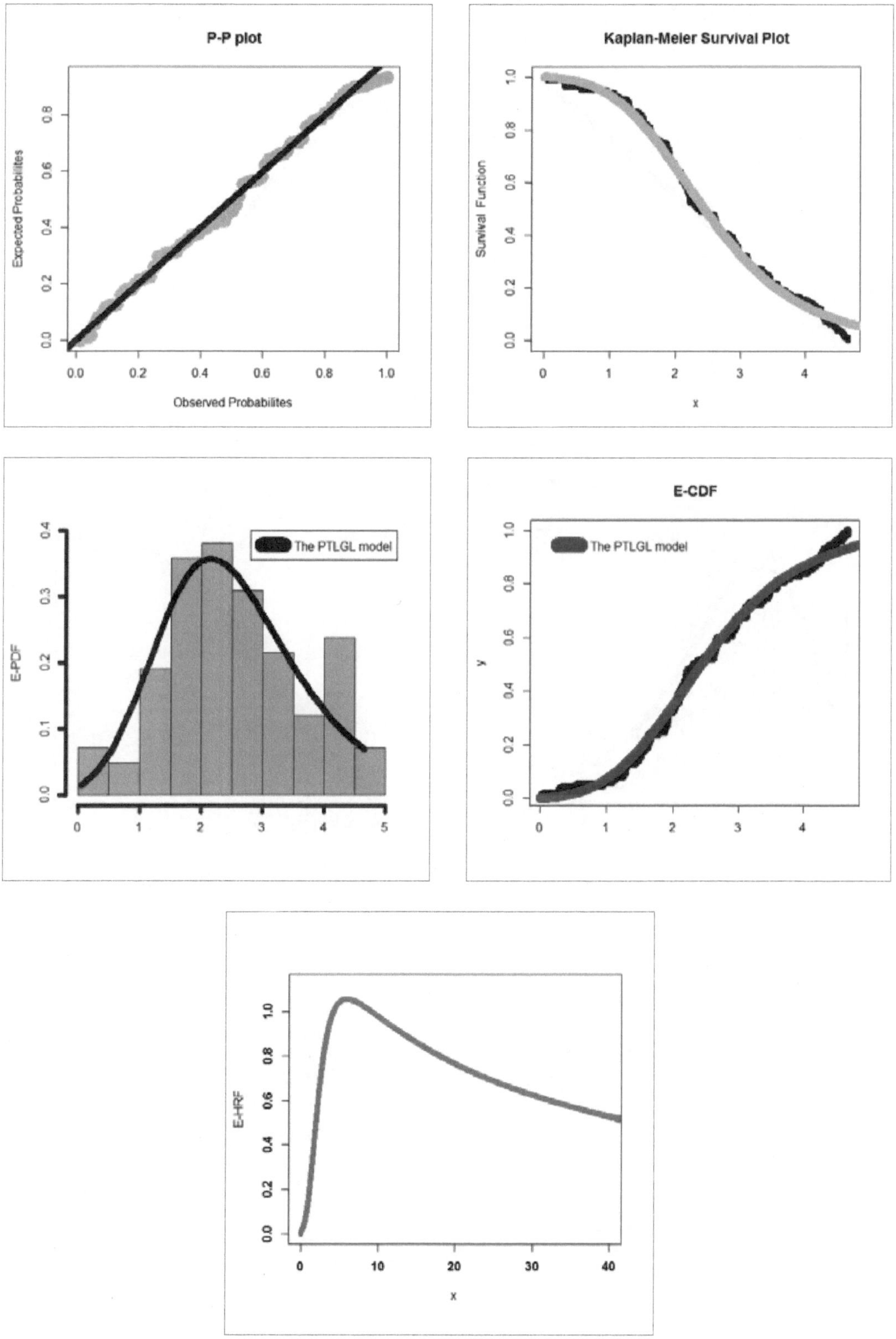

Figure 4: EPDF, EHRF, P-P, KMS plots for the 1st data set.

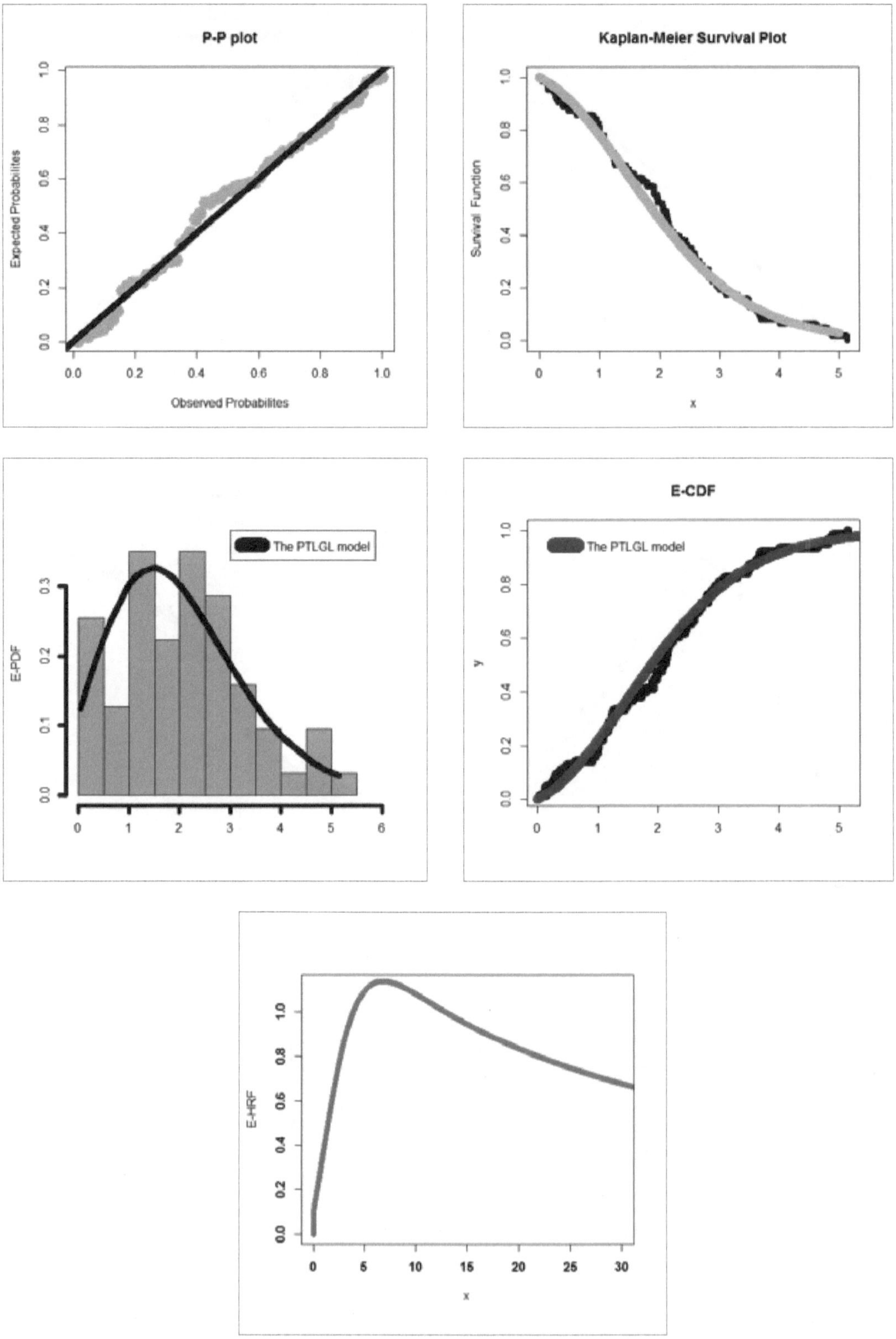

Figure 5: EPDF, EHRF, P-P, KMS plots for the 2nd data set.

9. Conclusions

A new compound G family of distributions called the Poisson Topp Leone generated-G (PTLG-G) family is defined and studied. The PTLG-G family is constructed by compounding the Poisson, and the Topp Leone generated G families. A special case based on the Lomax model called the Poisson Topp Leone generated Lomax (PTLGL) model is studied and analyzed. The density function of the PTLGL model can be "asymmetric right-skewed" and "symmetric" with many useful shapes. The hazard rate of the PTLGL model can be "constant", "decreasing", "upside-down", "increasing" and " increasing-constant". Relevant properties of the PTLGL model, including moment of the residual life, ordinary moments, moment of the reversed residual life, incomplete moments, probability weighted moments, order statistics and mean deviation, are derived and numerically analyzed. Several new bivariate PTLG-G families using the "Clayton copula", "Farlie-Gumbel-Morgenstern copula", "modified Farlie-Gumbel-Morgenstern copula", "Ali-Mikhail-Haq copula" and "Renyi's entropy copula" are investigated. Certain characterizations based on two truncated moments and the reverse hazard function are presented. We briefly describe seven classical estimation methods: the maximum likelihood, Cramér-von-Mises, ordinary least squares, weighted least square, right tail Anderson Darling, and left tail Anderson Darling methods used in the estimation process. Monte Carlo simulation experiments are performed to compare the performances of the proposed estimation methods for both small and large samples. These methods are used in the estimation process of the unknown parameters. Monte Carlo simulation experiments are performed to compare the performances of the proposed estimation methods for both small and large samples. Two different applications to real-life datasets are presented to illustrate the applicability and importance of the PTLG-G family. For the two real datasets: The "initial density shapes" are explored by the nonparametric Kernel density function, the "normality condition" is checked by the "Quantile-Quantile plot", the shape of the hazard rates is discovered by the "total time in test" graphical tool, the "box plots explore the extremes". Based on the two applications, the PTLGL distribution gives the lowest values for all test statistics with AICr = 269.8712, CAICr = 270.6404, BICr = 282.0253 and HQICr = 274.7570 for the failure times data, and AICr = 208.582, CAICr = 209.6347, BICr = 218.2977 and HQICr = 212.7966 for the service times data among all fitted competitive models.

As a future work, we can apply many new useful goodness-of-fit tests for right censored validation such as the Nikulin-Rao-Robson goodness-of-fit test and Bagdonavičius-Nikulin goodness-of-fit test as performed by Ibrahim et al. (2019), Goual et al. (2019, 2020), Mansour et al. (2020a,b,c,d,e,f), Yadav et al. (2020), Goual and Yousof (2020), Aidi et al. (2021) Yadav et al. (2022) and Ibrahim et al. (2022), among others.

References

Aboraya, M., M. Yousof, H. M., Hamedani, G. G. and Ibrahim, M. (2020). A new family of discrete distributions with mathematical properties, characterizations, Bayesian and non-Bayesian estimation methods. Mathematics, 8: 1648.

Aidi, K., Butt, N. S., Ali, M. M., Ibrahim, M., Yousof, H. M. et al. (2021). A modified chi-square type test statistic for the Double Burr X model with applications to right censored medical and reliability data. Pakistan Journal of Statistics and Operation Research, 17(3): 615–623.

Ali, M. M., Mikhail, N. N. and Haq, M. S. (1978). A class of bivariate distributions including the bivariate logistic. J. Multivar. Anal, 8: 405–412.

Alizadeh, M., Jamal, F., Yousof, H. M., Khanahmadi, M., Hamedani, G. G. et al. (2020a). Flexible Weibull generated family of distributions: characterizations, mathematical properties and applications. University Politehnica of Bucharest Scientific Bulletin-Series A-Applied Mathematics and Physics, 82(1): 145–150.

Alizadeh, M., Yousof, H. M., Jahanshahi, S. M. A., Najibi, S. M., Hamedani, G. G. et al. (2020b). The transmuted odd log-logistic-G family of distributions. Journal of Statistics and Management Systems, 23(4): 1–27.

Altun, E., Yousof, H. M. and Hamedani, G. G. (2021). The Gudermannian generated family of distributions with characterizations, regression models and applications, Studia Scientiarum Mathematicarum Hungarica, forthcoming.

Aryal, G. R. and Yousof, H. M. (2017). The exponentiated generalized-G Poisson family of distributions. Economic Quality Control, 32(1): 1–17.

Balakrishnan, N. and Lai, C. D. (2009). Continuous Bivariate Distributions; Springer Science & Business Media: Berlin/ Heidelberg, Germany.

Brito, E., Cordeiro, G. M., Yousof, H. M., Alizadeh, M. Silva, G. O. et al. (2017). Topp-leone odd log-logistic family of distributions. Journal of Statistical Computation and Simulation, 87(15): 3040–3058.

Chesneau, C. and Yousof, H. M. (2021). On a special generalized mixture class of probabilistic models. Journal of Nonlinear Modeling and Analysis, 3(1): 71–92.

Chesneau, C., Yousof, H. M., Hamedani, G. and Ibrahim, M. (2022). A New One-parameter Discrete Distribution: The Discrete Inverse BurrDistribution: Characterizations, Properties, Applications, Bayesian and Non-Bayesian Estimations. Statistics, Optimization & Information Computing, forthcoming.

Cordeiro, G. M., Ortega, E. M. and Popovic, B. V. (2015). The gamma-Lomax distribution. Journal of Statistical computation and Simulation, 85(2): 305–319.

Cordeiro, G. M., Yousof, H. M., Ramires, T. G. and Ortega, E. M. M. (2018). The Burr XII system of densities: properties, regression model and applications. Journal of Statistical Computation and Simulation, 88(3): 432–456.

Elgohari, H. and Yousof, H. M. (2020). A Generalization of Lomax Distribution with Properties, Copula and Real Data Applications. Pakistan Journal of Statistics and Operation Research, 16(4): 697–711.

El-Morshedy, M., Alshammari, F. S., Hamed, Y. S., Eliwa, M. S., Yousof, H. M. et al. (2021). A New Family of Continuous Probability Distributions. Entropy, 23: 194.

Eugene, N., Lee, C. and Famoye, F. (2002). Beta-normal distribution and its applications. Commun. Stat. Theory Methods, 31: 497–512.

Farlie, D.J.G. (1960). The performance of some correlation coefficients for a general bivariate distribution. Biometrika, 47: 307–323.

Glänzel, W. (1987). A characterization theorem based on truncated moments and its application to some distribution families, Mathematical Statistics and Probability Theory (Bad Tatzmannsdorf, 1986), Vol. B, Reidel, Dordrecht, 1987: 75–84.

Goual, H., Yousof, H. M. and Ali, M. M. (2019). Validation of the odd Lindley exponentiated exponential by a modified goodness of fit test with applications to censored and complete data. Pakistan Journal of Statistics and Operation Research, 15(3): 745–771.

Goual, H. and Yousof, H. M. (2020). Validation of Burr XII inverse Rayleigh model via a modified chi-squared goodness-of-fit test. Journal of Applied Statistics, 47(3): 393–423.

Goual, H., Yousof, H. M. and Ali, M. M. (2020). Lomax inverse Weibull model: properties, applications, and a modified Chi-squared goodness-of-fit test for validation. Journal of Nonlinear Sciences & Applications, 13(6): 330–353.

Gumbel, E. J. (1960). Bivariate exponential distributions. J. Am. Stat. Assoc., 55: 698–707.

Gumbel, E. J. (1961). Bivariate logistic distributions. J. Am. Stat. Assoc., 56: 335–349.

Gupta, R. C., Gupta, P. L. and Gupta, R. D. (1998). Modeling failure time data by Lehman alternatives. Communications in Statistics-Theory and Methods, 27(4): 887–904.

Hamedani, G. G. Yousof, H. M., Rasekhi, M., Alizadeh, M., Najibi, S. M. et al. (2017). Type I general exponential class of distributions. Pak. J. Stat. Oper. Res., XIV (1): 39–55.

Hamedani, G. G., Altun, E, Korkmaz, M. C., Yousof, H. M., Butt, N. S et al. (2018). A new extended G family of continuous distributions with mathematical properties, characterizations and regression modeling. Pak. J. Stat. Oper. Res., 14(3): 737–758.

Hamedani, G. G. Rasekhi, M., Najib, S. M., Yousof, H. M., Alizadeh, M. et al. (2019). Type II general exponential class of distributions. Pak. J. Stat. Oper. Res., XV (2): 503–523.

Hamedani, G. G., Korkmaz, M. C., Butt, N. S. and Yousof, H. M. (2021). The Type I Quasi Lambert Family: Properties, Characterizations and Different Estimation Methods. Pakistan Journal of Statistics and Operation Research, 17(3): 545–558.

Hamedani, G. G., Korkmaz, M. Ç., Butt, N. S. and Yousof H. M. (2022). The Type II quasi lambert g family of probability distributions. Pakistan Journal of Statistics and Operation Research, Forthcoming.

Ibrahim, M., Yadav, A. S., Yousof, H. M., Goual, H., Hamedani, G. G. et al. (2019). A new extension of Lindley distribution: modified validation test, characterizations and different methods of estimation. Communications for Statistical Applications and Methods, 26(5): 473–495.

Ibrahim, M., Ali, M. M. and Yousof, H. M. (2021). The discrete analogue of the Weibull G family: properties, different applications, Bayesian and non-Bayesian estimation methods. Annals of Data Science, forthcoming.

Ibrahim, M., Hamedani, G. G., Butt, N. S., Yousof, H. M. et al. (2022). Expanding the Nadarajah Haghighi Model: Characterizations, Properties and Application. Pakistan Journal of Statistics and Operation Research, forthcoming.

Johnson, N. L. and Kotz, S. (1975). On some generalized Farlie-Gumbel-Morgenstern distributions. Commun. Stat. Theory, 4: 415–427.

Johnson, N. L. and Kotz, S. (1977). On some generalized Farlie-Gumbel-Morgenstern distributions-II: Regression, correlation and further generalizations. Commun. Stat. Theory, 6: 485–496.

Karamikabir, H., Afshari, M., Yousof, H. M., Alizadeh, M., Hamedani, G. et al. (2020). The Weibull topp-leone generated family of distributions: statistical properties and applications. Journal of The Iranian Statistical Society, 19(1): 121–161.

Korkmaz, M. C. Yousof, H. M. and Hamedani, G. G. (2018a). The exponential Lindley odd log-logistic G family: properties, characterizations and applications. Journal of Statistical Theory and Applications, 17(3): 554–571.

Korkmaz, M. C., Yousof, H. M., Hamedani, G. G. and Ali, M. M. (2018b). The Marshall–Olkin generalized G Poisson family of distributions. Pakistan Journal of Statistics, 34(3): 251–267.

Korkmaz, M. Ç., Altun, E., Yousof, H. M. and Hamedani, G. G. (2020). The Hjorth's IDB generator of distributions: properties, characterizations, regression modeling and applications. Journal of Statistical Theory and Applications, 19(1): 59–74.

Lemonte, A. J. and Cordeiro, G. M. (2013). An extended Lomax distribution. Statistics, 47(4): 800–816.

Lomax, K.S. (1954). Business failures: Another example of the analysis of failure data, Journal of the American Statistical Association, 49: 847–852.

Mansour, M. M., Ibrahim, M., Aidi, K., Shafique Butt, N., Ali, M. M. et al. (2020a). A new log-logistic lifetime model with mathematical properties, copula, modified goodness-of-fit test for validation and real data modeling. Mathematics, 8(9): 1508.

Mansour, M. M., Butt, N. S., Ansari, S. I., Yousof, H. M., Ali, M. M. et al. (2020b). A new exponentiated Weibull distribution's extension: copula, mathematical properties and applications. Contributions to Mathematics, 1: 57–66. DOI: 10.47443/cm.2020.0018.

Mansour, M., Korkmaz, M. C., Ali, M. M., Yousof, H. M., Ansari, S. I. and Ibrahim, M. (2020c). A generalization of the exponentiated Weibull model with properties, Copula and application. Eurasian Bulletin of Mathematics, 3(2): 84–102.

Mansour, M., Rasekhi, M., Ibrahim, M., Aidi, K., Yousof, H. M. et al. (2020d). A New Parametric Life Distribution with Modified Bagdonavičius–Nikulin Goodness-of-Fit test for censored validation, properties, applications, and different estimation methods. Entropy, 22(5): 592.

Mansour, M., Yousof, H. M., Shehata, W. A. and Ibrahim, M. (2020e). A new two parameter Burr XII distribution: properties, copula, different estimation methods and modeling acute bone cancer data. Journal of Nonlinear Science and Applications, 13(5): 223–238.

Mansour, M. M., Butt, N. S., Yousof, H. M., Ansari, S. I., Ibrahim, M. et al. (2020f). A generalization of reciprocal exponential model: clayton copula, statistical properties and modeling skewed and symmetric real data sets. Pakistan Journal of Statistics and Operation Research, 16(2): 373–386.

Marshall, A. W. and Olkin, I. (1997). A new method for adding a parameter to a family of distributions with application to the Exponential and Weibull families. Biometrika, 84: 641–652.

Merovci, F., Alizadeh, M., Yousof, H. M. and Hamedani, G. G. (2017). The exponentiated transmuted-G family of distributions: theory and applications. Communications in Statistics-Theory and Methods, 46(21): 10800–10822.

Merovci, F., Yousof, H. M. and Hamedani, G. G. (2020). The poisson topp leone generator of distributions for lifetime data: theory, characterizations and applications. Pakistan Journal of Statistics and Operation Research, 16(2): 343–355.

Morgenstern, D. (1956). Einfache beispiele zweidimensionaler verteilungen. Mitteilingsbl. Math. Stat., 8: 234–235.

Murthy, D.N.P., Xie, M.and Jiang, R. (2004). Weibull Models; John Wiley & Sons: Hoboken, NJ, USA.

Nascimento, A. D. C., Silva, K. F., Cordeiro, G. M., Alizadeh, M., Yousof, H. M. et al. (2019). The odd Nadarajah-Haghighi family of distributions: properties and applications. Studia Scientiarum Mathematicarum Hungarica, 56(2): 1–26.

Nelsen, R. B. (2007). An Introduction to Copulas; Springer Science & Business Media: Berlin/Heidelberg, Germany.

Nofal, Z. M., Afify, A. Z., Yousof, H. M. and Cordeiro, G. M. (2017). The generalized transmuted-G family of distributions. Communications in Statistics-Theory and Method, 46: 4119–4136.

Pougaza, D. B. and Djafari, M. A. (2010). Maximum entropies copulas. In Proceedings of the 30th international workshop on Bayesian inference and maximum Entropy methods in Science and Engineering, Chamonix, France, 4–9 July 2010; pp. 329–336.

Rezaei, S., B. B. Sadr, M. Alizadeh and S. Nadarajah. (2017). Topp-Leone generated family of distributions: Properties and applications. Communications in Statistics: Theory and Methods 46(6): 2893–2909.

Shehata, W. A. M. and Yousof, H. M. (2021a). The four-parameter exponentiated Weibull model with Copula, properties and real data modeling. Pakistan Journal of Statistics and Operation Research, 17(3): 649–667.

Shehata, W. A. M. and Yousof, H. M. (2021b). A novel two-parameter Nadarajah-Haghighi extension: properties, copulas, modeling real data and different estimation methods. Statistics, Optimization & Information Computing, forthcoming.

Shehata, W. A. M., Yousof, H. M. and Aboraya, M. (2021). A novel generator of continuous probability distributions for the asymmetric left-skewed bimodal real-life data with properties and copulas. Pakistan Journal of Statistics and Operation Research, 17(4): 943–961.

Yadav, A. S., Goual, H., Alotaibi, R. M., Ali, M. M., Yousof, H. M. et al. (2020). Validation of the Topp-Leone-Lomax model via a modified Nikulin-Rao-Robson goodness-of-fit test with different methods of estimation. Symmetry, 12(1): 57.

Yadav, A. S., Shukl, S., Goual, H., Saha, M., Yousof, H. M et al. (2022). Validation of xgamma exponential model via nikulin-rao-robson goodness-of- fit-test under complete and censored sample with different methods of estimation. Statistics, Optimization & Information Computing, 10(2): 457–483.

Yousof, H. M., Afify, A. Z., Alizadeh, M., Butt, N. S., Hamedani, G. G. et al. (2015). The transmuted exponentiated generalized-G family of distributions. Pak. J. Stat. Oper. Res., 11: 441–464.

Yousof, H. M., Afify, A. Z., Hamedani, G. G. and Aryal, G. (2017a). The Burr X generator of distributions for lifetime data. Journal of Statistical Theory and Applications, 16: 288–305.

Yousof, H. M., Alizadeh, M., Jahanshahi, S. M. A., Ramires, T. G., Ghosh, I. and Hamedani, G. G. (2017b). The transmuted Topp-Leone G family of distributions: theory, characterizations and applications. Journal of Data Science, 15(4): 723–740.

Yousof, H. M., Altun, E., Ramires, T. G., Alizadeh, M. and Rasekhi, M. et al. (2018c). A new family of distributions with properties, regression models and applications. Journal of Statistics and Management Systems, 21(1): 163–188.

Yousof, H. M., Mansoor, M. Alizadeh, M., Afify, A. Z., Ghosh, I. et al. (2020). The Weibull-G Poisson family for analyzcheckeding lifetime data. Pak. J. Stat. Oper. Res., 16(1): 131–148.

Yousof, H. M., Chesneau, C., Hamedani, G. and Ibrahim, M. (2021). A New Discrete Distribution: Properties, Characterizations, Modeling Real Count Data, Bayesian and Non-Bayesian Estimations. Statistica, 81(2): 135–162.

Chapter 2

A Novel Family of Continuous Distributions

Properties, Characterizations, Statistical Modeling and Different Estimation Methods

Haitham M Yousof,[1,*] *M Masoom Ali,*[2] *Gauss M Cordeiro,*[3] *GG Hamedani*[4] and *Mohamed Ibrahim*[5]

1. Introduction

Statistical literature contains various G families of distributions which were generated either by compounding well-known existing G families or by adding one (or more) parameters to the existing classes. These novel families were employed for modeling real data in many applied areas such as engineering, insurance, demography, medicine, econometrics, biology and environmental sciences; see Cordeiro and de Castro (2011) (Kumaraswamy G family), Cordeiro et al. (2014) (the Lomax generator), Afify et al. (2016a) (transmuted geometric-G family), Afify et al. (2016b) (complementary geometric transmuted family), Aryal and Yousof (2017) (exponentiated generalized Poisson family), Brito et al. (2017) (Topp Leone odd log-logistic family), Yousof et al. (2017) (Burr X family), Cordeiro et al. (2018) (Burr XII family), Korkmaz et al. (2018) (exponential-Lindley odd log-logistic family) and Karamikabir et al. (2020) (Weibull Topp Leone generated family), Hamedani et al. (2021) and Hamedani et al. (2022) (type I quasi Lambert family and type II quasi Lambert family) among others. For other useful G families see Hamedani et al. (2017, 2018) and Hamedani et al. (2019).

We propose and study a new family of distributions called the geometric generated Rayleigh (GcGR) family with a strong physical motivation. Let $g_{\underline{\mathbb{V}}}$ (x) and $G_{\underline{\mathbb{V}}}(x)$ denote the probability density function (PDF) and cumulative distribution function (CDF) of an arbitrary baseline model with parameter vector $\underline{\mathbb{V}}$ and consider the CDF of the generated Rayleigh (GR) family

$$H_{\beta,\underline{\mathbb{V}}}(x) = 1 - \exp[-\nabla^2_{\beta,\underline{\mathbb{V}}}(x)]|_{x\in\mathbf{R},\,\beta>0}, \tag{1}$$

where, $\nabla_{\beta,\underline{\mathbb{V}}}(x) = \dfrac{G^{\beta}_{\underline{\mathbb{V}}}(x)}{1-G^{\beta}_{\underline{\mathbb{V}}}(x)}|_{x\in\mathbf{R},\beta>0}$ and $k_{\beta,\underline{\mathbb{V}}}(x) = dH_{\beta,\underline{\mathbb{V}}}(x)/dx$ is the PDF corresponding to (1). For any arbitrary baseline random variable (RV) having CDF $G_{\underline{\mathbb{V}}}(x)$, the CDF of the geometric G (Gc-G) family is defined by,

[1] Department of Statistics, Mathematics and Insurance, Faculty of Commerce, Benha University, Benha 13518, Egypt
[2] Department of Mathematical Sciences, Ball State University, Muncie, IN 47306 USA
[3] Universidade Federal de Pernambuco, Departamento de Estatística, Brazil.
[4] Department of Mathematical and Statistical Sciences, Marquette University, USA
[5] Department of Applied, Mathematical and Actuarial Statistics, Faculty of Commerce, Damietta University, Damietta, Egypt.
Emails: mali@bsu.edu; gauss@de.ufpe.br; gholamhoss.hamedani@marquette.edu; mohamed_ibrahim@du.edu.eg
* Corresponding author: haitham.yousof@fcom.bu.edu.eg

$$F_{\theta,\underline{V}}(x) = \frac{\theta G_{\underline{V}}(x)}{1-(1-\theta)G_{\underline{V}}(x)}|_{x\in\mathbf{R},\theta>0.} \tag{2}$$

By combining (1) and (2), we propose and study a new extension of the well-known Gc-G family to provide more flexibility to the generated family. The new family has a strong physical motivation as given in Section 2. The new family is derived based on expanding the geometric Rayleigh family with the generated odd ratio $\nabla_{\beta,\underline{V}}(x)$. After a quick study of its properties, different classical estimation methods under uncensored schemes are considered, such as the maximum likelihood (ML), Anderson–Darling (AD), ordinary least squares (OLS), Cramér-von Mises (CVM), weighted least squares (WLS), left-tail Anderson–Darling (LTAD), and right-tail Anderson–Darling (RTAD) methods. Numerical simulations are performed for comparing the estimation methods using different sample sizes for three different combinations of parameters.

In fact, the GcGR family is motivated by its flexibility in applications which is important. By means of three applications, we show that the GcGR class provides better fits than many other families. The new family could be useful in modeling real data with an "asymmetric monotonically increasing hazard rate function (HRF)" as illustrated in Figure 1, Figure 4 and Figure 7 (1st row right panels); the real data which has some extreme values as shown in Figure 1 and Figure 4 (2nd row right and left panels); the real data which has no extreme values as shown in Figure 7 (2nd row right and left panels); the real data for which its nonparametric Kernel density is asymmetric bimodal and heavy tail as illustrated in Figure 1, Figure 4 (1st row left panels); the real data for which its nonparametric Kernel density is symmetric and unimodal as illustrated in Figure 7 (1st row left panels); the real data which cannot be fitted by the common theoretical distributions such as normal, uniform, exponential, logistic, beta, lognormal and Weibull distributions as illustrated in Figure 1, Figure 4 and Figure 7 (3rd row right panels).

The rest of the paper is organized as follows. In Section 2, we define the GcGR family and give a useful representation of its density function. In Section 3, we derive some of its mathematical properties. In Section 4, some characterization results are addressed. In Section 5, we present a special model corresponding to the baseline Fréchet distribution. Different classical estimation methods under uncensored schemes are addressed in Section 6. Numerical simulations are performed for comparing the estimation methods under different scenarios in Section 7. In Section 8, we provide three applications to real data to illustrate the flexibility of the new family. Finally, some concluding remarks are addressed in Section 9.

2. The new family

We use (1) and (2) to construct a new two-parameter family of continuous probability distributions called the GcGR family by taking (1) as the baseline CDF of (2). The CDF of the GcGR family can be defined by,

$$F_{\underline{\varsigma}}(x) = \frac{\theta-\theta\, exp\left[-\nabla^2_{\beta,\underline{V}}(x)\right]}{1-(1-\theta)\left\{1-exp\left[-\nabla^2_{\beta,\underline{V}}(x)\right]\right\}}|_{x\in\mathbf{R},\theta>0,\beta>0,} \tag{3}$$

where, $\underline{\varsigma} = (\theta,\beta,\underline{V})$ refers to the parameter vector. The new CDF in (3) can be used for presenting a new discrete G family for modeling the count data (refer to Aboraya et al. (2020), Ibrahim et al. (2021), Chesneau et al. (2021) and Yousof et al. (2021) for more details). Then, the PDF corresponding to (3) is,

$$f_{\underline{\varsigma}}(x) = 2\theta\beta \frac{g_{\underline{V}}(x)G^{\beta-1}_{\underline{V}}(x)[1-G^{\beta}_{\underline{V}}(x)]^{-3}\, exp\left[-\nabla^2_{\beta,\underline{V}}(x)\right]}{\left(1-(1-\theta)\left\{1-exp\left[-\nabla^2_{\beta,\underline{V}}(x)\right]\right\}\right)^2}|_{x\in\mathbf{R},\theta>0,\beta>0.} \tag{4}$$

The GcGR family is motivated by the following points. Suppose a system is made up N of independent components in series, where N is a RV with a standard geometric distribution and probability mass function (PMF)

$$\mathcal{p}(N = n;\, \theta) = \theta(1-\theta)^{n-1}\,|_{n\in\mathcal{N} \text{ and } \theta\in(0,1)},$$

where, $\mathcal{N} = \{1,2,\ldots\}$. Suppose that RVs $X_1, X_2, \ldots, X_n$ represent the lifetimes of the component of the GR family. Then,

$$Y = \min(X_1, X_2, \ldots, X_n)$$

represents the time for the first failure with CDF (3) given $\theta \in (0,1)$. In a similar manner, consider now a parallel system with N independent components and suppose that a RV N has a geometric distribution with the PMF

$$\mathcal{p}(N = n;\, \theta) = \frac{1}{\theta}\left(1-\frac{1}{\theta}\right)^{n-1}\,|_{n\in\mathcal{N} \text{ and } \theta>2}.$$

Let $X_1, X_2, \ldots, X_n$ be the lifetimes of the GR family components. Then,

$$T = \max(X_1, X_2, \ldots, X_n)$$

represents the lifetime of the system. Therefore, the RV X follows (4) given $\theta > 1$.

Some power series expansions for Equations (3) and (4) can be derived using the concept of exponentiated G (Exp-G) family of distributions. Hereafter, for an arbitrary baseline CDF $G_{\underline{\mathbb{V}}}(x)$, we define a RV Y_β having the Exp-G distribution with power parameter $\beta > 0$, say $Y \sim$ Exp-G$(\beta, \underline{\mathbb{V}})$, if its PDF and CDF are given by $\pi_{\beta,\underline{\mathbb{V}}}(x) = \beta g_{\underline{\mathbb{V}}}(x) G_{\underline{\mathbb{V}}}(x)^{\beta-1}$ and $\mathbf{\Pi}_{\beta,\underline{\mathbb{V}}}(x) = G_{\underline{\mathbb{V}}}(x)^\beta$, respectively. Using the generalized binomial expansion and the power series, the PDF in (4) can be expressed as,

$$f_{\underline{\varsigma}}(x) = 2\theta\beta g_{\underline{\mathbb{V}}}(x) \sum_{i,j,k=0}^{\infty} \frac{(1-\theta)^i(-1)^k}{k!(1+i)^{-k}} \frac{G_{\underline{\mathbb{V}}}^{2\beta(k+1)-1}(x)]}{[1-G_{\underline{\mathbb{V}}}^{\beta}(x)]^{3+2k}} \binom{-2}{i}\binom{i}{j}.$$

Using the Taylor expansion, we can write,

$$f_{\underline{\varsigma}}(x) = \sum_{k,m=0}^{\infty} \mathcal{M}_{k,m}\, \pi_{\beta*,\underline{\mathbb{V}}}(x)|_{\beta=\beta[2(k+1)+m]>0}, \tag{5}$$

where,

$$\mathcal{M}_{k,m} = 2\theta\beta \sum_{i=0}^{\infty}\sum_{j=0}^{i} \frac{(1-\theta)^i(-1)^k(1+)^k}{k!\beta*} \binom{-2}{j}\binom{i}{j}\binom{-3-2k}{m}.$$

Thus, some mathematical properties of the GcGR family can be obtained simply from the properties of the Exp-G family. Equation (5) is the main result of this section. The CDF of the GcGR family can also be expressed as a mixture of Exp-G densities. By integrating (5), we obtain the same mixture representation,

$$F_{\underline{\varsigma}}(x) = \sum_{k,m=0}^{\infty} \mathcal{M}_{k,m}\, \mathbf{\Pi}_{\beta*,\underline{\mathbb{V}}}(x)|_{\beta*>0}, \tag{6}$$

where $\mathbf{\Pi}_{\beta*,\underline{\mathbb{V}}}(x)$ is the CDF of the Exp-G family with power parameter $\beta* > 0$.

3. Properties

The rth moment of X, say $\mu'_{r,X}$, follows from (5) as $\mu'_{r,X} = E(X^r) = \sum_{k,m=0}^{\infty} \mathcal{M}_{k,m}\, E(Y^r_{\beta*})$.

Henceforth, $Y_{\beta*}$ denotes the Exp-G distribution with positive power parameter $\beta*$. We have $E(Y^r_{\beta*}) = \beta* \int_{-\infty}^{\infty} x^r\, g_{\underline{V}}(x)\, G_{\underline{V}}(x)^{\beta*-1}\, dx$, which can be computed numerically in terms of the baseline quantile function (QF) $Q_{G,\underline{V}}(u) = G^{-1}_{\underline{V}}(u)$ as $E(Y^n_{\gamma}) = \beta* \int_0^1 Q_{G,\underline{V}}(u)^n\, u^{\beta*-1} du$. The variance, skewness, and kurtosis measures can now be calculated through simple relations. Then, the moment generating function (MGF) $M_X(t) = E(exp(tX))$ of X can be derived from Equation (5) as,

$$M_X(t) = \sum_{k,m=0}^{\infty} \mathcal{M}_{k,m}\, M_{\beta*}(t),$$

where $M_{\beta*}(t)$ is the MGF of $Y_{\beta*}$. Hence, $M_{\beta*}(t)$ can be determined from the Exp-G generating function. The sth incomplete moment, say $\phi_{s,X}(t)$, of X can be expressed from (5) as

$$\phi_{s,X}(t) = \int_{-\infty}^{t} x^s f_{\underline{\varsigma}}(x)dx = \sum_{k,m=0}^{\infty} \mathcal{M}_{k,m}\, \mathbf{I}^{t}_{-\infty}(x^s;\beta*),$$

where $\mathbf{I}^{t}_{-\infty}(x^s;\beta*) = \int_{-\infty}^{t} x^s\, \pi_{\beta*}(x)dx$. The nth moment of the residual life, say $m_{n,X}(t) = E[(X-t)^n \mid X > t]$, $n = 1,2,\ldots$, uniquely determines $F_{\underline{\varsigma}}(x)$. The nth moment of the residual life of X is given by $m_{n,X}(t) = \frac{1}{\boldsymbol{R}_{\varsigma}(t)} \int_t^{\infty} (x-t)^n\, dF(x)$. Then,

$$m_{n,X}(t) = \frac{1}{\boldsymbol{R}_{\varsigma}(t)} \sum_{k,m=0}^{\infty} \sum_{r=0}^{n} \mathcal{M}_{k,m} \binom{n}{r} (-t)^{n-r}\, \mathbf{I}^{\infty}_{t}(x^n;\beta*),$$

where $\mathbf{I}^{\infty}_{t}(x^n;\beta*) = \int_t^{\infty} x^n\, \pi_{\beta*}(x)dx$. The nth moment of the reversed residual life, say $M_{n,X}(t) = E[(t-X)^n \mid X \le t]$ for $t > 0$ and $n = 1,2,\ldots$, follows as $M_{n,X}(t) = \frac{1}{F_{\varsigma}(t)} \int_0^t (t-x)^n\, dF(x)$. Therefore, the nth moment of the reversed residual life of X becomes

$$M_n(t) = \frac{1}{F_{\varsigma}(t)} \sum_{k,m=0}^{\infty} \sum_{r=0}^{n} \mathcal{M}_{k,m} (-1)^r \binom{n}{r} t^{n-r}\, \mathbf{I}^{t}_{0}(x^n;\beta*),$$

where, $\mathbf{I}^{t}_{0}(x^n;\beta*) = \int_0^t x^n\, \pi_{\beta*}(x)dx$.

4. Characterizations of the GcGR family

To understand the behavior of the data obtained through a given process, we need to be able to describe this behavior through its approximate probability law. This, however, requires establishing conditions which govern the required probability law. In other words, we need to have certain conditions under which we may be able to recover the data probability law. So, the characterization of a distribution is important in applied sciences, where an investigator is vitally interested in finding out if the model follows the selected distribution. Therefore, the investigator relies on conditions under which the model would follow a specified distribution. A probability distribution can be characterized in different directions one of which is based on the truncated moments. This type of characterization pioneered by Galambos and Kotz (1978) and followed by other authors such as Kotz and Shanbhag (1980), Glänzel et al. (1984), Glänzel (1987), Glänzel and Hamedani (2001) and Kim and Jeon (2013), to name a few. For example, Kim and Jeon (2013) proposed a credibility theory based on the truncation of the loss data to estimate conditional mean loss for a given risk function. It should also be mentioned that characterization results are mathematically challenging. In this section, we present certain characterizations of the GcGR distribution based on: (i) conditional expectation (truncated moment) of certain functions of a RV and (ii) reverse hazard function.

4.1 Characterizations based on two truncated moments

This subsection is devoted to the characterizations of the GcGR distribution in terms of a simple relationship between two truncated moments. We will recall the Theorem of Glänzel (1987). As shown in Glänzel (1990), this characterization is stable in the sense of weak convergence. The first characterization given below can also be employed when the CDF does not have a closed form.

Proposition 4.1.1. Let $X : \Omega \rightarrow \boldsymbol{R}$ a continuous RV and let

$$q_1(x) = [1-(1-\theta)(1-exp\{-\nabla^2_{\beta,\underline{\varphi}}(x)\})]^2$$

and

$$q_2(x) = q_1(x)\ exp\{-\nabla^2_{\beta,\underline{\varphi}}(x)\} \text{ for } x \in \boldsymbol{R}.$$

for

Then X has PDF (4) if and only if the function ξ defined in Theorem 1 in is of the form

Proof. If X has PDF, then,

$$(1-F(x))E[q_1(X) \mid X \geq x] = \theta\ exp\{-\nabla^2_{\beta,\underline{\varphi}}(x)\} \text{ for } x \in \boldsymbol{R},$$

and

$$(1-F(x))E[q_2(X) \mid X \geq x] = \frac{\theta}{2} exp\{-2\nabla^2_{\beta,\underline{\varphi}}(x)\},\ x \in \boldsymbol{R},$$

and hence

$$\xi(x) = \frac{1}{2} exp\{-\nabla^2_{\beta,\underline{\varphi}}(x)\},\ x \in \boldsymbol{R}.$$

We also have

$$\xi(x)q_1(x) - q_2(x) = -\frac{1}{2} q_1(x)\ exp\{-\nabla^2_{\beta,\underline{\varphi}}(x)\} < 0, \mathit{for}\ x \in \boldsymbol{R}.$$

Conversely, if ξ is of the above form, then

$$s'(x) = \frac{\xi'(x)q_1(x)}{\xi(x)q_1(x)-q_2(x)} = 2\beta g_{\underline{\varphi}}(x)G_{\underline{\varphi}}^{2\beta-1}(x)[1-G_{\underline{\varphi}}^{\beta}(x)]^{-3},\ x \in \boldsymbol{R}.$$

Now, according to Theorem 1, X has density (4).

Corollary 4.1.1. Suppose X is a continuous RV. Let $q_1(x)$ be as in Proposition 4.1.1. Then X has density (4) if and only if there exist functions $q_2(x)$ and $\xi(x)$ defined in Theorem 1 for which the following first order differential equation holds

$$\frac{\xi'(x)q_1(x)}{\xi(x)q_1(x)-q_2(x)} = 2\beta g_{\underline{\varphi}}(x)G_{\underline{\varphi}}^{2\beta-1}(x)[1-G_{\underline{\varphi}}^{\beta}(x)]^{-3},\ x \in \boldsymbol{R}.$$

Corollary 4.1.2. The differential equation in Corollary 4.1.1 has the following general solution

$$\xi(x) = exp[\nabla^2_{\beta,\underline{\varphi}}(x)]\begin{bmatrix} -\int 2\beta g_{\underline{\varphi}}(x)G_{\varphi}^{2\beta-1}(x)[1-G_{\underline{\varphi}}^{\beta}(x)]^{-3} \\ \times exp\{-\nabla^2_{\beta,\underline{\varphi}}(x)\}(q_1(x))^{-1}q_2(x)+D \end{bmatrix},$$

where D is a constant. A set of functions satisfying the above differential equation is given in Proposition 4.1.1 with $D = 0$. Clearly, there are other triplets ($q_1(x)$, $q_2(x)$, $\xi(x)$) satisfying the conditions of Theorem 1.

4.2 Characterization in terms of the reverse hazard function

The reverse hazard function, $r_F(x)$, of a twice differentiable distribution function, F, is defined as,

$$r_F(x) = \frac{f(x)}{F(x)},\ x \in \text{support of } F.$$

We present a characterization of the GcGR distribution in terms of the reverse hazard function.

Proposition 4.2.1. Let $X : \Omega \rightarrow \boldsymbol{R}$ be a continuous RV. The RV X has PDF (4) if and only if its reverse hazard function $r_F(x)$ satisfies the following differential equation,

$$r_F'(x) - \frac{(2\beta-1)g_{\underline{\varphi}}(x)}{G_{\underline{\varphi}}(x)} r_F(x) = 2\beta G_{\underline{\varphi}}^{2\beta-1}(x)\frac{d}{dx}\mathcal{A}_{\underline{\varphi}}^{\theta,\beta}(x), x \in \boldsymbol{R},$$

with boundary condition where,

$$\mathcal{A}_{\underline{\varphi}}^{\theta,\beta}(x) = g_{\underline{\varphi}}^{\beta}(x)\frac{[1-G_{\underline{\varphi}}^{2\beta-1}(x)]^{-3}\left(1-exp\{-\nabla_{\beta,\underline{\varphi}}^{2}(x)\}\right)^{-1}}{\left[1-(1-\theta)\left(1-exp\{-\nabla_{\beta,\underline{\varphi}}^{2}(x)\}\right)\right]}.$$

5. Special case

The Fréchet (Fr) model is one of the most important distributions in modeling extreme values. The Fr model was originally proposed by Fréchet (1927). It has many applications, for example, accelerated life testing, earthquakes, floods, wind speed, horse racing, rainfall, queues in supermarkets, and sea waves (see Von Mises (1964) and Kotz and Johnson (1992)). One can find more details about the Fr model in the literature, for example, Nadarajah and Kotz (2003) investigated the exponentiated Fr distribution. Moreover, Jahanshahi et al. (2019) defined and applied a new version of the Fr distribution, called the Burr X Fréchet (BX-Fr) model for relief times and survival times data, Krishna et al. (2013) proposed some applications of the Marshall–Olkin Fr (MO-Fr) distribution, Al-Babtain et al. (2020) investigated a new three parameter Fr model called the generalized odd generalized exponential Fréchet (GOFE-Fr) with mathematical properties and applications, among others.

A RV X is said to have the Fr distribution if its PDF and CDF are given by $g_{\alpha_1,\alpha_2}(x) = \alpha_2\, \alpha_1^{\alpha_2} x^{\alpha_2-1} \exp\left[-\left(\frac{\alpha_1}{x}\right)^{\alpha_2}\right]|_{x\geq 0}$, and $G_{\alpha_1,\alpha_2}(x) = \exp\left[-\left(\frac{\alpha_1}{x}\right)^{\alpha_2}\right]|_{x\geq 0}$, where $\alpha_1 > 0$ is a scale parameter and $\alpha_2 > 0$ is a shape parameter. Based on (3), the CDF of the geometric generated Rayleigh Fréchet (GcGR-Fr) model can be defined by

$$F_{\underline{\varsigma}}(x) = \frac{\theta - \theta\left(-\left\{\exp\left[\beta\left(\frac{\alpha_1}{x}\right)^{\alpha_2}\right]-1\right\}^{2}\right)}{1-(1-\theta)\left[1-exp\left(-\left\{\exp\left[\beta\left(\frac{\alpha_1}{x}\right)^{\alpha_2}\right]-1\right\}^{-2}\right)\right]}|_{x\geq 0,\theta>0,\beta>0},$$

where $\underline{\varsigma} = (\theta, \beta, \alpha_1, \alpha_2)$ refers to the parameters vector. For the GcGR-Fr model, we obtain the following results:

$$\mu_{r,X}' = E(X^r) = \alpha_1^r \sum_{k,m=0}^{\infty} \mathcal{M}_{k,m}\beta^{*\frac{r}{\alpha_2\Gamma}}\left(1-\frac{r}{\alpha_1}\right)|_{\alpha_2>r},$$

$$\phi_{1,X}(t) = \alpha_1 \sum_{k,m=0}^{\infty} \mathcal{M}_{k,m}\beta^{*\frac{r}{\alpha_2\gamma}}\left(1-\frac{1}{\alpha_2}, \left(\frac{\alpha_1}{t}\right)^{\alpha_2}\right)|_{\alpha_2>1},$$

$$m_{n,X}(t) = \frac{1}{R_{\underline{\varsigma}}(t)} \sum_{k,m=0}^{\infty} \sum_{r=0}^{n} \mathcal{M}_{k,m} \binom{n}{r} (-t)^{n-r} \alpha_1^n \beta * \frac{n}{\alpha_2 \Gamma} \left(1 - \frac{n}{\alpha_2}, \left(\frac{\alpha_1}{t}\right)^{\alpha_2}\right) | \alpha_2 > n$$

$$M_n(t) = \frac{1}{F_{\underline{\varsigma}}(t)} \sum_{k,m=0}^{\infty} \sum_{r=0}^{n} \mathcal{M}_{k,m} (-1)^r \binom{n}{r} t^{n-r} \alpha_1^n \beta * \frac{n}{\alpha_2 \gamma} \left(1 - \frac{n}{\alpha_2}, \left(\frac{\alpha_1}{t}\right)^{\alpha_2}\right) | \alpha_2 > n,$$

where $\Gamma(\delta)|_{\rho>0} = \int_0^\infty z^{\delta-1} \, exp(-z)dz$ and $\gamma(\delta,\rho)$ refers to the lower incomplete gamma function $\gamma(\delta,\rho)|_{(\delta\neq 0,-1,-2,\ldots)} = \int_0^\rho z^{\delta-1} \, exp(-z)dz = \sum_{\kappa=0}^{\infty} \rho^{\delta+\kappa} \frac{(-1)^\kappa}{\kappa!(\delta+\kappa)}$. The function $\Gamma(\delta,\rho)|_{\rho>0} = \int_\rho^\infty z(\delta^{-1}) \, exp(-z)dz$ and $\Gamma(\delta,\rho) = \Gamma(\delta) - \gamma(\delta,\rho)$ refer to the upper incomplete gamma function.

6. Classical estimation under uncensored scheme

We discuss seven methods to estimate the parameters of the GcGR family which can be implemented using the "Adequacy Model" script in "R" software, which provides a general meta-heuristic optimization technique for maximizing or minimizing an arbitrary objective function (Marinho et al. (2019)).

6.1 The ML method

The maximum likelihood estimates (MLEs) enjoy desirable properties and can be used when constructing confidence intervals. The normal approximation for these estimates in large sample theory is easily handled either analytically or numerically. Here, we determine the MLEs of the parameters of the new family of distributions from complete samples only. Let $\underline{\varsigma} = (\theta,\beta,\underline{\mathbb{V}}^T)^T$ be the $p \times 1$ parameter vector. To obtain the MLE of $\underline{\varsigma}$, the log-likelihood function can be expressed as

$$\ell(\underline{\varsigma}) = \hbar\log(2) + \hbar\log(\beta) + \hbar\log(\theta) + \sum_{i=0}^{\hbar} \log g_{\underline{\mathbb{V}}}(x_i) + (2\beta-1)\sum_{i=0}^{\hbar} \log G_{\underline{\mathbb{V}}}(x_i)$$

$$-3\sum_{i=0}^{\hbar} \log\left[1 - G_{\underline{\mathbb{V}}}^{\beta}(x_i)\right] - \sum_{i=0}^{\hbar} s_i^2 - 2\sum_{i=0}^{\hbar} \log z_i,$$

where, $s_i = \frac{G_{\underline{\mathbb{V}}}^{\beta}(x_i)}{1 - G_{\underline{\mathbb{V}}}^{\beta}(x_i)}$ and $z_i = \{1 - (1-\theta)[1 - exp(-s_i^2)]\}$. The function $\ell(\underline{\varsigma})$ can be maximized either directly by using the R (optim function), SAS (PROC NLMIXED) or Ox program (sub-routine MaxBFGS) or by solving the nonlinear likelihood equations obtained by differentiating $\ell(\underline{\varsigma})$. For interval estimation of the model parameters, we require the observed information matrix. Under standard regularity conditions when $n \to \infty$, the distribution of $\hat{\underline{\varsigma}}$ can be approximated by a multivariate normal distribution to construct approximate confidence intervals for the parameters.

6.2 The CVM method

The Cramér-von Mises estimates (CVMEs) of θ, β and $\underline{\mathbb{V}}$ are obtained by minimizing the following expression with respect to these parameters

$$CVM_{(\underline{\varsigma})} = \frac{1}{12}\hbar^{-1} + \sum_{i=1}^{\hbar} \left[F_{\underline{\varsigma}}(x_{[i,\hbar]}) - c_{(i,\hbar)}^{[1]}\right]^2,$$

where, $c_{(i,\hbar)}^{[1]} = \frac{2i-1}{2\hbar}$ and

$$CVM_{(\underline{\varsigma})} = \sum_{i=1}^{\hbar} \left[\frac{\theta - \theta\, exp\left[-V_{\beta,\underline{\mathbb{V}}}^2(x_{[i,\hbar]})\right]}{1 - (1-\theta)\left\{exp\left[-V_{\beta,\underline{\mathbb{V}}}^2(x_{[i,\hbar]})\right]\right\}} - c_{(i,\hbar)}^{[1]}\right]^2.$$

6.3 OLS method

Let $F_{\varsigma}(x_{[i,k]})$ denote the CDF of the new family and let $x_{[1,k]} < x_{[2,k]} < \cdots < x_{[k,k]}$ be the k ordered observations. The ordinary least squares estimates (OLSEs) are obtained by minimizing

$$OLSE(\underline{\varsigma}) = \sum_{i=1}^{k} \left\{ \frac{\theta - \theta\, exp\left[-\nabla^2_{\beta,\underline{V}}(x_{[i,k]}) \right]}{1-(1-\theta)\left\{1-exp\left[-\nabla^2_{\beta,\underline{V}}(x_{[i,k]}) \right]\right\}} - c^{[2]}_{(i,k)} \right\}^2,$$

where, $c^{[2]}_{(i,k)} = \dfrac{i}{k+1}$.

6.4 WLS method

The weighted least square estimates (WLSEs) are obtained by minimizing the function $WLSE(\underline{\varsigma})$ with respect to θ, β and $\underline{V}$,

$$WLSE(\underline{\varsigma}) = \sum_{i=1}^{k} c^{[3]}_{(i,k)} \left[F_{\underline{\varsigma}}(x_{[i,k]}) - c^{[2]}_{(i,k)} \right]^2,$$

where, $c^{[3]}_{(i,k)} = [(1+k)^2(2+k)]/[i(1+k-i)]$.

6.5 The AD method

The Anderson-Darling estimates (ADEs) are obtained by minimizing the function,

$$ADE(\underline{\varsigma}) = -k - k^{-1}\sum_{i=1}^{k}(2i-1)\left\{logF_{\underline{\varsigma}}(x_{[i,k]}) + log\overline{F}_{\underline{\varsigma}}(x_{[-i+1+k:k]})\right\}.$$

where, $\overline{F}_{\underline{\varsigma}}(x_{[-i+1+k:k]}) = 1 - F_{\underline{\varsigma}}(x_{[-i+1+k:k]})$

6.6 The RTAD method

The right-tail Anderson–Darling estimates (RTADEs) are determined by minimizing

$$\boldsymbol{RT}_{(ADE)}(\underline{\varsigma}) = \frac{1}{2}k - 2\sum_{i=1}^{k} F_{\underline{P}}(x_{[i,k]}) - \frac{1}{k}\sum_{i=1}^{k}(2i-1)\left[log\overline{F}_{\underline{\varsigma}}(x_{[-i+1+k:k]})\right].$$

6.7 The LTAD method

The left-tail Anderson–Darling estimates (LTADEs) are obtained by minimizing

$$LT_{(ADE)}(\underline{\varsigma}) = \frac{3}{2}k + 2\sum_{i=1}^{k} F_{\varsigma}(x_{[i,k]}) - \frac{1}{k}\sum_{i=1}^{k}(2i-1)\log F_{\underline{\varsigma}}(x_{[i,k]}).$$

7. Comparing estimation methods

7.1 A simulation study for comparing estimation methods

A numerical simulation is performed to compare the classical estimation methods under the GcGR-Fr model. The simulation study is based on N = 1,000 generated data from the GcGR-Fr model with n = 50,100,150, and 300, and under the following three scenarios:

Scenario I: $(\theta = 0.5, \beta = 0.6, \alpha_1 = 0.9, \alpha_2 = 0.8)$,

Scenario II: $(\theta = 0.7, \beta = 0.7, \alpha_1 = 0.7, \alpha_2 = 0.7)$,

Scenario III: $(\theta = 0.9, \beta = 0.8, \alpha_1 = 0.5, \alpha_2 = 0.3)$

The estimates are compared in terms of the root of the mean-standard error ($RMSE_{(\varsigma)}$). The numbers in Table 1 indicate that the $RMSE_{(\varsigma)}$ for each parameter tends to as increases, which means incidence of the consistency property.

Table 1: Simulation results for comparing methods.

Scenario I					
Results →		**RMSE**			
Methods ↓		$\hat{\theta}$	$\hat{\beta}$	$\widehat{\alpha_1}$	$\widehat{\alpha_2}$
ML	20	0.05475	0.00190	0.00668	0.02412
OLS		0.05072	0.00216	0.00760	0.04411
WLS		0.05523	0.00206	0.00726	0.04263
CVM		0.05045	0.00215	0.00753	0.05006
AD		0.04892	0.00196	0.00690	0.02474
RTAD		0.04642	0.00184	0.00645	0.03007
LTAD		0.06649	0.00266	0.00929	0.05512
ML	50	0.01625	0.00077	0.00272	0.00786
OLS		0.01641	0.00091	0.00319	0.01643
WLS		0.01796	0.00083	0.00294	0.01413
CVM		0.01639	0.00090	0.00318	0.01741
AD		0.01619	0.00083	0.00292	0.01004
RTAD		0.01734	0.00080	0.00282	0.01232
LTAD		0.01778	0.00105	0.00370	0.01821
ML	100	0.00787	0.000352	0.00123	0.00366
OLS		0.00755	0.00044	0.00155	0.00761
WLS		0.00823	0.00040	0.00140	0.00527
CVM		0.00754	0.00044	0.00154	0.00788
AD		0.00734	0.00039	0.00140	0.00469
RTAD		0.00793	0.00038	0.00134	0.00570
LTAD		0.00782	0.00051	0.00179	0.00811
ML	200	0.00386	0.00018	0.00063	0.00176
OLS		0.00408	0.00024	0.00085	0.00418
WLS		0.00438	0.00020	0.00071	0.00237
CVM		0.00408	0.00024	0.00085	0.00421
AD		0.00404	0.00022	0.00077	0.00248
RTAD		0.00424	0.00020	0.00071	0.00259
LTAD		0.00439	0.00029	0.00102	0.00451
Scenario II					
ML	20	0.10719	0.00289	0.00592	0.01545
OLS		0.10205	0.00355	0.00723	0.02306
WLS		0.10174	0.00328	0.00673	0.03052
CVM		0.10601	0.00364	0.00743	0.02443
AD		0.10162	0.00329	0.00672	0.01638
RTAD		0.10917	0.00322	0.00659	0.02587
LTAD		0.11015	0.00401	0.00816	0.03624

Table 1 contd. ...

...Table 1 contd.

Scenario II					
Results →		**RMSE**			
Methods ↓		$\hat{\theta}$	$\hat{\beta}$	$\widehat{\alpha_1}$	$\widehat{\alpha_2}$
ML	50	0.03390	0.00118	0.00240	0.00518
OLS		0.03442	0.00150	0.00307	0.00901
WLS		0.03563	0.00123	0.00251	0.00721
CVM		0.03504	0.00146	0.00297	0.00815
AD		0.03423	0.00131	0.00267	0.00607
RTAD		0.03472	0.00122	0.00250	0.00837
LTAD		0.03982	0.00173	0.00352	0.01241
ML	100	0.01526	0.00058	0.00119	0.00241
OLS		0.01393	0.00064	0.00132	0.00416
WLS		0.01726	0.00066	0.00134	0.00335
CVM		0.01622	0.00074	0.00150	0.00407
AD		0.03423	0.00131	0.00267	0.00607
RTAD		0.01710	0.00064	0.00131	0.00431
LTAD		0.01725	0.00085	0.00174	0.00611
ML	200	0.00786	0.00029	0.00059	0.00123
OLS		0.00737	0.00035	0.00072	0.00188
WLS		0.00860	0.00031	0.00064	0.00167
CVM		0.00746	0.00036	0.00073	0.00198
AD		0.00727	0.00032	0.00065	0.00150
RTAD		0.00798	0.00032	0.00065	0.00224
LTAD		0.00786	0.00041	0.00084	0.00288
Scenario III					
ML	20	0.17067	0.00403	0.01824	0.00235
OLS		0.15867	0.00504	0.02388	0.00420
WLS		0.17851	0.00473	0.02240	0.00653
CVM		0.16833	0.00531	0.02534	0.00409
AD		0.16039	0.00473	0.02271	0.00240
RTAD		0.16203	0.00454	0.02189	0.00465
LTAD		0.20638	0.00623	0.02941	0.00479
ML	50	0.05045	0.00152	0.00661	0.00086
OLS		0.05758	0.00201	0.00876	0.00145
WLS		0.05479	0.00185	0.00857	0.00143
CVM		0.05262	0.00199	0.00899	0.00147
AD		0.04995	0.00184	0.00852	0.00094
RTAD		0.05634	0.00178	0.00796	0.00170
LTAD		0.05582	0.00227	0.01028	0.00174

Table 1 contd. ...

...Table 1 contd.

Scenario III					
Results →		RMSE			
Methods ↓		$\hat{\theta}$	$\hat{\beta}$	$\widehat{\alpha_1}$	$\widehat{\alpha_2}$
ML	100	0.02503	0.00085	0.00376	0.00037
OLS		0.02532	0.00104	0.00471	0.00075
WLS		0.03098	0.00095	0.00418	0.00057
CVM		0.02476	0.00096	0.00416	0.00073
AD		0.02431	0.00091	0.00406	0.00050
RTAD		0.02610	0.00086	0.00376	0.00075
LTAD		0.02718	0.00111	0.00486	0.00087
ML	200	0.01208	0.00040	0.00172	0.00019
OLS		0.01282	0.00052	0.00227	0.00035
WLS		0.01447	0.00046	0.00201	0.00024
CVM		0.01242	0.00050	0.00219	0.00035
AD		0.01251	0.00047	0.00206	0.00025
RTAD		0.01308	0.00045	0.00196	0.00037
LTAD		0.01317	0.00057	0.00248	0.00043

7.2 Applications for comparing estimation methods

In order to compare the estimation methods, we consider the Cramér-von Mises (C*) statistic, the Anderson-Darling (A*) statistic, the Kolmogorov-Smirnov (KS) statistic and its corresponding p-value (P_υ). These four statistics are widely used to determine how closely a specific CDF fits the empirical distribution of a given data. The following data are considered: The 1st uncensored data set consists of 100 observations on breaking stress of carbon fibers (in Gba) given by Nichols and Padgett (2006). The 2nd data set refers to the strengths of glass fibers reported by Smith and Naylor (1987). The 3rd data set called "Wingo data" represents a complete sample from a clinical trial described as relief times (in hours) for 50 arthritic patients. Table 2 gives the results for all estimation methods using these three real data sets.

The numbers in Table 2 indicate that the ML method is the best method for estimating the unknown parameters for the 1st data set with C* = 0.05923, A* = 0.43370, KS = 0.05832 and p-value = 0.88574. For the 2nd data set, the ML method is the best for estimating the unknown parameters with C* = 0.04481 and p-value = 0.82587. However, the OLS method is the best with KS = 0.06959 and C* = 0.04473. For the 3rd data set, the LTAD method is the best method for estimating the unknown parameters with C* = 0.04740 and A* = 0.40493. However, the OLS method is the best with KS = 0.06406 and p-value = 0.06406.

8. Comparing the competitive models

For illustrating the wide flexibility of the GcGR-Fr model, we consider the previous statistics for model comparison. Table 3 reports some competitive models.

Exploring real data can be done either numerically or graphically or with both techniques. We will consider many graphical techniques such as the skewness-kurtosis plot (or the Cullen and Frey plot) for exploring initial fit to the theoretical distributions such as normal, uniform, exponential, logistic, beta, lognormal and Weibull . Bootstrapping is applied and plotted for more accuracy. The Cullen and Frey plot compares distributions in the space of squared skewness, kurtosis which provides a summary of the properties of a distribution. So, many other graphical techniques are considered such as the "nonparametric Kernel

Table 2: Comparing estimation methods.

	For the 1st data set							
Results →	Estimates				Statistics			
Methods ↓	$\hat{\theta}$	$\hat{\beta}$	$\widehat{\alpha_1}$	$\widehat{\alpha_2}$	C*	A*	KS	p-value
ML	910.24494	0.05233	105.84380	0.98874	0.05923	0.43370	0.05832	0.88574
OLS	215.77420	0.08418	34.39883	1.12237	0.05961	0.44788	0.07278	0.66462
WLS	480.03974	0.09176	41.35466	1.07116	0.06032	0.43905	0.06115	0.84866
CVM	570.58613	0.09413	48.92754	1.01985	0.05918	0.43658	0.06405	0.80656
ADE	652.03622	0.15174	30.77710	1.02497	0.05953	0.43567	0.05987	0.86597
RTAD	372.49960	0.10113	43.95148	1.01228	0.05836	0.44189	0.07851	0.56863
LTAD	1840.81247	0.20254	40.22368	0.89669	0.05791	0.43099	0.06059	0.85643
For the 2nd data set								
ML	839.20873	0.14677	18.62822	1.25341	0.04481	0.34819	0.07906	0.82587
OLS	573042862	0.38244	8.35374	1.24981	0.04473	0.35496	0.06959	0.92046
WLS	650.61291	0.08583	45.75462	1.06944	0.04569	0.37974	0.09692	0.59487
CVM	434.59422	0.16560	14.05591	1.31446	0.04503	0.35521	0.07253	0.89471
ADE	1232.04744	0.02387	123.55314	1.13980	0.04433	0.35320	0.06973	0.91929
RTAD	538.16030	0.08137	36.83841	1.14559	0.04535	0.36853	0.10302	0.51565
LTAD	1283.60868	0.20206	18.12401	1.16273	0.04438	0.35001	0.07511	0.86934
For the 3rd data set								
ML	0.37760	0.00839	21.01780	1.21289	0.04850	0.40140	0.08191	0.89060
OLS	0.41411	0.01257	20.65118	1.10861	0.04774	0.40555	0.06406	0.98645
WLS	0.38930	0.00786	20.74389	1.23507	0.04861	0.40168	0.09496	0.75806
CVM	0.44939	0.01095	20.25422	1.15889	0.04747	0.40348	0.06913	0.97067
ADE	0.39734	0.00096	22.72119	1.15070	0.04790	0.40287	0.07341	0.95043
RTAD	0.37284	0.00916	23.51166	1.15051	0.04826	0.40318	0.07312	0.95201
LTAD	0.48905	0.01033	20.70115	1.17315	0.04740	0.40493	0.06861	0.97266

Table 3: Some competitive models.

Competitive models	Abbreviation	Author(s)
Fréchet	Fr	Fréchet (1927)
Exponentiated-Fréchet	E-Fr	Nadarajah and Kotz (2003)
Beta- Fréchet	Beta-Fr	Barreto-Souza et al. (2011)
Marshal-Olkin-Fréchet	MO-Fr	Krishna et al. (2013)
Transmuted-Fréchet	T-Fr	Mahmoud and Mandouh (2013)
Kumaraswamy-Fréchet	Kum-Fr	Mead and Abd-Eltawab (2014)
McDonald- Fréchet	Mc-Fr	Shahbaz et al. (2012)
odd log-logistic-inverse Rayleigh	OLL-IR	-
odd log-logistic exponentiated-Fréchet	OLLE-Fr	-
odd log-logistic exponentiated IR	OLLE-IR	-
generalized odd log-logistic- IR	GOLL-IR	-

density estimation (NKDE)" approach for exploring initial density shape, the "Quantile-Quantile (Q-Q)" plot for exploring "normality" of the data, the "total time in test (TTT)" plot for exploring the initial shape of the empirical HRFs, the "box plot" and scattergrams for exploring the extremes.

8.1 Comparing the competitive models under stress data

The 1st uncensored data set consists of 100 observations on the breaking stress of carbon fibers (in Gba) given by Nichols and Padgett (2006). Figure 1 gives the NKDE plot (1st row left panel), the TTT plot (1st row right panel), box plot (2nd row left panel), the Q-Q plot (2nd row right panel), scattergram plot (3rd row left panel), and the skewness-kurtosis plot (3rd row right panel). Based on Figure 1 (1st row left panel), it is noted that the breaking stress of carbon fibers data is asymmetric bimodal and right heavy tail. Based on Figure 1 (1st row right panel), it is clear that the HRF of the current data is monotonically increasing. Based on Figure 1 (2nd row left panel and 2nd row right panel), it is noted that this data includes some extreme values. Based on Figure 1 (3rd row right panel), it is noted that the current data cannot be explained by the theoretical distributions such as normal, uniform, exponential, logistic, beta, lognormal and Weibull distributions.

The statistics C*, A*, K-S and . for all fitted models are presented in Table 4. The MLEs and corresponding standard errors (SEs) are reported in Table 5. From Table 4, the GcGR-Fr model gives the lowest values C* = 0.0612, A* = 0.4467, K-S = 0.05887 and P_{υ} = 0.8789 as compared to the other models. Therefore, the GcGR-Fr can be chosen as the best model. Figure 2 gives the estimated PDF and CDF. Figure 3 gives the Probability-Probability (P-P) plot and estimated HRF for the current data. From Figure 2 and Figure 3, we note that the new GcGR-Fr model provides adequate fits to the empirical functions.

8.2 Comparing the competitive models with glass fibers data

The 2nd data set refers to the strengths of glass fibers as given by Smith and Naylor (1987). Figure 4 gives the NKDE plot (1st row left panel), the TTT plot (1st row right panel), box plot (2nd row left panel), the Q-Q plot (2nd row right panel), scattergram plot (3rd row left panel), and the skewness-kurtosis plot (3rd row right panel). Figure 4 (1st row left panel) indicates that the glass fibers data is asymmetric bimodal and right heavy tail. Figure 4 (1st row right panel) indicates that the HRF of the glass fibers data is monotonically

Table 4: C*, A*, K-S and for the breaking stress of carbon fibers data.

Criteria	**Goodness of fit criteria**			
Model	**C***	**A***	**K-S**	P_{υ}
GcGR-Fr	0.0612	0.4467	0.05887	0.8789
OB-Fr	0.0664	0.4706	0.0630	0.8220
OLLE-Fr	0.1203	0.9639	0.5561	< 0.0001
OLLE-IR	0.1553	1.2120	0.6550	< 0.0001
OLL-IR	0.1553	1.2120	0.6550	< 0.0001
Fr	0.1090	0.7657	0.0874	0.4282
Kum-Fr	0.0812	0.6217	0.0759	0.6118
E-Fr	0.1091	0.7658	0.0874	0.4287
Beta-Fr	0.0809	0.6207	0.0757	0.6147
T-Fr	0.0871	0.6209	0.0782	0.5734
MO-Fr	0.0886	0.6142	0.0763	0.5168
Mc-Fr	0.1333	1.0608	0.0807	0.5332

Table 5: MLEs and SEs for the breaking stress of carbon fibers data.

Estimates Model	Estimates				
	$\hat{\theta}$	$\hat{\beta}$	$\hat{c}$	$\widehat{\alpha_1}$	$\widehat{\alpha_2}$
GcGR-Fr	250.215 (393.53)	0.02987 (0.0175)		74.1258 (68.922)	1.17293 (0.2228)
GcGR-Fr	5.1954 (0.001)	0.5990 (0.032)		1.0404 (0.044)	1.2324 (0.003)
OLLE-Fr	0.1351 (0.011)		3.7216 (0.0034)	0.9296 (0.0033)	21.319 (0.0034)
OLLE-IR	0.49460 (0.0414)		0.06743 (0.7195)	1.74262 (9.3007)	
OLL-IR	0.49459 0.04135		0.45242 0.03869		
Fr				1.3968 (0.0336)	4.3724 (0.3278)
Kum-Fr		0.8489 (16.083)	1.6239 (0.6979)	1.6341 (9.049)	3.4208 (0.7635)
E-Fr		0.9395 (3.543)		1.4169 (2.568)	0.9395 (0.3278)
Beta-Fr		0.7346 (1.5290)	1.5830 (0.7132)	1.6684 (0.7662)	3.5112 (0.9683)
T-Fr	–0.7166 (0.2616)			1.2656 (0.0579)	4.7121 (0.3657)
MO-Fr		0.0033 (0.0009)		6.2296 (1.0134)	1.2419 (0.1181)
Mc-Fr	0.8503 (0.1353)	44.423 (25.100)	19.859 (6.706)	0.0203 (0.0060)	46.974 (21.871)

increasing. Figure 4 (2nd row left panel and 2nd row right panel) shows that the glass fibers data includes some extreme values. Figure 4 (3rd row right panel) indicates that the glass fibers data cannot be explained by the theoretical distributions such as normal, uniform, exponential, logistic, beta, lognormal and Weibull distributions.

The statistics C*, A*, K-S and for all fitted models are reported in Table 6. The MLEs and corresponding SEs are given in Table 7. From Table 6, the GcGR-Fr model gives the lowest values C* = 0.11304, A* = 0.89752, K-S = 0.12348 and = 0.2691 as compared to other models. Therefore, the GcGR-Fr distribution can be chosen as the best model. Figure 5 gives the estimated PDF and CDF. Figure 6 gives the P-P plot and estimated HRF for the glass fibers data. Based on Figure 5 and Figure 6, it is clear that the GcGR-Fr model provides adequate fits to the empirical functions.

8.3 Comparing the competitive models with the relief times data

The 3rd data set is called "Wingo data" and represents a complete sample from a clinical trial described as relief times (in hours) for 50 arthritic patients. Figure 7 gives the NKDE plot (1st row left panel), the TTT plot (1st row right panel), box plot (2nd row left panel), the Q-Q plot (2nd row right panel), scattergram plot (3rd row left panel), and the skewness-kurtosis plot (3rd row right panel). Figure 7 (1st row left panel) indicates that the relief times data can be considered as symmetric data. Based on Figure 7 (1st row right panel), it is noted that the HRF of these data is monotonically increasing. Based on Figure 7 (2nd row left panel and 2nd row right panel), it is clear that the relief times do not include any extreme values. Based

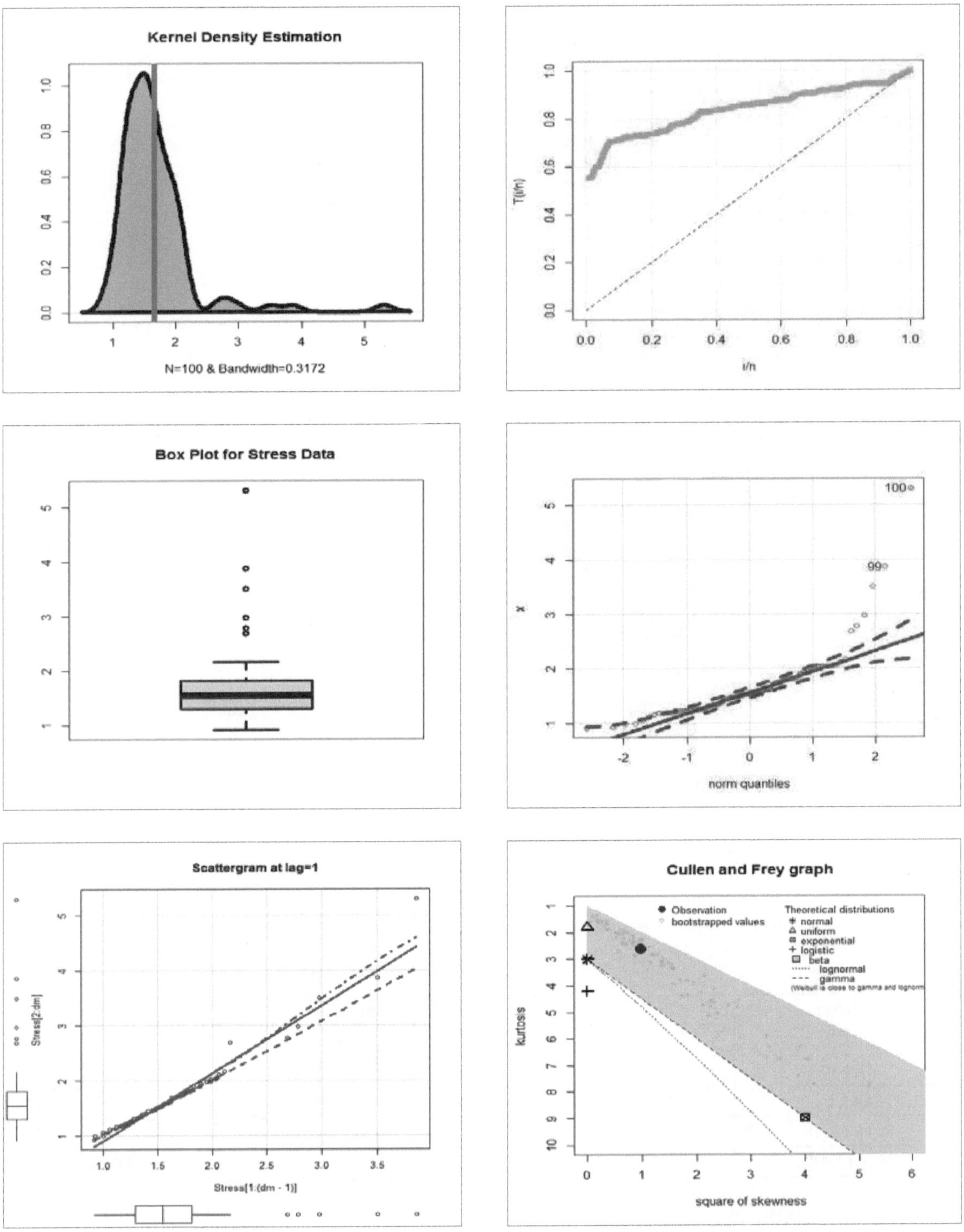

Figure 1: Graphical description for the breaking stress of carbon fibers data.

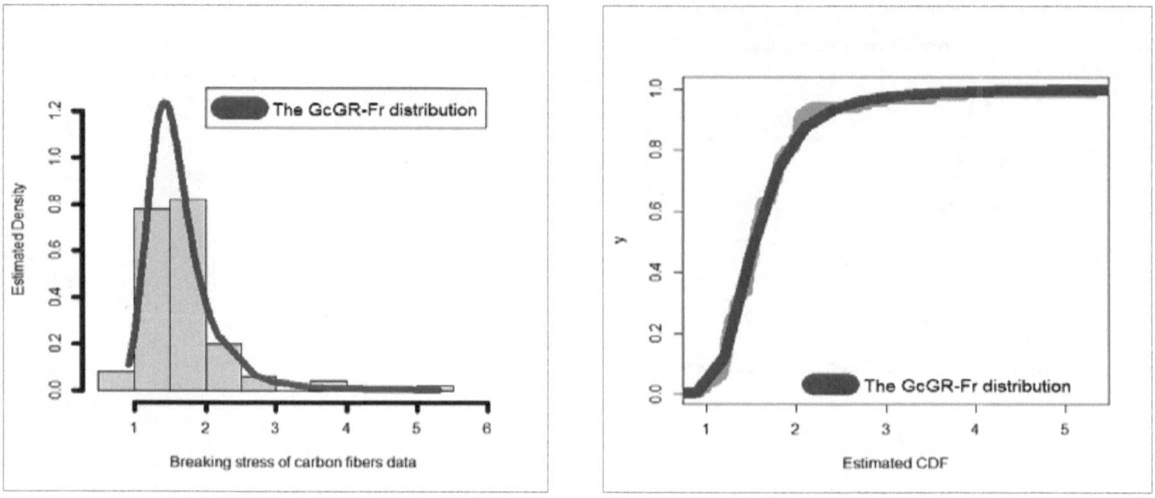

Figure 2: Estimated PDF and estimated CDF for the breaking stress of carbon fibers data.

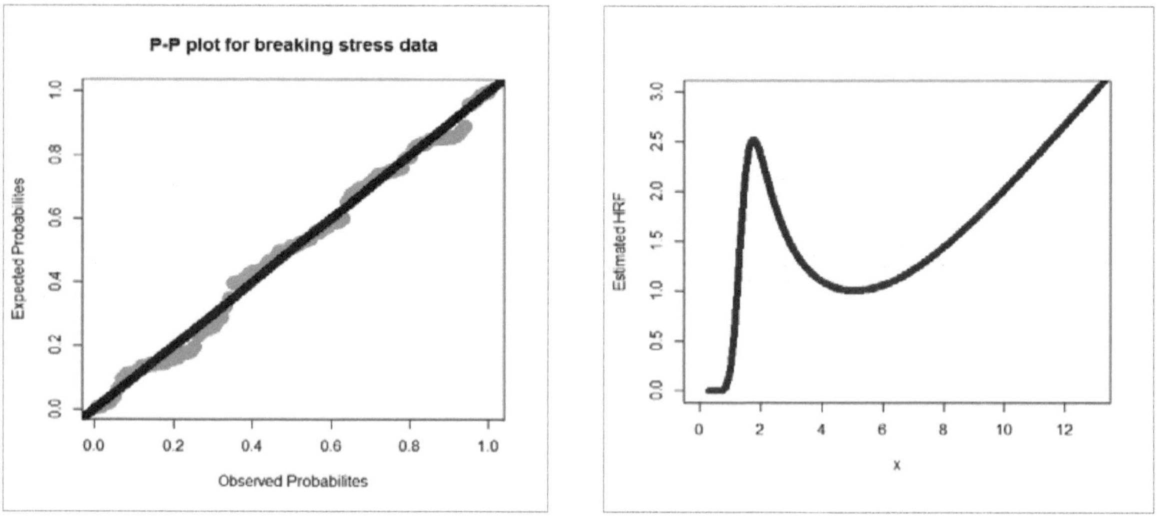

Figure 3: P-P plot and estimated HRF for the breaking stress of carbon fibers data.

Table 6: C*, A*; K-S and for the glass fibers data.

Criteria → Model ↓	Goodness of fit criteria			
	C*	A*	K-S	P_v
GcGR-Fr	0.11304	0.89752	0.12348	0.2691
OLLE-Fr	0.10487	0.83250	0.55196	< 0.0001
OLLE-IR	0.15020	1.14697	0.67949	< 0.0001
OLL-IR	0.15021	1.14697	0.67951	< 0.0001

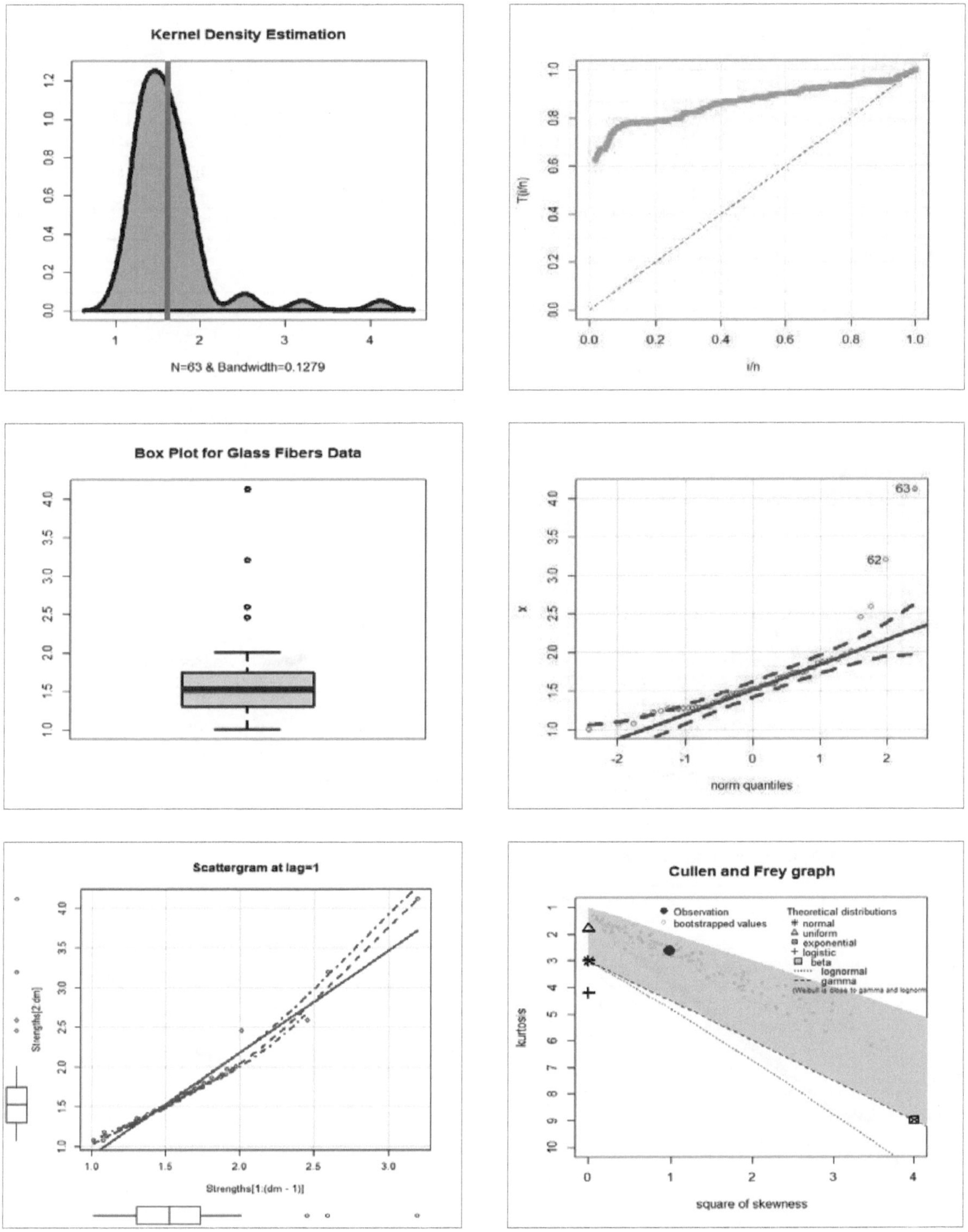

Figure 4: Graphical description for the glass fibers data.

Table 7: MLEs and SEs for the glass fibers data.

Estimates → **Model ↓**	**Estimates**				
	$\hat{\theta}$	$\hat{\beta}$	$\hat{c}$	$\widehat{\alpha_1}$	$\widehat{\alpha_2}$
GcGR-Fr	14.89934 (6.2264)	0.00567 (0.00085)		53.55753 (11.9327)	1.5947538 (0.097052)
OLLE-Fr	0.1449 (0.0129)		0.00879 (0.000)	1.2997 (0.000)	24.878 (0.000)
OLLE-IR	0.5025 (0.0529)		0.0716 (1.13062)	1.7048 (13.47)	
OLL-IR	0.50251 0.052946		0.45599 0.048652		

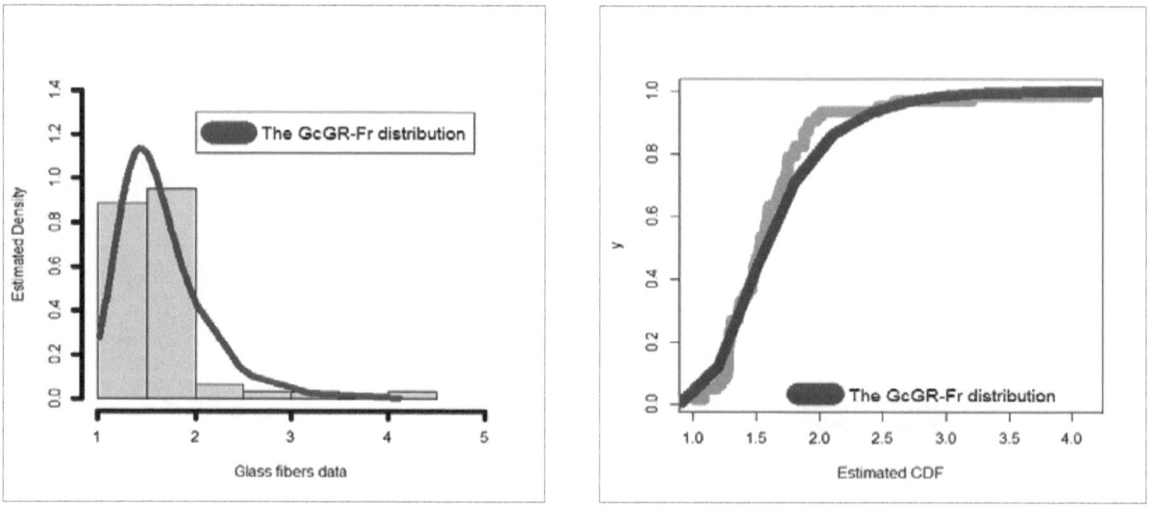

Figure 5: Estimated PDF and CDF for the glass fibers data.

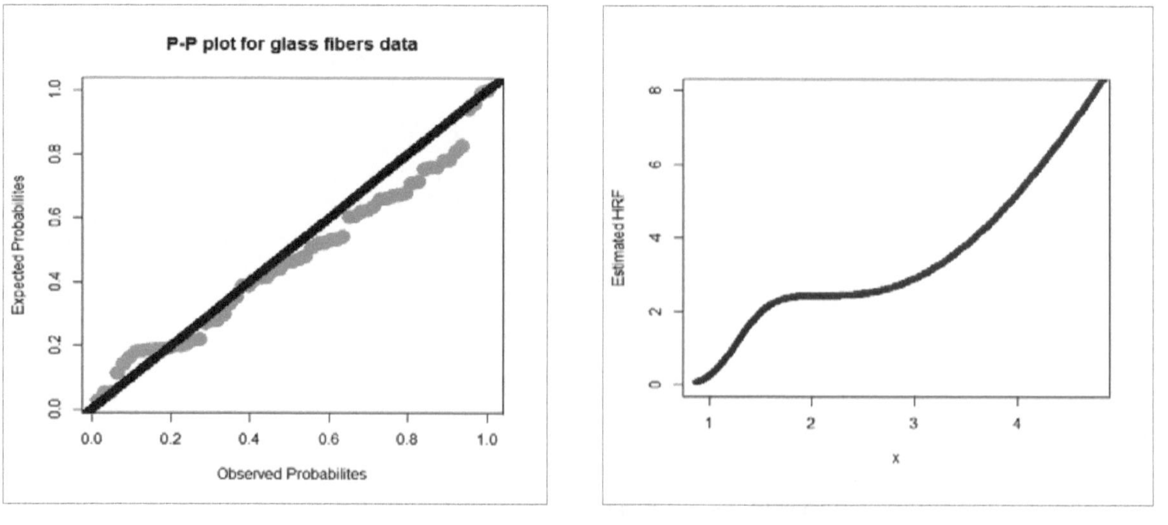

Figure 6: P-P plot and estimated HRF for the glass fibers data.

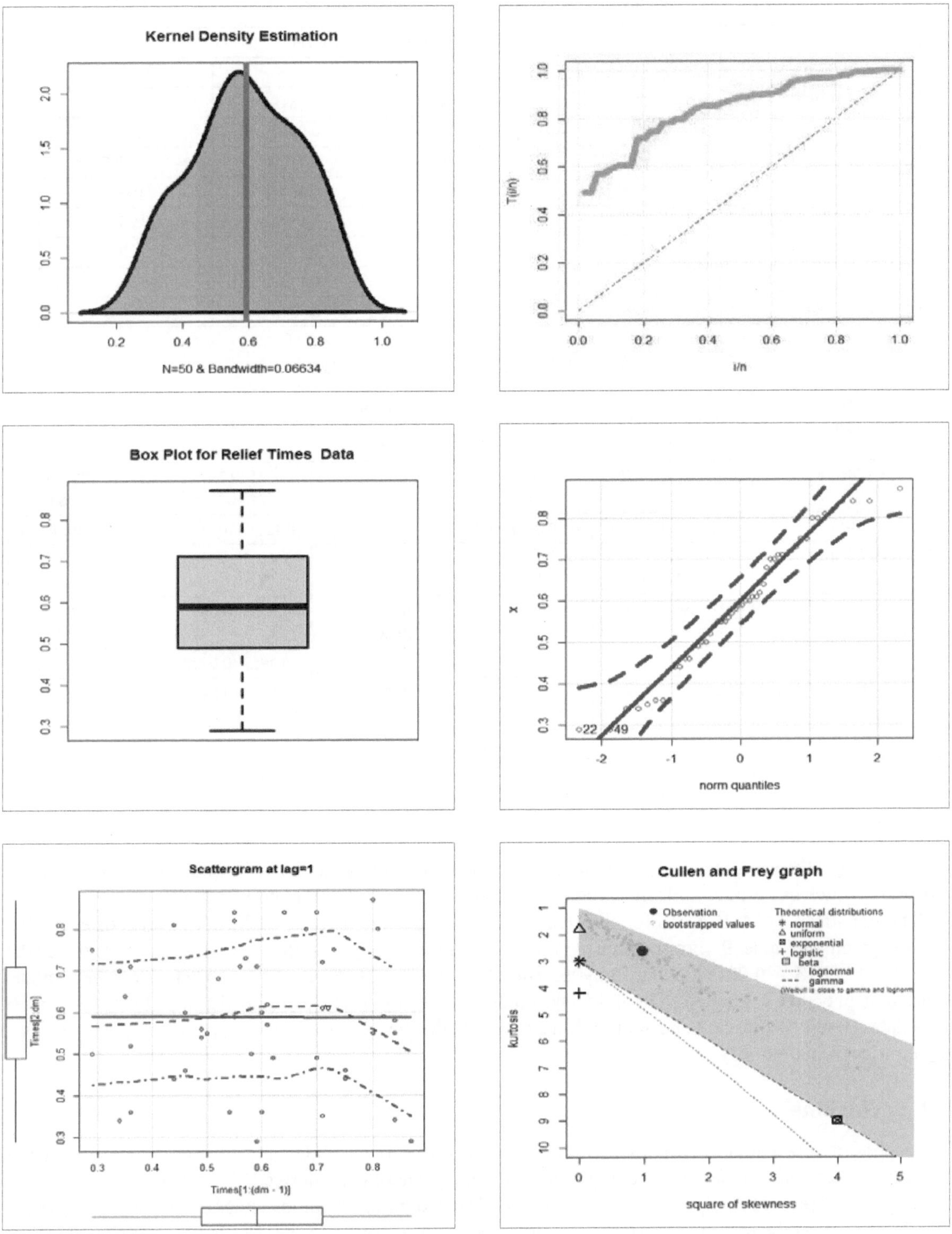

Figure 7: Graphical description for the relief times data.

Table 8: C*, A*, K-S and P_{υ} for the relief times data.

Criteria Model	Goodness of fit criteria			
	C*	A*	K-S	
GcGR-Fr	0.0485	0.4014	0.08191	0.8906
OB-Fr	0.0490	0.4208	0.09124	0.7994
GOLL-IR	0.1955	1.3498	0.11008	0.5797
OLLE-Fr	0.1577	1.0988	0.53498	< 0.0001
Fr	0.3233	2.0301	0.15062	0.2066
E-Fr	0.3233	2.0301	0.15061	0.2064
Beta-Fr	0.3611	2.5131	0.14334	0.3601
T-Fr	0.2823	1.8152	0.13701	0.3045

Table 9: MLEs and SEs for the relief times data.

Estimates → Model ↓	Estimates				
	$\hat{\theta}$	$\hat{\beta}$	$\hat{c}$	$\widehat{\alpha_1}$	$\widehat{\alpha_2}$
GcGR-Fr	0.377602 (0.60784)	0.008388 (0.00152)		21.01781 (21.46850)	1.2128934 (0.456084)
OB-Fr	17.7905 (0.0001)	6.9955 (4.0355)		0.12686 (0.0002)	0.17843 (0.0004)
GOLL-IR	1.96132 (0.2340)	0.1112 (0.001)		1.41232 (0.005)	
OLLE-Fr	0.0669 (0.0076)		0.00459 (0.0028)	0.3558 (0.0047)	32.561 (0.006)
Fr				0.4859 (0.0227)	3.2078 (0.3263)
E-Fr			0.9047 (18.784)	0.5013 (3.2444)	3.2077 (0.3263)
Beta-Fr		4.015 (0.111)	1.3349 (0.147)	2.0022 (0.321)	0.87017 (0.0033)
T-Fr	0.5816 (0.2787)			0.4400 (0.0290)	3.4974 (0.3527)

on Figure 7 (3rd row right panel), it is noted that the relief times cannot be explained from theoretical distributions such as normal, uniform, exponential, logistic, beta, lognormal and Weibull .

The statistics C*, A*, K-S and for all fitted models are reported in Table 8. The MLEs and corresponding SEs are given in Table 9. From Table 8, the GcGR-Fr model gives the lowest values C* = 0.0485, A* = 0.4014, K-S = 0.08191 and = 0.8906. Therefore, the GcGR-Fr may be chosen as the best model. Figure 8 displays the estimated PDF and CDF. Figure 9 gives the P-P plot and estimated HRF for the relief times data. Based on Figure 8 and Figure 9, we note that the new GcGR-Fr model provides adequate fits to the empirical functions.

9. Conclusions

This paper presents a novel two-parameter G family of distributions. Relevant statistical properties such as the ordinary moments, incomplete moments and generating function are derived. Special attention is devoted to the standard Fréchet baseline model. Different classical estimation methods under uncensored schemes are considered such as the maximum likelihood, Anderson–Darling, ordinary least squares, Cramér-von Mises, weighted least squares, left-tail Anderson–Darling, and right-tail Anderson–Darling. Numerical simulations are performed for comparing the estimation methods. Moreover, all methods of estimation are compared by

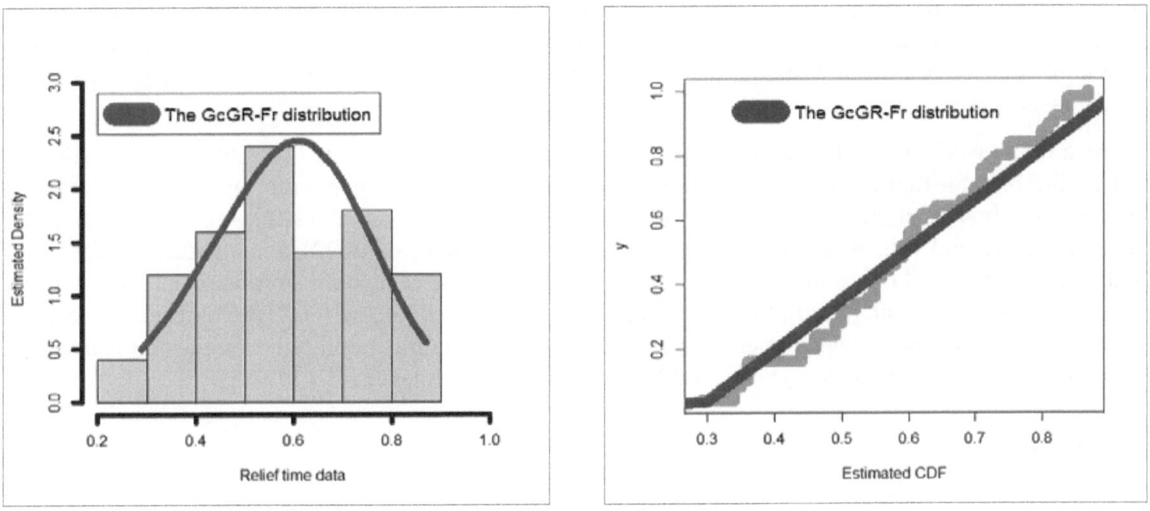

Figure 8: Estimated PDF and estimated CDF for the relief times data.

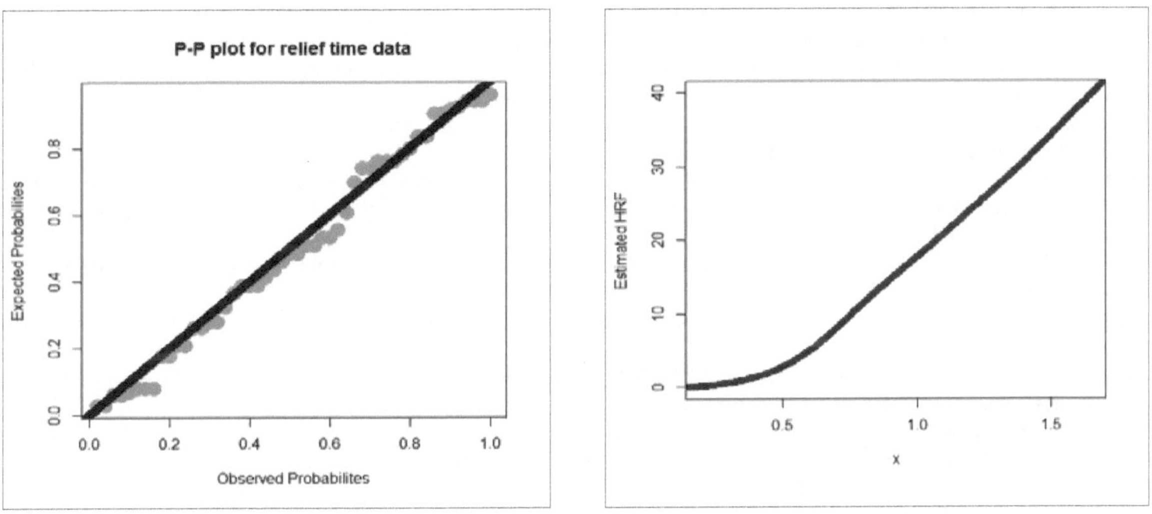

Figure 9: P-P plot and estimated HRF for the relief times data.

means of three real data sets. The usefulness and flexibility of the purposed family are illustrated by means of three applications to real data. The new family proved its superiority against many well-known G families as shown below:

I. In modeling the breaking stress of carbon fibers, the new family is better than the odd Burr G family, the odd log-logistic G family, the odd log-logistic exponentiated G family, the transmuted G family, the Kumaraswamy G family, exponentiated G family, Beta G family, the McDonald G family and the Marshall-Olkin G family under the Cramér-von Mises statistic, the Anderson-Darling statistic, the Kolmogorov-Smirnov test statistic, and its corresponding p-value.

II. In modeling the glass fibers, the purposed family is better than the odd log-logistic G family and the odd log-logistic exponentiated G family under the Cramér-von Mises statistic, the Anderson-Darling statistic, the Kolmogorov-Smirnov test statistic, and its corresponding p-value.

III. In modeling the relief times, the new family is better than the odd Burr G family, the generalized odd log-logistic G family, the odd log-logistic exponentiated G family, exponentiated G family, Beta G family and the transmuted G family under the Cramér-von Mises statistic, the Anderson-Darling statistic, the Kolmogorov-Smirnov test statistic, and its corresponding p-value.

As a future work, we will consider many new useful goodness-of-fit tests for right censored validation such as the Nikulin-Rao-Robson goodness-of-fit test and Bagdonavičius-Nikulin goodness-of-fit test as performed by Ibrahim et al. (2019), Goual et al. (2019, 2020), Mansour et al. (2020a–f), Yadav et al. (2020 and 2022), Goual and Yousof (2020), Aidi et al. (2021) and Ibrahim et al. (2022), among others. Some useful reliability studies based on multicomponent stress-strength and the remaining stress-strength concepts can be presented (Rasekhi et al. (2020), Saber et al. (2022a,b), Saber and Yousof (2022)). Some new acceptance sampling plans based on the complementary geometric Weibull-G family or on some special members can be presented in separate articles (see Ahmed and Yousof (2022) and Ahmed et al. (2022)).

References

Aboraya, M., M. Yousof, H. M., Hamedani, G. G. and Ibrahim, M. (2020). A new family of discrete distributions with mathematical properties, characterizations, Bayesian and non-Bayesian estimation methods. Mathematics, 8: 1648.

Afify, A. Z., Alizadeh, M., Yousof, H. M., Aryal, G., Ahmad, M. et al. (2016a). The transmuted geometric-G family of distributions: theory and applications. Pak. J. Statist, 32: 139–160.

Afify, A. Z., Cordeiro, G. M., Nadarajah, S., Yousof, H. M., Ozel, G. et al. (2016b). The complementary geometric transmuted-G family of distributions: model, properties and application. Hacettepe Journal of Mathematics and Statistics, 47: 1348–1374.

Ahmed, B. and Yousof, H. M. (2022). A New Group Acceptance Sampling Plans based on Percentiles for the Weibull Fréchet Model. Statistics, Optimization & Information Computing, forthcoming.

Ahmed, B., Ali, M. M. and Yousof, H. M. (2022). A Novel G Family for Single Acceptance Sampling Plan with Application in Quality and Risk Decisions. Annals of Data Science, forthcoming.

Aidi, K., Butt, N. S., Ali, M. M., Ibrahim, M., Yousof, H. M. et al. (2021). A modified chi-square type test statistic for the double burr X model with applications to right censored medical and reliability data. Pakistan Journal of Statistics and Operation Research, 17(3): 615–623.

Al-Babtain, A. A., Elbatal, I. and Yousof, H. M. (2020). A new three parameter Fréchet model with mathematical properties and applications. Journal of Taibah University for Science, 14: 265–278.

Aryal, G. R. and Yousof, H. M. (2017). The exponentiated generalized-G Poisson family of distributions. Stochastics and Quality Control, 32: 7–23.

Barreto-Souza, W., Cordeiro, G. M. and Simas, A. B. (2011). Some results for beta Fréchet distribution. Communications in Statistics—Theory and Methods, 40(5): 798–811.

Brito, E., Cordeiro, G. M., Yousof, H. M., Alizadeh, M., Silva, G.O. et al. (2017). The Topp–Leone odd log-logistic family of distributions. Journal of Statistical Computation and Simulation, 87: 3040–3058.

Cordeiro, G. M. and de Castro, M. (2011). A new family of generalized distributions. Journal of Statistical Computation and Simulation, 81: 883–898.

Cordeiro, G. M., Ortega, E. M., Popović, B. V. and Pescim, R. R. (2014). The Lomax generator of distributions: Properties, minification process and regression model. Applied Mathematics and Computation, 247: 465–486.

Cordeiro, G. M., Yousof, H. M., Ramires, T. G. and Ortega, E. M. (2018). The Burr XII system of densities: properties, regression model and applications. Journal of Statistical Computation and Simulation, 88: 432–456.

Fréchet, M. (1927). Sur la loi de probabilité de lécart maximum. Ann. Soc. Pol. Math., 6: 93–116.

Galambos, J. and Kotz, S. (1978). Characterizations of probability distributions. A unified approach with emphasis on exponential and related models, Lecture Notes in Mathematics, p. 675. Springer, Berlin.

Glänzel, W., Telcs, A. and Schubert, A. (1984). Characterization by truncated moments and its application to Pearson-type distributions, Z. Wahrsch. Verw. Gebiete, 66: 173–182.

Glänzel, W. (1987). A characterization theorem based on truncated moments and its application to some distribution families, Mathematical Statistics and Probability Theory (Bad Tatzmannsdorf, 1986), Vol. B, Reidel, Dordrecht, 75–84.

Glänzel, W. (1990). Some consequences of a characterization theorem based on truncated moments, Statistics: A Journal of Theoretical and Applied Statistics, 21: 613–618.

Glänzel, W. and Hamedani, G. G. (2001). Characterizations of the univariate distributions, Studia Scien. Math. Hung., 37: 83–118.

Goual, H., Yousof, H. M. and Ali, M. M. (2019). Validation of the odd Lindley exponentiated exponential by a modified goodness of fit test with applications to censored and complete data. Pakistan Journal of Statistics and Operation Research, 15(3): 745–771.

Goual, H. and Yousof, H. M. (2020). Validation of Burr XII inverse Rayleigh model via a modified chi-squared goodness-of-fit test. Journal of Applied Statistics, 47(3): 393–423.

Goual, H., Yousof, H. M. and Ali, M. M. (2020). Lomax inverse Weibull model: properties, applications, and a modified Chi-squared goodness-of-fit test for validation. Journal of Nonlinear Sciences & Applications, 13(6): 330–353.

Hamedani, G. G., Altun, E., Korkmaz, M. Ç., Yousof, H. M., Butt, N. S. et al. (2018). A new extended G family of continuous distributions with mathematical properties, characterizations and regression modeling. Pakistan Journal of Statistics and Operation Research, 737–758.

Hamedani, G. G., Yousof, H. M., Rasekhi, M., Alizadeh, M., Najibi, S. M. et al. (2017). Type I general exponential class of distributions. Pakistan Journal of Statistics and Operation Research, 14(1): 39–55.

Hamedani, G. G., Rasekhi, M., Najibi, S., Yousof, H. M., Alizadeh, M. et al. (2019). Type II general exponential class of distributions. Pakistan Journal of Statistics and Operation Research, 15(2): 503–523.

Hamedani, G. G., Korkmaz, M. C., Butt, N. S. and Yousof, H. M. (2021). The Type I Quasi Lambert Family: Properties, Characterizations and Different Estimation Methods. Pakistan Journal of Statistics and Operation Research, 17(3): 545–558.

Hamedani, G. G., Korkmaz, M. Ç., Butt, N. S. and Yousof H. M. (2022). The Type II Quasi Lambert G Family of Probability Distributions. Pakistan Journal of Statistics and Operation Research, forthcoming.

Ibrahim, M., Yadav, A. S., Yousof, H. M., Goual, H., Hamedani, G. G. et al. (2019). A new extension of Lindley distribution: modified validation test, characterizations and different methods of estimation. Communications for Statistical Applications and Methods, 26(5): 473–495.

Ibrahim, M., Ali, M. M. and Yousof, H. M. (2021). The discrete analogue of the Weibull G family: properties, different applications, Bayesian and non-Bayesian estimation methods. Annals of Data Science, forthcoming.

Ibrahim, M., Hamedani, G. G., Butt, N. S. and Yousof, H. M. (2022). Expanding the Nadarajah Haghighi Model: Characterizations, Properties and Application. Pakistan Journal of Statistics and Operation Research, forthcoming.

Jahanshahi, S. M. A., Yousof, H. M. and Sharma, V. K. (2019). The Burr X Fréchet Model for Extreme Values: Mathematical Properties, Classical Inference and Bayesian Analysis. Pakistan Journal of Statistics and Operation Research, 15: 797.

Karamikabir, H., Afshari, M., Yousof, H. M., Alizadeh, M., Hamedani, G. et al. (2020). The Weibull Topp-Leone Generated Family of Distributions: Statistical Properties and Applications. Journal of The Iranian Statistical Society, 19: 121–161.

Kim, J. H. and Jeon, Y. (2013). Credibility theory based on trimming, Insur. Math. Econ., 53: 36–47.

Korkmaz, M. C. Yousof, H. M. and Hamedani G. G. (2018). The exponential Lindley odd log-logistic G family: properties, characterizations and applications. Journal of Statistical Theory and Applications, 17: 554–571.

Kotz, S. and Shanbhag, D. N. (1980). Some new approach to probability distributions, Adv. Appl. Probab., 12: 903–921.

Kotz, S. and Johnson, N.L. (1992). Breakthroughs in Statistics: Foundations and Basic Theory; Springer-Verlag, Volume 1.

Krishna, E., Jose, K. K., Alice, T. and Risti, M. M. (2013). The Marshall-Olkin Fréchet distribution. Commun. Stat. Theory Methods, 42: 4091–4107.

Mansour, M. M., Ibrahim, M., Aidi, K., Shafique Butt, N., Ali, M. M. et al. (2020a). A new log-logistic lifetime model with mathematical properties, copula, modified goodness-of-fit test for validation and real data modeling. Mathematics, 8(9): 1508.

Mansour, M. M., Butt, N. S., Ansari, S. I., Yousof, H. M., Ali, M. M. et al. (2020b). A new exponentiated Weibull distribution's extension: copula, mathematical properties and applications. Contributions to Mathematics, 1: 57–66. DOI: 10.47443/cm.2020.0018.

Mansour, M., Korkmaz, M. Ç., Ali, M. M., Yousof, H. M., Ansari, S. I. and Ibrahim, M. (2020c). A generalization of the exponentiated Weibull model with properties, Copula and application. Eurasian Bulletin of Mathematics, 3(2): 84–102.

Mansour, M., Rasekhi, M., Ibrahim, M., Aidi, K., Yousof, H. M. et al. (2020d). A new parametric life distribution with modified bagdonavičius–nikulin goodness-of-fit test for censored validation, properties, applications, and different estimation methods. Entropy, 22(5): 592.

Mansour, M., Yousof, H. M., Shehata, W. A. and Ibrahim, M. (2020e). A new two parameter Burr XII distribution: properties, copula, different estimation methods and modeling acute bone cancer data. Journal of Nonlinear Science and Applications, 13(5): 223–238.

Mansour, M. M., Butt, N. S., Yousof, H. M., Ansari, S. I., Ibrahim, M. et al. (2020f). A generalization of reciprocal exponential model: clayton copula, statistical properties and modeling skewed and symmetric real data sets. Pakistan Journal of Statistics and Operation Research, 16(2): 373–386.

Marinho, P. R. D., Silva, R. B., Bourguignon, M., Cordeiro, G. M., Nadarajah, S. et al. (2019). Adequacy Model: An R package for probability distributions and general-purpose optimization. PloS One, 14: e0221487.

Mead, M. E. A. and Abd-Eltawab, A. R. (2014). A note on Kumaraswamy Fréchet distribution. Australia, 8: 294-300.

Nadarajah, S. and Kotz, S. (2003). The exponentiated Fréchet distribution. Interstate Electronic Journal, 14: 1–7.

Rasekhi, M., Saber, M. M. and Yousof, H. M. (2020). Bayesian and classical inference of reliability in multicomponent stress-strength under the generalized logistic model. Communications in Statistics-Theory and Methods, 50(21): 5114–5125.

Saber, M. M. and Yousof, H. M. (2022). Bayesian and Classical Inference for Generalized Stress-strength Parameter under Generalized Logistic Distribution, Statistics, Optimization & Information Computing, forthcoming.

Saber, M. M., Marwa M., Mohie El-Din and Yousof, H. M. (2022a). Reliability estimation for the remained stress-strength model under the generalized exponential lifetime distribution, Journal of Probability and Statistics, 20a21, 1–10.

Saber, M. M., Rasekhi, M. and Yousof, H. M. (2022b). Generalized Stress-Strength and Generalized Multicomponent Stress-Strength Models, Statistics, Optimization & Information Computing, forthcoming.

Shahbaz, M. Q., Shahbaz, S. and Butt, N. S. (2012). The Kumaraswamy-inverse Weibull distribution. Pak. J. Stat. Oper. Res., 8: 479–489.

Smith, R. L. and Naylor, J. C. (1987). A comparison of maximum likelihood and Bayesian estimators for the three-parameter Weibull distribution. Appl. Statist., 36: 358–369.

Von Mises, R. (1964). Selected papers of Richard von Mises. American Mathematical Society, 1.

Yadav, A. S., Goual, H., Alotaibi, R. M., Ali, M. M., Yousof, H. M. et al. (2020). Validation of the Topp-Leone-Lomax model via a modified Nikulin-Rao-Robson goodness-of-fit test with different methods of estimation. Symmetry, 12(1): 57.

Yadav, A. S., Shukl, S., Goual, H., Saha, M., Yousof, H. M. et al. (2022). Validation of Xgamma Exponential Model via Nikulin-Rao-Robson Goodness-of- Fit-Test under Complete and Censored Sample with Different Methods of Estimation. Statistics, Optimization & Information Computing, 10(2): 457–483.

Yousof, H. M., Afify, A. Z., Hamedani, G. G. and Aryal, G. (2017). The Burr X generator of distributions for lifetime data. Journal of Statistical Theory and Applications, 16: 288–305.

Yousof, H. M., Chesneau, C., Hamedani, G. G. and Ibrahim, M. (2021). A new discrete distribution: properties, characterizations, modeling real count data, bayesian and non-bayesian estimations. Statistica, 81(2): 135–162.

Chapter 3

On the use of Copulas to Construct Univariate Generalized Families of Continuous Distributions

Christophe Chesneau[1,*] and *Haitham M Yousof*[2]

1. Introduction

Many scientific disciplines are interested in modeling random multivariate events. However, the derivation of multivariate probability distributions that accurately model the variables at play is difficult. Copulas are a useful concept for dealing with this problem. As a first definition, a bivariate (or 2-dimensional) copula is a bivariate function $\mathcal{C}(u,v)$, $(u,v) \in [0,1]^2$, satisfying, the following properties: For any $(u,v) \in [0,1]^2$, $\mathcal{C}(u,0) = \mathcal{C}(0,v) = 0$, $\mathcal{C}(u,1) = u$, $\mathcal{C}(1,v) = v$, and for any $(u_1, u_2, v_1, v_2) \in [0,1]^4$, such that $u_1 \leq u_2$ and $v_1 \leq v_2$, $\mathcal{C}(u_2, v_2) - \mathcal{C}(u_2, v_1) - \mathcal{C}(u_1, v_2) + \mathcal{C}(u_1, v_1) \geq 0$.

The major result on the concept of copulas is the Sklar theorem established in Sklar (1959). This concept is quite flexible; over time, a variety of copulas have been proposed. In particular, there are those of the families of Archimedean copulas, elliptical copulas and extreme value copulas. They contain various copulas, depending on one or more parameters, which have found a place of choice in many applications. In addition, some copulas of interest are independent of these families and have been at the origin of important innovations in multivariate modeling. It is the case with the famous Farlie-Gumbel-Morgensten (FGM) copula. For the theoretical aspects, we may refer to Nelsen (2006), Yong-Quan (2008), Georges et al. (2001), Coles et al. (1999), Bekrizadeh et al. (2015), Bekrizadeh and Jamshidi (2017), Bekrizadeh et al. (2012), Trivedi and Zimmer (2005), Chesneau (2021a) and Chesneau (2021b). Diverse application evidence can be found in References Frees and Valdez (1998), Georges et al. (2001), Kazianka and Pilz (2009), Zhang et al. (2011), Thompson and Kilgore (2011) and Shiau et al. (2011).

In a completely different branch of probability and statistics, there is a high demand for new and original families of univariate distributions. These families are used for the analysis of data sets in various applied fields, as well as the construction of various models, such as regression models, clustering models, and so on. Through various mathematical functions, the compounding, integral, and mixing techniques are the most commonly used for defining such families. In this regard, we may refer the reader to references, Bekrizadeh

[1] Université de Caen Normandie, LMNO, Campus II, Science 3, 14032, Caen, France.
[2] Department of Statistics, Mathematics and Insurance, Faculty of Commerce, Benha University, Benha 13518, Egypt.
Email: haitham.yousof@fcom.bu.edu.eg
* Corresponding author: christophe.chesneau@gmail.com

et al. (2012), Bekrizadeh et al. (2015), Brito (2017), Chesneau (2021), Casella and Berger (1990), Chesneau and Yousof (2021), Coles et al. (1999), Cordeiro et al. (2018), Jin and Shitan (2014), Korkmaz et al. (2018a,b and 2020), Fischer and Klein (2007), Frees and Valdez (1998), Domma (2009), Karamikabir et al. (2020), Kazianka and Pilz (2009), Chesneau (2021), Eugene et al. (2002), Georges (2001) and Bekrizadeh and Jamshidi (2017), Aryal and Yousof (2017), Hamedani et al. (2017, 2018 and 2019), Nascimento et al. (2019), Yousof et al. (2017, 2018 and 2020), Merovci et al. (2017 and 2020), Alizadeh et al. (2020a,b), Altun et al. (2022) and El-Morshedy et al. (2021), among others.

In this article, we explore a new research approach consisting of using the flexible analytical properties of copulas to create new general families of univariate distributions. These families are called Copula-G families. As a primary remark, they have the feature of possibly depending on two different baseline distributions, parametric or not, and one independent tuning parameter, or more, inherent to the definition of the copula. We thus transpose some flexible dependence features of the copula to a new perspective of modeling in the univariate case. We develop this idea in a comprehensive manner. Some theoretical results are given in full generality. Then, by selecting some families of particular interest, as well as interesting baselines, we present two special members and determine some of its practical properties by a graphical approach. We highlight the fact that it can be applied to analyze real-life data, among other members of the family.

The plan of the paper is divided into the following sections. Section 2 lists several famous copulas. In Section 3, we develop our idea and show how these copulas can serve as flexible generators of new families of univariate distributions. Some comments conclude the paper in Section 4.

2. A list of known copulas

In the literature, there is a plethora of copulas presenting different features and properties. The main aim of this section is to present the general forms of some copulas depending on only one parameter, denoted by θ, often encountered in applications. The majority of them are cited in Nelsen (2006). We refer to this referenced table for the possible values of θ, which differ from one copula to another; it will be precise later only for the copulas taken into account in our applications.

1. The Archimedean copula 1 of [Nelsen (2006) Table 4.1]:

$$\mathcal{C}(u, v) = [(u^{-\theta} + v^{-\theta} - 1)_{+}]^{-1/\theta}.$$

2. The Archimedean copula 2 of [Nelsen (2006) Table 4.1]:

$$\mathcal{C}(u, v) = (1 - [(1 - u)^{\theta} + (1 - v)^{\theta}]^{-1/\theta})_{+}.$$

3. The Archimedean copula 3 of [Nelsen (2006) Table 4.1]:

$$\mathcal{C}(u, v) = \frac{uv}{1 - \theta(1 - u)(1 - v)}.$$

4. The Archimedean copula 4 of [Nelsen (2006) Table 4.1]:

$$\mathcal{C}(u, v) = exp\,(-[(-\,log\,u)^{\theta} + (-\,log\,v)^{\theta}]^{1/\theta}).$$

5. The Archimedean copula 5 of [Nelsen (2006) Table 4.1]:

$$\mathcal{C}(u, v) = -\frac{1}{\theta}\,log\left[1 + \frac{(e^{-\theta u} - 1)(e^{-\theta v} - 1)}{e^{-\theta} - 1}\right].$$

6. The Archimedean copula 6 of [Nelsen (2006) Table 4.1]:

$$\mathcal{C}(u,v) = 1 - [(1 - u)^{\theta} + (1 - v)^{\theta} - (1 - u)^{\theta}(1 - v)^{\theta}]^{1/\theta}.$$

7. The Archimedean copula 7 of [Nelsen (2006) Table 4.1]:

$$\mathcal{C}(u,v) = (\theta uv + (1-\theta)(u+v-1))_+.$$

8. The Archimedean copula 8 of [Nelsen (2006) Table 4.1]:

$$\mathcal{C}(u,v) = \left(\frac{\theta^2 uv - (1-u)(1-v)}{\theta^2 - (\theta-1)^2(1-u)(1-v)}\right)_+.$$

9. The Archimedean copula 9 of [Nelsen (2006) Table 4.1]:

$$\mathcal{C}(u,v) = uv\, exp[-\theta(log\, u)(log\, v)].$$

10. The Archimedean copula 10 of [Nelsen (2006) Table 4.1]:

$$\mathcal{C}(u,v) = \frac{uv}{[1+(1-u^\theta)(1-v^\theta)]^{1/\theta}}$$

11. The Archimedean copula 11 of [Nelsen (2006) Table 4.1]:

$$\mathcal{C}(u,v) = [(u^\theta v^\theta - 2(1-u^\theta)(1-v^\theta))_+]^{1/\theta}.$$

12. The Archimedean copula 12 of [Nelsen (2006) Table 4.1]:

$$\mathcal{C}(u,v) = (1 + [(u^{-1}-1)^\theta + (v^{-1}-1)^\theta]^{1/\theta})^{-1}.$$

13. The Archimedean copula 13 of [Nelsen (2006) Table 4.1]:

$$\mathcal{C}(u,v) = exp(1 - [(1-log\, u)^\theta + (1-log\, v)^\theta - 1]^{1/\theta}).$$

14. The Archimedean copula 14 of [Nelsen (2006) Table 4.1]:

$$\mathcal{C}(u,v) = (1 + [(u^{-1/\theta}-1)^\theta + (v^{-1/\theta}-1)^\theta]^{1/\theta})^{-\theta}.$$

15. The Archimedean copula 15 of [Nelsen (2006) Table 4.1]:

$$\mathcal{C}(u,v) = ((1 - [(1-u^{-1/\theta})^\theta + (1-v^{-1/\theta})^\theta]^{1/\theta})_+)^\theta.$$

16. The FGM copula:

$$\mathcal{C}(u,v) = uv[1 + \theta(1-u)(1-v)]. \tag{1}$$

17. The simple polynomial-sine (SPS) copula:

$$\mathcal{C}(u,v) = uv + \theta\frac{1}{\pi^2} sin\,(\pi u)\, sin(\pi v). \tag{2}$$

18. The power cosine copula (see Chesneau (2021a)):

$$\mathcal{C}(u,v) = uv\left[1+\theta\, cos\left(\frac{\pi}{2}u\right)cos\left(\frac{\pi}{2}v\right)\right].$$

19. The ratio extended FGM copula (see Chesneau (2021b)):

$$\mathcal{C}(u,v) = uv\left[1+\frac{1}{1+\theta uv}(1-u)(1-v)\right].$$

We recall that $(u,v)\in[0,1]^2$ in all cases, that $(a)_+ = max(a,0)$ and that the domain of definition for θ can change from one copula to another. Most of these copulas are involved in applications to describe dependence structures in two conjoint random phenomena, such as those illustrated in the references mentioned in the introductory section. All the copulas above are applicable to our idea of the Copula-G family, an idea which is developed in the next section.

3. Copula-G families

We now describe a new generator of distribution strategies based on the notion of copulas, which yields the Copula-G families.

3.1 Main idea

Starting with a copula $\mathcal{C}(u,v)$, the idea of creating a new generalized family is as follows. Let X and Y be two random variables defined on the domain [0,1] with the copula $\mathcal{C}(u,v)$ as a joint cumulative distribution function (CDF), and $\mathcal{W}(x)$ and $\mathcal{H}(x)$ as the CDFs of continuous univariate distributions. Let us introduce the univariate random variable

$$Z = max(\mathcal{W}^{-1}(X), \mathcal{H}^{-1}(Y)). \tag{3}$$

Then the CDF of Z is given by

$$F(x) = \mathcal{C}(\mathcal{W}(x), \mathcal{H}(x)), \ x \in R.$$

Thus defined, $F(x)$ is any univariate CDF, depending on $\mathcal{C}(u,v)$, $\mathcal{W}(x)$ and $\mathcal{H}(x)$. The Copula-G families are defined by $F(x)$. They depend on the three functions $\mathcal{C}(u,v)$, $\mathcal{W}(x)$ and $\mathcal{H}(x)$, and their choices drive the modeling ability of the Copula-G family from the modeling sense. We thus link the univariate CDFs, $\mathcal{W}(x)$ and $\mathcal{H}(x)$ through a copula strategy governed by $\mathcal{C}(u,v)$.

The probability density function (PDF) related to $F(x)$ is obtained via the differentiation of $F(x)$ with respect to x. After some algebra, it is obtained as

$$f(x) = w(x)\frac{\partial}{\partial u}\mathcal{C}(\mathcal{W}(x), \mathcal{H}(x)) + h(x)\frac{\partial}{\partial v}\mathcal{C}(\mathcal{W}(x), \mathcal{H}(x)), \ x \in R,$$

where $w(x)$ and $h(x)$ denote the PDFs related to $\mathcal{W}(x)$ and $\mathcal{H}(x)$, respectively. We emphasize that $\partial\mathcal{C}(\mathcal{W}(x), \mathcal{H}(x))/\partial u$ must be read as $\partial\mathcal{C}(u,v)/\partial u$ composed by $(u,v) = (\mathcal{W}(x), \mathcal{H}(x))$.

The hazard rate function (HRF) is given as $r(x) = f(x)/[1 - F(x)]$, $x \in R$, which can be expressed via $\mathcal{C}(u,v)$, $\mathcal{W}(x)$ and $\mathcal{H}(x)$ as

$$r(x) = \frac{1}{1 - \mathcal{C}(\mathcal{W}(x), \mathcal{H}(x))}\left[w(x)\frac{\partial}{\partial u}\mathcal{C}(\mathcal{W}(x), \mathcal{H}(x)) + h(x)\frac{\partial}{\partial v}\mathcal{C}(\mathcal{W}(x), \mathcal{H}(x))\right], \ x \in R.$$

The reversed hazard rate function is given as $r_*(x) = f(x)/F(x)$, $x \in R$, which can be expressed as,

$$r_*(x) = \frac{1}{\mathcal{C}(\mathcal{W}(x), \mathcal{H}(x))}\left[w(x)\frac{\partial}{\partial u}\mathcal{C}(\mathcal{W}(x), \mathcal{H}(x)) + h(x)\frac{\partial}{\partial v}\mathcal{C}(\mathcal{W}(x), \mathcal{H}(x))\right], \ x \in R.$$

These functions are the basis for a deep reliability study for given baseline distributions.

The quantile function is theoretically defined as $Q(x) = F^{-1}(x)$, which has no closed-form in our general setting. For some simple selected baseline distributions, it can certainly be expressed simply in some cases.

The use of two baseline distributions may be too arbitrary a choice in practice. To simplify the situation and fix the idea, a possible simple choice is $\mathcal{W}(x) = \mathcal{H}(x)$. In this case and if the copula $\mathcal{C}(u,v)$ is symmetric, the main functions are reduced to:

$$F(x) = \mathcal{C}(\mathcal{W}(x), \mathcal{H}(x)),$$

$$f(x) = 2w(x)\frac{\partial}{\partial u}\mathcal{C}(\mathcal{W}(x), \mathcal{H}(x)),$$

$$r(x) = \frac{2w(x)\partial\mathcal{C}(\mathcal{W}(x), \mathcal{W}(x))/\partial u}{1-\mathcal{C}(\mathcal{W}(x), \mathcal{H}(x))}.$$

These functions are quite manageable from the analytical viewpoint.

3.2 General theory

This part is devoted to some theoretical results of the Copula-G families, under some realistic assumptions. First, assume that $\mathcal{C}(u,v)$ can be expanded into a power series expansion as

$$\mathcal{C}(u,v) = \sum_{j,k=0}^{+\infty} a_{j,k}u^j v^k.$$

Such an expansion can be derived from the direct definition of the used copula (with possibly a lot of vanishing coefficients producing finite sums), or from the Taylor theorem. Then, under this assumption, some series expansion on important aspects of the Copula-G family can be given. In particular, it is clear that

$$F(x) = \sum_{j,k=0}^{+\infty} a_{j,k}\mathcal{W}(x)^j\,\mathcal{H}(x)^k.$$

Upon the condition that the derivative under the sums is mathematically valid, we immediately get

$$f(x) = \sum_{j,k=0}^{+\infty} a_{j,k}[jw(x)\mathcal{W}(x)^{j-1}\,\mathcal{H}(x)^k + kh(x)\mathcal{W}(x)^j\,\mathcal{H}(x)^{k-1}].$$

Some standard moment measures related to the Copula-G family can be derived from this formula. In particular, with the derivation under the sign sum condition, the rth moment of a random variable X with the CDF in Equation (3) is given as,

$$m_X(r) = E(X^r) = \sum_{j,k=0}^{+\infty} a_{j,k}[\theta_{j,k}(r) + \gamma_{j,k}(r)],$$

where,

$$\theta_{j,k}(r) = j\int_{-\infty}^{+\infty} x^r w(x)\mathcal{W}(x)^{j-1}\,\mathcal{H}(x)^k\,dx$$

and

$$\gamma_{j,k}(r) = k\int_{-\infty}^{+\infty} x^r h(x)\mathcal{W}(x)^j\,\mathcal{H}(x)^{k-1}\,dx.$$

The following approximation result is thus valid, provided the integer N chosen is large enough:

$$m_X(r) \approx \sum_{j,k=0}^{+\infty} a_{j,k}[\theta_{j,k}(r) + \gamma_{j,k}(r)].$$

The mean $m_X(1) = E(X)$, variance $V(X)$, skewness $S(X)$ and kurtosis $K(X)$ can be derived based on well-known relationships. More generally, the rth incomplete moment is given as,

$$m_X(r;t) = \sum_{j,k=0}^{+\infty} a_{j,k}[\theta_{j,k}(r;t) + \gamma_{j,k}(r;t)],$$

where.

$$\theta_{j,k}(r;t) = j\int_{-\infty}^{t} x^r\, w(x)\mathcal{W}(x)^{j-1}\,\mathcal{H}(x)^k\,dx$$

and

$$\gamma_{j,k}(r;t) = k \int_{-\infty}^{t} x^r\, h(x)\mathcal{W}(x)^j\, \mathcal{H}(x)^{k-1}\, dx.$$

In particular, the first incomplete moment is involved in the definition of key probabilistic functions, such as the mean residual life and mean inactivity time. In economics, life insurance, health sciences, demographics, product quality control, and product technology, these functions are useful (Lai and Xie 2006).

3.3 Examples of Copula-G family and members

Two Copula-G families are described below.

I. As a first example, we can consider the FGM copula defined in Equation (1) with $\theta \in [-1,1]$. Then, the following CDF is the result of the proposed strategy:

$$F(x) = \mathcal{W}(x)\, \mathcal{H}(x)[1 + \theta\,(1 - \mathcal{W}(x))\,(1 - \mathcal{H}(x)), x \in R. \tag{4}$$

The PDF related to $F(x)$ is given by,

$$f(x) = \mathcal{H}(x)w(x)[\theta\, \bar{\mathcal{H}}(x)\,(\bar{\mathcal{W}}(x) - \mathcal{W}(x)) + 1]$$
$$+ \mathcal{W}(x)\, h(x)[\theta\, \bar{\mathcal{W}}(x)\,(\bar{\mathcal{H}}(x) - \mathcal{H}(x)) + 1], x \in R,$$

where, $\bar{\mathcal{W}}(x) = 1 - \mathcal{W}(x)$ and $\bar{\mathcal{H}}(x) = 1 - \mathcal{H}(x)$. We recall that that $h(x)$ and $w(x)$ denote the PDFs related to $\mathcal{H}(x)$ and $\mathcal{W}(x)$, respectively. The family defined by this CDF $F(x)$ is called the FGM-G family. The HRF is obtained as,

$$r(x) = \frac{\mathcal{H}(x)w(x)[\theta\, \bar{\mathcal{H}}(x)\,(\bar{\mathcal{W}}(x) - \mathcal{W}(x)) + 1] + \mathcal{W}(x)\, h(x)[\theta\, \bar{\mathcal{W}}(x)\,(\bar{\mathcal{H}}(x) - \mathcal{H}(x)) + 1]}{1 - \mathcal{W}(x)\, \mathcal{H}(x)[1 + \theta\,(1 - \mathcal{W}(x))\,(1 - \mathcal{H}(x))]}, x \in R.$$

II. As a second example, we can consider the SPS copula defined in Equation (2) with $\theta \in [-1,1]$. Then, the following CDF is the result of the proposed strategy:

$$F(x) = \mathcal{W}(x)\, \mathcal{H}(x) + \theta\frac{1}{\pi^2} sin\,[\pi\mathcal{W}(x)] sin[\pi\mathcal{H}(x)], x \in R. \tag{5}$$

The PDF related to $F(x)$ is given by,

$$f(x) = \theta\frac{1}{\pi}\,(w(x)\, cos[\pi\mathcal{W}(x)]\, sin[\pi\mathcal{H}(x)] + h(x)\, sin\,[\pi\mathcal{W}(x)] cos[\pi\mathcal{H}(x)])$$
$$+ \mathcal{H}(x)w(x) + \mathcal{W}(x)h(x), x \in R.$$

The related family is called the SPS-G family. The corresponding HRF is indicated as,

$$r(x) = \frac{(\theta/\pi)(w(x)\, cos[\pi\mathcal{W}(x)]\, sin[\pi\mathcal{H}(x)] + h(x)\, sin\,[\pi\mathcal{W}(x)]\, cos[\pi\mathcal{H}(x)]) + \mathcal{H}(x)w(x) + \mathcal{W}(x)h(x)}{1 - \mathcal{W}(x)\, \mathcal{H}(x) + (\theta/\pi^2)\, sin\,[\pi\mathcal{W}(x)]\, sin[\pi\mathcal{H}(x)]}.$$

For the choices of $\mathcal{W}(x)$ and $\mathcal{H}(x)$, they mainly depend on the context in which we want to apply the copula-G family. In the field of reliability, one can think of considering the two main lifetime distributions in literature: The exponential and Lindley distributions.

Thus, for $\mathcal{W}(x)$, we consider the CDF of the exponential distribution with parameter $\alpha > 0$ given as,

$$\mathcal{W}(x) = 1 - e^{-\alpha x},\ x > 0,$$

and $\mathcal{W}(x) = 0$ for $x \leq 0$, with $\alpha > 0$. The corresponding PDF is specified by

$$w(x) = \alpha e^{-\alpha x},\ x > 0,$$

and $w(x) = 0$ for $x \leq 0$. For $\mathcal{H}(x)$, we consider the CDF of the Lindley distribution with parameter $\beta > 0$ given by,

$$\mathcal{H}(x) = 1 - \left(1+\frac{\beta}{\beta+1}x\right)e^{-\beta x},\ x > 0,$$

and $\mathcal{H}(x) = 0$ for $x \leq 0$, with $\beta > 0$. The corresponding PDF is specified by

$$h(x) = \frac{\beta^2}{\beta+1}(1+x)e^{-\beta x},\ x > 0,$$

and $h(x) = 0$ for $x \leq 0$. The Lindley distribution is defined as a special mixture of the exponential distribution with parameter β and gamma distribution with parameters 2 and β. These distributions are lifetime distributions which differ a bit in modeling properties. In particular, the HRF of the exponential distribution is constant, whereas the HRF of the Lindley distribution is not. For details on the Lindley distribution, we refer to Tomy (2018). With these baseline distributions, the two following two-parameter distributions can be introduced.

I. As a first example, based on the FGM-G family described by the CDF in Equation (4), and the exponential and Lindley distributions, we consider the following CDF:

$$F(x) = (1 - e^{-\alpha x})\left[1-\left(1+\frac{\beta}{\beta+1}x\right)e^{-\beta x}\right]\left[1+\theta e^{-\alpha x}\left(1+\frac{\beta}{\beta+1}x\right)e^{-\beta x}\right],\ x > 0,$$

and $F(x) = 0$ for $x \leq 0$, with $\theta \in [-1,1]$, $\alpha > 0$ and $\beta > 0$. We call this three-parameter model as the FGM-EL distribution. The related PDF is given by

$$f(x) = \left[1-\left(1+\frac{\beta}{\beta+1}x\right)e^{-\beta x}\right]\alpha e^{-\alpha x}\left[\theta\left(1+\frac{\beta}{\beta+1}x\right)e^{-\beta x}(2e^{-\alpha x}-1)+1\right],$$

$$+(1-e^{-\alpha x})\frac{\beta^2}{\beta+1}(1+x)e^{-\beta x}\left(\theta e^{-\alpha x}\left[2\left(1+\frac{\beta}{\beta+1}x\right)e^{-\beta x}-1\right)\right]+1\right), x > 0,$$

and $f(x) = 0$ for $x \leq 0$.

II. As a first example, based on the SPS-G family described by the CDF in Equation (5), and the exponential and Lindley distributions, we consider the following CDF:

$$F(x) = (1-e^{-\alpha x})\left[1-\left(1+\frac{\beta}{\beta+1}x\right)e^{-\beta x}\right]$$

$$+\theta\frac{1}{\pi^2}sin(\pi e^{-\alpha x})sin\left[\pi\left(1+\frac{\beta}{\beta+1}x\right)e^{-\beta x}\right], x > 0,$$

and $F(x) = 0$ for $x \leq 0$, with $\theta \in [-1,1]$, $\alpha > 0$ and $\beta > 0$. We call this three-parameter model as the SPS-EL distribution. The related PDF is given by

$$f(x) = -\theta\frac{1}{\pi}\left\{\begin{array}{l}\alpha e^{-\alpha x}cos(\pi e^{-\alpha x})sin\left[\pi\left(1+\frac{\beta}{\beta+1}x\right)e^{-\beta x}\right]\\ +\frac{\beta^2}{\beta+1}(1+x)e^{-\beta x}sin(\pi e^{-\alpha x})cos\left[\pi\left(1+\frac{\beta}{\beta+1}x\right)e^{-\beta x}\right]\end{array}\right\}$$
$$+\left[1-\left(1+\frac{\beta}{\beta+1}x\right)e^{-\beta x}\right]\alpha e^{-\alpha x}+(1-e^{-\alpha x})\frac{\beta^2}{\beta+1}(1+x)e^{-\beta x}, x>0,$$

and $f(x) = 0$ for $x \leq 0$.

Table 1 presents a numerical analysis of moment-type measures, $m_X(1)$, $V(X)$, $S(X)$ and $K(X)$, of the FGM-EL distribution.

Based on the numerical result given in Table 1, it is noted that $m_X(1)$ decreases as θ increases and increases as α and β increase. On the other hand, $S(X)$ is positive and can range in the interval (1.991305,66.15859), and the spread for its $K(X)$ is ranging from 8.190794 to 5785.361 for the extreme case, so the distribution is mainly leptokurtic. If data from a lifetime-type phenomenon is available, the maximum likelihood method can be used to calculate the parameters of the FGM-EL and SPS-EL distributions. Then, by substituting the obtained estimates into the CDF or PDF, we get the estimated CDF or PDF, respectively. We can visualize their fits as a suitable graphical representation of the data, such as the empirical CDF for the estimated CDF and the shape of the histogram for the estimated PDF.

Table 1: $m_X(1)$, $V(X)$, $S(X)$ and $K(X)$ of the FGM-EL distribution for various values of the parameters.

θ	α	β	$m_X(1)$	$V(X)$	$S(X)$	$K(X)$
0.99	0.5	50	0.01459327	0.00046747	1.991305	8.190794
0.50			0.01231556	0.00039155	2.299100	10.03958
0			0.00999137	0.00030340	2.707796	13.08657
0.50			0.00766718	0.00020445	3.250578	18.50450
0.90			0.00580783	0.00011751	3.623670	25.06420
0.05	0.001	5	0.00022747	0.00010259	66.15859	5785.361
	0.15		0.03208547	0.01303870	5.293313	40.01863
	0.75		0.12675900	0.03360885	2.296001	10.31271
	1		0.15412670	0.03402075	2.045883	9.017326
	2		0.22020870	0.02162654	2.608773	10.51440
	3		0.24663160	0.00727922	12.34179	26.00328
0.85	0.5	0.1	0.07437125	0.3514299	10.36599	135.0803
		0.5	0.21458330	0.4795370	4.118848	23.51379
		10	0.06303644	0.0103301	2.172377	9.068581
		50	0.01394249	0.0004468	2.071928	8.640574
		100	0.00704930	0.0001124	2.054335	8.609646

4. Conclusion

This paper is devoted to a new idea for generating a family of flexible distributions. The goal is to apply the well-known properties of copulas and multidimensional flexibility to the univariate situation to create new families of univariate distributions. These families can depend on two baseline distributions. Some theoretical results are provided, with discussions. Finally, some of the families are defined, and some members of interest are highlighted, to show how the proposed methodology can be used for modeling purposes. This work opens the horizons for the creation of a plethora of Copula-G families that can inspire statisticians all over the world.

References

Alizadeh, M., Jamal, F., Yousof, H. M., Khanahmadi, M., Hamedani, G. G. et al. (2020a). Flexible Weibull generated family of distributions: characterizations, mathematical properties and applications. University Politehnica of Bucharest Scientific Bulletin-Series A-Applied Mathematics and Physics, 82(1): 145—50.

Altun, E., Alizadeh, M., Yousof, H. M. and Hamedani, G. G. (2022). The Gudermannian generated family of distributions with characterizations, regression models and applications. Studia Scientiarum Mathematicarum Hungarica, 59(2): 93–115.

Aryal, G. R. and Yousof, H. M. (2017). The exponentiated generalized-G Poisson family of distributions. Economic Quality Control, 32(1): 1–17.

Bekrizadeh, H. and Jamshidi, B. (2017). A new class of bivariate copulas: dependence measures and properties. Metron, 75: 31–50.

Bekrizadeh, H., Parham, G. A. and Zadkarami, M. R. (2012). The new generalization of Farlie-Gumbel-Morgenstern copulas, Applied Mathematical Sciences, 6: 3527–3533.

Bekrizadeh, H., Parham, G. A. and Zadkarami, M. R. (2015). An asymmetric generalized FGM copula and its properties, Pakistan Journal of Statistics, 31: 95–106.

Brito, E., Cordeiro, G. M., Yousof, H. M., Alizadeh, M., Silva, G. O. et al. (2017). Topp-leone odd log-logistic family of distributions. Journal of Statistical Computation and Simulation, 87(15): 3040–3058.

Casella, G. and Berger, R. L. (1990). Statistical Inference, Brooks/Cole Publishing Company: Bel Air, CA, USA.

Chesneau, C. (2021a). A study of the power-cosine copula. Open Journal of Mathematical Analysis, 5, 1, 85–97.

Chesneau, C. (2021b). A new two-dimensional relation copula inspiring a generalized version of the Farlie-Gumbel-Morgenstern copula. Research and Communications in Mathematics and Mathematical Sciences, 13, 2, 99–128.

Chesneau, C. and Yousof, H. M. (2021). On a special generalized mixture class of probabilistic models. Journal of Nonlinear Modeling and Analysis, 3: 71–92.

Coles, S., Currie, J. and Tawn, J. (1999). Dependence measures for extreme value analyses, Extremes, 2: 339–365.

Cordeiro, G. M., Yousof, H. M., Ramires, T. G. and Ortega, E. M. M. (2018). The Burr XII system of densities: properties, regression model and applications. Journal of Statistical Computation and Simulation, 88(3): 432–456.

Domma, F. (2009). Some properties of the bivariate Burr type III distribution. Statistics, 44: 203–215.

El-Morshedy, M., Alshammari, F. S., Hamed, Y. S., Eliwa, M. S., Yousof, H. M et al. (2021). A New Family of Continuous Probability Distributions. Entropy, 23: 194.

Eugene, N., Lee, C. and Famoye, F. (2002). Beta-normal distribution and its applications. Commun. Stat. Theory Methods, 31: 497–512.

Fischer, M. and Klein, I. (2007). Constructing generalized fgm copulas by means of certain univariate distributions, Metrika, 65: 243–260.

Frees, E. W. and Valdez, E. A. (1998). Understanding relationship using copulas, North American Actuarial Journal, 2: 1–25.

Georges, P., Lamy, A. G., Nicolas, E., Quibel, G., Roncalli, T. et al. (2001). Multivariate Survival Modelling: A Unified Approach with Copulas, SSRN Electron. J.

Hamedani, G. G., Altun, E., Korkmaz, M. C., Yousof, H. M., Butt, N. S. et al. (2018). A new extended G family of continuous distributions with mathematical properties, characterizations and regression modeling. Pak. J. Stat. Oper. Res., 14: 737–758.

Hamedani, G. G., Yousof, H. M., Rasekhi, M., Alizadeh, M., Najibi, S. M. et al. (2017). Type I general exponential class of distributions. Pak. J. Stat. Oper. Res., 14: 39–55.

Hamedani, G. G., Rasekhi, M., Najib, S. M., Yousof, H. M., Alizadeh, M. et al. (2019). Type II general exponential class of distributions. Pak. J. Stat. Oper. Res., 15: 503–523.

Jin, P. W. M. and Shitan, M. (2014). Construction of a new bivariate copula based on Ruschendorf method, Applied Mathematical Sciences, 8: 7645–7658.

Karamikabir, H., Afshari, M., Yousof, H. M., Alizadeh, M., Hamedani, G. et al. (2020). The weibull topp-leone generated family of distributions: statistical properties and applications. Journal of The Iranian Statistical Society, 19: 121–161.

Kazianka, H. and Pilz, J. (2009). Copula-based geostatistical modeling of continuous and discrete data including covariates, Stochastic Environmental Research and Risk Assessment, 24: 661–673.

Korkmaz, M. C., Yousof, H. M. and Hamedani G. G. (2018a). The exponential Lindley odd log-logistic G family: properties, characterizations and applications. Journal of Statistical Theory and Applications, 17: 554–571.

Korkmaz, M. C., Yousof, H. M., Hamedani, G. G. and Ali, M. M. (2018b). The Marshall-Olkin generalized G Poisson family of distributions, Pakistan Journal of Statistics, 34: 251–267.

Korkmaz, M. C., Altun, E., Yousof, H. M. and Hamedani, G. G. (2020). The Hjorth's IDB generator of distributions: properties, characterizations, regression modeling and applications. Journal of Statistical Theory and Applications, 19: 59–74.

Lai, C. D. and Xie, M. (2006). Stochastic Ageing and Dependence for Reliability, Springer Science and Business Media, Berlin, Germany.

Merovci, F., Alizadeh, M., Yousof, H. M. and Hamedani, G. G. (2017). The exponentiated transmuted-G family of distributions: theory and applications, Communications in Statistics-Theory and Methods, 46: 10800–10822.

Merovci, F., Yousof, H. M. and Hamedani, G. G. (2020). The poisson topp leone generator of distributions for lifetime data: theory. Characterizations and Applications. Pakistan Journal of Statistics and Operation Research, 16(2): 343–355.

Nascimento, A. D. C., Silva, K. F., Cordeiro, G. M., Alizadeh, M., Yousof, H. M. et al. (2019). The odd Nadarajah-Haghighi family of distributions: properties and applications. Studia Scientiarum Mathematicarum Hungarica, 56: 1–26.

Nelsen, R. (2006). An Introduction to Copulas, Springer Science+Business Media, Inc. second edition.

Shiau, J. T., Wang, H. Y. and Tsai, C. T. (2011). Bivariate frequency analysis of floods using copulas, Journal of the American Water Resources Association, 46: 1549–1564.

Sklar, M. (1959). Fonctions de repartition a n dimensions et leurs marges, Publ. L'Institut Stat. L'Universite Paris, 8: 229–231.

Tomy, L. (2018). A retrospective study on Lindley distribution, Biometrics & Biostatistics International Journal, 7(2): 163–169.

Thompson, D. and Kilgore, R. (2011). Estimating joint flow probabilities at stream confluences using copulas. Transportation Research Record, 2262: 200–206.

Trivedi, P. K. and Zimmer, D. M. (2005). Copula modelling: an introduction to practitioners, Foundations and Trends in Econometrics, 1: 1–111.

Yong-Quan, D. (2008). Generation and prolongation of FGM copula, Chinese Journal of Engineering Mathematics, 25: 1137–1141.

Yousof, H. M., Alizadeh, M., Jahanshahi, S. M. A., Ramires, T. G., Ghosh, I. et al. (2017). The transmuted Topp-Leone G family of distributions: theory, characterizations and applications. Journal of Data Science, 15: 723–740.

Yousof, H. M., Altun, E., Ramires, T. G., Alizadeh, M., Rasekhi, M. et al. (2018). A new family of distributions with properties, regression models and applications, Journal of Statistics and Management Systems, 21: 163–188.

Yousof, H. M., Mansoor, M. Alizadeh, M., Afify, A. Z., Ghosh, I. et al. (2020). The Weibull-G Poisson family for analyzing lifetime data. Pak. J. Stat. Oper. Res., 16: 131–148.

Zhang, Q., Chen, Y.D., Chen, X. and Li, J. (2011). Copula-based analysis of hydrological extremes and implications of hydrological behaviors in Pearl river basin, China. Journal of Hydrologic Engineering, 16: 598–607.

Chapter 4

A Family of Continuous Probability Distributions

Theory, Characterizations, Properties and Different Copulas

Mohammad Mehdi Saber,[1] *GG Hamedani,*[2] *Haitham M Yousof,*[3,*] *Nadeem Shafique Butt,*[4] *Basma Ahmed*[5] and *Mohamed Ibrahim*[6]

1. Introduction

Statistical literature contains various G families of distributions which were generated either by compounding well-known existing G families or by adding one (or more) parameters to the existing classes. These novel families were employed for modeling real data in many applied areas such as engineering, insurance, demography, medicine, econometrics, biology, environmental sciences, and others; refer to Aryal and Yousof (2017) for exponentiated generalized Poisson-G family, Brito et al. (2017) for the Topp Leone odd log-logistic-G family, Yousof et al. (2017) for the Burr type X-G family, Cordeiro et al. (2018) for the Burr XII-G family, Korkmaz et al. (2018a and 2018b) for the exponential-Lindley odd log-logistic-G family and the Marshall–Olkin generalized G Poisson family of distributions, Karamikabir et al. (2020) for the Weibull Topp Leone generated-G family, Yousof et al. for the extended odd Fréchet-G family, Abouelmagd et al. (2019a and 2019b) for the Poisson Burr X-G family and the Topp-Leone Poisson-G family, Nascimento et al. (2019) for the odd Nadarajah-Haghighi-G family, Merovci et al. (2017 and 2020) for the exponentiated transmuted-G family and the Poisson Topp Leone-G family, Korkmaz et al. (2020) for the Hjorth's IDB generator of distributions, Alizadeh et al. (2020a and 2020b) for flexible Weibull generated-G family of distributions and the transmuted odd log-logistic-G family, Hamedani et al. (2017, 2018, 2019, 2021) the type I general exponential class of distributions, the new extended-G family of continuous distributions, the type II general exponential class of distributions and the type I quasi-Lambert family, Altun et al. (2021) for the Gudermannian generated family of distributions, Chesneau and Yousof (2021) for the special generalized

[1] Department of Statistics, Higher Education Center of Eghlid, Eghlid, Iran.
[2] Department of Mathematical and Statistical Sciences, Marquette University, USA.
[3] Department of Statistics, Mathematics and Insurance, Faculty of Commerce, Benha University, Benha, Egypt.
[4] Department of Family and Community Medicine, King Abdul Aziz University, Jeddah, Kingdom of Saudi Arabia.
[5] Department of Information System, Higher Institute for Specific Studies, Giza, Egypt.
[6] Department of Applied, Mathematical and Actuarial Statistics, Faculty of Commerce, Damietta University, Damietta, Egypt.
Emails: mmsaber@eghlid.ac.ir; gholamhoss.hamedani@marquette.edu; nshafique@kau.edu.sa; dr. basma13@gmail.com; mohamed_ibrahim@du.edu.eg
* Corresponding author: haitham.yousof@fcom.bu.edu.eg

mixture class of probabilistic models and El-Morshedy et al. (2021) for the Poisson generalized exponential-G family, among others.

Let $g_{\underline{V}}(x)$ and $G_{\underline{V}}(x)$ denote the density and cumulative distribution functions of the baseline model with the parameter vector $\underline{V}$ and consider the Weibull CDF,

$$F_{\beta_2,\beta_3}(x) = 1 - exp\left(-\frac{1}{\beta_3} x\beta_2\right) | \; x > 0,$$

with positive parameters β_2 and β_3. Based on this density and using the argument,

$$Q_{\underline{V}}(x) = \frac{1}{\frac{1}{G_{\underline{V}}(x)-1}},$$

Bourguignon et al. (2014) defined the CDF of their Weibull-G class by,

$$H_{\beta_2,\beta_3,\underline{V}}(x) = 1 - exp\left(-\frac{1}{\beta_3} Q_{\underline{V}}^{\beta_2}(x)\right) |_{\beta_2,\beta_3>0 \text{ and } x\in R}. \quad (1)$$

The Weibull-G density function is given by,

$$h_{\beta_2,\beta_3,\underline{V}}(x) = \frac{\beta_2}{\beta_3} g_{\underline{V}}(x) \frac{G_{\underline{V}}(x)^{\beta_2-1}}{G_{\underline{V}}(x)^{\beta_2+1}} exp\left(-\frac{1}{\beta_3} Q_{\underline{V}}^{\beta_2}(x)\right) |_{\beta_2,\beta_3>0 \text{ and } x\in R}. \quad (2)$$

where, $\bar{G}_{\underline{V}}(x) = 1 - G_{\underline{V}}(x)$. For a baseline random variable with probability density function (PDF) $h_{\beta_2,\beta_3,\underline{V}}(x)$ and CDF $H_{\beta_2,\beta_3,\underline{V}}(x)$, the complementary geometric-G (CGc-G) family is defined by the CDF,

$$F_{\beta_1,\underline{V}}(x) = \frac{\beta_1 G_{\underline{V}}(x)}{1-\bar{\beta}_1 G_{\underline{V}}(x)} |_{\bar{\beta}_1 = 1-\beta_1 \text{ and } x\in R}, \quad (3)$$

and the PDF is given by,

$$f_{\beta_1,\underline{V}}(x) = \frac{\beta_1 g_{\underline{V}}(x)}{\left[1-\bar{\beta}_1 G_{\underline{V}}(x)\right]^2} |_{\bar{\beta}_1 = 1-\beta_1 \text{ and } x\in R}, \quad (4)$$

where, $\beta_1 > 0$. In this paper, we propose and study a new extension of the CGc-G family to provide more flexibility to the generated family. We construct a new generator called the complementary geometric Weibull-G (CGcW-G) family by taking the Weibull-G CDF in (1) as the baseline CDF in equations (3) and (4). Further, we give a comprehensive description of the mathematical properties of the new family. In fact, the CGcW-G family is motivated by its flexibility in applications which has importance.

2. The new family

The CDF of the CGcW-G family is defined by,

$$F_{\underline{\Phi}}(x) = \frac{\beta_1 \left\{1 - exp\left(-\frac{1}{\beta_3} Q_{\underline{V}}^{\beta_2}(x)\right)\right\}}{1-\bar{\beta}_1 \left\{1 - exp\left(-\frac{1}{\beta_3} Q_{\underline{V}}^{\beta_2}(x)\right)\right\}} |_{\beta_1,\beta_2,\beta_3>0 \text{ and } x\in R}. \quad (5)$$

where, $\underline{\Phi} = (\beta_1, \beta_2, \beta_3, \underline{V})$ is the vector of parameters for the baseline $G_{\underline{V}}(x)$. The corresponding PDF is given by,

$$f_{\underline{\Phi}}(x) = \beta_1\beta_2\beta_3^{-1}\frac{g_{\underline{V}}(x)G_{\underline{V}}(x)^{\beta_2-1}exp-\left[\frac{1}{\beta_3}\mathcal{Q}_{\underline{V}}^{\beta_2}(x)\right]}{\overline{G}_{\underline{V}}(x)^{\beta_2+1}\left(1-\overline{\beta}_1\left\{1-exp\left[-\frac{1}{\beta_3}\mathcal{Q}_{\underline{V}}^{\beta_2}(x)\right]\right\}\right)^2}|_{\beta_1,\beta_2,\beta_3>0 \text{ and } x\in R}. \quad (6)$$

For $\beta_1 = 1$, the CGcW-G family reduces to the Weibull-G family (Bourguignon et al. (2014)). For $\beta_1 = \beta_3 = 1$, the CGcW-G family reduces to the one parameter Weibull-G family (Bourguignon et al. (2014)). For $\beta_1 = 1$ and $\beta_2 = 2$, the CGcW-G family reduces to the Rayleigh-G family (Bourguignon et al. (2014)). For $\beta_1 = \beta_2 = 1$, the CGcW-G family reduces to the odd exponential-G family (Bourguignon et al. (2014)). For $\beta_3 = 1$, the CGcW-G family reduces to the two parameter CGcW-G family.
Using the Taylor and generalized binomial expansions, the PDF in (6) can be expressed as,

$$f_{\underline{\Phi}}(x) = \sum_{k,m=0}^{\infty}\varpi_{k,m}h_{\beta_2^*}(x)|_{\beta_1,\beta_2,\beta_3>0 \text{ and } x\in R}, \quad (7)$$

where, $\beta_2^* = \beta_2(k+1)+m$ and

$$\varpi_{k,m} = \beta_2, \beta_1, \beta_3^{k-1}(1+i)\frac{(-1)^{k+m}}{k!\beta_2^*}\binom{-\beta_2(k+1)-1}{m}$$
$$\times\sum_{i=0}^{\infty}\sum_{j=0}^{i}(-1)^{i+j}(i+1)(\beta_1-1)^i\binom{i}{j}(j+1)^k,$$

and $\pi_{\nabla}(x) = \nabla g_{\underline{V}}(x)G_{\underline{V}}(x)^{\nabla-1}$ is the Exp-G PDF with power parameter $\nabla > 0$. Thus, several mathematical properties of the CGcW-G family can be obtained from those of the Exp-G family. Equation (7) is the main result of this section. The CDF of the CGcW-G family can also be expressed as a mixture of the Exp-G densities. By integrating (7), we obtain the same mixture representation

$$F_{\underline{\Phi}}(x) = \sum_{k,m=0}^{\infty}w_{k,m}\Pi_{\beta_2^*}(x)|_{\beta_1,\beta_2,\beta_3>0 \text{ and } x\in R}, \quad (8)$$

where $\Pi_{\nabla}(x)$ is the CDF of the Exp-G family with the power parameter ∇.

3. Characterizations of the CGcW-G distribution

Characterization of a distribution is important in applied sciences, where an investigator is vitally interested to find out if their model follows the selected distribution. Therefore, the investigator relies on conditions under which their model would follow a specified distribution. A probability distribution can be characterized in different directions one of which is based on the truncated moments. It should also be mentioned that characterization results are mathematically challenging and elegant. In this section, we present certain characterizations of the CGcW-G distribution based on: (i) conditional expectation (truncated moment) of a certain function of a random variable and (ii) reverse hazard function.

3.1 Characterizations based on a simple relationship between two truncated moments

This subsection presents characterizations of the CGcW-G distribution in terms of a simple relationship between two truncated moments. We employ a Theorem by Glänzel (1987) given in Appendix A. As shown in Glänzel (1990), this characterization is stable in the sense of weak convergence. The first characterization given below can also be employed when the CDF does not have a closed form.

Proposition 3.1.1. Let $X: \Omega \rightarrow R$ be a continuous random variable and let

$$\Upsilon_1(x) = \left(1 + \overline{\beta_1}\left\{1 - exp\left[-\frac{1}{\beta_3} Q_{\underline{V}}^{\beta_2}(x)\right]\right\}\right)^2$$

and

$$\Upsilon_2(x) = \Upsilon_1(x) exp\left[-\frac{1}{\beta_3} Q_{\underline{V}}^{\beta_2}(x)\right] \text{for } x \in R.$$

Then X has PDF (6) if and only if the function ξ defined in Theorem 1 is of the form

$$\xi(x) = \frac{1}{2} exp\left\{-\frac{1}{\beta_3} Q_{\underline{V}}^{\beta_2}(x)\right\} |_{x \in R}.$$

Proof. If X has PDF (6), then,

$$(1 - F(x)) E[\Upsilon_1(X) \mid X \geq x] = \beta_1 \, exp\left\{-\frac{1}{\beta_3} Q_{\underline{V}}^{\beta_2}(x)\right\} |_{x \in R},$$

and

$$(1 - F(x)) E[\Upsilon_2(X) \mid X \geq x] = \frac{1}{2} \beta_1 \, exp\left\{-\frac{2}{\beta_3} Q_{\underline{V}}^{\beta_2}(x)\right\} |_{x \in R},$$

and hence,

$$\xi(x) = \frac{1}{2} exp\left\{-\frac{1}{\beta_3} Q_{\underline{V}}^{\beta_2}(x)\right\} |_{x \in R}.$$

We also have,

$$\xi(x)\Upsilon_1(x) - \Upsilon_2(x) = -\frac{1}{2} \Upsilon_1(x) \, exp\left\{-\frac{1}{\beta_3} Q_{\underline{V}}^{\beta_2}(x)\right\} < 0 \, |_{x \in R}.$$

Conversely, if $\xi(x)$ is of the above form, then

$$s'(x) = \frac{\xi'(x)\Upsilon_1(x)}{\xi(x)\Upsilon_1(x) - \Upsilon_2(x)} = \beta_2 \beta_3^{-1} g_{\underline{V}}(x) [G_{\underline{V}}(x)]^{\beta_2 - 1} [1 - G_{\underline{V}}(x)]^{-(\beta_2 + 1)} |_{x \in R}.$$

Now, according to Theorem 1, X has density (6)

Corollary 3.1.1. Let X be a continuous random variable and $\Upsilon_1(x)$ be as in Proposition 3.1.1. The PDF of X is (6) if and only if there exist functions $\Upsilon_1(x)$ and $\xi(x)$ defined in Theorem 1 for which the following first order differential equation holds:

$$s'(x) = \frac{\xi'(x)\Upsilon_1(x)}{\xi(x)\Upsilon_1(x) - \Upsilon_2(x)} = \beta_2 \beta_3^{-1} g_{\underline{V}}(x) [G_{\underline{V}}(x)]^{\beta_2 - 1} [1 - G_{\underline{V}}(x)]^{-(\beta_2 + 1)} |_{x \in R}.$$

Corollary 3.1.2. The differential equation in Corollary 3.1.1 has the following general solution:

$$\xi(x) = exp\left\{\frac{1}{\beta_3} Q_{\underline{V}}^{\beta_2}(x)\right\} \times \left[\begin{array}{c} -\int \beta_2 \beta_3^{-1} g_{\underline{V}}(x)[G_{\underline{V}}(x)]^{\beta_2 -1}[1-G_{\underline{V}}(x)]^{-(\beta_2+1)} \\ \times exp\left[-\frac{1}{\beta_3} Q_{\underline{V}}^{\beta_2}(x)\right][\Upsilon_1(x)]^{-1}\Upsilon_2(x) + D \end{array}\right],$$

where D is a constant. A set of functions satisfying the above differential equation is given in Proposition 3.1.1 with $D = 0$. Clearly, there are other triplets $(\Upsilon_1(x), \Upsilon_2(x), \xi(x))$ satisfying the conditions of Theorem 1.

3.2 Characterization based on reverse hazard function

The reverse hazard function, r_F, of a twice differentiable distribution function, F, is defined as,

$$r_F(x) = \frac{f(x)}{F(x)},\ x \in \text{support of } F.$$

In this subsection we present a characterization of the CGcW-G distribution in terms of the reverse hazard function.

Proposition 3.2.1. Let $X : \Omega \rightarrow R$ be a continuous random variable. The random variable X has PDF (6) if and only if its reverse hazard function $r_F(x)$ satisfies the following differential equation,

$$r_F'(x) - \frac{(\beta_2 - 1)\, g_{\underline{V}}(x)}{G_{\underline{V}}(x)} r_F(x)$$

$$\beta_2 \beta_3^{-1} G_{\underline{V}}^{\beta_2 -1}(x) \frac{d}{dx}\left\{\frac{g_{\underline{V}}(x)(1-G_{\underline{V}}(x))^{-(\beta_2+1)} exp\left\{-\frac{1}{\beta_3} Q_{\underline{V}}^{\beta_2}(x)\right\}}{1-\overline{\beta}_1\left(1-exp\left\{-\frac{1}{\beta_3} Q_{\underline{V}}^{\beta_2}(x)\right\}\right)}\right\}\Bigg|_{x\in R},$$

with boundary condition $lim_{x\rightarrow -\infty} r_F(x) = 0$ for $\beta_2 > 1$.

4. Properties

In this section, we study some general properties of the CGcW-G family of distributions.

4.1 General properties

The r th moment of X, say $\mu_{r,X}'$, follows from (7) as,

$$\mu_{r,X}' = E(X^r) = \sum_{k,m=0}^{\infty} \varpi_{k,m} E(Y_{\beta_2^*}^r). \tag{9}$$

Henceforth, $Y_{\beta_2^*}$ denotes the Exp-G distribution with power parameter β_2^*. For $\beta_2^* > 0$, and we have,

$$E(Y_{\beta_2^*}^r) = \beta_2^* \int_{-\infty}^{\infty} x^r g_{\underline{V}}(x) G_{\underline{V}}(x)^{\beta_2^* -1} dx,$$

which can be computed numerically in terms of the baseline quantile function (qf) $Q_{G,\underline{V}}(u) = G_{\underline{V}}(u)^{-1}$ as,

$$E(Y^n_{\beta_2^*}) = \beta_2^* \int_0^1 Q_{G,\underline{V}}(u)^n \quad u^{\beta_2^*-1}\, du.$$

The variance, skewness, and kurtosis measures can now be calculated. Then, the MGF $M_X(t) = E(exp(tX))$ of can be derived from Equation (7) as,

$$M_X(t) = \sum_{k,m=0}^{\infty} \varpi_{k,m} M_{\beta_2^*}(t),$$

where $M_X(t)$ is the MGF of Y_k.

Hence, $M_X(t)$ can be determined from the Exp-G generating function. The nth central moment of X, say M_n, is given by,

$$M_{n,X} = E(X - \mu'_{1,X})n = \sum_{r=0}^{n} \sum_{k,m=0}^{\infty} \varpi_{k,m}(-1)^{n-r}\binom{n}{r}(\mu'_{r,X})^{n-r} E(Y^r_{\beta_2^*}).$$

The nth descending factorial moment of X (for $n = 1,2,\ldots$) is,

$$\mu'_{[1]} = E[X(X-1)\times\ldots\times(X-n+1)] = \sum_{j=0}^{n} s(n,j)\mu'_j = E(X^{[1]}),$$

where, $s(n,j) = \frac{1}{j!}\left[\frac{d^j}{dx^j}(j^{(n)})\right]_{x=0}$ is the Stirling number of the first kind. The th incomplete moment, say $\varpi_{s,X}(t)$, of X can be expressed from (10) as,

$$\varpi_{s,X}(t) = \int_{-\infty}^{t} x^s f_{\underline{\Phi}}(x)dx = \sum_{k,m=0}^{\infty} \varpi_{k,m} \int_{-\infty}^{t} x_s \pi_{\beta_2^*}(x)dx.$$

The mean deviations about the mean $[b_X = E(|X - \mu_1'|)]$ and about the median $[b_X = E(|X - M|)]$ of X are given by,

$$b_X(\mu'_{1,X}) = 2\mu'_{1,X})\, F(\mu'_{1,X} - 2\varpi_{1,X}(\mu'_{1,X})$$

and

$$b_X(M) = \mu'_{1,X} - 2\varpi_{1,X}(M),$$

respectively, where $\mu'_{1,X} = E(X)$, $M = Median(X) = Q(\frac{1}{2})$ is the median, $F(\mu'_{1,X})$ is easily calculated and $\varpi_{1,X}(t)$ is the first incomplete moment given by (11) with $s = 1$. Now, the general equation for $\varpi_{1,X}(t)$ can be derived from $\varpi_{s,X}(t)$ as,

$$\varpi_{1,X}(t) = \sum_{k,m=0}^{\infty} \varpi_{k,m} V_{\beta_2^*}(x),$$

where, $V(x) = \int_{-\infty}^{t} x h_k(x)\, dx$ is the first incomplete moment of the Exp-G distribution. This equations for $\varpi_{1,X}(t)$ can be applied to construct Bonferroni and Lorenz curves defined for a given probability π by $B(\pi) = \varpi_1(q)/(\pi\mu'_{1,X})$ and $L(\pi) = \varpi_1(q)/\mu'_{1,X}$, respectively, where $\mu'_{1,X} = E(X)$ and $q = Q(\pi)$ is the qf of X at π.

4.2 Probability weighted moments

The $(s,r)^{th}$ probability weighted moments of X following the CGcW-G family of distribution, say $\rho_{s,r}$, is formally defined by,

$$\rho_{s,r} = E\{X^s F(X)^r\} = \int_{-\infty}^{\infty} x^s F_{\underline{\Phi}}(x)^r f_{\underline{\Phi}}(x)dx.$$

From Equation (6) and (8), we can write,

$$f_{\underline{\Phi}}(x)F_{\underline{\Phi}}(x)^r = \sum_{k,m=0}^{\infty} C_{k,m}\pi_{\beta_2^*}(x)$$

where,

$$C_{k,m} = \frac{\beta_2\beta_1^{r+1}}{\beta_3}\frac{(-1)^{k+m}[(1+i)\beta_3]^k}{k!(j+1)^{-k}}\binom{-\beta_2(k+1)-1}{m}$$
$$\times\sum_{i,j=0}^{\infty}\frac{(-1)^j}{\overline{\beta_1}^{-i}[\beta_2(k+1)+m]}\binom{-r-2}{i}\binom{i+r}{j}$$

Finally, the $(s,r)^{\text{th}}$ PWM of X can be obtained from an infinite linear combination of Exp-G moments given by,

$$\rho_{s,r} = \sum_{k,m=0}^{\infty} C_{k,m}E(Y_{\beta_2^*}^s).$$

4.3 Order statistics

Order statistics make their appearance in many areas of statistical theory and practice. Let $X_1,\ldots,X_n$ be a random sample from the CGcW-G family of distributions. The PDF of the th order statistic, say _$X_{i:n}$, can be written as,

$$f_{i:n}(x) = \frac{f_{\underline{\Phi}}(x)}{B(i,n-i+1)}\sum_{j=0}^{n-i}(-1)^j\binom{n-i}{j}F_{\underline{\Phi}}(x)^{j+i-1}. \tag{10}$$

Using (5) and (6),
we get,

$$f_{\underline{\Phi}}(x)F_{\underline{\Phi}}(x)^{j+i-1} = \sum_{k,m=0}^{\infty} t_{k,m}\pi_{\beta_2^*}(x), \tag{11}$$

where $\pi_k(x)$ is the Exp-G density with power parameter k and

$$t_{k,m} = \beta_2\beta_1^{j+i}\beta_3^{-1}\frac{(-1)^{k+m}}{k!(w+1)^{-k}}[(1+i)\beta_3^{-1}]^k\binom{-\beta_2(k+1)-1}{m}$$
$$\times\sum_{h,w=0}^{\infty}\frac{(-1)^w\overline{\beta_2}^{-h}}{\beta_2(k+1)+m]}\binom{-(j+i+1)}{h}\binom{h+j+i-1}{w}.$$

Substituting (13) into Equation (12), the PDF of $X_{i:n}$ can be expressed as,

$$f_{i:n}(x) = \sum_{h,w=0}^{\infty}\sum_{j=0}^{n-i}\frac{(-1)^j t_{k,m}}{B(i,n-i+1)}\binom{n-i}{j}\pi_{\beta_2^*}(x), \tag{12}$$

Then, the density function of the CGcW-G order statistics is a mixture of Exp-G densities. Based on the last Equation, we note that the properties of $X_{i:n}$ follow from the properties of Y_{a+k}. For example, the moments of $X_{i;n}$ can be expressed as

$$E(X_{i:n}^q) = \sum_{h,w=0}^{\infty}\sum_{j=0}^{n-i}\frac{(-1)^j t_{k,m}}{B(i,n-i+1)}\binom{n-i}{j}E\pi_{\beta_2^*}(Y_{\beta_2^*}^q). \tag{13}$$

5. Estimation and inference

Several approaches for parameter estimation were proposed in the literature but the maximum likelihood method is the most commonly employed. The maximum likelihood estimators (MLEs) enjoy desirable properties and can be used when constructing confidence intervals with test statistics. The normal approximation for these estimators in large sample theory is easily handled either analytically or numerically. So, we consider the estimation of the unknown parameters for this family from complete samples only by maximum likelihood. Here, we determine the MLEs of the parameters of the new family of distributions from complete samples only. Let be a random sample from the CGcW-G family with parameters β_1, β_2, β_3 and $\underline{V}$. Then, the log-likelihood function for $\underline{\Phi}$, say $\ell = \ell(\underline{\Phi})$, is given by,

$$\ell = n\log\beta_1 + n\log\beta_2 + n\log\beta_3^{-1} + \sum_{i=0}^{n} \log g_{\underline{V}}(x_i) + (\beta_2 - 1)\sum_{i=0}^{n} \log G_{\underline{V}}(x_i)$$

$$-(\beta_2+1)\sum_{i=0}^{n} \log \overline{G}_{\underline{V}}(x_i) - \sum_{i=0}^{n} s_i^{\beta_2} - 2\sum_{i=0}^{n} \log z_i,$$

where,

$$s_i = G_{\underline{V}}(x_i)/\overline{G}_{\underline{V}}(x_i)$$

and

$$z_i = 1 + \overline{\beta}_1\left\{1 - exp\left[-\frac{1}{\beta_3} s_i^{\beta_2}\right]\right\}.$$

The log-likelihood function can be maximized either directly by using the R (optim function), SAS (PROC NLMIXED) or Ox program (sub-routine MaxBFGS) or by solving the nonlinear likelihood equations obtained by differentiating (17). The score vector components are given by,

$$U_{\beta_1} = \frac{n}{\beta_1} - 2\sum_{i=0}^{n}\frac{w_i}{z_i}$$

$$U_{\beta_2} = \frac{n}{\beta_2} + \sum_{i=0}^{n} \log G_{\underline{V}}(x_i) - \sum_{i=0}^{n} \log \overline{G}_{\underline{V}}(x_i) - \sum_{i=0}^{n} s_i^{\beta_2}\log s_i - 2\sum_{i=0}^{n}\frac{m_i}{z_i}$$

$$U_{\beta_3} = n\beta_2 - 2\sum_{i=0}^{n}\frac{\overline{\beta}_1}{z_i}s_i^{\beta_2} exp\left[-\frac{1}{\beta_3}s_i^{\beta_2}\right]\left[1 - exp\left(-\frac{1}{\beta_3}s_i^{\beta_2}\right)\right],$$

and

$$U_{V_r} = \sum_{i=0}^{n}\frac{g'_{r,\underline{V}}(x_i)}{g_{\underline{V}}(x_i)} + (\beta_2 - 1)\sum_{i=0}^{n}\frac{G'_{r,\underline{V}}(x_i)}{G_{\underline{V}}(x_i)} + (\beta_2+1)\sum_{i=0}^{n}\frac{G'_{r,\underline{V}}(x)}{\overline{G}_{\underline{V}}(x)} - \beta_2\sum_{i=0}^{n} p_{i,r}s_i^{\beta_2-1} - 2\sum_{i=0}^{n}\frac{d_i}{z_i},$$

where,

$$g'_{r,\underline{V}}(x) = \frac{\partial g_{\underline{V}}(x_i)}{\partial \underline{V}_k}, G'_{r,\underline{V}}(x_i) = \frac{\partial G_{\underline{V}}(x_i)}{\partial \underline{V}_k},$$

$$p_{i,r} = [\overline{G}_{\underline{V}}(x_i)G'_{\underline{V}}(x_i) + G_{\underline{V}}(x_i)G'_{\underline{V}}(x_i)]\overline{G}_{\underline{V}}(x_i)^{-2}, w_i = 1 - exp\left(-\frac{1}{\beta_3}s_i^{\beta_2}\right),$$

$$m_i = \overline{\beta}_1 s_i^{\beta_2} exp\left(-\frac{1}{\beta_3}s_i^{\beta_2}\right)\log s_i,$$

and

$$d_i = \overline{\beta}_1 \beta_2 p_{i,r} s_i^{\beta_2 - 1} exp\left(-\frac{1}{\beta_3} s_i^{\beta_2}\right).$$

Setting the nonlinear system of equations $U_{\beta1} = U_{\beta2} = U_{\beta3} = U_{Vk} = 0$ and solving them simultaneously yields the MLEs. These equations cannot be solved analytically, and statistical software can be used to solve them numerically using iterative methods such as the Newton-Raphson type algorithms. For interval estimation of the model parameters, we require the observed information matrix whose elements are easily derived. Under standard regularity conditions when $n \to \infty$, the distribution $\hat{\underline{\Phi}}$ of the estimated can be approximated by a multivariate normal distribution to construct approximate confidence intervals for the parameters. Here, $\mathbf{J}(\hat{\underline{\Phi}})$ is the total observed information matrix evaluated at . The method of the re-sampling bootstrap can be used for correcting the biases of the MLEs of the model parameters. Interval estimates may also be obtained using the bootstrap percentile method. Likelihood ratio tests can be performed for the proposed family of distributions in the usual way.

6. Copulas

For modeling of the bivariate real data sets, we shall derive some new bivariate CGCW-G (Bv-CGCW-G) type distributions using "Farlie-Gumbel-Morgenstern copula" (FGMC) copula, modified FGMC, "Clayton copula", "Renyi's entropy copula (REC)" and "Ali-Mikhail-Haq copula (AMHC)". The multivariate CGCW-G (Mv-CGCW-G) type can be easily derived based on the Clayton copula. However, future works may be allocated to study these new models. For more recent applications of some probability models see Al-babtain et al. (2020), Salah et al. (2020), Elgohari and Yousof (2020a and 2020b), Ali et al. (2021a and 2021b), Shehata and Yousof (2021a and 2021b), Elgohari and Yousof (2021), Elgohari et al. (2021) and Shehata et al. (2022).

6.1 BvCGCW-G type via *Clayton copula*

Let $X_1 \sim$ PTLG – G($\underline{\Phi}^1$) and $X_2 \sim$ PTLG – G($\underline{\Phi}^2$). Depending on the continuous marginals $\overline{\mathcal{U}} = 1 - \mathcal{U}$ and $\overline{\mathcal{V}} = 1 - \mathcal{V}$, the Clayton copula can be considered as,

$$C_{\mathscr{b}}(\overline{\mathcal{U}}, \overline{\mathcal{V}}) = [\max(\overline{\mathcal{U}}^{-\mathscr{b}}) + \overline{\mathcal{V}}^{-\mathscr{b}} - 1); 0]^{-\frac{1}{\mathscr{b}}}),$$

where,

$$\mathscr{b} \in [-1, \infty) - \{0\},\ \overline{\mathcal{U}} \in (0,1) \text{ and } \overline{\mathcal{V}} \in (0,1)$$

Let

$$\overline{\mathcal{U}} = 1 - F_{\underline{\Phi}_1}(x_1)|_{\underline{\Phi}_1},\ \overline{\mathcal{V}} = 1 - F_{\underline{\Phi}_2}(x_2)\,|_{\underline{\Phi}_2}$$

and

$$F_{\underline{\Phi}_i}(x_i)|_{i=1,2} = \frac{1}{1 - e^{-a_i}} \{1 - exp(-a_i\ \{G_{\underline{\Psi}}(x)^{\beta_i}\ [2 - G_{\underline{\Psi}}(x)^{\beta_i}]\}^{\alpha_i})\}.$$

Then, the BvCGCW-G type distribution can be obtained from $C_{\mathscr{b}}(\overline{\mathcal{U}}, \overline{\mathcal{V}})$.

6.2 BvCGCW-G type via REC

The REC can be derived using the continuous marginal functions $\mathcal{U} = 1 - \overline{\mathcal{U}} = F_{\underline{\Phi}_1}(x_1) \in (0,1)$ and $\mathcal{V} = 1 - \overline{\mathcal{V}} = F_{\underline{\Phi}_1}(x_2) \in (0,1)$ as follows:

$$F(x_1, x_2) = C(F_{\underline{\Phi}_1}(x_1), F_{\underline{\Phi}_1}(x_2)) = x_2\mathcal{U} + x_1\mathcal{V} - x_1x_2.$$

6.3 BvCGCW-G type via FGMC

Considering the FGMC, the joint CDF can be written as,

$$C_{\mathcal{B}}(\mathcal{U},\mathcal{V}) = \mathcal{U},\mathcal{V} + \mathcal{U},\mathcal{V}\,\mathcal{B}\,\bar{\mathcal{U}},\bar{\mathcal{V}},$$

where, the continuous marginal function $\mathcal{U} \in (0,1)$, $\mathcal{V} \in (0,1)$ and $\mathcal{B} \in [-1,1]$.
Setting

$$\bar{\mathcal{U}} = \bar{\mathcal{U}}_{\Phi_1}|_{\Phi_1>0} \text{ and } \bar{\mathcal{V}} = \bar{\mathcal{V}}_{\Phi_2}|_{\Phi_2>0},$$

we then have,

$$F(x_1, x_2) = \mathcal{U}\mathcal{V}(1 + \mathcal{B}\,\bar{\mathcal{U}},\bar{\mathcal{V}}).$$

Then, the joint PDF can be expressed as,

$$c_{\mathcal{B}}(\bar{\mathcal{U}},\bar{\mathcal{V}}) = 1 + \mathcal{B}\,\mathcal{U}^*\mathcal{V}^*,$$

where,

$$\mathcal{U}^* = 1 - 2\mathcal{U} \text{ and } \mathcal{V}^* = 1 - 2\mathcal{V},$$

or

$$f_{\mathcal{B}}(x_1, x_2) = f_{\Phi_1}(x_1)\,f_{\Phi_2}(x_2)\;C(F_{\Phi_1}(x_1), F_{\Phi_1}(x_2)),$$

where the two function $f_{\mathcal{B}}(x_1, x_2)$ and $c_{\mathcal{B}}(\mathcal{U},\mathcal{V})$ are PDFs corresponding to the joint CDFs and $f_{\mathcal{B}}(x_1, x_2)$ and $c_{\mathcal{B}}(\mathcal{U},\mathcal{V})$.

6.4 BvCGCW-G type via modified FGMC

The modified formula of the modified FGMC can be expressed as $C_{\mathcal{B}}(\mathcal{U},\mathcal{V}) = \mathcal{B}\;\mathcal{O}(\mathcal{U})^*\;\mathcal{J}(\mathcal{V})^* + \mathcal{U}\mathcal{V}$, with $\mathcal{U}\mathcal{O}(\mathcal{U})^* = \mathcal{J}(\mathcal{V})$ and $\mathcal{J}(\mathcal{V})^* = \mathcal{V}\mathcal{J}(\mathcal{V})$ where $\mathcal{O}(\mathcal{U}) \in (0,1)$ and $\mathcal{J}(\mathcal{V}) \in (0,1)$ are two continuous functions and $\mathcal{O}(\mathcal{U}=0) = \mathcal{O}(\mathcal{U}=1) = \mathcal{J}(\mathcal{V}=0) = \mathcal{J}(\mathcal{V}=1) = 0$. The, the following four types can be derived and considered:

6.4.1 Type I

The new bivariate version via modified FGMC type I can be written as

$$C_{\mathcal{B}}(\mathcal{U},\mathcal{V}) = \mathcal{B}\,\mathcal{O}(\mathcal{U})^*\,\mathcal{J}(\mathcal{V})^* + \mathcal{U}\mathcal{V}.$$

6.4.2 Type II

Consider $\mathbf{A}(\mathcal{U};\mathcal{B}_1)$ and $\mathbf{B}(\mathcal{V};\mathcal{B}_2)$ which satisfy the above conditions where,

$$\mathbf{A}(\mathcal{U};\mathcal{B}_1)|_{(\mathcal{B}1>0)} = \mathcal{U}^{\mathcal{B}_1}(1-\mathcal{U})^{1-\mathcal{B}_1}.$$

and

$$\mathbf{B}(\mathcal{V};\mathcal{B}_2)|_{(\mathcal{B}2>0)} = \mathcal{V}^{\mathcal{B}_2}(1-\mathcal{V})^{1-\mathcal{B}_2}.$$

Then, the corresponding bivariate version (modified FGMC Type II) can be derived as

$$C_{\theta 0,\theta 1,\theta 2}(\mathcal{U},\mathcal{V}) = \mathcal{U}\mathcal{V} + \theta_0\,\mathcal{U}\mathcal{V}\mathbf{A}(\mathcal{U};\theta_1)\mathbf{B}(\mathcal{V};\theta_2).$$

6.4.3 Type III

Let $\widetilde{\mathbf{A}(\mathcal{U})} = \mathcal{U}\,[log(1+\bar{\mathcal{U}})]_{(\bar{\mathcal{U}}=1-\mathcal{U})}$ and $\widetilde{\mathbf{B}(\mathcal{V})} = \mathcal{V}\,[log(1+\bar{\mathcal{V}})]_{(\bar{\mathcal{V}}=1-\mathcal{V})}$. Then, the associated CDF of the BvCGCW-G-FGM (modified FGMC type III) is

$$C_{\theta}(\mathcal{U},\mathcal{V}) = \mathcal{U}\mathcal{V} + \mathcal{U}\mathcal{V}\,\theta\widetilde{\mathbf{A}(\mathcal{U})}\,\widetilde{\mathbf{B}(\mathcal{V})}.$$

6.4.4 Type IV

Using the quantile concept, the CDF of the BvCGCW-G-FGM (modified FGMC type IV) model can be obtained as,

$$C(\mathcal{U},\mathcal{V}) = \mathcal{U}F^{-1}(\mathcal{U}) + \mathcal{V}F^{-1}(\mathcal{V}) - F^{-1}(\mathcal{U})\,F^{-1}(\mathcal{V})$$

where, $F^{-1}(\mathcal{U}) = Q(\mathcal{U})$ and $F^{-1}(\mathcal{V}) = Q(\mathcal{V})$.

6.5 BvCGCW-G type via AMHC

Under the "stronger Lipschitz condition", the joint CDF of the Archimedean Ali-Mikhail-Haq copula can be written as,

$$C_{\theta}(\upsilon,\mathcal{V}) = \frac{1}{1-\bar{\upsilon}\bar{\mathcal{V}}}\,\upsilon\mathcal{V}|_{\theta\in(-1,1)},$$

the corresponding joint PDF of the Archimedean Ali-Mikhail-Haq copula can be expressed as,

$$C_{\theta}(\upsilon,\mathcal{V}) = \frac{1}{[1-\theta\bar{\upsilon}\bar{\mathcal{V}}]^2}\left(1-\theta+2\theta\frac{\upsilon\mathcal{V}}{1-\theta\bar{\upsilon}\bar{\mathcal{V}}}\right)|_{\theta\in(-1,1)},$$

and then for any $\bar{\upsilon} = 1 - f_{\Phi_1}(x_1) = |_{[\bar{\upsilon}=(1-\upsilon)\in(0,1)]}$ and $\bar{\mathcal{V}} = 1 - F_{\Phi_2}(x_2)|_{[\bar{\mathcal{V}}=(1-\mathcal{V})\in(0,1)]}$ we have,

$$C_{\theta}(x_1,x_2) = \frac{1}{1-\theta[1-F_{\Phi_1}(x_1)][1-F_{\Phi_2}(x_2)]}\,[F_{\Phi_1}(x_1)\,F_{\Phi_2}(x_2)]|_{\theta\in(-1,1)},$$

and

$$C_{\theta}(x_1,x_2) = \frac{1}{\{1-\theta[1-F_{\Phi_1}(x_1)][1-F_{\Phi_2}(x_2)]\}^2}$$
$$\times\left(1-\theta+2\theta\left\{\frac{F_{\Phi_1}(x_1)\,F_{\Phi_2}(x_2)}{1-\theta[1-F_{\Phi_1}(x_1)][1-F_{\Phi_2}(x_2)]}\right\}\right)|_{\theta\in(-1,1)},$$

7. Conclusions

The present paper studied a new three-parameter compound family of probability distributions called the complementary geometric Weibull-G family. The relevant mathematical properties such as the ordinary moments, probability weighted moments and order statistics are derived and analyzed. The probability density function of the complementary geometric Weibull-G family is expressed as a mixture of the exponentiated-G densities. We presented certain characterizations of the new family based on: (*i*) conditional expectation (truncated moment) of certain functions of a random variable and (*ii*) reverse hazard function. For facilitating the mathematical modeling of the bivariate real data, we derive some new bivariate type extensions using

Farlie-Gumbel-Morgenstern, modified Farlie-Gumbel-Morgenstern, Clayton, "Renyi and "Ali-Mikhail-Haq copulas.

As future potential works, we can apply many new useful goodness-of-fit tests for right censoring distributional validity such as the Nikulin-Rao-Robson goodness-of-fit test, modified Nikulin-Rao-Robson goodness-of-fit test, Bagdonavicius-Nikulin goodness-of-fit test, modified Bagdonavicius-Nikulin goodness-of-fit test, to the new family as performed by Ibrahim et al. (2019), Goual et al. (2019, 2020), Mansour et al. (2020a–f), Yadav et al. (2020), Goual and Yousof (2020), Ibrahim et al (2021) and Yousof et al. (2021), Aidi et al. (2021) and Yousof et al. (2021a), among others.

Some new acceptance sampling plans based on the complementary geometric Weibull-G family or based on some special members can be presented in separate articles (refer to Ahmed and Yousof (2022) and Ahmed et al. (2022)).

Some useful reliability studies based on multicomponent stress-strength and the remaining stress-strength concepts can be presented (Rasekhi et al. (2020) and Saber et al. (2022a,b), Saber and Yousof (2022)).

References

Abouelmagd, T. H. M., Hamed, M. S., Hamedani, G. G., Ali, M. M., Goual, H. et al. (2019a). The zero truncated Poisson Burr X family of distributions with properties, characterizations, applications, and validation test. Journal of Nonlinear Sciences and Applications, 12(5): 314–336.

Abouelmagd, T. H. M., Hamed, M. S., Handique, L., Goual, H., Ali, M. M. et al. (2019b). A new class of distributions based on the zero truncated Poisson distribution with properties and applications, Journal of Nonlinear Sciences & Applications (JNSA), 12(3): 152–164.

Ahmed, B. and Yousof, H. M. (2022). A New Group Acceptance Sampling Plans based on Percentiles for the Weibull Fréchet Model. Statistics, Optimization & Information Computing, forthcoming.

Ahmed, B., Ali, M. M. and Yousof, H. M. (2022). A Novel G Family for Single Acceptance Sampling Plan with Application in Quality and Risk Decisions, Annals of Data Science, forthcoming.

Al-babtain, A. A., Elbatal, I. and Yousof, H. M. (2020). A new flexible three-parameter model: properties, clayton copula, and modeling real data. Symmetry, 12(3): 440.

Aidi, K., Butt, N. S., Ali, M. M., Ibrahim, M., Yousof, H. M. et al. (2021). A modified chi-square type test statistic for the double burr x model with applications to right censored medical and reliability data. Pakistan Journal of Statistics and Operation Research, 17(3): 615–623.

Ali, M. M., Yousof, H. M. and Ibrahim, M. (2021a). A new version of the generalized Rayleigh distribution with copula, properties, applications and different methods of estimation. Optimal Decision Making in Operations Research & Statistics: Methodologies and Applications, 1: 1–25.

Ali, M. M., Ibrahim, M. and Yousof, H. M. (2021b). Expanding the Burr X model: properties, copula, real data modeling and different methods of estimation. Optimal Decision Making in Operations Research & Statistics: Methodologies and Applications, 1: 26–49.

Alizadeh, M., Jamal, F., Yousof, H. M., Khanahmadi, M., Hamedani, G. G. et al. (2020a). Flexible Weibull generated family of distributions: characterizations, mathematical properties and applications. University Politehnica of Bucharest Scientific Bulletin-Series A-Applied Mathematics and Physics, 82(1): 145–150.

Alizadeh, M., Yousof, H. M., Jahanshahi, S. M. A., Najibi, S. M., Hamedani, G. G. et al. (2020b). The transmuted odd log-logistic-G family of distributions. Journal of Statistics and Management Systems, 23(4): 1–27.

Altun, E., Yousof, H. M. and Hamedani, G. G. (2021). The Gudermannian generated family of distributions with characterizations, regression models and applications, Studia Scientiarum Mathematicarum Hungarica, forthcoming.

Aryal, G. R. and Yousof, H. M. (2017). The exponentiated generalized-G Poisson family of distributions. Stochastics and Quality Control, 32: 7–23.

Bourguignon, M., Silva, R. B. and Cordeiro, G. M. (2014). The Weibull-G family of probability distributions. Journal of Data Science, 12(1): 53–68.

Brito, E., Cordeiro, G. M., Yousof, H. M., Alizadeh, M., Silva, G. O et al. (2017). The Topp–Leone odd log-logistic family of distributions. Journal of Statistical Computation and Simulation, 87: 3040–3058.

Chesneau, C. and Yousof, H. M. (2021). On a special generalized mixture class of probabilistic models. Journal of Nonlinear Modeling and Analysis, 3(1): 71–92.

Cordeiro, G. M., Yousof, H. M., Ramires, T. G. and Ortega, E. M. (2018). The Burr XII system of densities: properties, regression model and applications. Journal of Statistical Computation and Simulation, 88: 432–456.

Elgohari, H. and Yousof, H. M. (2020a). A generalization of lomax distribution with properties, copula and real data applications. Pakistan Journal of Statistics and Operation Research, 16(4): 697–711.

Elgohari, H. and Yousof, H. M. (2020b). New extension of weibull distribution: copula, mathematical properties and data modeling. Statistics, Optimization & Information Computing, 8(4): 972–993.

Elgohari, H. and Yousof, H. M. (2021). A New Extreme Value Model with Different Copula, Statistical Properties and Applications. Pakistan Journal of Statistics and Operation Research, 17(4): 1015–1035.

Elgohari, H., Ibrahim, M. and Yousof, H. M. (2021). A new probability distribution for modeling failure and service times: properties, copulas and various estimation methods. Statistics, Optimization & Information Computing, 8(3): 555–586.

El-Morshedy, M., Alshammari, F. S., Hamed, Y. S., Eliwa, M. S., Yousof, H. M. et al. (2021). A New Family of Continuous Probability Distributions. Entropy, 23: 194.

Goual, H., Yousof, H. M. and Ali, M. M. (2019). Validation of the odd Lindley exponentiated exponential by a modified goodness of fit test with applications to censored and complete data. Pakistan Journal of Statistics and Operation Research, 15(3): 745–771.

Goual, H. and Yousof, H. M. (2020). Validation of Burr XII inverse Rayleigh model via a modified chi-squared goodness-of-fit test. Journal of Applied Statistics, 47(3): 393–423.

Goual, H., Yousof, H. M. and Ali, M. M. (2020). Lomax inverse Weibull model: properties, applications, and a modified Chi-squared goodness-of-fit test for validation. Journal of Nonlinear Sciences & Applications (JNSA), 13(6): 330–353.

Glänzel, W., Telcs, A. and Schubert, A. (1984). Characterization by truncated moments and its application to Pearson-type distributions. Z. Wahrsch. Verw. Gebiete 66: 173–182.

Glänzel, W. (1987). A characterization theorem based on truncated moments and its application to some distribution families, Mathematical Statistics and Probability Theory (Bad Tatzmannsdorf, 1986), Vol. B, Reidel, Dordrecht, 75–84.

Glänzel, W. (1990). Some consequences of a characterization theorem based on truncated moments, Statistics: A Journal of Theoretical and Applied Statistics, 21(4): 613–618.

Hamedani, G. G., Altun, E., Korkmaz, M. Ç., Yousof, H. M., Butt, N. S. et al. (2018). A new extended G family of continuous distributions with mathematical properties, characterizations and regression modeling. Pakistan Journal of Statistics and Operation Research, 737–758.

Hamedani, G. G., Yousof, H. M., Rasekhi, M., Alizadeh, M., Najibi, S. M. et al. (2017). Type I general exponential class of distributions. Pakistan Journal of Statistics and Operation Research, 14(1): 39–55.

Hamedani, G. G., Rasekhi, M., Najibi, S., Yousof, H. M., Alizadeh, M. et al. (2019). Type II general exponential class of distributions. Pakistan Journal of Statistics and Operation Research, 15(2): 503–523.

Hamedani, G. G., Korkmaz, M. C., Butt, N. S. and Yousof, H. M. (2021). The Type I Quasi Lambert Family: Properties, Characterizations and Different Estimation Methods. Pakistan Journal of Statistics and Operation Research, 17(3): 545–558.

Ibrahim, M., Yadav, A. S., Yousof, H. M., Goual, H., Hamedani, G. G. et al. (2019). A new extension of Lindley distribution: modified validation test, characterizations and different methods of estimation. Communications for Statistical Applications and Methods, 26(5): 473–495.

Ibrahim, M., Aidi, K., Ali, M. M. and Yousof, H. M. (2021). A Novel Test Statistic for Right Censored Validity under a new Chen extension with Applications in Reliability and Medicine. Annals of Data Science, 29(1): 681–705.

Korkmaz, M. C. Yousof, H. M. and Hamedani G. G. (2018a). The exponential Lindley odd log-logistic G family: properties, characterizations and applications. Journal of Statistical Theory and Applications, 17(3): 554–571.

Korkmaz, M. C., Yousof, H. M., Hamedani G. G. and Ali, M. M. (2018b). The Marshall–Olkin generalized G Poisson family of distributions, Pakistan Journal of Statistics, 34(3): 251–267.

Karamikabir, H., Afshari, M., Yousof, H. M., Alizadeh, M., Hamedani, G. et al. (2020). The Weibull ToppLeone Generated Family of Distributions: Statistical Properties and Applications. Journal of The Iranian Statistical Society, 19: 121–161

Korkmaz, M. Ç., Altun, E., Yousof, H. M. and Hamedani, G. G. (2020). The Hjorth's IDB generator of distributions: properties, characterizations, regression modeling and applications. Journal of Statistical Theory and Applications, 19(1): 59–74.

Mansour, M. M., Ibrahim, M., Aidi, K., Shafique Butt, N., Ali, M. M. et al. (2020a). A New Log-Logistic Lifetime Model with Mathematical Properties, Copula, Modified Goodness-of-Fit Test for Validation and Real Data Modeling. Mathematics, 8(9); 1508.

Mansour, M. M., Butt, N. S., Ansari, S. I., Yousof, H. M., Ali, M. M., Ibrahim, M. et al. (2020b). A new exponentiated Weibull distribution's extension: copula, mathematical properties and applications. Contributions to Mathematics, 1 (2020) 57–66. DOI: 10.47443/cm.2020.0018.

Mansour, M., Korkmaz, M. C., Ali, M. M., Yousof, H. M., Ansari et al. (2020c). A generalization of the exponentiated Weibull model with properties, Copula and application. Eurasian Bulletin of Mathematics, 3(2): 84–102.

Mansour, M., Rasekhi, M., Ibrahim, M., Aidi, K., Yousof, H. M. et al. (2020d). A new parametric life distribution with modified bagdonavi?ius–nikulin goodness-of-fit test for censored validation, properties, applications, and different estimation methods. Entropy, 22(5): 592.

Mansour, M., Yousof, H. M., Shehata, W. A. and Ibrahim, M. (2020e). A new two parameter Burr XII distribution: properties, copula, different estimation methods and modeling acute bone cancer data. Journal of Nonlinear Science and Applications, 13(5): 223–238.

Mansour, M. M., Butt, N. S., Yousof, H. M., Ansari, S. I., Ibrahim, M. et al. (2020f). A generalization of reciprocal exponential model: clayton copula, statistical properties and modeling skewed and symmetric real data sets. Pakistan Journal of Statistics and Operation Research, 16(2): 373–386.

Merovci, F., Alizadeh, M., Yousof, H. M. and Hamedani, G. G. (2017). The exponentiated transmuted-G family of distributions: theory and applications. Communications in Statistics-Theory and Methods, 46(21): 10800–10822.

Merovci, F., Yousof, H. M. and Hamedani, G. G. (2020). The poisson topp leone generator of distributions for lifetime data: theory, characterizations and applications. Pakistan Journal of Statistics and Operation Research, 16(2): 343–355.

Nascimento, A. D., Silva, K. F., Cordeiro, G. M., Alizadeh, M., Yousof, H. M et al. (2019). The odd Nadarajah-Haghighi family of distributions: properties and applications. Studia Scientiarum Mathematicarum Hungarica, 56(2): 185–210.

Rasekhi, M., Saber, M. M. and Yousof, H. M. (2020). Bayesian and classical inference of reliability in multicomponent stress-strength under the generalized logistic model. Communications in Statistics-Theory and Methods, 50(21): 5114–5125.

Saber, M. M. and Yousof, H. M. (2022). Bayesian and Classical Inference for Generalized Stress-strength Parameter under Generalized Logistic Distribution, Statistics, Optimization & Information Computing, forthcoming.

Saber, M. M. Marwa M. Mohie El-Din and Yousof, H. M. (2022a). Reliability estimation for the remained stress-strength model under the generalized exponential lifetime distribution.Journal of Probability and Statistics, 2021: 1–10.

Saber, M. M., Rasekhi, M. and Yousof, H. M. (2022b). Generalized Stress-Strength and Generalized Multicomponent Stress-Strength Models, Statistics, Optimization & Information Computing, forthcoming.

Salah, M. M., El-Morshedy, M., Eliwa, M. S. and Yousof, H. M. (2020). Expanded Fréchet Model: mathematical properties, copula, different estimation methods, applications and validation testing. Mathematics, 8(11): 1949.

Shehata, W. A. M., Butt, N. S., Yousof, H. and Aboraya, M. (2022). A new lifetime parametric model for the survival and relief times with copulas and properties. Pakistan Journal of Statistics and Operation Research, 18(1): 249–272.

Shehata, W. A. M. and Yousof, H. M. (2021a). The four-parameter exponentiated Weibull model with Copula, properties and real data modeling. Pakistan Journal of Statistics and Operation Research, 17(3): 649–667.

Shehata, W. A. M. and Yousof, H. M. (2021b). A novel two-parameter Nadarajah-Haghighi extension: properties, copulas, modeling real data and different estimation methods. Statistics, Optimization & Information Computing, forthcoming.

Yadav, A. S., Goual, H., Alotaibi, R. M., Ali, M. M., Yousof, H. M. et al. (2020). Validation of the Topp-Leone-Lomax model via a modified Nikulin-Rao-Robson goodness-of-fit test with different methods of estimation. Symmetry, 12(1): 57.

Yousof, H. M., Afify, A. Z., Hamedani, G. G. and Aryal, G. (2017). The Burr X generator of distributions for lifetime data. Journal of Statistical Theory and Applications, 16: 288–305.

Yousof, H. M., Rasekhi, M., Altun, E. and Alizadeh, M. (2018). The extended odd Fréchet family of distributions: properties, applications and regression modeling. International Journal of Applied Mathematics and Statistics, 30(1): 1–30.

Yousof, H. M., Aidi, K., Hamedani, G. G. and Ibrahim, M. (2021a). A new parametric lifetime distribution with modified Chi-square type test for right censored validation, characterizations and different estimation methods. Pakistan Journal of Statistics and Operation Research, 17(2): 399–425.

Yousof, H. M., Al-nefaie, A. H., Aidi, K., Ali, M. M., Ibrahim, M. et al. (2021). A modified chi-square type test for distributional validity with applications to right censored reliability and medical data: a modified chi-square type test. Pakistan Journal of Statistics and Operation Research, 17(4): 1113–1121.

Appendix A

Theorem 1. Let (Ω, F, P) be a given probability space and let $H = [a, b]$ be an interval for some $d < b$ ($a = -\infty$, $b = \infty$ might as well be allowed). Let $X : \Omega \to H$ be a continuous random variable with the distribution function F and let $\Upsilon_1(x)$ and $\Upsilon_2(x)$ be two real functions defined on H such that,

$$E[\Upsilon_2(x)|X \geq x] = E[\Upsilon_2(x)|X \geq x]\ \xi(x),\ x \in H,$$

is defined with some real function ξ. Assume that $\Upsilon_1(x)$, $\Upsilon_2(x) \in C^1(H)$, $\xi(x) \in C^2(H)$ and F is twice continuously differentiable and strictly monotone function on the set H. Finally, assume that the equation

$\xi\Upsilon_1 = \Upsilon_2$ has no real solution in the interior of H. Then F is uniquely determined by the functions $\Upsilon_1(x)$, $\Upsilon_2(x)$ and $\xi(x)$, particularly,

$$F(x) = \int_a^x C\left|\frac{\xi'(u)}{\xi(u)\Upsilon_1(u) - \Upsilon_1(u)}\right| exp(-s(u))du,$$

where the function $s(u)$ is a solution of the differential equation,

$$s' = \frac{\xi'(x)\Upsilon_1(x)}{\xi(x)\Upsilon_1(x) - \Upsilon_1(x)}$$

and C is the normalization constant, such that $\int_H dF = 1$.

Note: The goal is to have the function $\xi(x)$ as simple as possible.

Chapter 5

New Odd Log-Logistic Family of Distributions

Properties, Regression Models and Applications

Emrah Altun,[1,*] *Morad Alizadeh,*[2] *Gamze Ozel*[3] and *Haitham M Yousof*[4]

1. Introduction

Statistical distributions are used to model and make predictions about the data in many applied sciences. However, known distributions, such as normal, gamma, and Weibull, are insufficient to provide conclusions about complex datasets. Therefore, many families of distributions have been proposed, especially in the last decade. Many complex data sets can be modeled with high accuracy thanks to these families of distributions. One of the most important generalizations of the log-logistic distribution was introduced by Gleaton and Lynch (2006) and described as the odd log-logistic (OLL) family of distributions. Several generalizations of the OLL family were studied by many authors such as Cordeiro et al. (2017), Alizadeh et al. (2017), Korkmaz et al. (2018), Alizadeh et al. (2018a), Alizadeh et al. (2018b), Korkmaz et al. (2019), Alizadeh et al. (2021), Rasekhi et al. (2021). Researchers continue to be interested in the generalizations of the known distributions. For instance, Kilai et al. (2022) studied a new generalization of the gull alpha power family which was originally introduced by Ijaz et al. (2020). Another generalization of the Gumbel-Weibull distribution was introduced by Osatohanmwen et al. (2022). An interesting work was done by Omair et al. (2022). The Whittaker function was used to define a new distribution.

The cumulative distribution function (cdf) of the OLL family is,

$$F(x;\alpha,\xi)=\frac{G(x;\xi)^{\alpha}}{G(x;\xi)^{\alpha}+\bar{G}(x;\xi)^{\alpha}}, \tag{1}$$

where, α is the shape parameter. The probability density function (pdf) of the cdf in (1) is,

$$f(x;\alpha,\xi)=\frac{\alpha g(x;\xi)G(x;\xi)^{\alpha-1}\bar{G}(x;\xi)^{\alpha-1}}{\left[G(x;\xi)^{\alpha}+\bar{G}(x;\xi)^{\alpha}\right]^{2}}. \tag{2}$$

[1] Department of Mathematics, Bartin University, Turkey.
[2] Department of Statistics, Faculty of Intelligent Systems Engineering and Data Science, Persian Gulf University, Bushehr, 75169, Iran
[3] Department of Statistics, Hacettepe University, Turkey.
[4] Department of Statistics, Mathematics and Insurance, Benha University, Benha, Egypt.
* Corresponding author: emrahaltun@bartin.edu.tr

This paper proposes a new odd log-logistic family of distributions (NOLL-G) using the idea of Alzaatreh et al. (2013). The cdf of the NOLL-G is derived by,

$$F(x;\alpha,\beta,\xi)=\int_0^{\frac{G(x;\xi)^\alpha}{\bar{G}(x;\xi)^\beta}}\frac{1}{(1+t)^2}dt=\frac{G(x;\xi)^\alpha}{G(x;\xi)^\alpha+\bar{G}(x;\xi)^\beta},\tag{3}$$

where, $\alpha > 0$ and $\beta > 0$ are two shape parameters, is the vector of parameters for parent distributions such as $G(\cdot)$ and $\bar{G}(x;\xi) = 1 - G(x;\xi)$. The pdf and hazard rate of NOLL-G are given by,

$$f(x;\alpha,\beta,\xi)=\frac{g(x;\xi)G(x;\xi)^{\alpha-1}\bar{G}(x;\xi)^{\beta-1}\left[\alpha+(\beta-\alpha)G(x;\xi)\right]}{\left[G(x;\xi)^\alpha+\bar{G}(x;\xi)^\beta\right]^2},\tag{4}$$

and

$$h(x;\alpha,\beta,\xi)=\frac{g(x;\xi)G(x;\xi)^{\alpha-1}\left[\alpha+(\beta-\alpha)G(x;\xi)\right]}{\bar{G}(x;\xi)\left[G(x;\xi)^\alpha+\bar{G}(x;\xi)^\beta\right]},\tag{4}$$

This family is denoted by X ~ NOLL-G (α,β,ξ). The NOLL-G contains the OLL-G as its sub-model. When the parameters $\alpha = \beta$ in the pdf of the NOLL-G, we have the OLL-G. When the parameters $\alpha = \beta = 1$, we have the parent distribution $G(x;\xi)$.
The below algorithm is given to generate random variables from the NOLL-G model.

Algorithm

I. Generate $U \sim U(0,1)$
II. Solve the non-linear equation: $U = G(x;\xi)^\alpha \big/ \left[G(x;\xi)^\alpha+\bar{G}(x;\xi)^\beta\right]$ below
III. Repeat steps 1 and 2, N times.

The main motivation of the study is to provide a more flexible G-class family of distributions for modeling the different types of data sets such as increasing, decreasing, upside-down as well as bathtub hazard rates. Also, the proposed family, NOLL-G is capable of the modeling of the left, right and symmetric and bimodal data sets. The NOLL-G family can be viewed as a generalization of the OLL-G distribution. Additionally, the NOLL-G family is a wider generalization of the parent distribution. More importantly, thanks to the flexibility of the NOLL-G family, we define a regression model for the censored dependent variable. In the next section of the study, special members of the NOLL family are introduced.

2. Special NOLL models

2.1 The new odd log-logistic weibull (NOLLW) model

The cdf of the Weibull distribution is $G(x,\xi) = 1 - \exp[-(x/b)^a]$ where where $\xi = (b,a)^T$, $a > 0$, is the shape parameter and $b > 0$ is the scale parameter. Inserting the cdf of the Weibull distribution in (4), we have the pdf of the NOLLW distribution. For the sake of simplicity, the pdf of the NOLLW distribution is omitted. The densities and hrf shapes of the NOLLW distribution are displayed in Figures 1 and 2. The NOLLW has right skewed and bimodal densities. Also, it has flexible hrf shapes such as increasing, decreasing, bathtub, constant and upside-down.

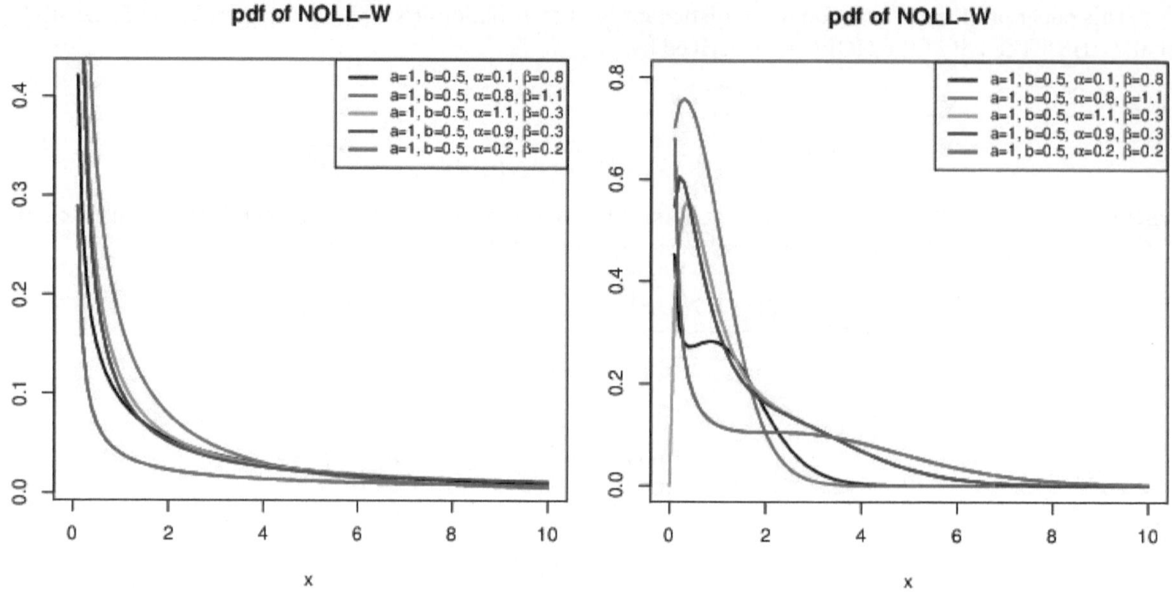

Figure 1: The pdf plots of NOLLW model.

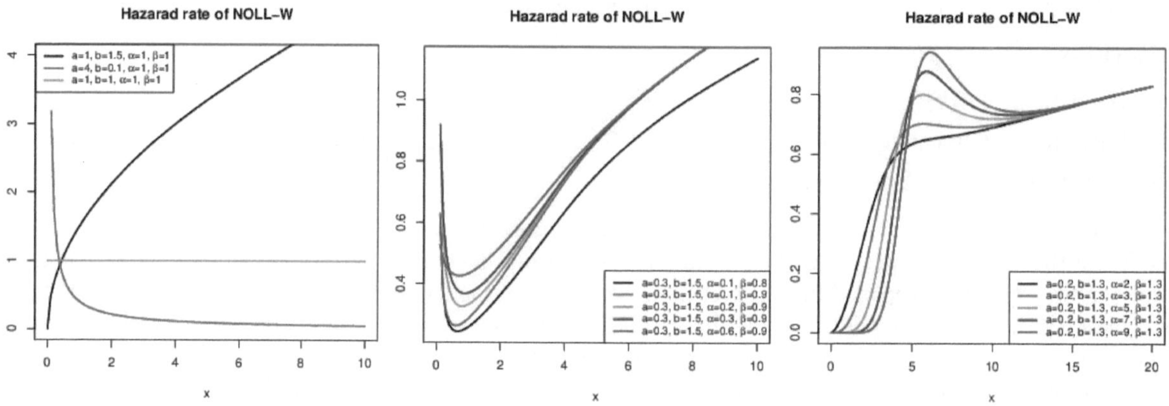

Figure 2: The hrf plots of NOLLW model.

2.2 The new odd log-logistic normal (NOLLN) model

Inserting the pdf and cdf of the normal distribution in (4), we have the following pdf for the NOLLN distribution:

$$f(x)=\frac{\varphi(z)\Phi(z)^{\alpha-1}\bar{\Phi}(z)^{\beta-1}\left[\alpha+(\beta-\alpha)\Phi(z)\right]}{\left[\Phi(z)^{\alpha}+\bar{\Phi}(z)^{\beta}\right]^{2}}, \tag{6}$$

where, $x \in R$ and $z = \frac{x-\mu}{\sigma}$. The mean, $\mu \in R$ is a location parameter and $\sigma > 0$ is a scale parameter, $\varphi(\cdot)$ and $\Phi(\cdot)$ are the pdf and cdf of the standard normal distribution and $\bar{\Phi}(z) = 1 - \Phi(z)$. Figure 3 displays the density shapes of the NOLLN distribution.

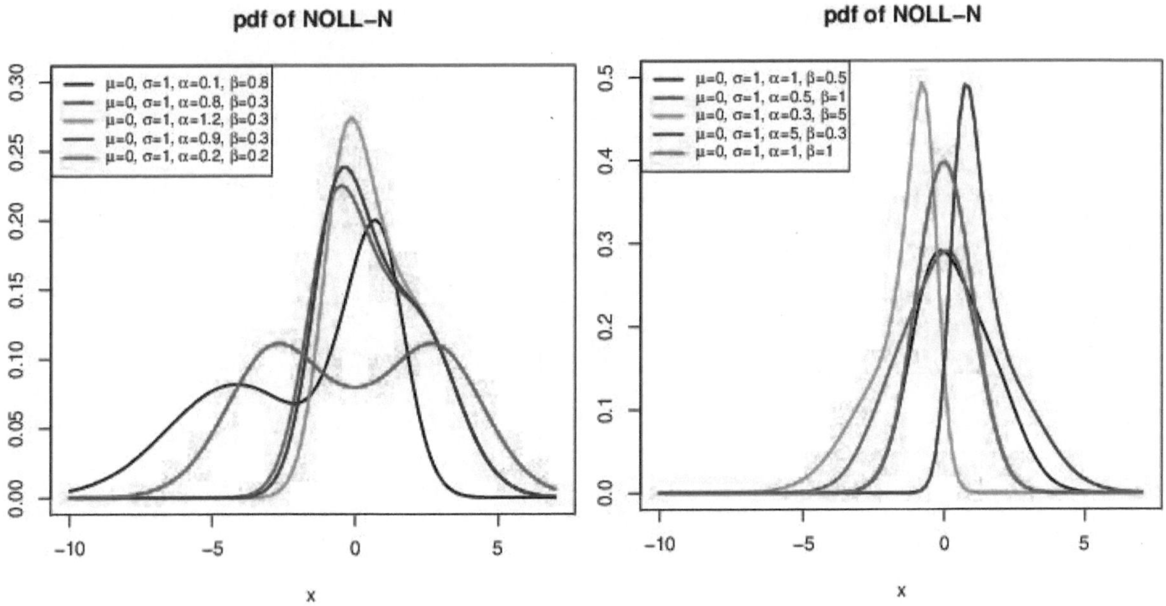

Figure 3: The pdf plots of NOLLN model.

2.3 The new odd log-logistic gamma (NOLLGa) model

The gamma distribution is another famous statistical distribution to model right skewed datasets. The cdf of the gamma distribution is $G_{a,b}(x) = 1 - \frac{\gamma\left(a, \frac{x}{b}\right)}{\Gamma(a)}$, where $\Gamma(a)$ is the gamma function and $\gamma\left(a, \frac{x}{b}\right)$ is the incomplete gamma function. The shape and scale parameters are $a > 0$ and $b > 0$, respectively. Inserting the cdf and pdf of the gamma distribution in (4), we have the pdf for the NOLLGa distribution which is omitted here for the sake of simplicity. Some density and hrf shapes of the NOLLGa are displayed in Figure 4. The NOLLGa distribution has right skewed and bimodal densities as well as increasing and upside-down hrf shapes. The new generalization of the gamma distribution opens new opportunities to model bimodal right skewed data sets.

3. General properties

Several statistical properties of the NOLL-G model are obtained in the rest of this section.

3.1 Useful expansions

Let $G(x)$ be a parent distribution and the exponentiated-G (Exp-G) model is defined by the cdf and pdf $H_c(x) = G(x)^c$ and $h_c(x) = cg(x)G(x)^{c-1}$, respectively. Using the exp-G model several properties of the NOLL-G model can be obtained. Initially, we provide an expansion for the cdf of the NOLL-G model using the power series for $G(x)^\alpha$ ($\alpha > 0$ real) as,

$$G(x)^\alpha = \sum_{k=0}^{\infty} a_k G(x)^k, \tag{7}$$

where,

$$a_k = a_k(\alpha) = \sum_{j=k}^{\infty} (-1)^{k+j} \binom{\alpha}{j} \binom{j}{k}. \tag{8}$$

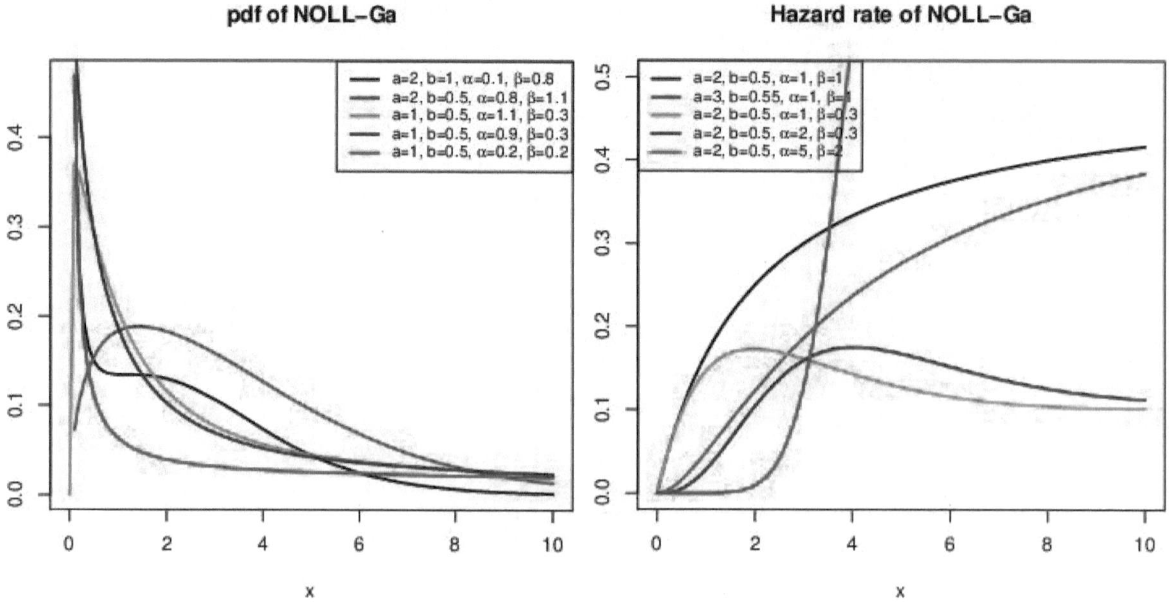

Figure 4: The pdf and hrf plots of NOLLGa model.

For any real $\alpha > 0$, consider the generalized binomial expansion

$$\left[1-G(x)^{\beta}\right]=\sum_{k=0}^{\infty}(-1)^{k}\binom{\beta}{k}G(x)^{k}. \tag{9}$$

Inserting (7) and (8) in Equation (3), we have,

$$F(x)=\frac{G(x)^{\alpha}}{\sum_{k=0}^{\infty}b_{k}G(x)^{k}}, \tag{10}$$

where, $b_k = a_k + (-1)^k \binom{\beta}{k}$. Using the inverse of power series, (10) can be written as follows:

$$F(x)=\sum_{k=0}^{\infty}c_{k}G(x)^{k+\alpha}, \tag{11}$$

where, $c_0 = \frac{1}{b_0}$ and for $k \geq 1$, c_k 's are determined from the recurrence equation,

$$c_{k}=-\frac{1}{b_{0}^{2}}\sum_{r=1}^{k}b_{r}c_{k-r}. \tag{12}$$

Differentiating the equation (11), we have the pdf of X,

$$f(x)=\sum_{k=0}^{\infty}c_{k}h_{k+\alpha}(x),$$

where, $h_{k+\alpha}(x) = (k+\alpha)G(x)^{k+\alpha-1}\, g(x)$ is the Exp-G density function with power parameter $k + \alpha$. So, equation (12) shows that NOLL-G model can be expressed as a linear combination of the exp-G densities. Under this fact, we obtain several properties of the NOLL-G model.

3.2 Moments

Here, we derive the moments of the NOLL-G model emphasizing them on the special case NOLLW model. Using mixture representation of the NOLL-G, the rth raw moment of X is defined by,

$$\mu_{r}'=E\left(X^{r}\right)=\sum_{k=0}^{\infty}c_{k}E\left(Y_{k+\alpha}^{r}\right), \tag{13}$$

where, $E(Y_\zeta^r) = \zeta \int_{-\infty}^{\infty} x^r\, g(x)G(x)^{\zeta-1}\, dx$ which can be calculated numerically using the quantile function (qf) $Q_G(u) = G^{-1}(u)$ as follows:

$$E(Y_\zeta^n) = \zeta \int_0^1 Q_G(u)^n\, u^{\zeta-1}\, du.$$

We have the mean of X for $r = 1$.
For the special NOLLW, we have

$$\mu_r' = \sum_{k,w=0}^{\infty} \delta_{k,w}^{(k+\alpha,r)} \Gamma\left(\frac{\gamma+r}{\gamma}\right), \forall -\gamma < r,$$

where, $\delta_{k,w}^{(k+\alpha,r)} = c_k \delta_w^{(k+\alpha,r)}$ and $\delta_w^{(k+\alpha,r)} = \dfrac{(k+\alpha)(-1)^w}{\lambda^{-r}(w+1)^{(r+\gamma)/\gamma}}\dbinom{k+\alpha-1}{w}$.

3.3 Incomplete moments and moment generation function

The rth incomplete moment of is defined by,

$$I_r(x) = \int_{-\infty}^{y} x^r f(x)dx,$$

which can be rewritten as,

$$I_r(x) = \sum_{k=0}^{\infty} c_k m_{r,k+\alpha}(y),$$

where, $I_{r,\zeta}(y) = \int_0^{G(y)} Q_G^r(u) u^{\zeta-1}\, du$. The integral $I_{r,\zeta}(y)$ can be determined analytically for special models using the qf. For the NOLLW model, we have,

$$I_r(x) = \sum_{k,w=0}^{\infty} \delta_{k,w}^{(k+\alpha,r)} \gamma\left(\frac{\gamma+r}{\gamma}, \left(\frac{1}{\lambda t}\right)^\gamma\right), \forall -\gamma < r,$$

The moment generating function (mgf) of X, say $M(t) = E(e^{tX})$, is determined from (12) as follows:

$$M(t) = \sum_{k=0}^{\infty} c_k M_{k+\alpha}(t),$$

where, $M_\zeta(t) = \zeta \int_{-\infty}^{\infty} e^{tx} G^{\zeta-1}(x) g(x) dx = \zeta \int_0^1 \exp\left[tQ_g(u)\right] u^{\zeta-1} du$ is the generation of Y_ζ.

4. Estimation

The maximum likelihood estimation (MLE) method is used. Let $\Phi = (\alpha,\beta,\xi)^T$ be an unknown parameter vector. The log-likelihood function of the NOLL family is,

$$\ell(\Phi) = \sum_{i=1}^{n} \log g(x_i,\xi) + (\alpha-1)\sum_{i=1}^{n} \log G(x_i,\xi) + (\beta-1)\sum_{i=1}^{n} \log \bar{G}(x_i,\xi)$$
$$+\sum_{i=1}^{n} \log\left[\alpha + (\beta-\alpha)G(x_i,\xi)\right] - 2\sum_{i=1}^{n} \log\left[G(x_i,\xi)^\alpha + \bar{G}(x_i,\xi)^\beta\right]$$

The given log-likelihood function is maximized to obtain the MLEs of the parameters of the NOLL family. This procedure is implemented with the optim function of the R software. The Hessian function is also used to obtain an observed information matrix to construct the confidence intervals of the parameters.

5. Simulation

Now, we discuss the efficiency of the MLE method for the NOLLN model. The simulation results are evaluated based on the estimated biases, mean square error (MSE), average length (AL) and coverage probability (CP).

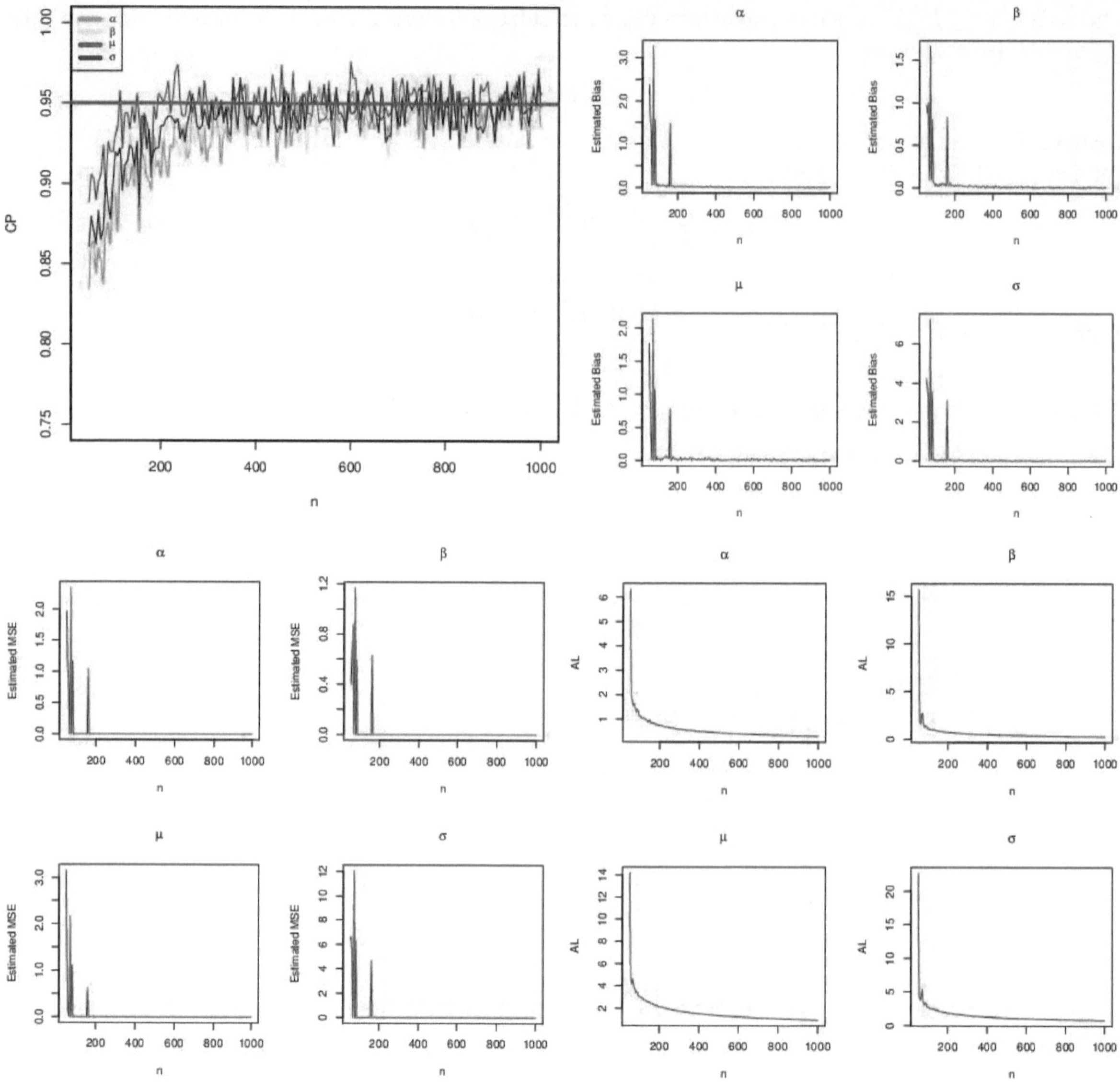

Figure 5: Simulation results of NOLLN model.

The simulation is repeated, $N = 10{,}000$ times. The selected parameter values are $\alpha = 0.5$, $\beta = 0.5$, $\mu = 2$ and $\sigma = 1$. The generated sample size is increased by 5 units started from $n = 50$ to $n = 1{,}000$. The simulation results are summarized in Figure 5. Our expectation is that when the sample size increases, the estimated biases, MSEs, ALs should be decreased. Also, the CP should be nearly, . The results verify the expectations, and it is concluded that the MLE is an appropriate method to obtain the parameters of the NOLL-G model.

6. Regression model

The aim of the regression models is to explain the variability of the dependent variable using some relevant covariates. In this section, we introduce a new regression model for the censored dependent variable. To do this, we benefit from the NOLLW density. First, we use the appropriate transformation on the NOLLW

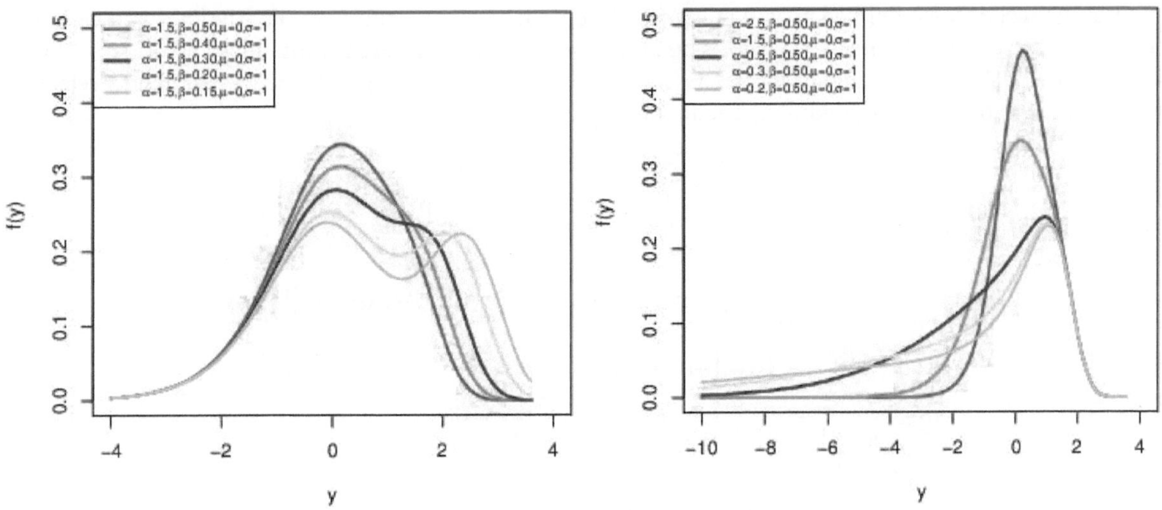

Figure 6: The pdf plots of the LNOLLW model.

density. Let X follow the NOLLW density and consider the random variable $Y = log(X)$. Substituting $a = 1/\sigma$ and $b = exp(\mu)$ in the density of Y (for $y \in \Re$), we have,

$$f(y) = \frac{1}{\sigma}\exp\left[\left(\frac{y-\mu}{\sigma}\right) - \exp\left(\frac{y-\mu}{\sigma}\right)\right]\left\{1-\exp\left[-\exp\left(\frac{y-\mu}{\sigma}\right)\right]\right\}^{\alpha-1}$$
$$\times\left(\exp\left[-\exp\left(\frac{y-\mu}{\sigma}\right)\right]\right)^{\alpha-1}\left[\alpha+(\beta-\alpha)\left\{1-\exp\left[-\exp\left(\frac{y-\mu}{\sigma}\right)\right]\right\}\right] \tag{14}$$
$$\times\left[\left\{1-\exp\left[-\exp\left(\frac{y-\mu}{\sigma}\right)\right]\right\}^{\alpha}+\left(\exp\left[-\exp\left(\frac{y-\mu}{\sigma}\right)\right]\right)^{\beta}\right]^{-2},$$

where, $\mu \in \Re$ is the location parameter, $\sigma > 0$ is the scale parameter, and $\alpha > 0, \beta > 0$ are the shape parameters. From now on, we denote the density in (14) as $Y \sim LNOLLW\,(\alpha,\beta,\sigma,\mu)$. The plots of the LNOLLW density is displayed in Figure 6. It is concluded that the new density can be used to model symmetric, left-skewed and bimodal lifetime datasets. The survival function of (14) is given by,

$$S(y) = \frac{\left(\exp\left[-\exp\left(\frac{y-\mu}{\sigma}\right)\right]\right)^{\beta}}{\left\{1-\exp\left[-\exp\left(\frac{y-\mu}{\sigma}\right)\right]\right\}^{\alpha}+\left(\exp\left[-\exp\left(\frac{y-\mu}{\sigma}\right)\right]\right)^{\beta}}, \tag{15}$$

Thanks to the flexible density of the LNOLLW, we propose a new regression model based on the following model:

$$y_i = x_i^T\beta + \sigma z_i, \quad i = 1,2,...,n,$$

where, y_i is the dependent variable following the density in (14) and $x_i^T = (1, x_{i1}, \ldots, x_{ip})$ are the vector covariates for the th individual. The vector $\beta = (\beta_1, \ldots, \beta_p)^T$ represents the unknown regression parameters. The scale parameter is given by $\sigma > 0$. We use the identity link function such as $\mu_i = x_1^T\beta$. The LNOLLW regression model contains the log-Weibull (LW) regression model as its sub-model (Lawless, 2011).

Let the response be defined as $y_i = \min\{log(x_i), log(c_i)\}$ where x_i and c_i are the lifetimes and censored times, respectively. We define two sets to assign the observations which are the lifetimes and censored times for the individuals involved in the study. These are F and C. Let $\tau = (\alpha, \beta, \sigma, \beta^T)^T$ be the parameter vector of the LNOLLW model. The log-likelihood function of the model is,

$$\ell = \sum_{i \in F} \ell_i(\tau) + \sum_{i \in C} \ell_i^{(c)}(\tau),$$

where, $l_i(\tau) = log[f(y_i)]$, $l_i^{(c)}(\tau) = log[S(y_i)]$. So, the total log-likelihood function is

$$\begin{aligned}\ell(\tau) &= r\log\left(\frac{1}{\sigma}\right) + \sum_{i \in F}(z_i - u_i) + (\alpha - 1)\sum_{i \in F}\log\{1 - \exp[-u_i]\} \\ &+ (\alpha - 1)\sum_{i \in F}\log(\exp[-u_i]) + \sum_{i \in F}\log\left[\alpha + (\beta - \alpha)\{1 - \exp[-u_i]\}\right] \\ &-2\sum_{i \in F}\log\left[\{1 - \exp[-u_i]\}^{\alpha} + (\exp[-u_i])^{\beta}\right] \\ &+\beta\sum_{i \in C}\log(\exp[-u_i]) - \sum_{i \in C}\log\left[\{1 - \exp[-u_i]\}^{\alpha} + (\exp[-u_i])^{\beta}\right]\end{aligned}, \quad (16)$$

where, $u_i = \exp(z_i)$, $z_i = (y_i - v_i^T\beta)/\sigma$ and r is the number of uncensored observations. The model parameters are estimated using the MLE method. The log-likelihood function in (16) is maximized using the optim function.

Two residuals are generally used to check the suitability of the regression model for the fitted data. The first residual is the Martingale residual of Fleming and Harrington (1994). The interpretation of the Martingale residual is problematic, so, the modified deviance residuals are preferred more (Therneau et al., 1990). Here, the modified deviance residuals are used to check the assumption on the residuals of the LNOLLW regression model.

7. Applications

In this section, several members of the NOLL-G family are compared with the existing models based on the real data modeling. The competitive models are listed in Table 1. Four applications of the NOLL-G model are presented in the rest of the section to show the importance and flexibility of the proposed models. The comparison of the models is done by using the following metrics and goodness-of-fit statistics: - log-likelihood function, Anderson-Darling (A*) and Cramer-von Mises (W*) test statistics.

7.1 Glass fibers dataset

The used data set is reported in the Smith and Naylor (1987) study. The data represents the strengths of 1.5 cm glass fibers. Table 2 shows the estimated model parameters and goodness-of-fit statistics. The NOLLN model has the lowest values of A* and W* statistics as well as the lowest value , $-\ell$. So, it is clear that the NOLLN distribution is the best model for the data set. Figure 7 shows the fitted densities and cdfs of the models. Figure 7 provides great evidence that the NOLLN model gives acceptable results for the data set.

Table 1: Competitive models and their abbreviations.

Models	Abbreviations	References
Odd Log-Logistic-G Family	OLL-G	Gleaton and Lynch (2006)
Kumaraswamy-G Family	KUM-G	Cordeiro and de Castro (2011)
Exponentiated Generalized-G Family	EG-G	Cordeiro et al. (2013)
Odd Burr-G Family	OBu-G	Alizadeh et al. (2017)
Generalized Odd Log-Logistic-G Family	GOLL-G	Cordeiro et al. (2017)

Table 2: The results of the fitted model for the glass fibers dataset.

Model	α	β	μ	σ	A*	W*	−ℓ
N			1.506	0.321	1.928	0.35	17.911
			0	0.028			
OLL-N	5.971		1.540	1.638	1.446	2.262	16.067
	10.841		0	2.941			
GOLL-N	1.559	0.012	2.295	0.074	1.139	0.205	14.626
	0.423	0.009	0.169	0.03			
NOLL-N	0.609	6.570	2.022	0.46	0.482	0.082	11.900
	0.257	6.975	0.381	0.226			
KUM-N	0.027	36.099	3.566	0.12	0.963	0.172	14.191
	0.015	92.947	1.288	0.062			
EG-N	13.800	0.582	2.376	0.43	0.969	0.173	14.201
	25.448	0.439	0.65	0.251			
OBu-N	242.460	3.409	1.868	83.768	0.754	0.135	13.183
	0.01	0.013	0.011	0.037			

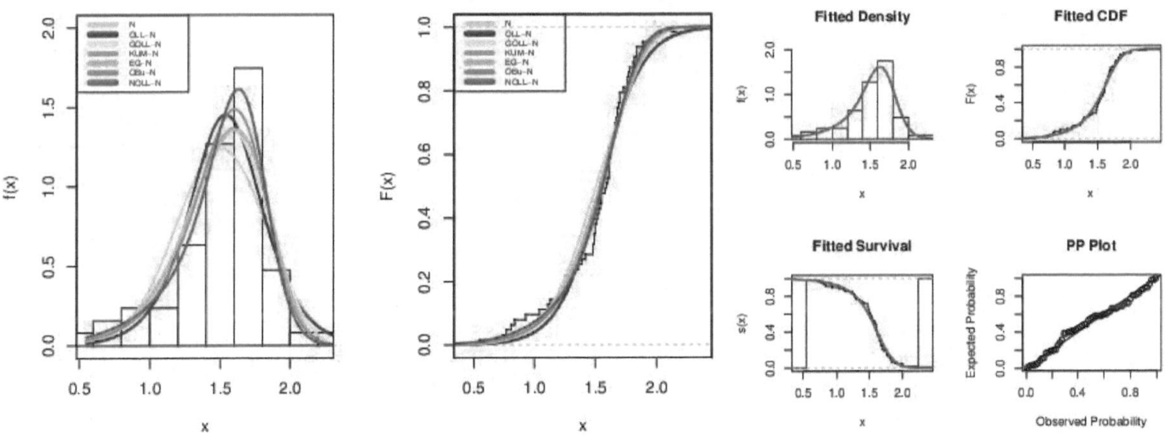

Figure 7: The graphical comparison of the fitted models for the glass fibers data set.

7.2 Daily ozone measurements

In the second application, the data comes from the daily ozone measurements in New York between the dates of May-September 1973. The data is fitted by NOLLW and other competitive models. The results are listed in Table 3. The NOLLW distribution has the lowest values for the goodness-of-fit statistics. Also, Figure 8 shows that the NOLLW distribution gives better results than other competitive models.

7.3 Failure times

In the third application, the data is about the 73 failure times (in hours) of unscheduled maintenance actions for the USS Halfbeak number 4 main propulsion diesel engine over 25.518 operating hours. The estimated parameters and goodness-of-fit statistics are listed in Table 4. Again, the NOLLGa distribution has the lowest values for these statistics. Therefore, the NOLLGa distribution outperforms the other competitive models. The suitability of the NOLLGa model for the data is checked graphically which is given in Figure 9.

Table 3: The results of the fitted model for the daily ozone dataset.

Model	α	β	a	b	A*	W*	–ℓ
W			1.340	46.074	0.966	0.17	542.610
			0.095	3.374			
OLL-W	1.308		1.067	47.653	0.822	0.139	542.284
	0.503		0.368	4.976			
GOLL-W	0.354	6.049	1.320	18.757	0.283	0.047	539.275
	0.157	2.796	0.319	5.031			
NOLL-W	1.395	0.233	1.714	24.744	0.14	0.019	537.933
	0.23	0.082	0.243	0.459			
KUM-W	3.643	0.225	0.958	6.870	0.357	0.048	541.019
	2.071	0.221	0.271	3.098			
E-W	0.284	2.587	0.834	4.688	0.54	0.085	541.202
	0.704	1.616	0.242	13.184			
Obu-W	1.797	0.148	1.175	14.010	0.28	0.041	539.442
	0.426	0.08	0.126	4.134			

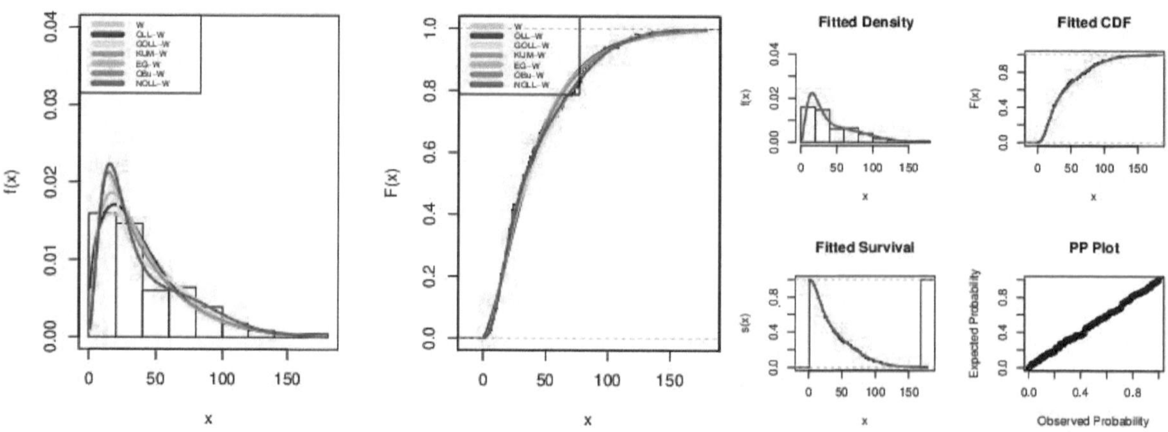

Figure 8: The graphical comparison of the fitted models for the daily ozone data set.

7.4 Heart transplant data

In the last application, we show the usefulness of the LNOLLW regression model. We use the Stanford Heart transplant data set which has the information of 103 individuals. The dependent variable is the survival times of the individuals which is modeled by the following covariates.

x_1 - year of acceptance to the program;

x_2 - age;

x_3 - surgery ($1 = yes, 0 = no$);

x_4 - transplant ($1 = yes, 0 = no$).

The same data set was modeled by Brito et al. (2017) using the log Topp-Leone odd log-logistic Weibull regression model. The model is shortly denoted as LTLOLLW. The data set is fitted by three models. These are LNOLLW, LTLOLLW and LW models.

Table 4: The results of the fitted model for the failure times data set.

Models	α	β	a	b	A*	W*	−ℓ
Ga			5.832	0.3	8.414	1.656	244.433
			0.952	0.051			
OLL-Ga	4.783		1	0.02	7.263	1.401	236.534
	0.915		0.167	0.007			
GOLL-Ga	4.576	0.27	2.717	0.041	7.171	1.382	235.883
	1.484	0.272	2.421	0.03			
NOLL-Ga	0.138	4567.849	11.260	0.143	0.43	0.07	194.090
	0.033	97.223	4.268	0.122			
KUM-Ga	1.021	6.910	4.699	0.118	7.374	1.435	236.211
	0.845	3.355	3.809	0.075			
EG-Ga	6.997	0.306	14.968	0.424	6.306	1.217	228.722
	3.704	0.053	0.002	0.002			
OBu-Ga	3.679	2.523	0.9019	0.023	6.263	1.199	228.442
	0.723	0.741	0.24	0.0095			

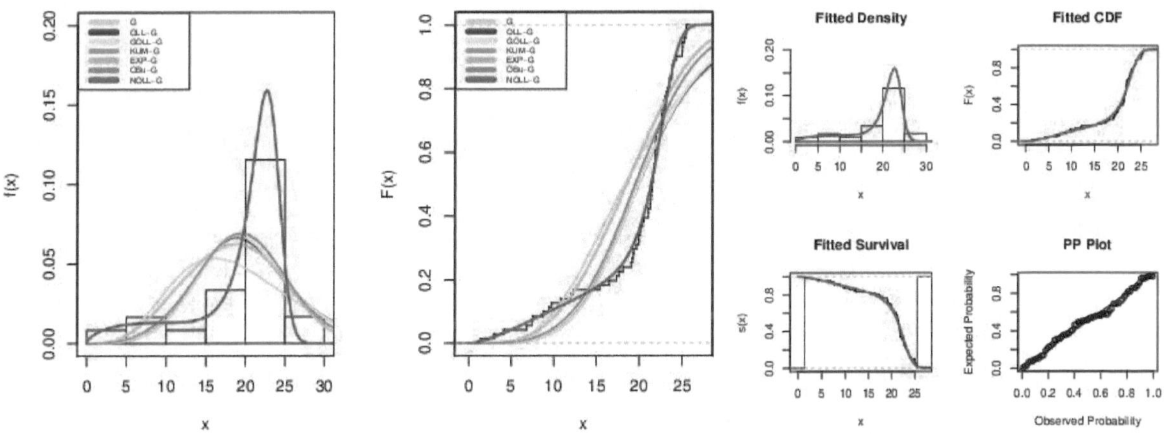

Figure 9: The graphical comparison of the fitted models for the failure times data set.

The results are given in Table 5. The best model is selected based on the Akaike Information Criteria (AIC) and Bayesian Information Criteria (BIC). The model having the lowest values of these statistics is the best model for the data set. As seen from Table 5, the proposed model has lower values of these statistics than those of the other two regression models. Therefore, the LNOLLW regression model produces more acceptable results then other models for the current data set. Additionally, the regression parameters β_1 and β_2 are statistically significant.

The validity of the LNOLLW model is checked by the residual analysis. Figure 10 displays the quantile-quantile plot of the modified deviance residuals and its index plot. From Figure 10, we conclude that none of the observations can be evaluated as possible outliers. Therefore, the fitted model is appropriate for the current data.

Table 5: The results of the fitted regression models.

	Models								
	LW			LTLOLLW			LNOLLW		
	MLEs	SE	p-value	MLEs	SE	p-value	MLEs	SE	p-value
α	-	-	-	2.340	3.546	-	4.674	10.120	-
β	-	-	-	24.029	3.015	-	3.815	17.608	-
σ	1.478	0.133	-	9.680	12.526	-	5.455	15.627	-
β_0	1.639	6.835	0.811	−0.645	8.459	0.939	3.777	11.725	0.747
β_1	0.104	0.096	0.279	0.074	0.097	0.448	0.214	0.096	0.026
β_2	−0.092	0.02	< 0.001	−0.053	0.020	0.009	−0.053	0.018	0.003
β_3	1.126	0.658	0.087	1.676	0.597	0.005	0.174	0.497	0.726
β_4	2.544	0.378	< 0.001	2.394	0.384	< 0.001	0.445	0.373	0.233
$-\ell$	171.240			164.684			159.360		
AIC	354.481			345.368			334.721		
BIC	370.2894			366.446			355.799		

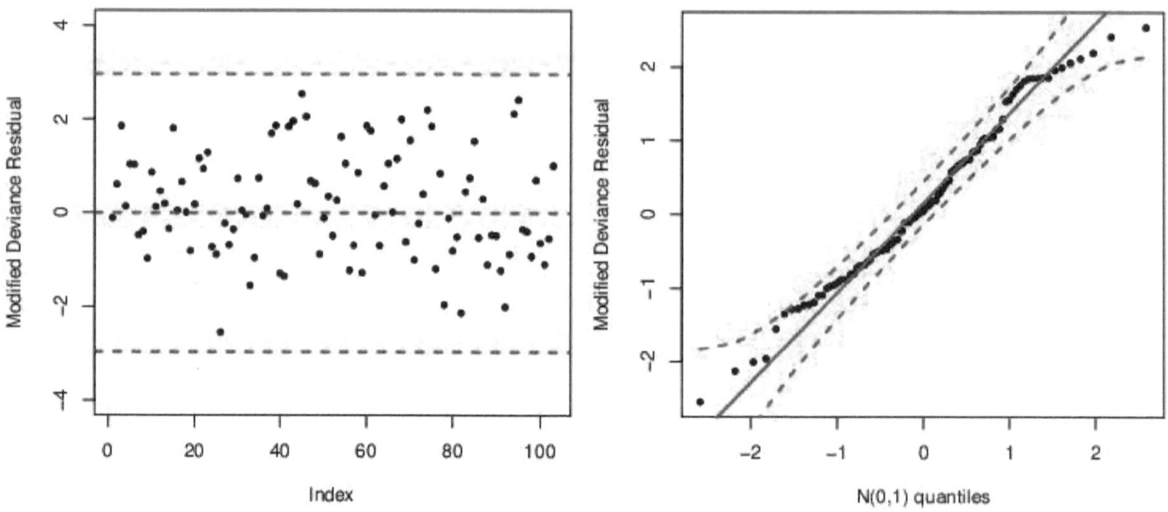

Figure 10: The plots of the modified deviance residual.

8. Conclusion and future work

This paper introduces the NOLL family of distributions. The special models of importance belonging to the NOLL family are implemented to data sets to convince the readers about the applicability of the proposed models. The regression model of the NOLLW distribution is defined based on the location-scale family. In the future work of the presented study, we plan to develop a new generalization of the Pareto distribution using the NOLL family to analyze extreme events by using peaks over threshold methodology.

References

Alizadeh, M., Afshariy, M., Karamikabir, H. and Yousof, H. M. (2021). The odd log-logistic burr-x family of distributions: properties and applications. Journal of Statistical Theory and Applications, 20(2): 228–241.

Alizadeh, M., Cordeiro, G. M., Nascimento, A. D., Lima, M. D. C. S., Ortega, E. M. et al. (2017). Odd-Burr generalized family of distributions with some applications. Journal of Statistical Computation and Simulation, 87(2): 367–389.

Alizadeh, M., Lak, F., Rasekhi, M., Ramires, T. G., Yousof, H. M. et al. (2018a). The odd log-logistic Topp–Leone G family of distributions: heteroscedastic regression models and applications. Computational Statistics, 33(3): 1217–1244.

Alizadeh, M., Yousof, H. M., Jahanshahi, S. M. A., Najibi, S. M., Hamedani, G. G. et al. (2020). The transmuted odd log-logistic-G family of distributions. Journal of Statistics and Management Systems, 23(4): 1–27.

Alizadeh, M., Yousof, H. M., Rasekhi, M. and Altun, E. (2018b). The odd log-logistic Poisson-G family of distributions. Journal of Mathematical Extension, 12(1): 81–104.

Alzaatreh, A., Lee, C. and Famoye, F. (2013). A new method for generating families of continuous distributions. Metron, 71(1): 63–79.

Brito, E., Cordeiro, G. M., Yousof, H. M., Alizadeh, M., Silva, G. O. et al. (2017). The Topp-Leone odd log-logistic family of distributions. Journal of Statistical Computation and Simulation, 87(15): 3040–3058.

Cordeiro, G. M. and de Castro, M. (2011). A new family of generalized distributions. Journal of Statistical Computation and Simulation, 81(7): 883–898.

Cordeiro, G. M., Ortega, E. M. and da Cunha, D. C. (2013). The exponentiated generalized class of distributions. Journal of Data Science, 11(1): 1–27.

Cordeiro, G. M., Alizadeh, M., Ozel, G., Hosseini, B., Ortega, E. M. M. et al. (2017). The generalized odd log-logistic family of distributions: properties, regression models and applications, Journal of Statistical Computation and Simulation, 87(5): 908–932.

Fleming, T. R. and Harrington, D. P. (1994). Counting process and survival analysis, John Wiley, New York.

Gleaton, J. U. and Lynch, J. D. (2006). Properties of generalized log-logistic families of lifetime distributions. Journal of Probability and Statistical Science, 4(1): 51–64.

Ijaz, M., Asim, S. M., Farooq, M., Khan, S. A., Manzoor, S. et al. (2020). A Gull Alpha Power Weibull distribution with applications to real and simulated data. Plos one, 15(6): e0233080.

Korkmaz, M. C., Yousof, H. M., Alizadeh, M. and Hamedani, G. G. (2019). The Topp-Leone generalized odd log-logistic family of distributions: properties, characterizations and applications. Communications Faculty of Sciences University of Ankara Series A1 Mathematics and Statistics, 68(2): 1506–1527.

Korkmaz, M. Ç., Yousof, H. M. and Hamedani, G. G. (2018). The exponential Lindley odd log-logistic-G family: properties, characterizations and applications. Journal of Statistical Theory and Applications, 17(3): 554–571.

Kilai, M., Waititu, G. A., Kibira, W. A., Alshanbari, H. M., El-Morshedy, M. et al. (2022). A new generalization of Gull Alpha Power Family of distributions with application to modeling COVID-19 mortality rates. Results in Physics, 105339.

Lawless, J. F. (2011). Statistical models and methods for lifetime data. John Wiley and Sons.

Osatohanmwen, P., Efe-Eyefia, E., Oyegue, F. O., Osemwenkhae, J. E., Ogbonmwan, S. M. et al. (2022). The Exponentiated Gumbel–Weibull {Logistic} Distribution with Application to Nigeria's COVID-19 Infections Data. Annals of Data Science, 1–35.

Omair, M. A., Tashkandy, Y. A., Askar, S. and Alzaid, A. A. (2022). Family of Distributions Derived from Whittaker Function. Mathematics, 10(7): 1058.

Rasekhi, M., Altun, E., Alizadeh, M. and Yousof, H. M. (2021). The odd log-logistic Weibull-G family of distributions with regression and financial risk models, Journal of the Operations Research Society of China, https://doi.org/10.1007/s40305-021-00349-6.

Smith, R. L. and Naylor, J. C. (1987). A comparison of maximum likelihood and Bayesian estimators for the three-parameter Weibull distribution. Applied Statistics, 358–369.

Therneau, T. M., Grambsch, P. M. and Fleming, T. R. (1990). Martingale-based residuals for survival models. Biometrika, 77(1): 147–160.

Chapter 6

On the Family of Generalized Topp-Leone Arcsin Distributions

Vikas Kumar Sharma,[1,*] *Komal Shekhawat*[2] and *Sanjay Kumar Singh*[1]

1. Introduction

The past decade was a prolific period in which most of the developments of the distributions took place. Various extended/modified families of the probability distributions were proposed for fitting lifetimes data, count data, and other random phenomenon from applied areas. Among many, we may mention some of the recent works here. Eliwa et al. (2021) introduced the exponentiated odd Chen-generated (G) family of distributions which can be served as a lifetime distribution for data modelling positively and negatively skewed data sets. El-Morshedy et al. (2021) also proposed an exponentiated type distribution which is suitable for fitting both symmeteric and asymmeteric data. Among others, Alzaatreh et al. (2013) presented a method for generating new classes of distributions. The method is described as follows. Let T be a RV of a generator distribution with PDF, $r(t)$ defined on $[a, b]$ and X be a continuous baseline RV with CDF, $G_X(x)$. The CDF of this family (called TX family) is given by,

$$F_{\text{TX}}(x) = \int_a^{W(G_X(x))} r(t)dt = \Psi\{W(G_X(x))\}, \tag{1}$$

where, $W(G_X(x)) \in [a, b]$ is a differentiable and monotonically increasing function in x and $\Psi(t)$ is the CDF of the RV T. This approach is widely used in statistical literature. This approach of unifying probability distributions was first utilized by Eugene et al. (2002) in which the authors introduced a four parameter beta-normal distribution while assuming $T \sim Beta(\alpha, \beta)$ and $X \sim Normal(\mu, \sigma^2)$ with $W(G_X(x)) = G_X(x)$. After Alzaatreh et al. (2013), various distributions were proposed with different choices of the distributions of X and T. For instance, Alzaatreh et al. (2014) and Sharma et al. (2017) introduced the gamma-normal and Maxwell-Weibull distributions, respectively.

[1] Department of Statistics, Institute of Science, Banaras Hindu University, Varanasi, India.
[2] Department of Basic Sciences, Institute of Infrastructure Technology Research and Management, Ahmedabad, India.
* Corresponding author: vikasstats@rediffmail.com

It is always good to have the simplest generator for developing flexible and parsimonious distributions. The TLD by Topp and Leone (1955) is considered to be a good choice for producing the extended distributions. A recent article on the TLD by Shekhawat and Sharma (2021) can be followed for associated theories, parameter estimation and application. Sangsanit and Bodhisuwan (2016) investigated the use of the TLD as a generator in (1) while they considered the generalized exponential distribution as the baseline distribution. Sharma (2018) introduced the Topp-Leone normal distribution for fitting skewed data sets that has increasing failure rate function.

Using the TLD, Chesneau et al. (2021) recently established a potential family of distributions that unifies the various known classes of the probability distributions. They defined the family by the following CDF,

$$F(x) = [G(x;\xi)]^{\lambda}[2 - G(x;\xi)]^{\lambda\beta}, \beta \leq 1, \lambda > 0, x \in \mathcal{R}, \tag{2}$$

where, ξ is the parameter(s) of the baseline CDF, $G(x;\xi)$. It is called the ETL-G family of distributions. Importantly, we can note here that this family is a wrapper of the various known families as follows:

- When $\beta = 0$, it reduces to the family of the Lehmann-type distributions, AL-Hussaini and Ahsanullah (2015).
- When $\beta = 1$, it reduces to the family of the Topp-Leone-G distributions, Sangsanit and Bodhisuwan (2016).
- When $\beta = -1$, it reduces to the exp-half-G family discussed by Bakouch (2020).
- When $\beta = -1$ and $\lambda = 1$, it can be seen as a particular member of the Marshal-Olkin family, Marshall and Olkin (1997).

Chesneau et al. (2021) illustrated this family with six well-known probability distributions and advocated the use of the ETL-Weibull distribution over the beta and Kumaraswamy type Weibull distributions for fitting the GAG concentration level in urine.

Recently, there have been high relevance and applicability of trignometric distributions for modelling various real-life phenomena. The most significant approach is to use trignometric transformations for introducing generalized families of distributions. As the TLD and ASD (see Feller (1967)) are simple and easily tractable, yet they hold impressive statistical properties and applications. During last few years, the TLD is employed by many authors to produce probability distributions Al-Babtain et al. (2020) developed Sin Topp-Leone-G family of distributions. Al-Shomrani et al. (2016); Brito et al. (2017); Yousof et al. (2017); Rezaei et al. (2017); Elgarhy et al. (2018); Hassan et al. (2019); Al-Babtain et al. (2020); Chipepa et al. (2020); Reyad et al. (2021); Moakofi et al. (2022); Adeyinka (2022); Oluyede et al. (2022) are among important literatures proposing the generalized family of distributions based on the TLD.

In this chapter, we consider the ASD as the baseline distribution. The ASD first appeared in Feller (1967) as a model of the random walk process. Various researchers has done enormous work based on the ASD due to its phenomenal properties. It is effectively useful in modelling unit range and symmeteric distributions. Schmidt and Zhigljavsky (2009), Arnold and Groeneveld (1980) and Ahsanullah (2015) discussed its characterizations. The ASD is one of the members of the McDonald distribution and so it is well-linked with the beta and Kumaraswamy distributions as a special case, see Cordeiro and Lemonte (2014); Cordeiro et al. (2016). The CDF of the ASD is $F(x) = \frac{2}{\pi}\arcsin\left(\sqrt{x}\right), x \in (0,1)$, which is quite simple and tractable. Cordeiro and Lemonte (2014) proposed the McDonald-arcsine distribution using the idea of the TX family.

The above discussion on the TLD and ASD motivates authors to combine both the distributions in the TX family given in (1) having the potential to produce a flexible distribution based on trigonometric functions. We further consider the ETL family with the ASD to introduced a flexible and parsimonious distribution for fitting unit range data sets. The CDF of the proposed distribution is given by,

$$F(x) = \left\{ \frac{2}{\pi} \arcsin\left(\sqrt{x}\right) \left[2 - \frac{2}{\pi} \arcsin\left(\sqrt{x}\right) \right]^{\beta} \right\}^{\lambda}, x \in (0,1), \lambda > 0, \beta \leq 1. \tag{3}$$

We call this distribution as the ETLASD. The ETLASD has the following PDF of the form,

$$\begin{aligned} f(x) &= \lambda[G(x)]^{\lambda-1}[2-(\beta+1)G(x)][2-G(x)]^{\lambda\beta-1}g(x), \qquad x \in (0,1). \\ &= \frac{\lambda}{\pi\sqrt{x(1-x)}} \left[\frac{2}{\pi} \arcsin\left(\sqrt{x}\right) \right]^{\lambda-1} \left[2 - (\beta+1)\frac{2}{\pi} \arcsin\left(\sqrt{x}\right) \right] \left[2 - \frac{2}{\pi} \arcsin\left(\sqrt{x}\right) \right]^{\lambda\beta-1}. \end{aligned} \tag{4}$$

Figure 1 illustrates various shapes of the PDF depending on the parameter values. The distribution has bathtub, increasing or decreasing, J and reversed J shaped frequency curves. Figure 2 shows the shapes of the HRF for various choices of the parameters. This Figure reveals that the ETLASD is capable of fitting increasing and bathtub data shaped HRF data sets. From these Figures, we conclude that the ETLASD may be found useful for fitting varieties of data sets. That justifies the investigation of this distribution over unit range distributions (see Section 9 on application).

The remaining parts of the chapter are organized into the following sections: In Section 2, we derive the significant expansions of the ETLASD. Section 3 is dedicated to the moments in which we derive central, raw and incomplete moments, cumulants and characteristic functions. In Section 4, measures of the quantiles, skewness and peakedness are obtained. Entropy measures are discussed in Section 5. In Section 6, the maximum likelihood estimation is utilized to estimate the unknown parameters. Stochastic ordering, stress strength reliability and identifiability are discussed in Sections 7 and 8, respectively. We provide an application of the ETLASD and compare it to its competing distributions in Section 9. Section 10 brings the chapter to a conclusion.

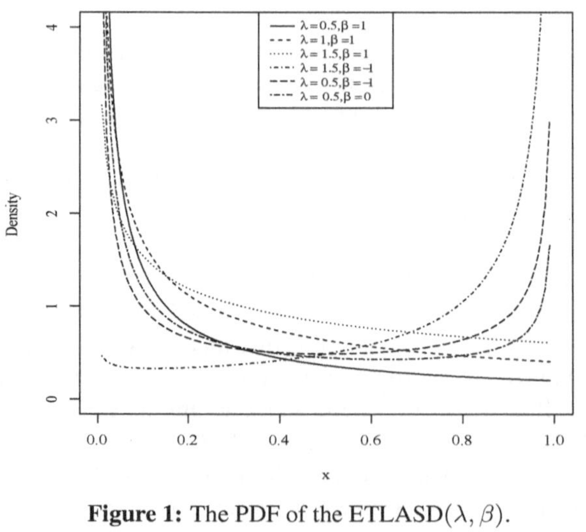

Figure 1: The PDF of the ETLASD(λ, β).

2. Expansion

In this section, we expand the ETLASD PDF and CDF in terms of the linear combination of the Lehmann type I distributions that are well studied in statistics literature, see Cordeiro et al. (2013). The expansion would

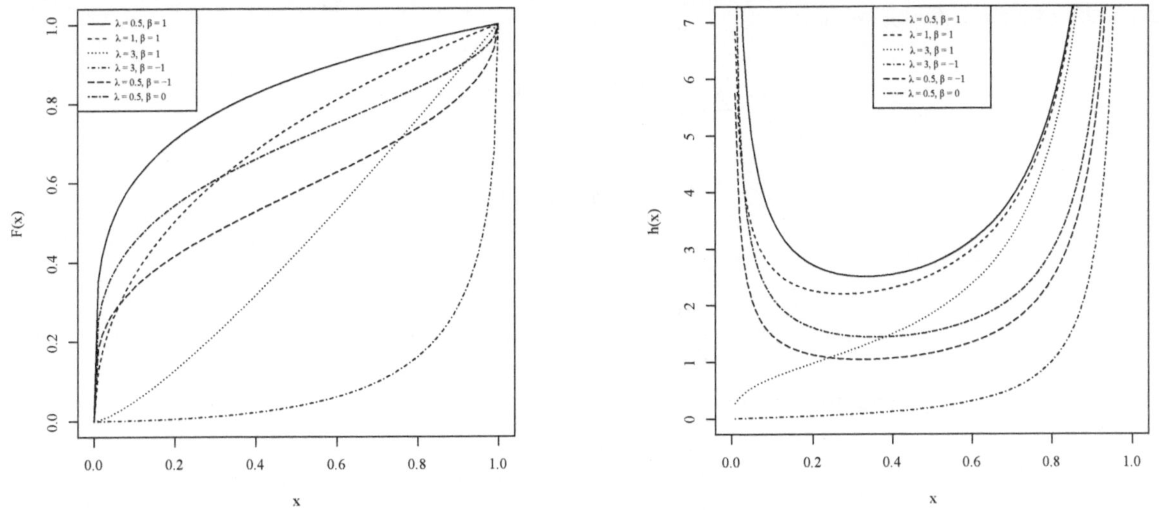

(a) CDF (b) HRF

Figure 2: The CDF and HRF of the ETLASD(λ, β).

provide the tractable properties such as moments and related measures. The expansion of the ETLASD can be achieved using the arcsin function expansion. It is defined by

$$\arcsin(\sqrt{x}) = \sum_{m=0}^{\infty} a_m x^{m+1/2}, \quad |x| \leq 1, \tag{5}$$

where, $a_m = (2m)!/[(2m+1)2^{2m}(m!)^2]$.

The power series raised to a positive integer r can be expanded as,

$$\left(\sum_{m=0}^{\infty} a_m z^m\right)^r = \sum_{m=0}^{\infty} p_{r,m} z^m, \tag{6}$$

where the coefficients $p_{r,m}(m = 1, 2, ...)$ can be calculated using the recursive equation (with $p_{r,0} = a_0^r$),

$$p_{r,m} = (ma_0)^{-1} \sum_{k=1}^{m} (rk - m + k) a_k p_{r,m-k}. \tag{7}$$

Further, we use exponentiated-G family to develop the linear combination of the beta distributions. The CDF and PDF of the exponentiated ASD are given by,

$$H_d(z) = \left[\frac{2}{\pi} arcsin(\sqrt{x})\right]^d \quad \text{and} \quad h_d(z) = \frac{d}{\pi(x - x^2)^{\frac{1}{2}}} \left[\frac{2}{\pi} arcsin(\sqrt{x})\right]^{d-1}, \text{respectively.} \tag{8}$$

Expanding $[2 - G(x)]^{\lambda\beta-1} = \sum_{i=0}^{\infty} \binom{n}{i} 2^{\lambda\beta-i-1}(-1)^i G(x)^i$ in equation (4), we have,

$$f(x) = \lambda g(x) \sum_{i=0}^{\infty} \binom{n}{i} 2^{\lambda\beta-i}(-1)^i G(x)^{\lambda+i-1} - \lambda g(x) \sum_{i=0}^{\infty} \binom{n}{i} 2^{\lambda\beta-i-1}(-1)^i G(x)^{\lambda+i},$$

$$f(x) = f_1(x)(\text{say}) - f_2(x)(\text{say}). \tag{9}$$

Expanding function $f_1(x)$, we have,

$$f_1(x) = \lambda \sum_{i=0}^{\infty} \binom{n}{i} 2^{\lambda\beta - i} (-1)^i g(x) G(x)^{\lambda+i-1}, \tag{10}$$

$$f_1(x) = w_i h_{\lambda+i}(x), \tag{11}$$

where $w_i = \sum_{i=0}^{\infty} \binom{n}{i} \frac{\lambda}{\lambda+i} 2^{\lambda\beta-i}$ and

$$h_{\lambda+i}(x) = \frac{\lambda+i}{\pi(x-x^2)^{1/2}} \left[\frac{2}{\pi} \arcsin(\sqrt{x})\right]^{\lambda+i-1}. \tag{12}$$

Using the expansions (5) and (6), we obtain,

$$h_{\lambda+i}(x) = \frac{\lambda+i}{\pi(x-x^2)^{1/2}} \left[\frac{2}{\pi}\right]^{\lambda+i-1} \sum_{m=0}^{\infty} p_{\lambda+i-1,m} x^{(2m+\lambda+i)/2-1}(1-x)^{1/2-1}, \tag{13}$$

and

$$f_1(x) = \sum_{i,r=0}^{\infty} s(i,r) g_B\left(x; \frac{2m+\lambda+i}{2}, \frac{1}{2}\right), \tag{14}$$

where, $s(i,r) = w_i 2^{\lambda+i-1} \frac{\lambda+i}{\pi^{\lambda+i}} p_{\lambda+i-1,m} B\left(\frac{2m+\lambda+i}{2}, \frac{1}{2}\right)$ and $g_B(.)$ denotes the beta density function. Similarly, we have,

$$f_2(x) = \sum_{i,r=0}^{\infty} s'(i,r) g_B\left(x; \frac{2m+\lambda+i+1}{2}, \frac{1}{2}\right), \tag{15}$$

where, $s'(i,r) = w'_i 2^{\lambda+i} \frac{\lambda+i+1}{\pi^{\lambda+i+1}} p_{\lambda+i,m} B\left(\frac{2m+\lambda+i+1}{2}, \frac{1}{2}\right)$, $s(i,r) = w_i 2^{\lambda+i-1} \frac{\lambda+i}{\pi^{\lambda+i}} p_{\lambda+i-1,m} B\left(\frac{2m+\lambda+i}{2}, \frac{1}{2}\right)$ and $w'_i = \sum_{i=0}^{\infty} (-1)^i \binom{n}{i} \frac{\lambda}{\lambda+i-1} 2^{\lambda\beta-i-1}$. Therefore, from the equation (9), we have,

$$f(x) = \sum_{i,r=0}^{\infty} s(i,r) g_B\left(x; \frac{2m+\lambda+i}{2}, \frac{1}{2}\right) - \sum_{i,r=0}^{\infty} s'(i,r) g_B\left(x; \frac{2m+\lambda+i+1}{2}, \frac{1}{2}\right); x \in (0,1), \lambda > 0, \beta \leq 1. \tag{16}$$

The corresponding CDF is given by,

$$F(x) = \sum_{i,r=0}^{\infty} s(i,r) G_B\left(x; \frac{2m+\lambda+i}{2}, \frac{1}{2}\right) - \sum_{i,r=0}^{\infty} s'(i,r) G_B\left(x; \frac{2m+\lambda+i+1}{2}, \frac{1}{2}\right); x \in (0,1), \lambda > 0, \beta \leq 1. \tag{17}$$

Since the PDF can be written as the linear combination of the beta distributions, the numerous statistical properties can be immediately derived similar to that of the beta distribution.

3. Moments

The moments are quantitative measures that are used for characterizing the distributions and to study their shapes. We measure the flatness of the distribution as well as distinctive degree of skewness. Here, we derive the expressions of the raw moments that are purposive to derive central moments, cumulants and other higher derivative moments. The kth central moment of the ETLASD is derived in the following Proposition.

Proposition 1 *For a RV X that follows the $ETLASD(\lambda, \beta)$, the kth moment about the origin is given by,*

$$E(X^k) = \sum_{i,r=0}^{\infty} \left(s(i,r) \frac{B\left(m+k+\frac{\lambda+i}{2}, \frac{1}{2}\right)}{B\left(m+\frac{\lambda+i}{2}, \frac{1}{2}\right)} - s^{'}(i,r) \frac{B\left(m+k+\frac{\lambda+i+1}{2}, \frac{1}{2}\right)}{B\left(m+\frac{\lambda+i+1}{2}, \frac{1}{2}\right)} \right). \tag{18}$$

Proof 1 *Using the expansion given in (16), the kth moment is defined as,*

$$\mu_k^{'} = E(X^k) = \sum_{i,r=0}^{\infty} \left(\frac{s(i,r)}{B\left(m+\frac{\lambda+i}{2}, \frac{1}{2}\right)} \int_0^1 x^{m+k+\frac{\lambda+i}{2}-1}(1-x)^{\frac{1}{2}-1}dx - \right.$$

$$\left. \frac{s^{'}(i,r)}{B\left(m+\frac{\lambda+i+1}{2}, \frac{1}{2}\right)} \int_0^1 x^{m+k+\frac{\lambda+i+1}{2}-1}(1-x)^{\frac{1}{2}-1}dx \right),$$

$$= \sum_{i,r=0}^{\infty} \left(s(i,r) \frac{B\left(m+k+\frac{\lambda+i}{2}, \frac{1}{2}\right)}{B\left(m+\frac{\lambda+i}{2}, \frac{1}{2}\right)} - s^{'}(i,r) \frac{B\left(m+k+\frac{\lambda+i+1}{2}, \frac{1}{2}\right)}{B\left(m+\frac{\lambda+i+1}{2}, \frac{1}{2}\right)} \right).$$

The central moments(μ_k) and cumulants (K_k) of X can be determined respectively from the equations given below:

$$\mu_k = \sum_{s=0}^{p} \binom{k}{s} (-1)^s \mu_1^{'k} \mu_{k-s}^{'} \quad \text{and} \quad K_k = \mu_k^{'} - \sum_{s=1}^{k-1} \binom{k-1}{s-1} K_s \mu_{k-s}^{'}, \tag{19}$$

where, $K_1 = \mu_1^{'}$, $K_2 = \mu_2^{'} - \mu_1^{'2}$, $K_3 = \mu_3^{'} - 3\mu_2^{'}\mu_1^{'} + 2\mu_1^{'3}$, $K_4 = \mu_4^{'} - 4\mu_3^{'}\mu_1^{'} - 3\mu_2^{'2} + 12\mu_2^{'}\mu_1^{'} - 6\mu_1^{'4}$. The measures of skewness ($\gamma_1 = K_3/K_2^{\frac{3}{2}}$) and kurtosis ($\gamma_2 = K_4/K_2^2$) can be calculated from the cumulants. We further derive the expressions of the expectations derived from the ETLASD RV. These are given by,

$$E(X^k(1-X)^{-k}) = \sum_{i,r=0}^{\infty} \left(s(i,r) \frac{B\left(m+k+\frac{\lambda+i}{2}, \frac{1}{2}-k\right)}{B\left(m+\frac{\lambda+i}{2}, \frac{1}{2}\right)} - s^{'}(i,r) \frac{B\left(m+k+\frac{\lambda+i+1}{2}, \frac{1}{2}-k\right)}{B\left(m+\frac{\lambda+i+1}{2}, \frac{1}{2}\right)} \right),$$

$$E(X) = \sum_{i,r=0}^{\infty} s(i,r) \left(1 - \frac{1}{2m+\lambda+i+1}\right) - s^{'}(i,r) \left(1 - \frac{1}{2m+\lambda+i+2}\right),$$

$$E(X(1-X)) = \sum_{i,r=0}^{\infty} \left(\frac{s(i,r)(2m+\lambda+i)}{(2m+\lambda+i+1)(2m+\lambda+i+3)} - \frac{s^{'}(i,r)(2m+\lambda+i+1)}{(2m+\lambda+i+2)(2m+\lambda+i+4)} \right),$$

$$E(\log(X)) = \sum_{i,r=0}^{\infty} \left(F\left(m+\frac{\lambda+i+1}{2}\right) + F\left(m+\frac{\lambda+i}{2}\right) \right) \left(s(i,r) - s^{'}(i,r) \right), -$$

$$\frac{2}{2m+\lambda+i} s^{'}(i,r),$$

where, $F(a) = \frac{d}{dx}\log(\gamma(a))$ is digamma function, and

$$E(X\log(X)) = \left(s(i,r)\frac{2m+\lambda+i}{2m+\lambda+i+1} - s^{'}(i,r)\frac{2m+\lambda+i+1}{2m+\lambda+i+2} \right) \times$$

$$\left(F\left(m+\frac{3}{2}+\frac{\lambda+i}{2}\right) - F\left(m+1+\frac{\lambda+i}{2}\right) \right) - \frac{2s^{'}(i,r)(2m+\lambda+i)}{(2m+\lambda+i+2)^2}.$$

3.1 Characteristic function

The CF is a significant statistical property that completely defines the probability distribution. The CF is the Fourier transform of the PDF. The CF of the ETLASD is given by,

$$\begin{aligned}
\psi(\lambda,\beta;t) = E(e^{itX}) &= \int_0^1 e^{itx} f_1(x)dx - \int_0^1 e^{itx} f_2(x)dx, \\
&= \sum_{i,r=0}^{\infty} s(i,r)\int_0^1 e^{itx} g_B\left(x; \frac{2m+\lambda+i}{2}, \frac{1}{2}\right) - s^{'}(i,r)\int_0^1 e^{itx} g_B\left(x; \frac{2m+\lambda+i+1}{2}, \frac{1}{2}\right) dx, \\
&= \sum_{i,r=0}^{\infty} s(i,r)_1F_1(m+\frac{\lambda+i}{2}, m+\frac{\lambda+i+1}{2}; it) - s^{'}(i,r)_1F_1(m+\frac{\lambda+i+1}{2}, m+1+\frac{\lambda+i}{2}; it),
\end{aligned}$$

where, ${}_1F_1(.,.;,)$ is Kummer's confluent hypergeometric function (of the first kind). It also follows that the moment generating function is given by,

$$\begin{aligned}
M_X(\lambda,\beta;t) &= \sum_{i,r=0}^{\infty} s(i,r)_1F_1(m+\frac{\lambda+i}{2}, m+\frac{\lambda+i+1}{2}; t) - s^{'}(i,r)_1F_1(m+\frac{\lambda+i+1}{2}, m+1+\frac{\lambda+i}{2}; t), \\
&= \sum_{i,r=0}^{\infty}\left(s(i,r)\sum_{n=0}^{\infty}\left(\frac{(2m+\lambda+i)^{(n)}}{(2m+\lambda+i+1)^{(n)}}\frac{t^n}{n!}\right) - s^{'}(i,r)\sum_{n=0}^{\infty}\left(\frac{(2m+\lambda+i+1)^{(n)}}{(2m+\lambda+i+2)^{(n)}}\frac{t^n}{n!}\right)\right), \\
&= \sum_{i,r=0}^{\infty}\sum_{n=0}^{\infty}\left(s(i,r)\left(\frac{(2m+\lambda+i)^{(n)}}{(2m+\lambda+i+1)^{(n)}}\frac{t^n}{n!}\right) - s^{'}(i,r)\left(\frac{(2m+\lambda+i+1)^{(n)}}{(2m+\lambda+i+2)^{(n)}}\frac{t^n}{n!}\right)\right)\frac{t^n}{n!}.
\end{aligned}$$

Using the moment generating function, the kth raw moment is given by,

$$E[X^k] = \sum_{i,r=0}^{\infty}\left(s(i,r)\left(\frac{(2m+\lambda+i)^{(k)}}{(2m+\lambda+i+1)^{(k)}}\right) - s^{'}(i,r)\left(\frac{(2m+\lambda+i+1)^{(k)}}{(2m+\lambda+i+2)^{(k)}}\right)\right),$$

where $(x)^k$ is a Pochhammer symbol representing rising factorial. The kth moment can also be expressed in terms of the Beta moments as given by,

$$E[X^k] = \frac{2m+\lambda+i+k-1}{2m+\lambda+i+k}\sum_{i,r=0}^{\infty} s(i,r)E(Y^{k-1}) + \frac{2m+\lambda+i+k}{2m+\lambda+i+k+1}\sum_{i,r=0}^{\infty} s^{'}(i,r)E(Z^{k-1}),$$

where, $Y \sim Beta\left(\frac{2m+\lambda+i}{2}, \frac{1}{2}\right)$ and $Z \sim Beta\left(\frac{2m+\lambda+i+1}{2}, \frac{1}{2}\right)$.

4. Quantiles, skewness and flatness

In this section, we derive the QF of the ETLASD and using it we study the shape of the distribution. The QF is an alternative to the moments. For some distributions, the moments are not expressible in closed forms but the quantiles are obtainable by inverting the CDF. In such cases, statistical measures can be obtained using the QF. The pth quantile (say, Q_p) is derived by solving the equation $F(Q_p) = p$, i.e., $Q_p = F^{-1}(p), p \in (0,1)$. For the ETLASD, the Q_p is obtained as,

$$\left\{\frac{2}{\pi}\arcsin\left(\sqrt{Q_p}\right)\left[2-\frac{2}{\pi}\arcsin\left(\sqrt{Q_p}\right)\right]^{\beta}\right\}^{\lambda} = p; p \in (0,1). \tag{20}$$

Taking $\frac{2}{\pi}\arcsin\left(\sqrt{Q_p}\right) = t^{\beta}$, we can obtain the pth quantile of the ETLASD by solving the following non linear equation,

$$t^{\beta+1} - 2t + p^{1/\lambda\beta} = 0. \tag{21}$$

The QF for some given β values are as follows:

I. For $\beta = 1$, $Q_p = \sin^2\left(\frac{\pi}{2}\left(1 - \sqrt{1 - p^{\frac{1}{\lambda}}}\right)\right)$.

II. For $\beta = 0$, $Q_p = \sin^2\left(\frac{\pi}{2}p^{\frac{1}{\lambda}}\right)$.

III. For $\beta = -1$, $Q_p = \sin^2\left(\frac{\pi p^{\frac{1}{\lambda}}}{1+p^{\frac{1}{\lambda}}}\right)$.

IV. For $\beta = -2$, $Q_p = \sin^2\left(\pi\tau(p)\right)$, where $\tau(p) = \frac{1+4p^{\frac{1}{\lambda}} - \sqrt{1+8p^{\frac{1}{\lambda}}}}{4p^{\frac{1}{\lambda}}}$.

V. For $\beta = -3$, $Q_p = \sin^2\left(\pi\tau(p)\right)$, where $\tau(p) = 1 - \frac{\left(9p^{\frac{1}{2\lambda}} + \sqrt{3(1+27p^{\frac{1}{\lambda}})}\right)^{\frac{2}{3}} 3^{\frac{2}{3}}}{2.3^{\frac{2}{3}}\left(9p^{\frac{2}{\lambda}} + \sqrt{3p^{\frac{3}{\lambda}}(1+27p^{\frac{1}{\lambda}})}\right)^{\frac{2}{3}}}$.

The median, Inter-quantile range, coefficients of skewness and kurtosis based on quantiles are given by,

$$\begin{aligned}
\text{(Median)}\ \ \mathrm{M} &= Q_{0.5},\\
\text{(Inter-quantile range)}\ \ \mathrm{IQR} &= Q_{0.75} - Q_{0.25},\\
\text{(Galton's cofficient)}\ \mathrm{S} &= \frac{Q_{.25} + Q_{.75} - 2M}{Q_{.75} - Q_{.25}} \text{ with } S \in [-1, 1],\\
\text{(Moors's cofficient)}\ \ \mathrm{T} &= \frac{Q_{.875} - Q_{.625} + Q_{.375} - Q_{.125}}{Q_{.75} - Q_{.25}} - 1.23.
\end{aligned}$$

The shapes of the distribution can be identified as,

- If $S = 0$, the distribution is normal, i.e., symmetric and if $S < 0 (S > 0)$, the distribution is left (right) skewed.
- If $T = 0$, the distribution is normal, i.e., mesokurtic and if $T < 0 (T > 0)$, the distribution is platykurtic(leptokurtic).

We sketch the curves for T, S, IQR and M with varying parameters in Figure 3. The ETLASD has the normal, mesokurtic and platykurtic shapes for its density depending upon its parameter values. It also provides symmeteric, positively skewed and negatively skewed shapes. The median increases with increasing values of the parameters. The IQR converges from highly skewed to a flat shape as parameter λ increases.

5. Shannon entropy measure

A classical measure of uncertainity for a given RV X is the differential entropy, known as Shannon entropy, which is defined by $I_s = E\left[-\log[f(X)]\right]$. For the ETLASD, the Shannon entropy is given by,

$$\begin{aligned}
I_s = \log(\lambda) - \log(\pi) - \frac{1}{2}E\left[\log(X(1-X))\right] + (\lambda - 1)E\left[\log\left(\frac{2}{\pi}arcsin(\sqrt{x})\right)\right] +\\
E\left[\log\left(2 - (\beta+1)\frac{2}{\pi}arcsin(\sqrt{x})\right)\right] + (\lambda\beta - 1)E\left[\log\left(2 - \frac{2}{\pi}arcsin(\sqrt{x})\right)\right].
\end{aligned} \tag{22}$$

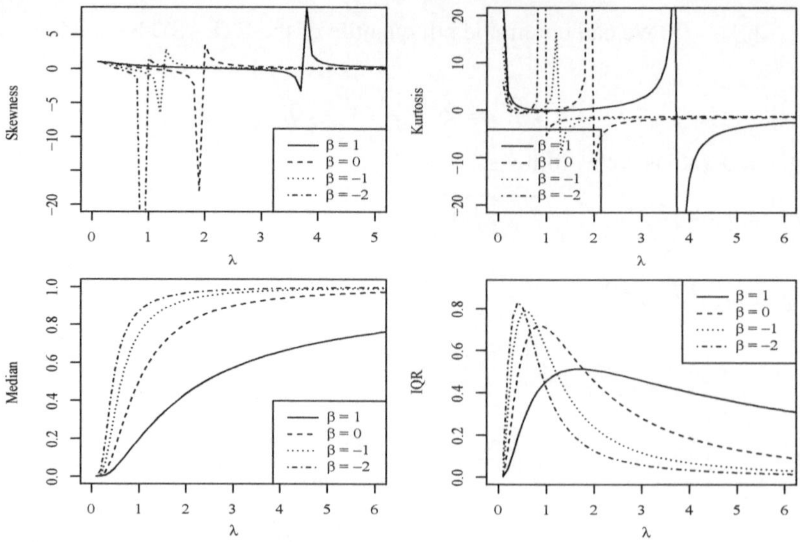

Figure 3: The skewness, kurtosis, median and IQR of the ETLASD(λ, β).

As a statistical average, I_s measures the expected uncertainty including the PDF and the predictability of the outcome X. It has numerous implementations in various fields such as data communication, physics, combinatorics and others. The elements appearing in (22) are separately derived, which are as follows:

$$E\left[\log\left(\frac{2}{\pi}arcsin(\sqrt{x})\right)\right] = \lambda\int_0^1 \lambda(2-(\beta+1)t)t^{\lambda-1}\log(t)(2-t)^{\beta\lambda-1}dt \tag{23}$$

$$\lambda 2^{\beta\lambda-1}\left(\frac{(\beta+1)\,{}_3F_2\left(\lambda+1,\lambda+1,1-\beta\lambda;\lambda+2,\lambda+2;\frac{1}{2}\right)}{(\lambda+1)^2} - \frac{2\,{}_3F_2\left(\lambda,\lambda,1-\beta\lambda;\lambda+1,\lambda+1;\frac{1}{2}\right)}{\lambda^2}\right),$$

$$E\left[\log\left(2-\frac{2}{\pi}arcsin(\sqrt{x})\right)\right] = \lambda\int_0^1 \beta\lambda(2-(\beta+1)t)t^{\lambda-1}\log(2-t)(2-t)^{\beta\lambda-1}dt \tag{24}$$
$$= 2^{(\beta+1)\lambda}B_{\frac{1}{2}}(\lambda+1,\beta\lambda),$$

$$E(\log(1-X)) = \sum_{i,r=0}^{\infty} F(\tfrac{1}{2})\left(s(i,r)-s^{'}(i,r)\right) - \sum_{i,r=0}^{\infty} s(i,r)F\left(m+\frac{\lambda+i+1}{2}\right)+ \tag{25}$$
$$\sum_{i,r=0}^{\infty} s^{'}(i,r)F\left(m+1+\frac{\lambda+i}{2}\right).$$

$$E(\log(X(1-X))) = -\frac{1}{2}\left(\sum_{i,r=0}^{\infty}\left(F\left(m+\frac{\lambda+i+1}{2}\right)+F\left(m+\frac{\lambda+i}{2}\right)+F(\tfrac{1}{2})\right)\left(s(i,r)-s^{'}(i,r)\right)+\right. \tag{26}$$
$$\left.\sum_{i,r=0}^{\infty} s(i,r)F\left(m+\frac{\lambda+i+1}{2}\right)+\sum_{i,r=0}^{\infty} s^{'}(i,r)\left(F\left(m+1+\frac{\lambda+i}{2}\right)-\frac{2}{2m+\lambda+i}\right)\right).$$

$$E\left[\log\left(2-(\beta+1)\frac{2}{\pi}arcsin(\sqrt{x})\right)\right] = \int_0^1 \lambda(2-(\beta+1)t)t^{\lambda-1}(2-t)^{\beta\lambda-1}\log(2-(\beta+1)t)dt \tag{27}$$
$$= \frac{1}{2}\left(\frac{(\beta+1)2^{\beta\lambda}F_1\left(\lambda+1;-\beta\lambda,1;\lambda+2;\frac{1}{2},\frac{\beta+1}{2}\right)}{\lambda+1}+2\log(1-\beta)\right).$$

Combining the elements (24), (25), (26) and (27), the Shannon entropy is given by,

$$I_s = \log(\frac{\lambda(1-\beta)}{\pi}) + 2^{\beta\lambda-1}\left((\lambda-1)\lambda\Gamma(\lambda)^2\left((\beta+1)\lambda^2\,{}_3\tilde{F}_2\left(\lambda+1,\lambda+1,1-\beta\lambda;\lambda+2,\lambda+2;\frac{1}{2}\right)-\right.\right.$$
$$\left.\left.2\,{}_3\tilde{F}_2\left(\lambda,\lambda,1-\beta\lambda;\lambda+1,\lambda+1;\frac{1}{2}\right)\right)+\frac{(\beta+1)F_1\left(\lambda+1;-\beta\lambda,1;\lambda+2;\frac{1}{2},\frac{\beta+1}{2}\right)}{\lambda+1}+\beta2^{\lambda+1}(\beta\lambda-1)\times\right.$$
$$\left.B_{\frac{1}{2}}(\lambda+1,\beta\lambda)\right)-\frac{1}{2}\left(\sum_{i,r=0}^{\infty}\left(F\left(m+\frac{\lambda+i+1}{2}\right)+F\left(m+\frac{\lambda+i}{2}\right)+F(\frac{1}{2})\right)\left(s(i,r)-s'(i,r)\right)+\right.$$
$$\left.\sum_{i,r=0}^{\infty}s(i,r)F\left(m+\frac{\lambda+i+1}{2}\right)+\sum_{i,r=0}^{\infty}s'(i,r)\left(F\left(m+1+\frac{\lambda+i}{2}\right)-\frac{2}{2m+\lambda+i}\right)\right).$$

6. Maximum likelihood estimation

We use the MLE approach to estimate the unknown ETLASD parameters. This is attained by maximizing the likelihood function so that, under the assumed statistical model, the observed data is most probable. The logic of the MLE is both instinctive and flexible. Let $\mathbf{x} = x_1, x_2, ..., x_n$ be a random sample of size n drawn from the ETLASD. The log-likelihood function, based on sample **x**, is given by,

$$l(\Theta|\mathbf{x}) = n\log(\lambda) - n\log(\pi) - \frac{1}{2}\sum_{i=1}^{n}[\log(x_i(1-x_i))] + (\lambda-1)\sum_{i=1}^{n}\left[\log\left(\frac{2}{\pi}arcsin(\sqrt{x_i})\right)\right]$$
$$+\sum_{i=1}^{n}\left[\log\left(2-(\beta+1)\frac{2}{\pi}arcsin(\sqrt{x_i})\right)\right]+(\lambda\beta-1)\sum_{i=1}^{n}\left[\log\left(2-\frac{2}{\pi}arcsin(\sqrt{x_i})\right)\right]. \tag{28}$$

To obtain the MLEs $(\hat{\lambda},\hat{\beta})$, we maximize the log-likelihood function given in (28). The MLEs can be determined numerically solving the following log-likelihood equations,

$$\frac{\partial l(\Theta|x)}{\partial\lambda} = \frac{n}{\lambda} + \beta\sum_{i=1}^{n}\log\left(2-\frac{2\sin^{-1}(\sqrt{x_i})}{\pi}\right)+\sum_{i=1}^{n}\log\left(\frac{2\sin^{-1}(\sqrt{x_i})}{\pi}\right) = 0, \tag{29}$$

$$\frac{\partial l(\Theta|x)}{\partial\beta} = \sum_{i=1}^{n}\frac{\sin^{-1}(\sqrt{x_i})}{(\beta+1)\sin^{-1}(\sqrt{x_i})-\pi}+\lambda\sum_{i=1}^{n}\log\left(2-\frac{2\sin^{-1}(\sqrt{x_i})}{\pi}\right) = 0. \tag{30}$$

From equation (29), $\hat{\lambda}$ is obtained by

$$\hat{\lambda} = \frac{-n}{\sum_{i=1}^{n}\log\left(\frac{2}{\pi}\arcsin(\sqrt{x_i})\right)+\hat{\beta}\sum_{i=1}^{n}\log\left(2-\frac{2}{\pi}\arcsin(\sqrt{x_i})\right)},$$

where $\hat{\beta}$ can be uniquely determined by solving the following non-linear equation,

$$\hat{\beta} = \frac{1}{n-1}\left(\sum_{i=1}^{n}\frac{\arcsin(\sqrt{x_i})\left(n\log\left(2-\frac{2}{\pi}\arcsin(\sqrt{x_i})\right)-\log\left(\frac{2}{\pi}\arcsin(\sqrt{x_i})\right)\right)+n\pi\log\left(\frac{2}{\pi}\arcsin(\sqrt{x_i})\right)}{\sum_{i=1}^{n}\arcsin(\sqrt{x_i})\log\left(2-\frac{2}{\pi}\arcsin(\sqrt{x_i})\right)}\right);n>1.$$

For interval estimation, we compute the expected information matrix which is given by,

$$\mathbf{I}(\boldsymbol{\Theta}) = \begin{bmatrix} I_{\lambda\lambda} & I_{\lambda\beta} \\ I_{\beta\lambda} & I_{\beta\beta} \end{bmatrix},$$

where, $I_{\lambda\lambda} = -\frac{n}{\lambda^2}$, $I_{\lambda\beta} = \sum_{i=1}^{n} \log\left(2 - \frac{2\sin^{-1}(\sqrt{x_i})}{\pi}\right)$ and $I_{\beta\beta} = -\sum_{i=1}^{n} -\frac{\sin^{-1}(\sqrt{x_i})^2}{(\pi-(\beta+1)\sin^{-1}(\sqrt{x_i}))^2}$.

The Inverse Fisher information matrix is given by,

$$\mathbf{I}(\boldsymbol{\Theta})^{-1} = \frac{1}{I_{\lambda\lambda}I_{\beta\beta} - I_{\lambda\beta}I_{\beta\lambda}} \begin{bmatrix} I_{\beta\beta} & -I_{\beta\lambda} \\ -I_{\lambda\beta} & I_{\lambda\lambda} \end{bmatrix}.$$

It is well-studied that the asymptotic distribution of the MLE is normal, i.e., $\sqrt{n}(\hat{\Theta}-\Theta) \sim N(0, \mathbf{I}(\boldsymbol{\Theta})^{-1})$ and it can be used to construct the approximate confidence interval for the parameters λ and β. The asymptotic $100(1-\alpha)\%$ confidence intervals for λ and β are $\hat{\lambda} \pm z_{\alpha/2}\sqrt{[var(\hat{\lambda})]}$ and $\hat{\beta} \pm z_{\alpha/2}\sqrt{[var(\hat{\beta})]}$, where $z_{\alpha/2}$ is $(1-\alpha/2)th$ quantile of $N(0,1)$ and $var(.)$ is the diagonal element of $\mathbf{I}(\boldsymbol{\Theta})^{-1}$.

7. Stochastic ordering

In the case of proportion data, comparison of two independent variables may be of interest. Therefore, we study the ordering of two ETLASD RVs. Let X and Y be two non-identical ETLASD RVs. Stochastic ordering is defined in terms of various statistical functions such as CDF, HRF, mean residual life function and likelihood ratio. The basic definition of the stochastic ordering is as follows. A RV X is said to be stochastically greater than Y if $F_X(x) \geq F_Y(x) \forall x$ and it is denoted by $X \geq_{st} Y$. We discuss here the stochastic ordering using the likelihood ratio.

Proposition 2 *Suppose* $X \sim ETLASD(\lambda_1, \beta_1)$ *and* $Y \sim ETLASD(\lambda_2, \beta_2)$. *Then, we have the following cases of stochastic ordering in terms of the likelihood ratio.*

1. *When* $\beta_1 = \beta_2 = \beta \leq 1$ *and* $\lambda_1 > \lambda_2$, $Y \leq_{lr} X$.
2. *When* $\lambda_1 = \lambda_2 = \lambda$ *and* $\beta_2 > \beta_1$, $Y \leq_{lr} X$.

Proof The likelihood ratio for the ETLASD is given by,

$$\frac{f_X(x;\lambda_1,\beta_1)}{f_Y(x;\lambda_2,\beta_2)} = \frac{\lambda_1}{\lambda_2}\left(\frac{2}{\pi}\arcsin(\sqrt{x})\right)^{\lambda_1-\lambda_2} \frac{[2-(\beta_1+1)\left(\frac{2}{\pi}\arcsin(\sqrt{x})\right)]}{[2-(\beta_2+1)\left(\frac{2}{\pi}\arcsin(\sqrt{x})\right)]}\left[2-\left(\frac{2}{\pi}\arcsin(\sqrt{x})\right)\right]^{\lambda_1\beta_1-\lambda_2\beta_2}.$$

We first take $\beta_1 = \beta_2 = \beta$ and differentiate log-likelihood ratio with respect to x that gives,

$$\frac{d}{dx}\log\left(\frac{f_X(x;\lambda_1,\beta)}{f_Y(x;\lambda_2,\beta)}\right) = (\lambda_1-\lambda_2)\frac{\arcsin(\sqrt{x})}{\sqrt{x(1-x)}}\frac{(2-(\beta+1)\left(\frac{2}{\pi}\arcsin(\sqrt{x})\right))}{\left(\frac{2}{\pi}\arcsin(\sqrt{x})\right)(2-\left(\frac{2}{\pi}\arcsin(\sqrt{x})\right))}.$$

We have $\frac{d}{dx}\log\left(\frac{f_X(x;\lambda_1,\beta)}{f_Y(x;\lambda_2,\beta)}\right) > 0 \quad \forall x$. Therefore, $(Y \leq_{lr} X)$ when $\lambda_2 > \lambda_1$ and $\beta_1 = \beta_2 = \beta \leq 1$.

For $\lambda_1 = \lambda_2 = \lambda$ differentiating the log-likelihood ratio with respect to x gives,

$$\frac{d}{dx}\log\left(\frac{f_X(x;\lambda,\beta_1)}{f_Y(x;\lambda,\beta_2)}\right) = \lambda(\beta_2-\beta_1)\frac{1}{\pi(\sqrt{x(1-x)})(2-\left(\frac{2}{\pi}\arcsin(\sqrt{x})\right))},$$

which is greater than 0 if $\beta_2 > \beta_1$. Thus, $(Y \leq_{lr} X)$ when $\beta_2 > \beta_1$ and $\lambda_1 = \lambda_2 = \lambda > 0$.

The ETLASD also holds ordering in the HRF and mean residual life following the implications stated by Shaked and Shanthikumar (1994) as under:

$$X \geq_{lr} Y \Longrightarrow \begin{bmatrix} X \geq_{hr} Y \\ X \geq_{rh} Y \end{bmatrix} \Longrightarrow X \geq_{st} Y \Longrightarrow X \geq_{mrl} Y.$$

8. Identifiability and stress strength reliability

For any two RVs X and Y, the stress-strength reliability parameter is specified as $R = P[X > Y]$ where RV X represents the strength and Y represents the stress. It measures the probability that the system has enough strength to overcome the stress. Johnson (1988) discussed the three stress strength models which are applied to a vast number of applications in civil, mechanical and aerospace engineering.

If $X \sim ETLASD(\lambda_1, \beta_1)$ and $Y \sim ETLASD(\lambda_2, \beta_2)$, then we have,

$$R = \frac{\lambda_1\left(\beta_1 - \lambda_2(\beta_1 - \beta_2)2^{(\lambda_1(\beta_1+1)+\lambda_2(\beta_2+2))}\text{Beta}\left[\frac{1}{2}, \lambda_1 + \lambda_2, 1 + \beta_1\lambda_1 + \beta_2\lambda_2\right]\right)}{\beta_1\lambda_1 + \beta_2\lambda_2}.$$

Identifiability is an important property which a model must hold for precise inferences. A model is said to be identifiable when it is theoretically possible to find the true value of the parameter when we obtain an infinite number of observations. Let $F = \{F_\theta; \theta \in \Theta\}$ be a statistical model where parameter(Θ) is said to be identifiable if $\theta \to F_\theta$ mapping is one to one, i.e.,

$$F_{\theta_1} = F_{\theta_2} \implies \theta_1 = \theta_2 \quad \forall \quad \theta_1, \theta_2 \in \Theta.$$

To prove the identification property, we use Theorem 1 of Basu and Ghosh (1980) which states that the density ratio $f_1(x; \Theta_1)/f_1(x; \Theta_2)$ of two different members of the family defined on the domain (a, b), converges to either 0 or ∞, when $x \to a$. For the ETLASD, we have

$$\lim_{x \to 0} \frac{f_1(x; \lambda_1, \beta_1)}{f_2(x; \lambda_2, \beta_2)} = \begin{cases} 0 & \text{if } \lambda_2 > \lambda_1, \beta_2 > \beta_1, \\ \infty & \text{if } \lambda_2 < \lambda_1, \beta_2 < \beta_1, \\ 1 & \text{if } \lambda_2 = \lambda_1, \beta_2 = \beta_1. \end{cases}$$

Therefore, the ETLASD is identifiable in parameters(λ, β).

9. Empirical study

In this section, an application to the real-life data is demonstrated to show the usefulness of the ETLASD over Beta, Kumaraswamy, GTL and UG distributions. All distributions are indexed by two-parameters and defined on a uni-interval. We consider a data set that studies the performance of an algorithm called SC16. The data set was first used by Caramanis et al. (1983). Altun and Hamedani (2018) also used this data set to demonstrate the application of the log-xgamma distribution.

From Table 1, it is seen that the coefficient of skewness $\beta > 0$, which reveals that the distribution is positively skewed. As coefficient of kurtosis $\gamma < 3$, the distribution is platykurtic. In order to identify the shape of the empirical hazard function, we use the TTT plot given by Aarset (1987). Let $t_{n:i}; i = 1, 2, ..., n$ denotes the ith ordered sample. The scaled empirical TTT transform is obtained by,

$$\Phi_n(s/n) = \left[\sum_{i=1}^{s} t_{n:i} + (n - s)t_{n:s}\right] / \sum_{i=1}^{n} t_{n:i}.$$

The linear interploation of consecutive points $(s/n, \Phi_n(s/n))$ $(s = 0, 1, 2, ..., n)$ are connected by straight lines to get the TTT plot. From Figure 4b, we can see that the SC16 data set accomodates the bathtub shaped hazard function as the TTT plot is initially convex then concave. Since the characteristics of the ETLASD match with that of the data, the distribution can be used for fitting this data set. We use the *goodness.fit()* command in R software along with the PSO method for the estimation of the parameters.

The PDFs of the beta, Kumaraswamy, UG and GTL distributions are given by,

$$f_{Beta}(y;\alpha,\beta) = \frac{1}{Beta(\alpha,\beta)}x^{\alpha-1}(1-x)^{\beta-1};\quad 0<y<1,\alpha,\beta>0,$$
$$f_{Kumar}(y;\alpha,\beta) = \alpha\beta x^{\alpha-1}(1-x^{\alpha})^{\beta-1};\quad 0<y<1,\alpha,\beta>0,$$
$$f_{UG}(y;v,r) = \frac{v^r y^{v-1}(-\log(y))^{r-1}}{\Gamma(r)};\quad 0<y<1,v,r>0,$$
$$f_{GTL}(y;\alpha,\beta) = 2\alpha\beta x^{\alpha\beta-1}\left(1-x^{\alpha}\right)\left(2-x^{\alpha}\right)^{\beta-1};\quad 0<x<1\ \alpha,\beta>0,$$

respectively.

Table 2 lists the estimated parameters and the goodness of fit statistics values for all the distributions for the SC16 data set. In order to prove that the ETLASD provides better fitting of the unit range data than its competing distributions, we use graphical and statistical measures such as CM, AD, KS, AIC, CAIC, BIC and HQAIC. The AIC, BIC, CAIC and HQIC quantify the relative information lost when the model is fitted to real data set. The ETLASD has the smallest information criterion value among other distributions and so it can be a reasonable choice for fitting the U-shaped data having a bathtub shaped hazard rate. Figures 4a and 5 display the fitted density and fitted HRF plots for the distributions respectively. The CM, K-S and AD measure the discrepancy between the empirical and hypothetical distributions. This shows that the ETLASD gives a better fit of the data set over other distributions.

Table 1: Descriptive statistics of SC16 data.

Statistic	SC16 data
Minimum	0.006
Maximum	0.866
Mean	0.2881
Median	0.1160
SD	0.318
Q_1	0.0325
Q_3	0.518
Skewness	0.767
Kurtosis	1.974

Table 2: MLEs of the parameters and goodness-of-fit statistics for the distributions based on the SC16 data.

Model	MLEs	W	AD	K-S	p-value	AIC	BIC	CAIC	HQIC
ETLASD	(0.917, 0.925)	0.102	0.643	0.148	0.692	-15.373	-13.102	-14.773	-14.802
UG	(2.505, 1.220)	8.271	46.118	0.998	0.000	-3.560	-1.289	-2.960	-2.988
Beta	(0.539,1.141)	0.114	0.712	0.161	0.586	-14.705	-12.434	-14.105	-14.134
Kumar	(0.602, 1.215)	0.112	0.700	0.243	0.130	-14.576	-12.305	-13.976	-14.000
GTL	(0.700,1.024)	0.115	0.718	0.264	0.080	-9.999	-7.728	-9.399	-9.427

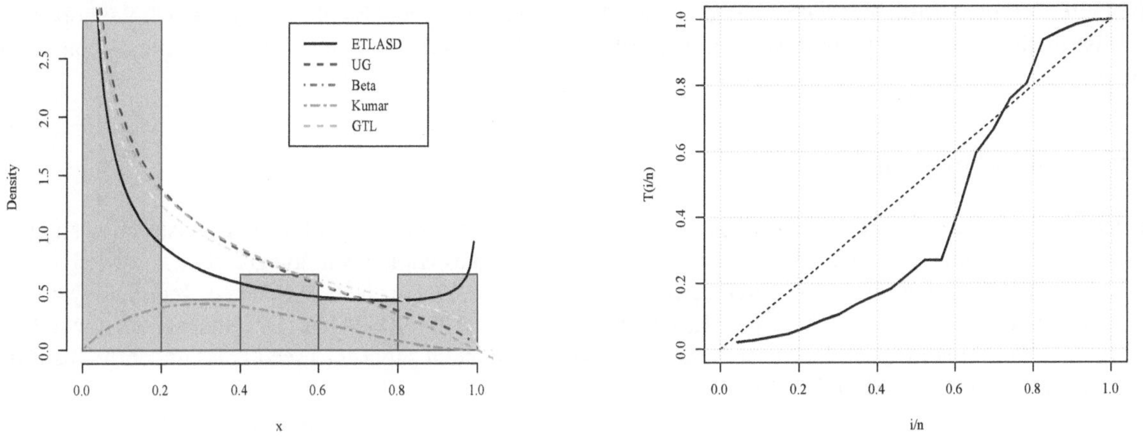

(a) Estimated densities

(b) TTT plot

Figure 4: Fitted densities of the distributions and TTT plot for the SC16 data.

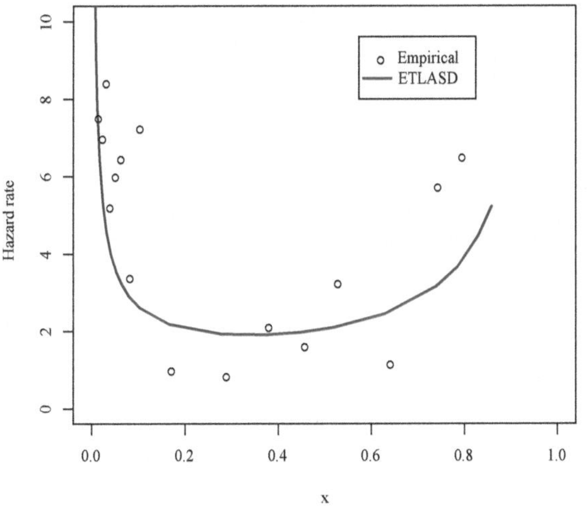

Figure 5: The empirical and fitted HRF of the ETLASD for the SC16 data set.

10. Conclusion

In this article, we propose a new two-parameter distribution called "extended Topp-Leone arcsine distribution" that accommodates four different major shapes for its density function such as J and revised J, increasing, decreasing and bathtub shaped. We discussed its fundamental properties that include shapes of the HRF and PDF, moments, quantile function, stochastic ordering, entropy measure, identifiability and stress-strength reliability. We also study the skewness and kurtosis of the distribution. It was observed that the ETLASD is positively and negatively skewed with varying degrees of kurtosis. We demonstrate empirically the importance of the proposed distribution and its flexibility in fitting U-shaped density data having a bathtub shaped HRF. It was seen that the ETLASD is a better fitted model as compared to other unit distributions based on various information measures. Summing up, it can be stated here that the ETLASD distribution is a better choice for fitting real-life unit range data that has a U-shaped density and bathtub shaped HRF. We hope that the distribution will be a stepping stone in the field of clinical studies, engineering, algorithms, economics and others.

Acknowledgments

Authors would like to thank all editors of the book and anonymous reviewers for their constructive suggestions that greatly improved the presentation of the paper. Dr. Vikas Kumar Sharma greatly acknowledges the financial support from Science and Engineering Research Board, Department of Science Technology, Govt. of India, under the scheme Early Career Research Award (ECR/2017/002416). Dr. Sharma acknowledges the Banaras Hindu University, Varanasi, India for providing financial support as seed grant under the Institute of Eminence Scheme (Scheme No. Dev. 6031).

Abbreviations

PDF Probability distribution function

CDF Cumulative distribution function

TLD Topp-Leone distribution

RV Random Variable

ETL Extended Topp-Leone

ASD Arcsin distribution

ETLASD Extended Topp-Leone arcsin distribution

HRF Hazard rate function

CF Characteristic function

QF Quantile function

MLE Maximum likelihood estimation

CM Cramér-von Misses

AD Anderson Darling

KS Kolmogorov-Smirnov

AIC Akaike Information Criterion

CAIC Consistent Akaike information criterion

BIC Bayesian information criterion

HQAIC Hannan-Quinn information criterion

GTL Generalized Topp-Leone

UG Unit gamma

PSO Particle swarm optimization

TTT Total time on test

References

Aarset, M. V. (1987). How to identify bathtub hazard rate. IEEE Trans. Reliab., 36: 106–108.

Adeyinka, F. S. (2022). On the tx class of topp leone-g family of distributions: Statistical properties and applications. European Journal of Statistics, 2: 2–2.

Ahsanullah, M. (2015). Some characterizations of arcsine distribution. Moroccan J. Pure and Appl. Anal., 1: 70–75.

Al-Babtain, A. A., Elbatal, I., Chesneau, C. and Elgarhy, M. (2020). Sine topp-leone-g family of distributions: Theory and applications. Open Physics, 18: 574–593.

AL-Hussaini, E. K. and Ahsanullah, M. (2015). Exponentiated Distributions. Atlantis Studies in Probability and Statistics, Atlantis Press.

Al-Shomrani, A., Arif, O., Shawky, A., Hanif, S., Shahbaz, M. Q. et al. (2016). Topp–leone family of distributions: Some properties and application. Pakistan Journal of Statistics and Operation Research, 443–451.

Altun, E. and Hamedani, G. (2018). The log-x gamma distribution with inference and application. Journal de la Société Française de Statistique, 159: 40–55.

Alzaatreh, A., Famoye, F. and Lee, C. (2014). The gamma-normal distribution: Properties and applications. Computational Statistics and Data Analysis, 69: 67–80.

Alzaatreh, A., Lee, C. and Famoye, F. (2013). A new method for generating families of continuous distributions. Metron, 71: 63–79.

Arnold, B. C. and Groeneveld, R. A. (1980). Some properties of the arcsine distribution. J. Amer. Statistical Assoc., 75: 173–175.

Bakouch, H. S. (2020). A new family of extended half-distributions: Theory and applications. FILOMAT, 34: 1–10.

Basu, A. P. and Ghosh, J. K. (1980). Identifiability of distributions under competing risks and complementary risks model. Communications in Statistics—Theory and Methods, 9: 1515–1525.

Brito, E., Cordeiro, G., Yousof, H., Alizadeh, M., Silva, G. et al. (2017). The topp–leone odd log-logistic family of distributions. Journal of Statistical Computation and Simulation, 87: 3040–3058.

Caramanis, M., Stremel, J., Fleck, W. and Daniel, S. (1983). Probabilistic production costing: An investigation of alternative algorithms. International Journal of Electrical Power and Energy Systems, 5: 75–86.

Chesneau, C., Sharma, V. K. and Bakouch, H. S. (2021). Extended topp–leone family of distributions as an alternative to beta and kumaraswamy type distributions: Application to glycosaminoglycans concentration level in urine. International Journal of Biomathematics, 14: 2050088.

Chipepa, F., Oluyede, B. and Makubate, B. (2020). The topp-leone-marshall-olkin-g family of distributions with applications. International Journal of Statistics and Probability, 9: 15–32.

Cordeiro, G., Ortega, E. and Cunha, D. (2013). The exponentiated generalized class of distributions. Journal of Data Science, 11: 1–27.

Cordeiro, G. M. and Lemonte, A. J. (2014). The McDonald arcsine distribution: A new model to proportional data. Statistics, 48: 182–199.

Cordeiro, G. M., Lemonte, A. J. and Campelo, A. K. (2016). Extended arcsine distribution to proportional data: Properties and applications. Studia Scientiarum Mathematicarum Hungarica, 53: 440–466.

El-Morshedy, M., El-Faheem, A. A., Al-Bossly, A. and El-Dawoody, M. (2021). Exponentiated generalized inverted gompertz distribution: Properties and estimation methods with applications to symmetric and asymmetric data. Symmetry, 13.

Elgarhy, M., Arslan Nasir, M., Jamal, F. and Ozel, G. (2018). The type II topp-leone generated family of distributions: Properties and applications. Journal of Statistics and Management Systems, 21: 1529–1551.

Eliwa, M. S., El-Morshedy, M. and Ali, S. (2021). Exponentiated odd chen-g family of distributions: Statistical properties, bayesian and non-bayesian estimation with applications. Journal of Applied Statistics, 48: 1948–1974.

Eugene, N., Lee, C. and Famoye, F. (2002). Beta-normal distribution and its applications. Communications in Statistics—Theory and Methods, 31: 497–512.

Feller, W. (1967). An Introduction to Probability Theory and its Applications (Vol. 1), (3rd ed.). New York: John Wiley and Sons.

Hassan, A. S., Elgarhy, M. and Ahmad, Z. (2019). Type II generalized topp-leone family of distributions: Properties and applications. Journal of Data Science, 17.

Johnson, R. A. (1988). 3 stress-strength models for reliability. pp. 27–54. *In*: Quality Control and Reliability. Elsevier. Volume 7 of Handbook of Statistics.

Marshall, A. W. and Olkin, I. (1997). A new method for adding a parameter to a family of distributions with application to the exponential and weibull families. Biometrika, 84: 641–652.

Moakofi, T., Oluyede, B. and Gabanakgosi, M. (2022). The topp-leone odd burr III-g family of distributions: Model, properties and applications. Statistics, Optimization & Information Computing, 10: 236–262.

Oluyede, B., Chamunorwa, S., Chipepa, F. and Alizadeh, M. (2022). The topp-leone gompertz-g family of distributions with applications. Journal of Statistics and Management Systems, 1–25.

Reyad, H., Korkmaz, M. Ç., Afify, A. Z., Hamedani, G., Othman, S. et al. (2021). The fréchet topp leone-g family of distributions: Properties, characterizations and applications. Annals of Data Science, 8: 345–366.

Rezaei, S., Sadr, B. B., Alizadeh, M. and Nadarajah, S. (2017). Topp–leone generated family of distributions: Properties and applications. Communications in Statistics—Theory and Methods, 46: 2893–2909.

Sangsanit, Y. and Bodhisuwan, W. (2016). The Topp-Leone generator of distributions: properties and inferences. Songklanakarin J. Sci. Technol., 38: 537–548.

Schmidt, K. M. and Zhigljavsky, A. (2009). A characterization of the arcsine distribution. Statistics & Probability Letters, 79: 2451–2455.

Shaked, M. and Shanthikumar, J. (1994). Stochastic Orders and their Applications. Boston: Academic Press.

Sharma, V. K. (2018). Topp–leone normal distribution with application to increasing failure rate data. Journal of Statistical Computation and Simulation, 88: 1782–1803.

Sharma, V. K., Bakouch, H. S. and Suthar, K. (2017). An extended maxwell distribution: Properties and applications. Communications in Statistics—Simulation and Computation, 46: 6982–7007.

Shekhawat, K. and Sharma, V. K. (2021). An extension of J-shaped distribution with application to tissue damage proportions in blood. Sankhya B, 83: 548–574.

Topp, C. W. and Leone, F. C. (1955). A family of J-shaped frequency functions. Journal of the American Statistical Association, 50: 209–219.

Yousof, H. M., Alizadeh, M., Jahanshahi, S., Ghosh, T. G. R. I., Hamedani, G. et al. (2017). The transmuted topp-leone g family of distributions: theory, characterizations and applications. Journal of Data Science, 15: 723–740.

Chapter 7

The Truncated Modified Lindley Generated Family of Distributions

Lishamol Tomy,[1,*] *Christophe Chesneau*[2] and *Jiju Gillariose*[3]

1. Introduction

Lifetime distributions are important for the understanding of lifetime phenomena in various fields of applied science. This is essentially due to the need for appropriate statistical models to analyze a variety of data. Because of its many applications in various areas, the exponential distribution is regarded as one of the most important one-parameter distributions. The modifications or generalizations of the exponential distribution have attracted much attention, especially since the Lindley (L) distribution was invented by Lindley (1958). The L and similar models have been used in a variety of applications to solve problems from several fields, including quality management, environmental studies, health sciences, ecology, marketing, finance and insurance. Consequently, the L distribution is used as an alternative to the exponential distribution in many statistical settings. The following one-parameter cumulative density function (cdf) governs it:

$$F_L(x;\lambda) = 1 - \left[1 + \frac{\lambda x}{1+\lambda}\right] e^{-\lambda x}, \quad x > 0,$$

where $\lambda > 0$, and $F_L(x; \lambda) = 0$ for $x \leq 0$. The probability density function (pdf) is then calculated as follows: $f_L(x; \lambda) = F_L'(x; \lambda)$. That is,

$$f_L(x;\lambda) = \frac{\lambda^2}{1+\lambda} e^{-\lambda x}, \quad x > 0,$$

and $f_L(x; \lambda) = 0$ for $x \leq 0$.

The modified L (ML) model has recently been presented by Chesneau et al. (2019) as a middle ground between the conventional exponential and the L distribution. It has a cdf that is specified by,

$$F_{ML}(x;\lambda) = 1 - \left[1 + \frac{\lambda x}{1+\lambda} e^{-\lambda x}\right] e^{-\lambda x}, \quad x > 0, \tag{1}$$

with $\lambda > 0$ and $F_{ML}(x; \lambda) = 0$ for $x \leq 0$, and the related pdf follows upon differentiation:

$$f_{ML}(x;\lambda) = \frac{\lambda}{1+\lambda}\left[(1+\lambda)e^{\lambda x} + 2\lambda x - 1\right] e^{-2\lambda x}, \quad x > 0,$$

[1] Department of Statistics, Deva Matha College, Kuravilangad, Kerala-686633, India.
[2] Université de Caen, LMNO, Campus II, Science 3, 14032, Caen, France.
[3] Department of Statistics, CHRIST (Deemed to be University), Bangalore, Karnataka–560029, India.
Emails: christophe.chesneau@gmail.com; jijugillariose@yahoo.com
* Corresponding author: lishatomy@gmail.com

and $f_{ML}(x;\lambda) = 0$ for $x \leq 0$. As the main reference, Chesneau et al. (2019) discussed the applicability of the ML model and demonstrated its usability using a variety of real-world data sets. The fact that its pdf may be represented as a linear combination of exponential and gamma pdfs is an important structural property of the ML distribution. Chesneau et al. (2020a), Chesneau et al. (2020b) and Chesneau et al. (2020c) proposed the inverted ML distribution, two expansions of the ML distribution, and the wrapped ML distribution, respectively, more recently. This paper explores two new facets of the ML distribution. First, we apply the simple truncation method to the ML distribution to offer a novel distribution with a range of (0,1), called the truncated-(0,1) ML (TML) distribution. Such kinds of distributions are ideal for the modeling of probabilities or percentages that occur in many applied areas. We provide the minimum theory and practice regarding the TML distribution. Second, based on the TML distribution, we create a novel generated family of distributions, called the TML generated (TML-G) family. We study its general theoretical properties. Then, based on some specific distribution, we show its effectiveness in data fitting, with the consideration of important data sets.

The following is a summary of the rest of the paper: We introduce and investigate a new TML distribution in Section 2. In Section 3, we offer a novel TML model-generated family of distributions as well as a particular example of the derived family, in addition to its probabilistic features. Section 4 concludes with some final thoughts.

2. The TML distribution

2.1 Presentation

The TML distribution is defined by the simple truncated version of the ML distribution on the interval (0,1). That is, based on Equation (1), it is governed by the cdf given as $F_{TML}(x;\lambda) = F_{ML}(x;\lambda)/F_{ML}(1;\lambda)$, $x \in (0,1)$, $F_{TML}(x;\lambda) = 0$ for $x \leq 0$, and $F_{TML}(x;\lambda) = 1$ for $x \geq 1$. Thus, for $x \in (0,1)$, we explicitly have,

$$F_{TML}(x;\lambda) = c_\lambda \left\{1 - \left[1 + \frac{\lambda x}{1+\lambda} e^{-\lambda x}\right] e^{-\lambda x}\right\}, \tag{2}$$

where, $c_\lambda = (1+\lambda)/\{1+\lambda-[1+\lambda(1+e^{-\lambda})]e^{-\lambda}\}$.
Thus, the corresponding pdf is defined by,

$$f_{TML}(x;\lambda) = d_\lambda\,[(1+\lambda)e^{\lambda x} + 2\lambda x - 1]\, e^{-2\lambda x}, \quad x \in (0,1),$$

where, $d_\lambda = \lambda/\{1+\lambda-[1+\lambda(1+e^{-\lambda})]e^{-\lambda}\}$ and $f_{TML}(x;\lambda) = 0$ for $x \notin (0,1)$. With the truncated construction, the analytical properties of the pdf of the ML distribution are transposed to the unit interval (0,1). Thus, we can directly say that $f_{TML}(x;\lambda)$ is unimodal. Here, the pdf of TML model is plotted in Figure 1 to see how its shape changes with variations in the values of the parameters.

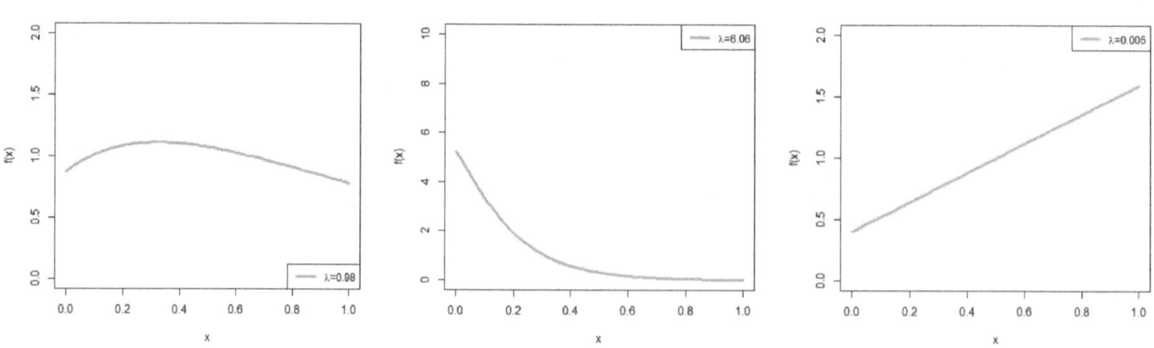

Figure 1: Examples of graphs of the pdf of the TML distribution.

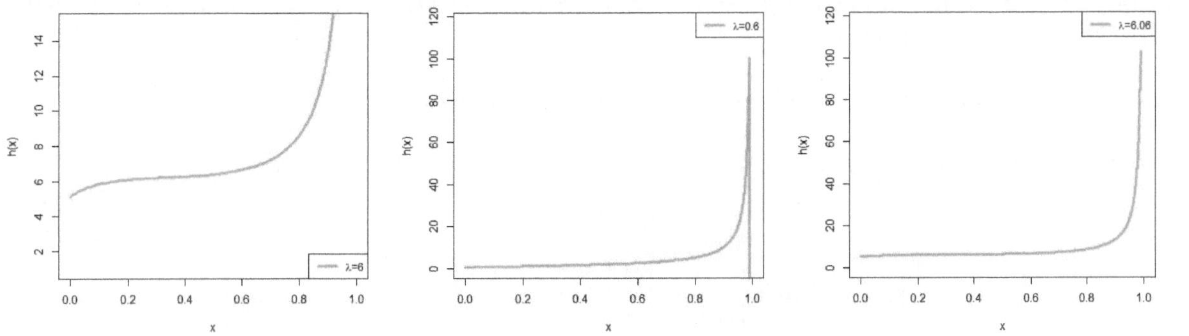

Figure 2: Examples of graphs of the pdf of the TML distribution.

Based on the two functions above, the hazard rate function (hrf) follows immediately:

$$h_{TML}(x;\lambda)=\frac{f_{TML}(x;\lambda)}{1-F_{TML}(x;\lambda)}$$
$$=\frac{d_{\lambda}\left[(1+\lambda)e^{\lambda x}+2\lambda x-1\right]e^{-2\lambda x}}{1-c_{\lambda}\left\{1-\left[1+\lambda xe^{-\lambda x}/(1+\lambda)\right]e^{-\lambda x}\right\}},\quad x\in(0,1),$$

and $h_{TML}(x;\lambda)=0$ for $x\notin(0,1)$. Figure 2 shows some possible shapes of the above Equation. We can see that the hrf is increasing and increasing-decreasing.

2.2 Some theory

The following power series expansion holds for $f_{TML}(x;\lambda)$.

Proposition 2.1 *The pdf of the TML distribution has the following power series expansion:*

$$f_{TML}(x;\lambda)=\sum_{k=0}^{+\infty}u_k x^k+\sum_{k=0}^{+\infty}v_k x^{k+1},$$

where,

$$u_k=d_{\lambda}\frac{(-\lambda)^k}{k!}[(1+\lambda)-2^k],\; v_k=-d_{\lambda}\frac{(-2\lambda)^{k+1}}{k!}. \tag{3}$$

Proof. We have,

$$f_{TML}(x;\lambda)=d_{\lambda}\,[(1+\lambda)e^{-\lambda x}+2\lambda xe^{-2\lambda x}-e^{-2\lambda x}],$$
$$=d_{\lambda}\left[(1+\lambda)\sum_{k=0}^{+\infty}\frac{(-\lambda)^k}{k!}x^k-\sum_{k=0}^{+\infty}\frac{(-2\lambda)^{k+1}}{k!}x^{k+1}-\sum_{k=0}^{+\infty}\frac{(-2\lambda)^k}{k!}x^k\right]$$
$$=d_{\lambda}\left[\sum_{k=0}^{+\infty}\frac{(-\lambda)^k}{k!}[(1+\lambda)-2^k]x^k-\sum_{k=0}^{+\infty}\frac{(-2\lambda)^{k+1}}{k!}x^{k+1}\right]$$
$$=\sum_{k=0}^{+\infty}u_kx^k+\sum_{k=0}^{+\infty}v_kx^{k+1}.$$

The proof of Proposition 2.1 ends. □

This series expansion will be useful in determining some important properties of the TML model and the proposed family.

Let us now introduce a random variable X with the TML distribution. The (raw) moments of are X examined in the next result.

Proposition 2.2 *Let us consider the incomplete gamma function given by* $\gamma(s,x) = \int_0^x t^{s-1} e^{-t} dt$, $x > 0$. *Then the* r^{th} *(raw) moment of* X *is given by,*

$$\mu_r = \mathbb{E}(X^r) = \frac{d_\lambda}{\lambda^r}\left(\gamma(r+1,\lambda) + \frac{1}{2^{r+1}(1+\lambda)}[\gamma(r+2,2\lambda) - \gamma(r+1,2\lambda)]\right).$$

Proof. From the definition of $f_{TML}(x;\lambda)$ and appropriate change of variables, we get

$$\mu_r = \int_0^1 x^r f_{TML}(x;\lambda)dx$$

$$= d_\lambda\left\{\int_0^1 x^r \lambda e^{-\lambda x} dx + \frac{1}{2(1+\lambda)}\left[\int_0^1 x^r (2\lambda)^2 x e^{-2\lambda x} dx - \int_0^1 x^r (2\lambda) e^{-2\lambda x} dx\right]\right\}$$

$$= \frac{d_\lambda}{\lambda^r}\left(\gamma(r+1,\lambda) + \frac{1}{2^{r+1}(1+\lambda)}[\gamma(r+2,2\lambda) - \gamma(r+1,2\lambda)]\right).$$

Hence the desired result. □

Based on Proposition 2.2, the mean of X is obtained as,

$$\mu_1 = \frac{d_\lambda}{\lambda}\left(\gamma(2,\lambda) + \frac{1}{4(1+\lambda)}[\gamma(3,2\lambda) - \gamma(2,2\lambda)]\right)$$

$$= \frac{d_\lambda}{\lambda^r}\left(1-(1+\lambda)e^{-\lambda} - \frac{1}{4(1+\lambda)}(4\lambda^2 e^{-2\lambda} + 2\lambda e^{-2\lambda} + e^{-2\lambda} - 1)\right).$$

The variance, standard deviation, moment-skewness and moment-kurtosis can be expressed in a similar manner.

The incomplete moments of X can be expressed in a similar manner; the above integral terms just need to be truncated at $t \in (0,1)$. The final result is clarified in the following proposition.

Proposition 2.3 *Let us consider the indicator function over an event B denoted by* 1_B. *Then, for* $t \in (0,1)$, *the* r^{th} *incomplete moment of* X *at* t *is given by,*

$$\mu_r^*(t) = \mathbb{E}(X^r 1_{\{X \le t\}})$$

$$= \frac{d_\lambda}{\lambda^r}\left(\gamma(r+1,\lambda t) + \frac{1}{2^{r+1}(1+\lambda)}[\gamma(r+2,2\lambda t) - \gamma(r+1,2\lambda t)]\right).$$

The proof is similar to the one of Proposition 2.2. It is thus omitted.

Based on Proposition 2.2, the incomplete mean of X at t is obtained as,

$$\mu_1 = \frac{d_\lambda}{\lambda}\left(\gamma(2,\lambda t) + \frac{1}{4(1+\lambda)}[\gamma(3,2\lambda t) - \gamma(2,2\lambda t)]\right)$$

$$= \frac{d_\lambda}{\lambda}\left(1-(1+\lambda t)e^{-\lambda t} - \frac{1}{4(1+\lambda)}[4\lambda^2 t^2 e^{-2\lambda t} + 2\lambda t e^{-2\lambda t} + e^{-2\lambda t} - 1]\right).$$

It is of particular importance, since it naturally appears in useful mean residual life functions and income curves.

2.3 Application

Here we turn out the TML distribution as a statistical model, assuming that is unknown.

2.3.1 *Estimation of the parameter λ*

We estimate the unique parameter λ by using the most popular method: the maximum likelihood method. Thus, the obtained estimate, say $\hat{\lambda}$, satisfies:

$$\hat{\lambda} = \text{argmax}_{\lambda}\, L(x_1,\ldots, x_n; \lambda)$$

where, $x_1,\ldots, x_n$ denote all realisations of independent random variables $X_1,\ldots, X_n$ with the common TML distribution, and $L(x_1,\ldots, x_n; \lambda)$ is the likelihood function defined by,

$$L(x_1,\ldots, x_n; \lambda) = \prod_{i=1}^{n} f_{TML}(x_i; \lambda)$$

$$= d_{\lambda}^{n}\, e^{-2\lambda \sum_{i=1}^{n} x_i} \prod_{i=1}^{n} [(1+\lambda)e^{\lambda x_i} + 2\lambda x_i - 1].$$

Equivalently, we can also use the logarithm of the likelihood function to have the following relationship: $\hat{\lambda} = \text{argmax}_{\lambda}\, \log L(x_1,\ldots, x_n; \lambda)$. The obtained estimates are known to have desirable probabilistic properties, which enable the determination of confidence intervals and various statistical tests on λ. The standard errors (SEs) and other basic properties of the estimates are also available, and provided numerically in all the statistical software, such as R or SAS.

2.3.2 Data Fitting

In this subsection, we show situations where the TML model is applicable. We consider a real dataset to ascertain the advantage of TML over the power distribution having the following cdf:

$$G(x;\alpha) = x^{\alpha}, \quad x \in (0,1),$$

and for $G(x;\alpha) = 0$ for $x \notin (0,1)$, where $\alpha > 0$ is a shape parameter. For $\alpha = 1$, it is a particular case of the uniform distribution bounded on the interval . Parameters are estimated using maximum likelihood estimation. The source of the data set is Klein and Moeschberger(2006). It gives the time it takes for kidney dialysis patients to become infected in months. For previous studies on this data set, see, Bantan et al. (2020) and Chesneau, et al. (2021). In this case, we perform a normalization operation by dividing the dataset by 30, to get data between 0 and 1. The transformed data lies between 0.08333333 to 0.91666667.

To determine whether the model is expedient, we derive the unknown parameters by the standard maximum likelihood method and then SE, estimated –log-likelihood (–logL), the values of the AIC (Akaike Information Criterion), BIC (Bayesian Information Criterion), Kolmogorov-Smirnov (K-S) statistic and the respective p-value and compare them.

Table 1 provides the fingings of the descriptive summary for the fitted TML and power models for the dataset. From the findings evaluated, the smallest –logL, AIC, BIC, K-S statistic and the highest p-values are obtained for the TML model. Moreover, Figures 3a and 3b present the estimated pdfs and estimated cdfs for the dataset. Therefore, it can be concluded that for the considered data set, the TML distribution is a better empirical model for fitting data than the power model.

Table 1: Comparison criterion for data set.

Distribution	**Estimates(SE)**	**–logL**	**AIC**	**BIC**	**K-S**	***p*-value**
TML	2.0165(0.5564)	–3.1959	–4.3917	–3.0595	0.1344	0.6930
power	0.8351(0.1578)	–0.4829	1.034179	2.366383	0.24084	0.07769

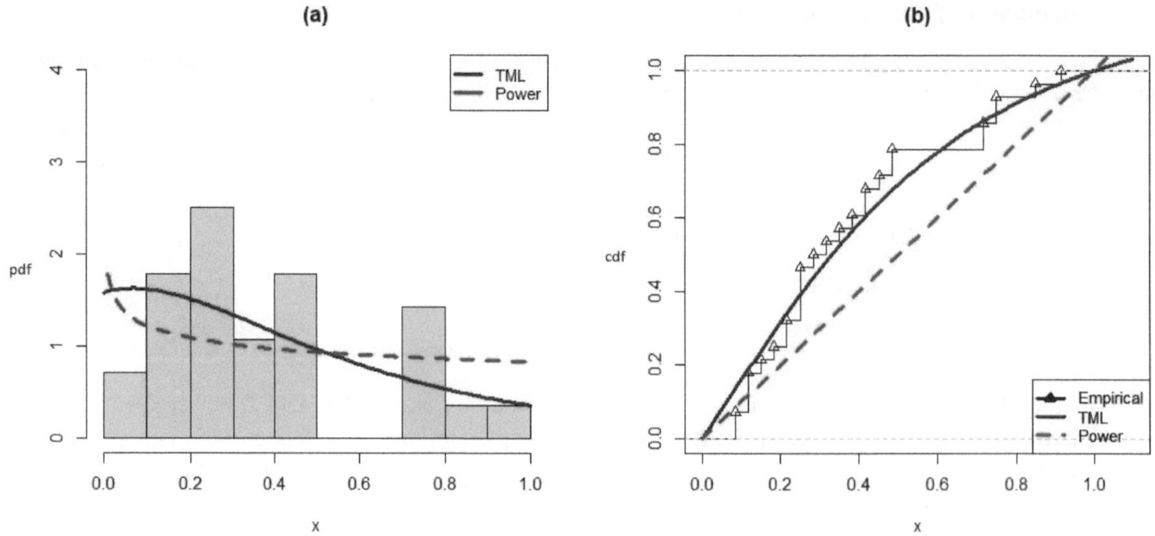

Figure 3: Fitted pdf plots for data set.

3. The TML-G family

3.1 Presentation

The TML distribution is the springboard for more in distribution theory, starting with the construction of the TML-G family of distributions. According to the cdf of a continuous distribution, say $G(x;\zeta)$ where ζ denotes a certain vector of parameters, and Equation (2) the TML-G family is defined with the given cdf:

$$F_{TML-G}(x;\lambda,\zeta) = c_\lambda\left\{1-\left[1+\frac{\lambda G(x;\zeta)}{1+\lambda}e^{-\lambda G(x;\zeta)}\right]e^{-\lambda G(x;\zeta)}\right\}, \quad x \in \mathbb{R}.$$

Thus, by considering the pdf corresponding to $G(x;\zeta)$, represented by $g(x;\zeta)$, the pdf of the TML-G family is given by,

$$f_{TML-G}(x;\lambda,\zeta) = d_\lambda\, g(x;\zeta) \times$$
$$[(1+\lambda)e^{\lambda G(x;\zeta)} + 2\lambda G(x;\zeta) - 1]\, e^{-2\lambda G(x;\zeta)}, \quad x \in \mathbb{R}.$$

Based on the two functions above, the hazard rate function (hrf) follows immediately:

$$h_{TML-G}(x;\lambda,\zeta) = \frac{f_{TML-G}(x;\lambda,\zeta)}{1-F_{TML-G}(x;\lambda,\zeta)}$$
$$= \frac{d_\lambda g(x;\zeta)\left[(1+\lambda)e^{\lambda G(x;\zeta)}+2\lambda G(x;\zeta)-1\right]e^{-2\lambda G(x;\zeta)}}{1-c_\lambda\left\{1-\left[1+\lambda G(x;\zeta)e^{-\lambda G(x;\zeta)}/(1+\lambda)\right]e^{-\lambda G(x;\zeta)}\right\}} \quad x \in \mathbb{R}.$$

The following series expansion holds for $f_{TML-G}(x;\lambda,\zeta)$.

Proposition 3.1 *The pdf of the TML-G distribution can be expressed via pdfs of the E-G family as,*

$$f_{TML-G}(x;\lambda,\zeta) = \sum_{k=0}^{+\infty} u_k^* f_{E-G}(x;k+1,\zeta) + \sum_{k=0}^{+\infty} v_k^* f_{E-G}(x;k+2,\zeta),$$

where, $f_{E-G}(x;k+1,\zeta) = kg(x;\zeta)G(x;\zeta)^k$ which is the pdf of the E-G family with power parameter $k+1$, $u_k^* = u_k/(k+1)$ and $v_k^* = v_k/(k+2)$, with u_k and v_k defined in (3).

This result is an immediate consequence of Proposition 2.1. It is thus omitted.
A possible functional approximation of $f_{TML-G}(x;\lambda,\zeta)$ is

$$f_{TML-G}(x;\lambda,\zeta) \approx \sum_{k=0}^{K(\to+\infty)} u_k^* f_{E-G}(x;k+1,\zeta) + \sum_{k=0}^{K(\to+\infty)} v_k^* f_{E-G}(x;\, k+2,\, \zeta).$$

Moment-type measures can be derived from Proposition 3.1. As a example, for a random variable X with a distribution into the TML-G family, the rth moment of X is given by,

$$\mu_r = \mathbb{E}(X^r) = \sum_{k=0}^{+\infty} u_k^* \mu_r^*(k) + \sum_{k=0}^{+\infty} v_k^* \mu_r^*(k+1),$$

where, $\mu_r^*(k) = \int_{-\infty}^{+\infty} x^r f_{E-G}(x;k+1,\zeta)dx$ denotes the rth moment of a random variable with distribution into the E-G family with power parameter $k + 1$, and can be approximated as,

$$\mu_r \approx \sum_{k=0}^{K(\to+\infty)} u_k^* \mu_r^*(k) + \sum_{k=0}^{K(\to+\infty)} v_k^* \mu_r^*(k+1).$$

The interest of this approximation is mainly computational: The moments $\mu_r^*(k)$ are available in the literature for a lot of parent distributions, and can be used to approximate μ_r which can be very complex in its former integral definition. Based on the series expansion technique, other moment-measures can be provided, such as incomplete moments, and diverse entropy.

3.2 TML-Exponential (TML-E) distribution

In this subsection, we explore one member of the TML-G family, namely, TML-exponential (TML-E) distribution and present its properties in detail. The exponential distribution was chosen because it is the simplest and widely used model. In practice, other distributions can be used to model real data.
Based on a cdf of a TML-G family, and TML-E distribution is defined with the following cdf:

$$F_{TML-E}(x;\lambda,\theta) = c_\lambda \left\{ 1 - \left[1 + \frac{\lambda\left(1-e^{-\frac{x}{\theta}}\right)}{1+\lambda} e^{-\lambda(1-e^{-\frac{x}{\theta}})} \right] e^{-\lambda(1-e^{-\frac{x}{\theta}})} \right\}, \quad x > 0,$$

$\lambda > 0$, $\theta > 0$ and $F_{TML-E}(x;\lambda,\theta) = 0$ for $x \le 0$. Thus, by considering the pdf corresponding to the considered exponential distribution, the pdf of the TML-G family is defined by,

$$f_{TML-E}(x;\lambda,\theta) = \frac{e^{-x/\theta}}{\theta} \times \left[(1+\lambda)e^{\lambda(1-e^{-x/\theta})} + 2\lambda(1-e^{-x/\theta}) - 1 \right] e^{-2\lambda(1-e^{-x/\theta})}, \quad x > 0,$$

and $f_{TML-E}(x;\lambda,\theta) = 0$ for $x \le 0$. According to the two functions above, the hrf follows immediately:

$$h_{TML-E}(x;\lambda,\theta) = \frac{f_{TML-E}(x;\lambda,\theta)}{1-F_{TML-E}(x;\lambda,\theta)}$$

$$= \frac{(d_\lambda/\theta \times e^{-x/\theta})\left[(1+\lambda)e^{\lambda(1-e^{-x/\theta})} + 2\lambda(1-e^{-x/\theta}) - 1 \right] e^{-2\lambda(1-e^{-x/\theta})}}{1 - c_\lambda \left\{ 1 - \left[1 + \lambda(1-e^{-x/\theta})e^{-\lambda(1-e^{-x/\theta})} / (1+\lambda) \right] e^{-\lambda(1-e^{-x/\theta})} \right\}}, \quad x > 0,$$

and $h_{TML-E}(x;\lambda,\theta) = 0$ for $x \le 0$.

3.3 Estimation and applications

By considering the TML-E model, we estimate the corresponding parameters collected into a vector, say $\Theta = (\lambda, \theta)$, by using the maximum likelihood method.

Our estimated vector, say $\hat{\Theta}$, satisfies

$$\hat{\Theta} = \text{argmax}_{\Theta}\, L(x_1, \ldots, x_n; \Theta)$$

where $x_1, \ldots, x_n$ denote realisations of independent random variables $X_1, \ldots, X_n$ with the TML-E distribution, and $L(x_1, \ldots, x_n; \Theta)$ is the likelihood function defined by

$$L(x_1, \ldots, x_n; \Theta) = \prod_{i=1}^{n} f_{TML-E}(x_i; \lambda, \theta)$$

Alternatively, the logarithm of the likelihood function can be used to obtain the following relationship: $\hat{\Theta} = \text{argmax}_{\Theta} \log L(x_1, \ldots, x_n; \Theta)$. The derived estimates are known to have good probabilistic features, allowing confidence ranges to be calculated and other statistical tests to be performed on Θ and its components. The SEs and other basic features of the estimates are likewise available in every statistical tool, such as R or SAS, and are supplied numerically.

3.4 Data fitting

In this section, we offer two real-world illustrations that demonstrate the TML-E model's versatility and compare the TML-E distribution to the Lindley and exponential distributions.

Data set 1: The first data set provides the time between failures for a repairable item, for more details see, Murthy et al. (2004).

Data set 2: The second real data set suggests 30 consecutive values of March precipitation (in inches) in Minneapolis/St Paul given by Hinkley (1977).

The findings of descriptive investigations for the fitted TML-E, Lindley, and exponential distributions for two data sets are presented in Tables 2 and 3. Since the smallest logL, AIC, BIC, K-S statistic and the highest p-values are obtained for the TML-E model, it can be considered as the best. The estimated pdfs for data sets 1 and 2 are presented in Figures 2a and 3a. In addition, the cdfs for each model are compared to the empirical distribution function in Figures 2b and 3b. Therefore, we conclude that the TML-E model provides good fits to these data.

Table 2: Comparison criterion for data set 1.

Distribution	**Estimates(SE)**	**−logL**	**AIC**	**BIC**	**K-S statistic**	***p*-value**
TML-E	$\hat{\lambda}$ = **0.9759 (0.6881)** $\hat{\theta}$ = 0.8794 (0.1875)	**39.8833**	**83.76661**	**86.569**	**0.0759**	**0.9952**
Lindley	$\hat{\theta}$ = 0.9762 (0.1345)	41.5473	85.0947	86.4958	0.1407	0.5928
exponential	$\hat{\theta}$ = 0.6482 (0.1183)	43.0054	88.0108	89.4120	0.1845	0.2590

Table 3: Comparison criterion for data set 2.

Distribution	**Estimates(SE)**	**−logL**	**AIC**	**BIC**	**K-S statistic**	***p*-value**
TML-E	$\hat{\lambda}$= **2.0202 (0.7861)** $\hat{\theta}$ = 0.7632 (0.1290)	**38.5082**	**81.0164**	**83.81879**	**0.0582**	**0.9900**
Lindley	$\hat{\theta}$= 0.9096 (0.1247)	43.1437	88.2875	89.6886	0.18823	0.2383
exponential	$\hat{\theta}$=0.5970149 0.1090	45.4744	92.9488	94.3499	0.2352	0.0724

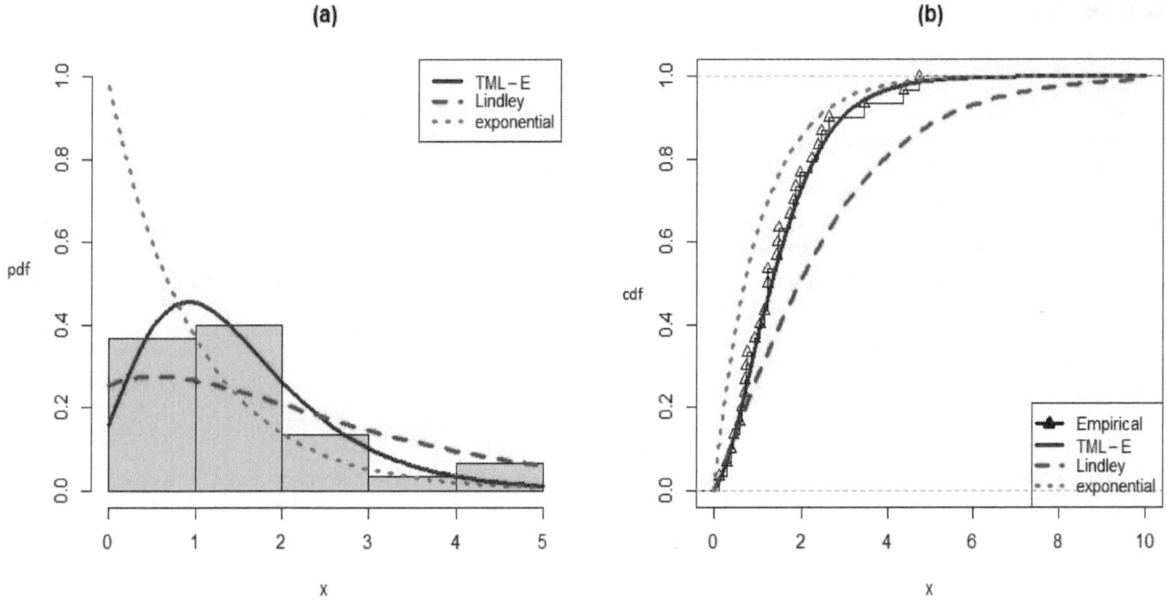

Figure 4: Fitted pdf plots for data set 1.

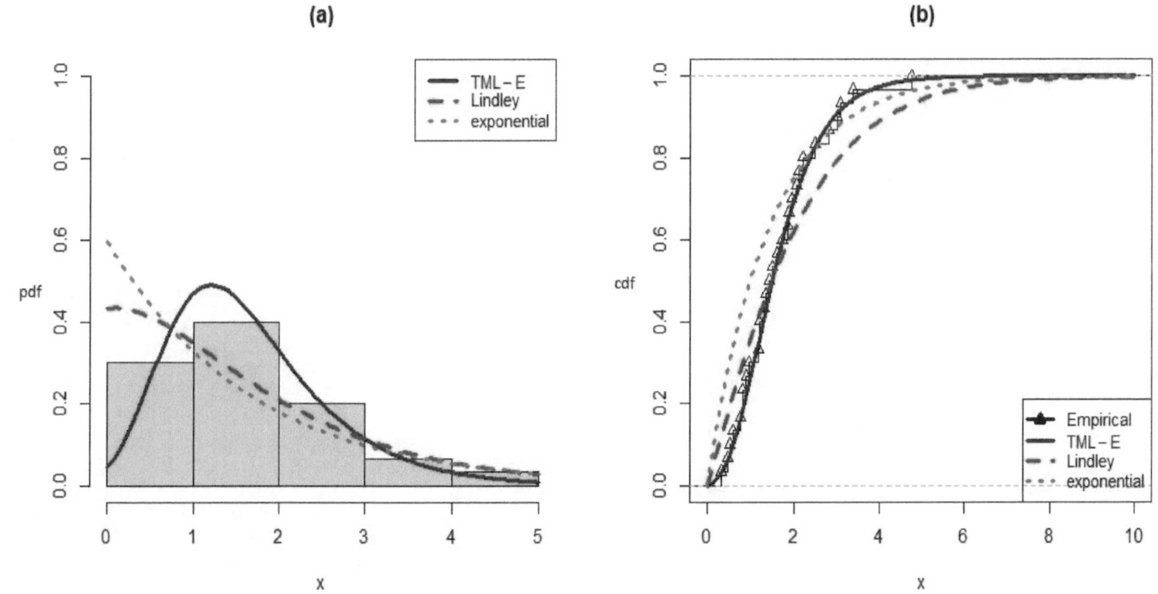

Figure 5: Fitted pdf plots for data set 2.

4. Conclusions

In this paper, we present a novel one-parameter model, the truncated-(0,1) ML (TML) distribution, and a new generated family, the TML-Generated (TML-G) family. We derive several structural properties for this new family, as well as investigate some properties of the new TML-Exponential (TML-E) model. Maximum likelihood is acclimated to estimate the model parameters. Two real-world examples show that the proposed TML-E distribution consistently outperforms other models in terms of fit.

References

Bantan, R. A. R., Chesneau, C., Jamal, F., Elgarhy, M., Tahir, M. H. et al. (2020). Some new facts about the unit-Rayleigh distribution with applications, Mathematics, 8, 11, 1954, 1–23.

Chesneau, C., Tomy, L. and Gillariose, J. (2019). A new modified Lindley distribution with properties and applications, Journal of Statistics and Management Systems, DOI: 10.1080/09720510.2020.1824727.

Chesneau, C., Tomy, L., Gillariose, J. and Jamal, F. (2020a). The inverted modified lindley distribution. Journal of Statistical Theory and Practice, 14: 1–17.

Chesneau, C., Tomy, L. and Gillariose, J. (2020b). On a sum and difference of two Lindley distributions: theory and applications, REVSTAT- Statistical Journal, 18: 673–695.

Chesneau, C., Tomy, L. and Jose, M. (2020c). Wrapped modified Lindley distribution. Journal of Statistics and Management Systems, DOI: 10.1080/09720510.2020.1796313.

Chesneau, C., Tomy, L. and Gillariose, J. (2021). On a new distribution based on the arccosine function, Arabian Journal of Mathematics, 10(3): https://doi.org/10.1007/s40065-021-00337-x.

Hinkley, D. (1977). On quick choice of power transformations. Applied Statistics, 26: 67–69.

Klein, J. P. and Moeschberger, M. L. (2006). Survival Analysis: Techniques for Censored and Truncated Data; Springer: Berlin/ Heidelberg, Germany.

Lindley, D. V. (1958). Fiducial distributions and Bayes theorem. Journal of the Royal Statistical Society, 20: 102–107.

Murthy, D. N. P., Xie, M. and Jiang, R. (2004). Weibull Models, Wiley series in probability and statistics, John Wiley & Sons, NJ.

Chapter 8

An Extension of the Weibull Distribution via Alpha Logarithmic G Family with Associated Quantile Regression Modeling and Applications

Yunus Akdoğan,[1] *Kadir Karakaya,*[1] *Mustafa Ç Korkmaz,*[2,*] *Fatih Şahin*[1] and *Aşır Genç*[3]

1. Introduction

The lifetime models are used to model data defined on $(0, \infty)$ and generally prefered in fields such as reliability engineering and life data analysis. In recent years, the need for data modeling has increased. A lot of distributions have been introduced recently by Alizadeh et al. (2020), Korkmaz (2020), Tanış et al. (2021), Rasekhi et al. (2019), and others. Also, several distribution families have been introduced using transformations of well-known existing distribution functions recently. For instance, the exponentiated family introduced by Mudholkar and Srivastava (1993) is presented as,

$$F(x) = F_0(x)^{\alpha}, \alpha > 0,$$

where, $F_0(x)$ is the cumulative distribution function (cdf) of the baseline model. Mahdavi and Kundu (2017) introduced another generalized family called α—power family. This family is specialized for the Exponential, Pareto and Weibull distribution in the studies of Mahdavi and Kundu (2017), Kinaci et al. (2019) and Nassar et al. (2017) respectively.

The α—logarithmic family is defined by Karakaya et al. (2017) and the exponential distribution is selected as a baseline distribution. The α—logarithmic family is as follows: Let $F_0(x)$ be the cdf of a baseline random variable , then the cdf of the α—logarithmic transformation for $x \in \mathbb{R}$, is given by,

$$F(x) = \frac{log(1+\alpha\{F_0(x)\})}{log(1+\alpha)}, \tag{1}$$

[1] Department of Statistics, Selcuk University, Konya, Turkey.
[2] Department of Measurement and Evaluation, Artvin Çoruh University, Artvin, Turkey.
[3] Department of Statistics, Necmettin Erbakan University, Konya, Turkey; Email: agenc@erbakan.edu.tr
Emails: yakdogan@selcuk.edu.tr; kkarakaya@selcuk.edu.tr; fatihsahin0644@gmail.com
* Corresponding author email: mustafacagataykorkmaz@gmail.com

where, $\alpha > -1$. This family is known as ALT and is easily seen that when $\alpha \to 0$ then $F_{ALT}(x) \to F_0(x)$. The probability distribution function (pdf) of the ALT family is given as follows:

$$F_{ALT}(x) = \frac{\alpha f_0(x)}{log(1+\alpha)(1+\alpha F_0(x))}, \tag{2}$$

where, f_0 is the pdf of the baseline distribution. In reliability research, the Weibull distribution is one of the most used lifetime distributions.This distribution is used to model data in fields such as medicine, biology, physics and economics. The Weibull distribution is lacking in data modeling for some applications. To overcome this, many new distributions based on the Weibull distribution have been proposed. In this study, we aim to bring a new perspective to the Weibull distribution and it is the baseline distribution is selected in the family of ALT introduced by Karakaya et al. (2017). This paper is organized as follows. In Section 2, some distributional properties of the new distribution are examined. The inference of the new model is examined by five different estimation methods discussed in Section 3. In Section 4, extensive Monte Carlo simulation studies are performed to investigate the efficiency of the five estimators. On the basis of the new model, a new quantile regression model is constructed, and model parameters are estimated using the maximum likelihood method in Section 5. Finally, Section 6 shows two data sets of the proposed model for modeling practical data sets.

2. Alpha logarithmic weibull distribution

In this section, the ALT method is applied to Weibull distribution and is called α—logarithmic Weibull distribution (ALWD). Let X follow the Weibull distribution with parameters β, $\theta > 0$. Recall that cdf associated with is given by,

$$F_W(x) = 1 - \exp\left(-\left(\frac{x}{\beta}\right)^{\theta}\right) \tag{3}$$

Using the Weibull distribution as $F_0(x)$ in (3), the cdf and pdf of ALWD are obtained, respectively, by,

$$F(x;\Xi) = \left\{\begin{matrix} 0, & x \le 0 \\ \dfrac{log\left(1+\alpha\left\{1-exp\left(-\left(\frac{x}{\beta}\right)^{\theta}\right)\right\}\right)}{2}, & x > 0 \end{matrix}\right\} \tag{4}$$

and

$$f(x;\Xi) = \frac{\alpha \frac{\theta}{\beta}\left(\frac{x}{\beta}\right)^{\theta-1} exp\left(-\left(\frac{x}{\beta}\right)^{\theta}\right)}{log(1+\alpha)\left(1+\alpha\left\{1-exp\left(-\left(\frac{x}{\beta}\right)^{\theta}\right)\right\}\right)} \mathbb{I}_{\mathbb{R}^+}(x), \tag{5}$$

where, $\alpha > -1$, $\beta > 0$ and $\theta > 0$ are parameters, $\Xi = (\alpha, \beta, \theta)$ and $\mathbb{I}_A(\cdot)$ is the indicator function on set A. When the random variable X has an ALWD with pdf given in (5), it is briefly shown as ALWD (Ξ). The ALWD includes some special-sub distributions as follows:

- The exponential distribution with scale parameter β (when $\theta = 1$ and $\alpha = 0$)
- The Weibull distribution with scale parameter β and shape parameter θ (when $\alpha = 0$)
- The gamma distribution with shape parameter 2 and scale parameter β (when $\theta = 1$ and $\alpha = 0$)

- The generalized gamma distribution with shape parameters 2, α and scale parameter λ (when $\alpha = 0$).
- The Rayleigh distribution with shape parameters 2, α and scale parameter $\lambda = \sigma\sqrt{2}$ (when $\alpha = 0$).

The quantile function of the ALWD can be easily obtained by inverting its cdf. Hence, th quantile function is obtained by,

$$x_u(\Xi) = -\beta\left(-log\left(\frac{\alpha+1-(1+\alpha)^u}{\alpha}\right)\right)^{1/\theta},$$

where, $0 < u < 1$. The hazard rate function (hrf) for ALWD is obtained by,

$$h(x;\Xi) = \frac{\alpha\frac{\theta}{\beta}\left(\frac{x}{\beta}\right)^{\theta-1} exp\left\{-\left(\frac{x}{\beta}\right)^{\theta}\right\}}{log\left(1+\alpha-\alpha\, exp\left\{-\left(\frac{x}{\beta}\right)^{\theta}\right\}\right)-log(1+a)}$$

$$\times\frac{1}{\alpha\, exp\left\{-\left(\frac{x}{\beta}\right)^{\theta}\right\}-\alpha-1}.$$

The pdf plot and hrf plot of ALWD are given in Figure 1 for some choices of parameter values α, β and θ.

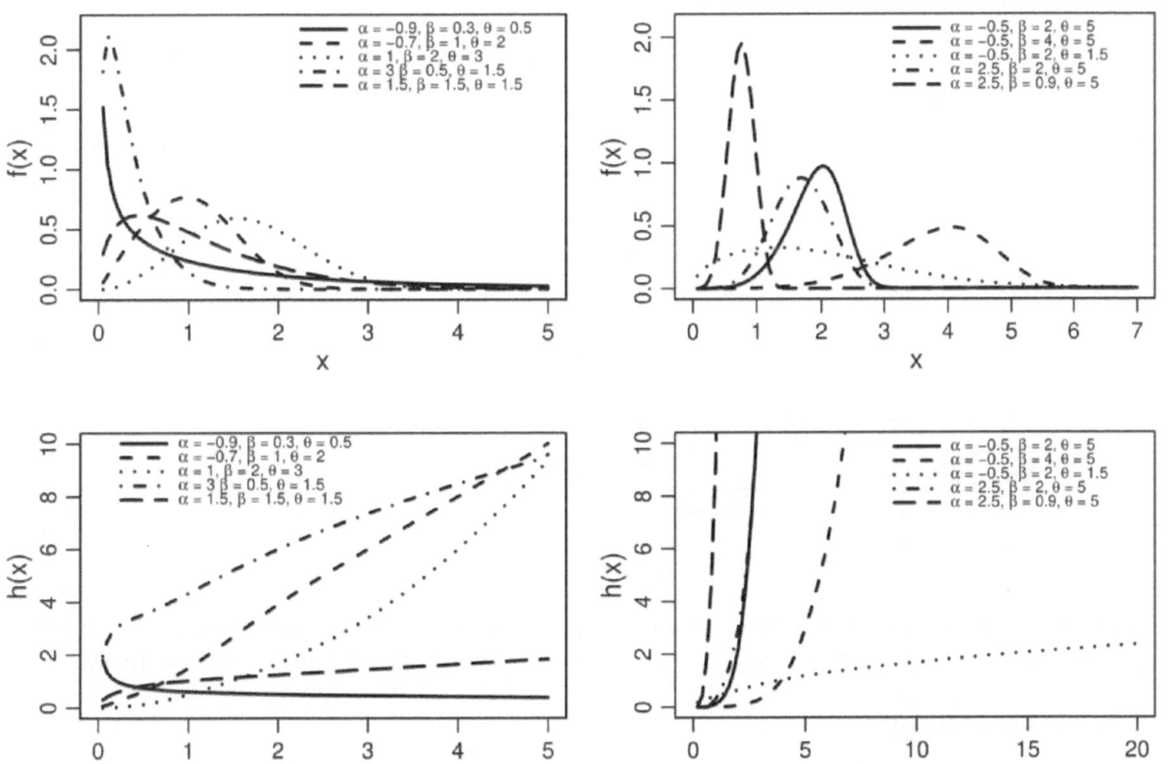

Figure 1: Pdf and hrf plots for different parameter values α, β and θ.

It is concluded from Figure 1 that the pdf of the ALWD is increasing, decreasing and unimodal. It is also concluded from Figure 1 that the hrf is increasing when $\alpha < 0$ and decreasing when $\alpha < 0$. It is also seen that the hrf is constant when $\theta = 1$, $\alpha \to 0$.
The *rth* moment of ALWD for $\alpha < 0$ is given as follows,

$$E(x^r) = \frac{\alpha\beta^r}{log(1+\alpha)} \Gamma\left(\frac{r}{\theta}+1\right) \Phi\left(\frac{\alpha}{1+\alpha}, \frac{r}{\alpha}+1, 1\right)$$

where, $\Phi(z,s,v) = \sum_{n=0}^{\infty} (v+n)^{-s} z_n$ is the Lerch function. The expected value is,

$$E(x) = \frac{\alpha\beta}{log(1+\alpha)} \Gamma\left(\frac{1}{\theta}+1\right) \sum_{n=0}^{\infty} (1+n)^{-\left(\frac{1}{\theta}+1\right)} \left(\frac{\alpha}{1+\alpha}\right)^n.$$

Note that when $\theta = 1$, $\alpha \to 0$ then $E(X) \to \beta$ and $\alpha \to \infty$ then $E(X) \to 0$. Moreover when $\theta = 1$, $\alpha \to 0$ then $Var(X) \to \beta^2$ and $\alpha \to \infty$ then $Var(X) \to 0$.

2.1 Order statistics

The order statistics of the ALWD (Ξ) distribution provided some interesting findings. Let be a random sample from ALWD (Ξ) distribution and $X_{(1)} \leq X_{(2)} \leq \cdots \leq X_{(n)}$ denote the corresponding order statistics. The cdf and pdf of *rth* order statistics, $X_{(r)}$, are given by,

$$F_{X_{(r)}}(x) = \sum_{k=r}^{n} \sum_{l=0}^{n-k} (-1)^l \binom{n}{k} \binom{n-k}{l} \left\{ \frac{log\left(1+\alpha\left(1-exp\left\{\left(-\frac{x}{\beta}\right)^\theta\right\}\right)\right)}{log(1+\alpha)} \right\}^{k+l}$$

and

$$f_{X_{(r)}}(x) = \frac{\alpha\theta}{\beta\, log(1+\alpha) B(r, n-r+1)} \sum_{k=0}^{n-r} (-1)^k \binom{n-r}{k} \frac{\left(\frac{x}{\beta}\right)^\theta exp\left\{-\left(\frac{x}{\beta}\right)^\theta\right\}}{\left(1+\alpha\left(1-exp\left\{\left(-\frac{x}{\beta}\right)^\theta\right\}\right)\right)}$$

$$\times \left\{ \frac{log\left(1+\alpha\left(1-exp\left\{\left(-\frac{x}{\beta}\right)^\theta\right\}\right)\right)}{log(1+\alpha)} \right\}^{r+k-1}$$

respectively, where $B(\cdot,\cdot)$ is the beta function and $r = 1,2,\ldots, n$. It is obviously seen that the cdf and pdf of $X_{(1)} = min\{X_1, X_2,\ldots, X_n\}$ and $X_{(n)} = max\{X_1, X_2,\ldots, X_n\}$ are obtained, for $r = 1$ and $r = n$ respectively.

2.2 Stochastic ordering

Shaked and Shantikumar (2007) obtained some stochastic ordering with many applications. Stochastic orders are an important criterion for the comparative determination of random variables. We provide some essential definitions. A random variable X is shown to be smaller than a random variable Y in the

- The stochastic order ($X <_{st} Y$) if $F_X(x) \geq F_Y(x)$.

- The mean residual life order ($X <_{mrl} Y$) if $m_X(x) \geq m_Y(y)$.
- The likelihood ratio order ($X <_{lr} Y$) if $\frac{f_x(x)}{f_Y(x)}$ if decreases in x.
- The hazard rate order ($X <_{hr} Y$) if $h_Y(x) \geq h_Y(x)$.

Theorem 1 *If $X \sim ALWD(\alpha_1,\beta,\theta)$ and $Y \sim ALWD(\alpha_2,\beta,\theta)$ and $\alpha_1 > \alpha_2$, then $X <_{lr} Y$.*

Corollary 2.1 *If $X \sim ALWD(\alpha_1,\beta,\theta)$ and $Y \sim ALWD(\alpha_2,\beta,\theta)$ and $\alpha_1 > \alpha_2$ then $\Rightarrow X <_{hr} Y \Rightarrow X <_{mrl} Y \Rightarrow X <_{st} Y$.*

For any $\alpha > 0 \rightleftharpoons$ and $x \in (0,\infty)$ get the $W(x)$ the likelihood ratio of ALWD is given by,

$$W(x) = \frac{\alpha_1 log(1+\alpha_2)\left(1+\alpha_2\left(1-exp\left(-\frac{x}{\beta}\right)^{\theta}\right)\right)}{\alpha_2 log(1+\alpha_1)\left(1+\alpha_1\left(1-exp\left(-\frac{x}{\beta}\right)^{\theta}\right)\right)}$$

Taking the derivative with respect to x,

$$W'(x) = -\frac{\overbrace{\alpha_1\theta\left(\frac{x}{\beta}\right)^{\theta} exp\left(-\frac{x}{\beta}\right)^{\theta} log(1+\alpha_1)log(1+\alpha_2)(\alpha_1+\alpha_2)}^{>0}}{\underbrace{x\alpha_2\left(-1-\alpha_1+\alpha_1 exp\left(-\frac{x}{\beta}\right)^{\theta}\right)^2}_{>0}}$$

for $\alpha_1 > \alpha_2$, $-(\alpha_1 - \alpha_2)$ is smaller than zero. So $W'(x) < 0 \rightleftharpoons$ when $\alpha_1 - \alpha_2$ is taken. $W(x)$ is a decreasing function in x. The proof is thus completed.

Theorem 2.1 shows that the ALWDs are ordered for stochastic orderings.

3. Parameter estimation

In this section, several methods of estimations are examined to estimate the three unknown parameters of ALWD. We discuss the maximum likelihood (ML), least-squares (LS), weighted least squares (WLS), Anderson-Darling (AD), and Cramer-von Mises (CvM) methods of estimation. Let $X_1, X_2, \ldots, X_n$ be a random sample from ALWD. $X_{(1)} < X_{(2)} < \cdots < X_{(n)}$ symbolizes the corresponding order statistics. Also, represents the observed value of $X_{(i)}$, $i = 1,2,\ldots, n$. Then the log-likelihood function is given by,

$$\ell(\Xi) = n\, log(\alpha) + n\, log(\theta) - n\, log(\beta) - \theta\sum_{j=1}^{n}\frac{x_j}{\beta} - n\, log\,(log(1+\alpha))$$

$$-\sum_{j=1}^{n} log\left(1+\alpha\left(1-exp\left\{-\left(\frac{x}{\beta}\right)^{\theta}\right\}\right)\right). \tag{6}$$

ML estimators of Ξ are obtained as,

$$\hat{\Xi}_{ML} = \max_{(\alpha,\theta,\beta)\in([-1,+\infty]-\{0\})\times(0,+\infty)\times(0,+\infty)} \{\ell(\Xi)\}. \tag{7}$$

Then using (4), the following equation can be written as,

$$F(x_{(j)}) = \frac{log\left(1+\alpha\left(1-exp\left\{-\left(\frac{x_{(j)}}{\beta}\right)^{\theta}\right\}\right)\right)}{log(1+\alpha)}, \quad j = 1, 2, \cdots, n.$$

Let us define the following four functions which are utilized to get the other estimates:

$$Q_{LS}(\Xi) = \sum_{j=1}^{n}\left(F(x_{(j)};\Xi) - \frac{j}{n+1}\right)^2,$$

$$Q_{WLS}(\Xi) = \sum_{j=1}^{n}\frac{(n+2)(n+1)^2}{j(n-j+1)}\left(F(x_{(j)};\Xi) - \frac{j}{n+1}\right)^2,$$

$$Q_{CvM}(\Xi) = \frac{1}{12n} + \sum_{j=1}^{n}\left(F(x_{(j)};\Xi) - \frac{2j-1}{2n}\right)^2,$$

and

$$Q_{AD}(\Xi) = -n - \frac{1}{n}\sum_{j=1}^{n}\left\{(2j-1)log\left(F(x_{(j)};\Xi)\right)\right\} + \frac{1}{n}\sum_{j=1}^{n}\left\{log\left(1-F(x_{(j)};\Xi)\right)\right\}.$$

Then, the LS, WLS, CvM and AD estimator of Ξ are given, respectively, by

$$\widehat{\Xi}_{LSE} = \underset{(\alpha,\theta,\beta)\in([-1,+\infty]-\{0\})\times(0,+\infty)\times(0,+\infty)}{\arg\min}\{Q_{LS}(\Xi)\}, \tag{8}$$

$$\widehat{\Xi}_{WLSE} = \underset{(\alpha,\theta,\beta)\in([-1,+\infty]-\{0\})\times(0,+\infty)\times(0,+\infty)}{\arg\min}\{Q_{WLS}(\Xi)\}, \tag{9}$$

$$\widehat{\Xi}_{CvME} = \underset{(\alpha,\theta,\beta)\in([-1,+\infty]-\{0\})\times(0,+\infty)\times(0,+\infty)}{\arg\min}\{Q_{CvM}(\Xi)\}, \tag{10}$$

$$\widehat{\Xi}_{ADE} = \underset{(\alpha,\theta,\beta)\in([-1,+\infty]-\{0\})\times(0,+\infty)\times(0,+\infty)}{\arg\min}\{Q_{AD}(\Xi)\}. \tag{11}$$

All optimization problems given in Equations (7, 8–11) can be carried out with some numerical methods such as BFGS or Nelder-Mead.

4. Simulation study

Consider, N = 1000 trials of size n = 50,55,...,1000 from the ALWD with true parameter values α = 1, $\beta = 10$ and $\theta = 2$. All estimates are obtained via **constrOptim** routine in the R. The bias and mean square errors (MSEs) are computed by (for $\Xi = \alpha, \beta, \theta$)

$$Bias_{\Xi}(n) = \frac{1}{N}\sum_{j=1}^{N}(\Xi_j - \hat{\Xi}j),$$

and

$$MSE_{\Xi}(n) = \frac{1}{N}\sum_{j=1}^{N}(\Xi_j - \hat{\Xi}j)^2$$

respectively. The results are served up in Figures 2–4. From Figures 2–4 it can be said that estimates are biased but asymptotically unbiased. Also, as n increases, the bias and MSE of the estimators decrease as expected.

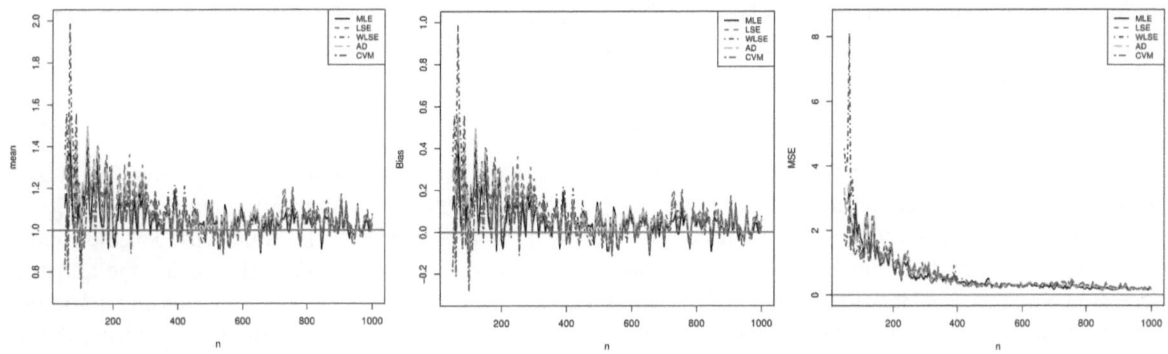

Figure 2: The empirical means, bias and MSEs of the parameter α.

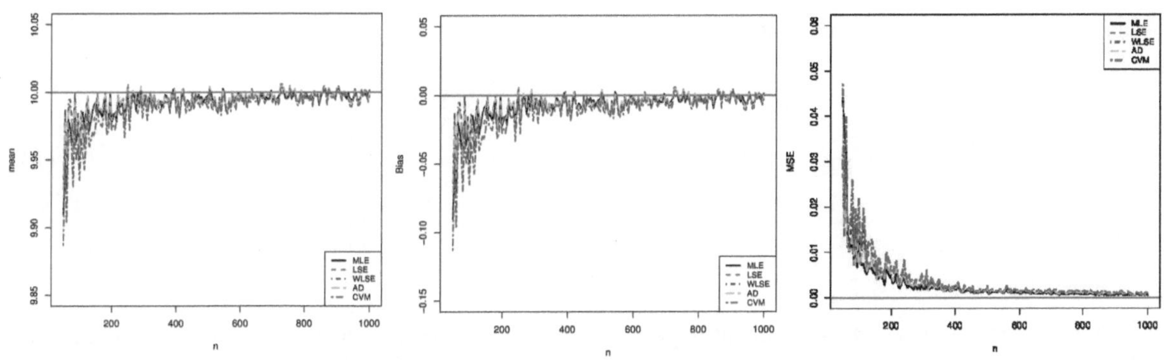

Figure 3: The empirical means, bias and MSEs of the parameter β.

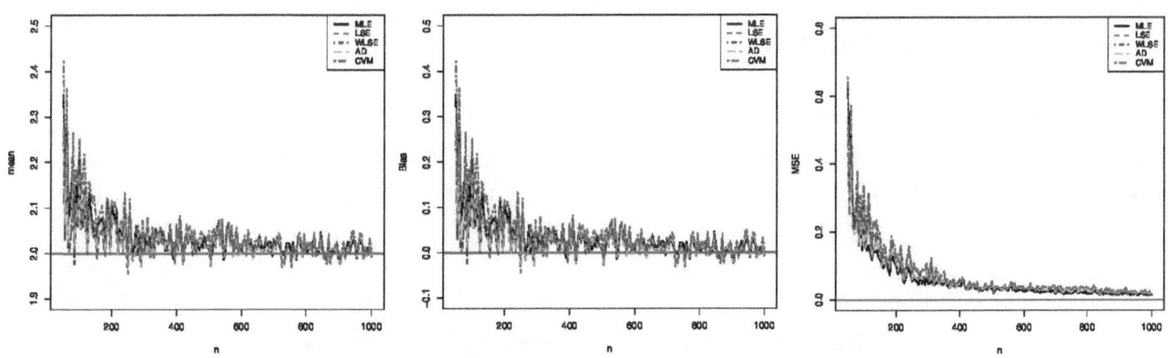

Figure 4: The empirical means, bias and MSEs of the parameter θ.

5. Alwd quantile regression model

When the data set is defined as the positive domain, the gamma regression model, proposed by Cuervo (2001), comes to mind to construct the linear relation between the response variable and independent variables (covariates, regressors). The gamma regression aims to model the conditional mean of a response variable via the covariates. On re-parameterizing the ordinary gamma distribution, the following pdf is obtained:

$$f(y,\alpha,\mu)=\frac{1}{y\Gamma(\alpha)}\left(\frac{\alpha y}{\mu}\right)^{\alpha}e^{-\alpha y/\mu}, y>0, \tag{12}$$

where, $\alpha > 0$ is the shape parameter and $\mu > 0$ is the expected value of the model. Now, using the appropriate link function the covariates can be linked to the mean of the model. In the gamma regression model, the regression structure is given by,

$$g(\mu_i) = x_i\, \gamma^T,$$

where, $\gamma = (\gamma_0, \gamma_1, \gamma_2, \ldots, \gamma_p)^T$ and $x_i = (1, x_{i1}, x_{i2}, x_{i3}, \ldots, x_{ip})$ are the unknown regression parameter vector and known i^{th} vector of the covariates. The function $g(z)$ is the link function. When the response variable has a skewed distribution or outliers, the gamma regression can be affected by these situations. So, the robust regression models will be more suitable than mean response regression model. For this reason, the quantile regression Koenker and Bassett (1978) can be seen as a good robust alternative model. Now, as an alternative to gamma mean response regression model, we propose the ALWD quantile regression model. Using the qf of the ALWD, let $\mu = x_u\ (\Xi)$ and $\beta = -\mu\ -\mu \left(\rightleftharpoons - log\left(\frac{\alpha + 1 - (1+\alpha)^u}{\alpha} \right) \right)^{-1/\theta}$ and be in (5). The following pdf is obtained based on this re-parametrization:

$$g(y,\alpha,\theta,\mu,u) = \frac{\alpha\theta\left(-log\left(\frac{\alpha+1-(1+\alpha)^u}{\alpha}\right)\right) y^{\theta-1} \left(\frac{\alpha+1-(1+\alpha)}{\alpha}\right)^{\left(\frac{y}{\mu}\right)^{\theta}}}{\mu^{\theta}\left(1+\alpha\left(1-\left(\frac{\alpha+1-(1+\alpha)^u}{\alpha}\right)^{\left(\frac{y}{\mu}\right)^{\theta}}\right)\right)\log(1+\alpha)} \tag{13}$$

where, the α, $\theta > 0$, $\mu > 0$, is the quantile parameter, and u is known. The random variable Y is denoted by $Y \sim ALWD(\alpha,\theta,\mu,u)$. This re-parametrization, in the next step, will link the covariates to the quantile of the ALWD random variable. Then, we use the log-link function to link the covariates to quantiles of the $ALWD(\alpha,\theta,\mu,u)$ model via for $log(\mu_i) = x_i\, \gamma^T$ for $i = 1,2\ldots n$. If the parameter is equal to 0.5, the covariates are linked to the conditional median of the response variable.

5.1 MLE method for the parameters of ALWD quantile regression model

Based on the ML estimates method, the log-likelihood function of the ALWD quantile regression model is given by (for $\Psi = (\alpha,\theta,\gamma)^T$ and $\mu_j = exp(x_j\, \gamma^T)$),

$$\ell(\Psi) = \text{n}\log\left[\frac{\alpha\theta\left(-log\left(\frac{\alpha+1-(1+\alpha)^u}{\alpha}\right)\right)}{\log(1+\alpha)}\right] + (\theta-1)\sum\nolimits_{j=1}^{n} log\, y_j - \theta\sum\nolimits_{j=1}^{n} log\, \mu_j$$
$$+ log\left(\frac{\alpha+1-(1+\alpha)^u}{\alpha}\right)\sum\nolimits_{j=1}^{n}\left(\frac{y_j}{\mu_j}\right)^{\theta} - \sum\nolimits_{j=1}^{n} log\left[1+\alpha\left(1+\alpha\left(1-\left(\frac{\alpha+1-(1+\alpha)^u}{\alpha}\right)^{\left(\frac{y}{\mu}\right)^{\theta}}\right)\right)\right] \tag{14}$$

Maximizing (14), the ML estimators can be obtained directly via the **maxLik** function in the R software. The standard errors of the estimated parameters can be obtained with this function also.

5.2 Model checking

After model fitting, the residual analysis plays an important role in model fitting. In order to do this, the randomized quantile residuals (rqrs) are focused on, Dunn and Smyth (1996). For $i = 1,\ldots, n$, the ith randomized quantile residual is defined by,

$$\hat{r}_j = \Phi^{-1}\,[G(y_j, \hat{\alpha}, \hat{\theta}, \hat{\mu}_j, u)],$$

where, the $G(y,\alpha,\theta,\mu,u)$ is the cdf of the re-parameterized ALWD and $\Phi^{-1}(x)$ is the quantile function of the standard normal distribution. If the model is valid then the rqrs have a $N(0,1)$ distribution.

6. Real data applications

In this section, a practical example is considered to observe the modeling capability of the ALWD. The number of failures for the air conditioning system of jet airplanes was examined. This data set was analyzed by Cordeiro and Lemonte (2011). Under ML estimates, we fit the ALWD to jet airplanes data and compare the ALWD with exponential (E), Weibull (W), gamma (G) and Rayleigh (R) based on the estimated log-likelihood values ($-2\hat{\ell}$), Akaike information criteria (AIC), Bayesian information criterion (BIC), Kolmogorov-Smirnov (K-S) goodness of fit statistic and related p-value. All calculations mentioned above are obtained by the BFGS command in R function optim. From Table 1, the ALWD can be selected as the best distribution because it has the lowest values of AIC, BIC, K-S and $\hat{\ell}$. These results show that the new proposed model has competitive power.

Second, we point out the applicability of the ALWD quantile regression model as an alternative to the gamma regression model. The used data set consists of the environment indicator of the Better Life Index (BLI) values of the OECD countries as well as Brazil, Russia and South Africa. The data set can be extracted from https://stats.oecd.org/index.aspx?DataSetCode=BLI2017.

The idea is to set the linear relation between water quality (WQ) and air pollution (AP). The regression structure has been set for μ_i given by,

$$logit(\mu_j) = \gamma_0 + \gamma_{1j} \quad \text{for } j = 1,\ldots,39$$

Table 2 indicates the ML estimates and their standard errors (SEs), and model selection criteria for the ALWD quantile and gamma regressions models.

From Table 2, for $u = 0.5$ quantile level, all coefficients of the models are statistically significant. Hence, the covariate has affected the response variable which is statistically significant at the usual significance levels. This affection has been seen as an opposite indication. Moreover, the log-likelihood values of the

Table 1. MLEs, $\hat{\ell}$ and goodness of fit statistics for the jet airplanes data (p-value given in (.)).

	Model				
Parameters	ALWD	E	W	G	R
$-2\hat{\ell}$	2070.0236	2076.4966	2073.5022	2075.2246	2382.5502
AIC	2076.0237	2078.4967	2077.5023	2079.2247	2384.5504
BIC	2085.7330	2081.7332	2083.9752	2085.6976	2387.7868
K-S	0.0547	0.0845	0.0572	0.0703	0.4003
	(0.6282)	(0.1368)	(0.5703)	(0.3114)	(0.0000)
$\hat{\alpha}$	2.0025			0.9049	
$\hat{\beta}$	74.6601		0.9109	0.0098	100.1541
$\hat{\lambda}$	0.6130	0.0108	87.7565		

Table 2. The results of ALWD quantile and Gamma regression models with model selection criteria.

	ALWD			Gamma		
Parameters	Estimate	SE	p-value	Estimate	SE	p-value
γ_0	–0.0631	0.0330	0.0555	–0.0392	0.0459	0.3930
γ_1	–0.0085	0.0025	0.0005	–0.0118	0.0031	0.0001
α	–0.9277	0.3222	0.0039	76.6523	0.7689	< 0.0001
θ	9.0136	2.9760	0.0025			
ℓ	40.2193			37.1693		
AIC	–72.4385			–68.3385		
BIC	–65.7843			–63.3478		

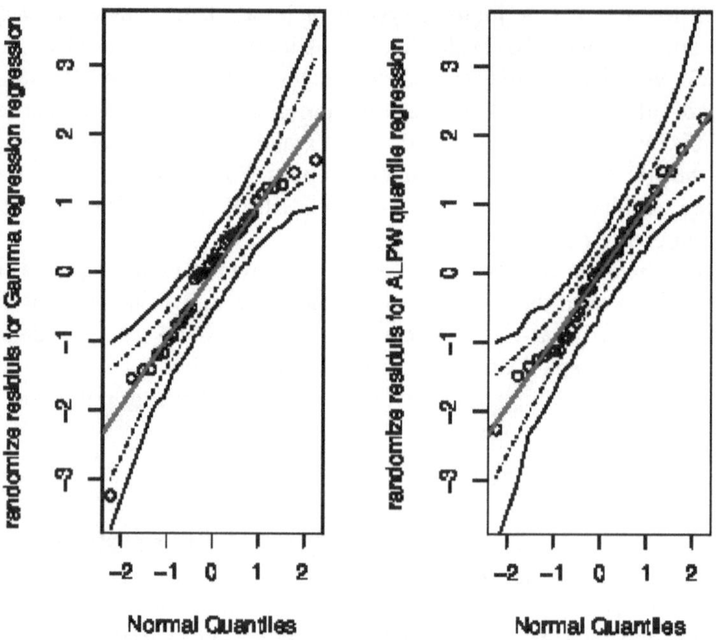

Figure 5: The Q-Q plot of the rqrs based on the fitted data set.

proposed quantile regression model are bigger than those of the gamma regression model with all the smallest likelihood-based statistics. These inferences are also supported by the half-normal plots for the Q-Q plots of the rqrs with simulated envelopes shown in Figure 5.

7. Conclusions

A new lifetime distribution based on the Weibull is proposed. The new distribution is derived from the ALT method. Many distributional properties of the new distribution have been studied. The unknown parameters of the new distribution have been estimated by five methods. A comprehensive simulation study was conducted to observe the performances of the five estimators. A new regression model has been proposed as an alternative to the Gamma regression model. Its usefulness in data modeling was shown via two applications to the real and regression data sets.

References

Alizadeh, M., Jamal, F., Yousof, H. M., M. Khanahmadi, G. G. Hamedani et al. (2020). Flexible Weibull generated family of distributions: characterizations, mathematical properties and applications. University Politehnica of Bucharest Scientific Bulletin-series A-Applied Mathematics and Physics, 82(1): 145–150.

Cordeiro, G. M. and Lemonte, A. J. (2011). The Birnbaum-Saunders distribution: An improved distribution for fatigue life modeling. Computational Statistics and Data Analysis, 55: 1445–1461.

Cuervo, E. C. (2001). Modelagem da variabilidade em modelos lineares generalizados. Doctoral dissertation, Tese de D. Sc., IM-UFRJ, Rio de Janeiro, RJ, Brasil.

Dunn, P. K. and Smyth, G. K. (1996). Randomized quantile residuals. Journal of Computational and Graphical Statistics, 5(3): 236–244.

Karakaya, K., Kinaci, I., Kus, C. and Akdogan, Y. (2017). A new family of distributions. Hacettepe Journal of Mathematics and Statistics, 46(2): 303–314.

Kinaci, İ., Kus, C., Karakaya, K. and Akdogan, Y. (2019). APT-Pareto Distribution and its Properties. Cumhuriyet Science Journal, 40(2): 378–387.

Koenker, R. and Bassett, G. Jr. (1978). Regression quantiles. Econometrica: Journal of the Econometric Society, 46(1): 33–50.

Korkmaz, M. C. (2020). The unit generalized half normal distribution: A new bounded distribution with inference and application. University Politehnica of Bucharest Scientific Bulletin-series A-Applied Mathematics and Physics, 82(2): 133–140.

Mahdavi, A. and Kundu, D. (2017). A new method for generating distributions with an application to exponential distribution. Communications in Statistics-Theory and Methods, 46(13): 6543–6557.

Mudholkar, G. S. and Srivastava, D. K. (1993). Exponentiated Weibull family for analyzing bathtub failure-rate data. IEEE Transactions on Reliability, 42(2): 299–302.

Shaked, M. and Shanthikumar, J. G. (2007). Stochastic Orders. Springer, New York, NY, USA.

Tanis, C., Saracoglu, B., Kus, C., Pekgor, A., Karakaya, K. et al. (2021). Transmuted lower record type frechet distribution with lifetime regression analysis based on Type I-censored data. Journal of Statistical Theory and Applications, 20(1): 86–96.

Nassar, M., Alzaatreh, A., Mead, M. and Abo-Kasem, O. (2017). Alpha power Weibull distribution: Properties and applications. Communications in Statistics-Theory and Methods, 46(20): 10236–10252.

Rasekhi, A., Rasekhi, M. and Hamedani, G.G. (2019). Incomplete Gamma Distribution: A new two Parameter Lifetime Distribution with Survival Regression Model. University Politehnica of Bucharest Scientific Bulletin-series A-Applied Mathematics and Physics, 81(2): 177–184.

Chapter 9

The Topp-Leone-G Power Series Distribution

Its Properties and Applications

Laba Handique,[1,*] *Subrata Chakraborty*[2] *and M Masoom Ali*[3]

MS Classification: 60E05; 62E15; 62F10.

1. Introduction

The Topp-Leone distribution is commonly used in many applied problems, particularly in lifetime data analysis that has attracted various statisticians as an alternative to the Beta distribution. A generalization of this distribution is the Topp-Leone-G family of distribution. The distribution function of the Topp-Leone-G family is given by Ali et al. (2016) as,

$$F^{\text{TLG}}(x;\alpha) = G(x)^{\alpha}[2-G(x)]^{\alpha} \;; \qquad x \in R,\ \alpha > 0 \tag{1}$$

The probability density function (pdf) of the corresponding (1) is given by,

$$f^{\text{TLG}}(x;\alpha) = 2\alpha\, g(x)\overline{G}(x)G(x)^{\alpha-1}[2-G(x)]^{\alpha-1} \tag{2}$$

where, $g(x)$ and $\overline{G}(x) = 1 - G(x)$ are the probability density function (pdf) and the survival function (sf) of the baseline distribution.

Roozegar and Nadaraja (2017) defined the Topp-Leone power series distribution with sf and pdf,

$$\overline{F}^{\text{TLPS}}(x;\lambda,\alpha,\beta) = \frac{\Phi(\lambda[1-G(x;\alpha,\beta)])}{\Phi(\lambda)} \text{ and } f^{\text{TLPS}}(x;\lambda,\alpha,\beta) = \frac{\lambda\, g(x;\alpha,\beta)\ \Phi'(\lambda[1-G(x;\alpha,\beta)])}{\Phi(\lambda)}$$

where, $g(x;\alpha,\beta)$ and $G(x;\alpha,\beta)$ are the pdf and cdf of the two parameter Topp-Leone distribution $\Phi(\lambda)$ and $\Phi'(\cdot)$ are defined below.

In this article, we introduce a new extension of the Topp-Leone-Power series family of distributions by compounding Topp-Leone-G by power series distributions, to get the Topp-Leone-G power series distribution.

Given $N = n$, let $X_i's(i = 1,2,\ldots, n)$ be independent and identically distributed (iid) random variables from the Topp-Leone-G family of distributions whose cdf is given by eq. (1). Consider N to be a discrete random

[1] Department of Statistics, Darrang College, Tezpur-784001, Assam, India.
[2] Department of Statistics, Dibrugarh University, Dibrugarh-784164, Assam, India.
[3] Department of Mathematical Sciences, Ball State University, Muncie, IN 47306 USA.
* Corresponding author: handiquelaba@gmail.com

variable independent of $X_i's$ from a power series distribution, truncated at zero, with the probability mass function given by,

$$P(N=n)=p_n=\frac{a_n\,\lambda^n}{\Phi(\lambda)}, \qquad n=1, 2, 3,... \tag{3}$$

where, $a_n \geq 0$ depends only on n and $\Phi(\lambda)=\sum_{n=1}^{\infty} a_n\,\lambda^n$, $\lambda \in (0,s)$ (s can be $+\infty$) such that $\Phi(\lambda)$ is finite and its first, second and third derivatives exist and are denoted by $\Phi(\cdot)$, $\Phi''(\cdot)$ and $\Phi'''(\cdot)$. Table 1 shows some useful quantities including a_n, $\Phi(\lambda)$, $\Phi'(\lambda)$ and $\Phi''(\lambda)$, $\Phi'''(\lambda)$ and $\Phi^{-1}(\lambda)$ for some power series distributions (truncated at zero) such as Poisson, geometric, logarithmic, binomial and negative binomial distributions.

Suppose that the failure time of each sub system has the cumulative distribution function (cdf) equation (1). Let Y_i denote the failure time of the ith subsystem and X denote the time to failure of the first out of the N functioning subsystems that is $X=\min\{Y_1, Y_2,..., Y_n\}$. Then the conditional cdf of X given N is $F(x/N=n)=1-\Pr(X>x/N=n)=1-P(Y_i>x)^n=1-[1-F^{\text{TLG}}(x;\alpha)]^n$. Observe that $X/N=n$ is a Kumaraswamy Tope-Leone-G (KwTLG) random variable with parameters α and n which we symbolized by $KwTLG(\alpha,n)$. So, the unconditional cdf of X (for $x>0$) can be expressed as,

$$F(x)=F^{\text{TLGPS}}(x;\alpha,\lambda;\Omega)=\sum_{n=1}^{\infty}\frac{a_n\,\lambda^n}{\Phi(\lambda)}\left(1-[1-F^{\text{TLG}}(x;\alpha)]^n\right)=1-\frac{\Phi(\lambda[1-F^{TLG}(x;\alpha)])}{\Phi(\lambda)} \tag{4}$$

This is the proposed Topp Leone-G power series (**in short,** TLGPS($\alpha,\lambda;\Omega$), **Ω is the parameter of baseline distribution *G*.**) distribution. The corresponding pdf of the TLGPS($\alpha,\lambda;\Omega$) family is given by,

$$f(x)=f^{\text{TLGPS}}(x;\alpha,\lambda;\Omega)=\frac{\lambda\, f^{\text{TLG}}(x;\alpha)\;\Phi'\left(\lambda[1-F^{\text{TLG}}(x;\alpha)]\right)}{\Phi\,(\lambda)} \tag{5}$$

The sf and hazard rate functions (hrf) of the TLGPS($\alpha,\lambda;\Omega$) distribution are,

$$\overline{F}^{\text{TLGPS}}(x;\alpha,\lambda;\Omega)=\frac{\Phi\left(\lambda[1-F^{\text{TLG}}(x;\alpha)]\right)}{\Phi\,(\lambda)} \text{ and } h^{\text{TLGPS}}(x;\alpha,\lambda;\Omega)=\frac{\lambda\, f^{\text{TLG}}(x;\alpha)\;\Phi'\left(\lambda[1-F^{\text{TLG}}(x;\alpha)]\right)}{\Phi\left(\lambda[1-F^{\text{TLG}}(x;\alpha)]\right)}$$

where, $\Phi(\lambda)$ and $\Phi'(\cdot)$ are defined above.
For $G(x)=x$, the TLGPS($\alpha,\lambda;\Omega$) reduces to TLPS(α,λ).

TLGPS($\alpha,\lambda;\Omega$) class of distributions contains several important families, including the Topp-Leone-G Poisson, Topp-Leone-G geometric, Topp-Leone-G logarithmic, Topp-Leone-G binomial and Topp-Leone-G negative binomial distributions. We derive some mathematical properties of this class.

Proposition 1: The $KwTLG(\alpha,c)$ distribution with parameters α and c is a limiting distribution of the TLGPS distribution.

Table 1: Useful quantities for some power series distributions.

Distribution	$\Phi\,(\lambda)$	$\Phi'(\lambda)$	$\Phi''(\lambda)$	$\Phi^{-1}(\lambda)$	a_n	s
Poisson	$e^{\lambda}-1$	e^{λ}	e^{λ}	$\log(1+\lambda)$	$(n!)^{-1}$	∞
Geometric	$\lambda(1-\lambda)^{-1}$	$(1-\lambda)^{-2}$	$2\,(1-\lambda)^{-3}$	$\lambda\,(1+\lambda)^{-1}$	1	1
Logarithmic	$-\log(1-\lambda)$	$(1-\lambda)^{-1}$	$(1-\lambda)^{-2}$	$1-e^{-\lambda}$	n^{-1}	1
Binomial	$(1+\lambda)^m-1$	$m(1+\lambda)^{m-1}$	$m(m-1)(1+\lambda)^{m-2}$	$(\lambda-1)^{\frac{1}{m}}-1$	$\binom{m}{n}$	∞
Negative Binomial	$\frac{\lambda^m}{(1-\lambda)^m}$	$\frac{m\,\lambda^{m-1}}{(1-\lambda)^{m+1}}$	$\frac{m(m+2\lambda-1)}{\lambda^{2-m}(1-\lambda)^{m+2}}$	$\frac{\lambda^{1/m}}{1+\lambda^{1/m}}$	$\binom{n-1}{m-1}$	1

i.e., $\lim_{x\to 0^+} F^{\text{TLGPS}}(x;\alpha,\lambda) = \lim_{x\to 0^+}\left\{1-\frac{\Phi(\lambda[1-F^{\text{TLG}}(x;\alpha)])}{\Phi(\lambda)}\right\} = 1-[1-F^{\text{TLG}}(x;\alpha)]^c$

where, $c = \min\{n \in N : a_n > 0\}$.

Proposition 2: The pdf of the TLGPS distribution can be written as an infinite mixture of the pdf of the *KwTLG*(α,*n*) distribution with parameters α and *n*.

Proof: By using eq. (5), we have,

$$f^{\text{TLGPS}}(x;\alpha,\lambda) = \frac{1}{\Phi(\lambda)}\sum_{n=1}^{\infty}\left\{na_n\lambda^n f^{\text{TLG}}(x;\alpha)[1-f^{\text{TLG}}(x;\alpha)^{n-1}]\right\} = \sum_{n=1}^{\infty} p_n h(x;\alpha,n), \tag{6}$$

since, $\Phi'(\lambda) = \sum_{n=1}^{\infty} n\,a_n\,\lambda^{n-1}$ where, $h(x;\alpha,n) = n\,f^{\text{TLG}}(x;\alpha)\;[1-F^{\text{TLG}}(x;\alpha)]^{n-1}$, is the pdf of *KwTLG*($\alpha$,*n*) and p_n is defined above.

Using the series expansion $(1-z)^k = \sum_{j=0}^{k}\binom{k}{j}(-z)^j$ we can, therefore, write,

$$f^{\text{TLGPS}}(x;\alpha,\lambda) = \sum_{n=1}^{\infty}\sum_{j=0}^{n-1} p_n\,\mu_j f^{\text{TLG}}(x;\alpha)\,[F^{\text{TLG}}(x;\alpha)]^j \tag{7}$$

$$= \sum_{n=1}^{\infty}\sum_{j=0}^{n-1}\frac{\xi_{n,j}}{j+1}\frac{d}{dx}[F^{\text{TLG}}(x;\alpha)]^{j+1} = \sum_{n=1}^{\infty}\sum_{j=0}^{n-1}\xi'_{n,j}\frac{d}{dx}[F^{\text{TLG}}(x;\alpha)]^{j+1}, \tag{8}$$

where, $\mu_j = n(-1)^j\binom{n-1}{j}$, $\xi'_{n,j} = \frac{1}{j+1}p_n\,\mu_j$ and $\xi_{n,j} = (j+1)\,\xi'_{n,j}$.

2. Special cases of the TLGPS distribution

Here, we study basic distributional properties of the Topp-Leone-G Poisson (TLGP), Topp-Leone-G geometric (TLGG), Topp-Leone-G logarithmic (TLGL), Topp-Leone-G binomial (TLGB) and Topp-Leone-G negative binomial (TLGNB) distributions as special cases of the TLGPS distribution. Table 2 expresses the pdf, sf, hrf and reversed hazard rate function (rhrf) of the TLGP, TLGG, TLGL, TLGB and TLGNB distributions. $g^{\text{TLG}}(x;\alpha)$ and $G^{\text{TLG}}(x;\alpha)$ denote the pdf and cdf of the Topp-Leone-G family of distributions, respectively. To illustrate the flexibility of the distributions, graphs of the pdf and hazard rate function for some selected values of the parameters are presented in Figures 1 and 2.

Special case of the TLGP and TLGG families of distributions with their pdf and cdf are listed along with their representative graphs for some selected values of the parameters are presented in Figures 1 to 4.

- **Topp-Leone-Weibull Poisson (TLWP) distribution**

Let the base line distribution be Weibull (Weibull, 1951) with parameters $\beta > 0$ and $\theta > 0$ having pdf and cdf, $g(x) = \beta\theta x^{\theta-1}e^{-\beta x^\theta}$ and $G(x) = 1-e^{-\beta x^\theta}$, $x > 0$, respectively. Then we get the pdf and hrf of TLWP(λ, α, β, θ) distribution, respectively, as,

$$f^{\text{TLWP}}(x;\lambda,\alpha,\beta,\theta) = \frac{2\lambda\alpha\beta\theta x^{\theta-1}e^{-2\beta x^\theta}(1-e^{-2\beta x^\theta})^{\alpha-1}\exp[\lambda\{1-(1-e^{-2\beta x^\theta})^\alpha\}]}{e^\lambda-1},$$

$$\text{and } h^{\text{TLWP}}(x;\lambda,\alpha,\beta,\theta) = \frac{2\lambda\alpha\beta\theta x^{\theta-1}e^{-2\beta x^\theta}(1-e^{-2\beta x^\theta})^{\alpha-1}\exp[\lambda\{1-(1-e^{-2\beta x^\theta})^\alpha\}]}{\exp[\lambda\{1-(1-e^{-2\beta x^\theta})^\alpha\}]-1}.$$

Table 2: pdf, sf, hrf and rhrf of the five special distributions.

Distribution	pdf	sf
TLGP	$\frac{\lambda g^{\text{TLG}}(x;\alpha)\exp[\lambda\{1-G^{\text{TLG}}(x;\alpha)\}]}{e^{\lambda}-1}$	$\frac{\exp[\lambda\{1-G^{\text{TLG}}(x;\alpha)\}]-1}{e^{\lambda}-1}$
TLGG	$\frac{(1-\lambda)g^{\text{TLG}}(x;\alpha)}{[1-\lambda\{1-G^{\text{TLG}}(x;\alpha)\}]^2}$	$\frac{(1-\lambda)[1-G^{\text{TLG}}(x;\alpha)]}{1-\lambda[1-G^{\text{TLG}}(x;\alpha)]}$
TLGL	$\frac{-\lambda g^{\text{TLG}}(x;\alpha)}{\{\ln(1-\lambda)\}[1-\lambda\{1-G^{\text{TLG}}(x;\alpha)\}]}$	$\frac{\ln[1-\lambda\{1-G^{\text{TLG}}(x;\alpha)\}]}{\ln(1-\lambda)}$
TLGB	$\frac{m\lambda g^{\text{TLG}}(x;\alpha)}{(1+\lambda)^m-1}[1+\lambda\{1-G^{\text{TLG}}(x;\alpha)\}]^{m-1}$	$\frac{[1+\lambda\{1-G^{\text{TLG}}(x;\alpha)\}]^m-1}{(1+\lambda)^m-1}$
TLGNB	$\frac{m(1-\lambda)^m g^{\text{TLG}}(x;\alpha)[1-G^{\text{TLG}}(x;\alpha)]^{m-1}}{[1-\lambda\{1-G^{\text{TLG}}(x;\alpha)\}]^{m+1}}$	$\frac{[(1-\lambda)\{1-G^{\text{TLG}}(x;\alpha)\}]^m}{[1-\lambda\{1-G^{\text{TLG}}(x;\alpha)\}]^m}$
Distribution	**hrf**	**rhrf**
TLGP	$\frac{\lambda g^{\text{TLG}}(x;\alpha)\exp[\lambda\{1-G^{\text{TLG}}(x;\alpha)\}]}{\exp[\lambda\{1-G^{\text{TLG}}(x;\alpha)\}]-1}$	$\frac{\lambda g^{\text{TLG}}(x;\alpha)\exp[\lambda\{1-G^{\text{TLG}}(x;\alpha)\}]}{e^{\lambda}-\exp[\lambda\{1-G^{\text{TLG}}(x;\alpha)\}]}$
TLGG	$\frac{g^{\text{TLG}}(x;\alpha)}{[1-\lambda\{1-G^{\text{TLG}}(x;\alpha)\}]\{1-G^{\text{TLG}}(x;\alpha)\}}$	$\frac{(1-\lambda)g^{\text{TLG}}(x;\alpha)}{1-\lambda[1-G^{\text{TLG}}(x;\alpha)]}$
TLGL	$\frac{-\lambda g^{\text{TLG}}(x;\alpha)[1-\lambda\{1-G^{\text{TLG}}(x;\alpha)\}]^{-1}}{\ln[1-\lambda\{1-G^{\text{TLG}}(x;\alpha)\}]}$	$\frac{\lambda g^{\text{TLG}}(x;\alpha)\ln(1-\lambda)}{[1-\lambda\{1-G^{\text{TLG}}(x;\alpha)\}](\ln[1-\lambda\{1-G^{\text{TLG}}(x;\alpha)\}])}$
TLGB	$\frac{m\lambda g^{\text{TLG}}(x;\alpha)[1+\lambda\{1-G^{\text{TLG}}(x;\alpha)\}]^{m-1}}{[1+\lambda\{1-G^{\text{TLG}}(x;\alpha)\}]^m-1}$	$\frac{m\lambda g^{\text{TLG}}(x;\alpha)[1+\lambda\{1-G^{\text{TLG}}(x;\alpha)\}]^{m-1}}{(1+\lambda)^m-[1+\lambda\{1-G^{\text{TLG}}(x;\alpha)\}]^m}$
TLGNB	$\frac{m\,g^{\text{TLG}}(x;\alpha)[1-G^{\text{TLG}}(x;\alpha)]^{-1}}{[1-\lambda\{1-G^{\text{TLG}}(x;\alpha)\}]}$	$\frac{m(1-\lambda)^m g^{\text{TLG}}(x;\alpha)[1-G^{\text{TLG}}(x;\alpha)]^{m-1}[1-\lambda\{1-G^{\text{TLG}}(x;\alpha)\}]^{-1}}{[1+\lambda\{1-G^{\text{TLG}}(x;\alpha)\}]^m-[1-G^{\text{TLG}}(x;\alpha)]^m(1-\lambda)^m}$

- Taking $\theta=1$, in TLWP(λ, α, β, θ) we get the TLEP(λ, α, β) with pdf and hrf, respectively, as,

$$f^{\text{TLEP}}(x;\lambda,\alpha,\beta)=\frac{2\lambda\alpha\beta e^{-2\beta x}(1-e^{-2\beta x})^{\alpha-1}\exp[\lambda\{1-(1-e^{-2\beta x})^{\alpha}\}]}{e^{\lambda}-1},$$

and $$h^{\text{TLEP}}(x;\lambda,\alpha,\beta)=\frac{2\lambda\alpha\beta e^{-2\beta x}(1-e^{-2\beta x})^{\alpha-1}\exp[\lambda\{1-(1-e^{-2\beta x})^{\alpha}\}]}{\exp[\lambda\{1-(1-e^{-2\beta x})^{\alpha}\}]-1}.$$

Obviously a large number of particular cases of TLGP distributions can be generated by assuming different baseline distributions.

Special cases of the TLGG family of distributions and list of their main distributional characteristics follow:

- **Topp-Leone-Weibull Geometric (TLWG) distribution**

Let the baseline distribution be Weibull (Weibull, 1951) with parameters $\beta>0$ and $\theta>0$ having pdf and cdf $g(x)=\beta\theta x^{\theta-1}e^{-\beta x^{\theta}}$ and $G(x)=1-e^{-\beta x^{\theta}}, x>0$, respectively. Then we get the pdf and hrf of the TLWG(λ, α, β, θ) distribution respectively as,

$$f^{\text{TLWG}}(x;\lambda,\alpha,\beta,\theta)=\frac{(1-\lambda)2\alpha\beta\theta x^{\theta-1}e^{-2\beta x^{\theta}}(1-e^{-2\beta x^{\theta}})^{\alpha-1}}{[1-\lambda\{1-(1-e^{-2\beta x^{\theta}})^{\alpha}\}]^2},$$

and $$h^{\text{TLWG}}(x;\lambda,\alpha,\beta,\theta)=\frac{2\alpha\beta\theta x^{\theta-1}e^{-2\beta x^{\theta}}(1-e^{-2\beta x^{\theta}})^{\alpha-1}}{[1-\lambda\{1-(1-e^{-2\beta x^{\theta}})^{\alpha}\}]\,[1-(1-e^{-2\beta x^{\theta}})^{\alpha}]}.$$

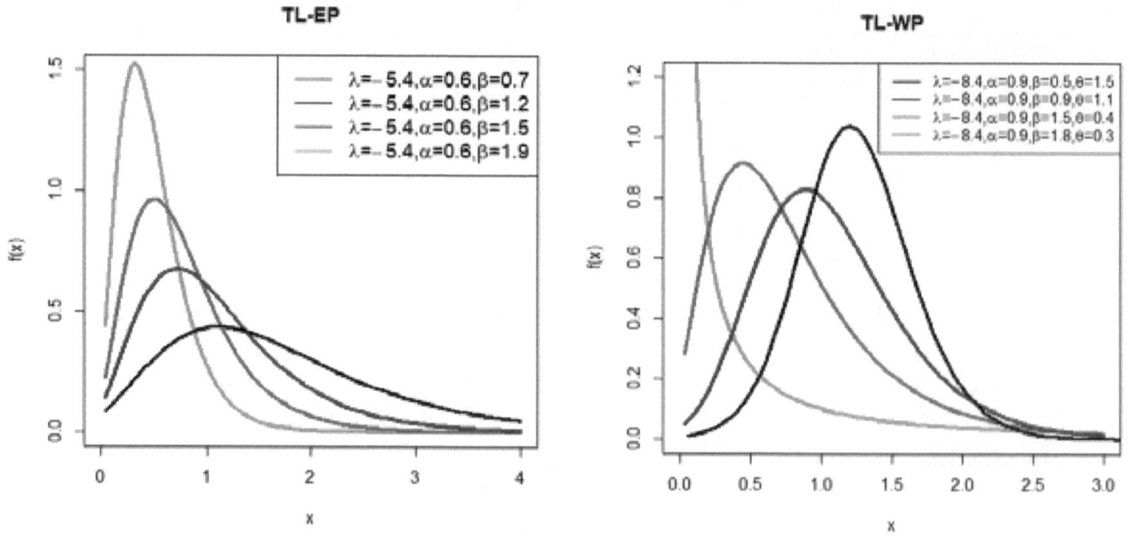

Figure 1: pdf plots of the TLEP and TLWP distribution

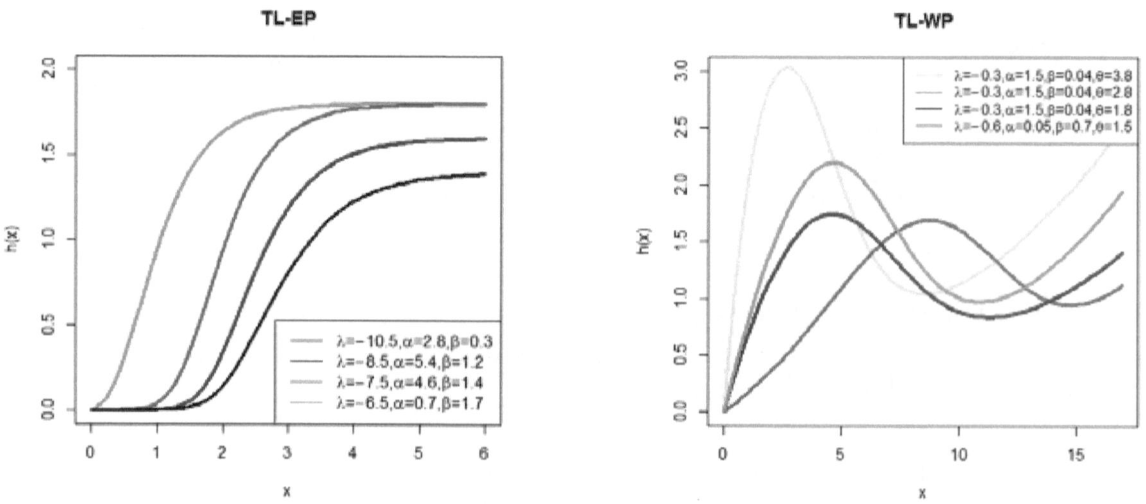

Figure 2: hrf plots of the TLEP and TLWP distribution

- Taking $\theta = 1$ in TLWG(λ, α, β, θ) we get the TLEG(λ, α, β, θ) with pdf and hrf, respectively, as,

$$f^{\text{TLEG}}(x;\lambda,\alpha,\beta) = \frac{(1-\lambda)2\alpha\,\beta\, e^{-2\beta x}(1-e^{-2\beta x})^{\alpha-1}}{[1-\lambda\{1-(1-e^{-2\beta x})^{\alpha}\}]^2},$$

$$\text{and } h^{\text{TLEG}}(x;\lambda,\alpha,\beta) = \frac{2\alpha\,\beta\, e^{-2\beta x}(1-e^{-2\beta x})^{\alpha-1}}{[1-\lambda\{1-(1-e^{-2\beta x})^{\alpha}\}]\ [1-(1-e^{-2\beta x})^{\alpha}]}.$$

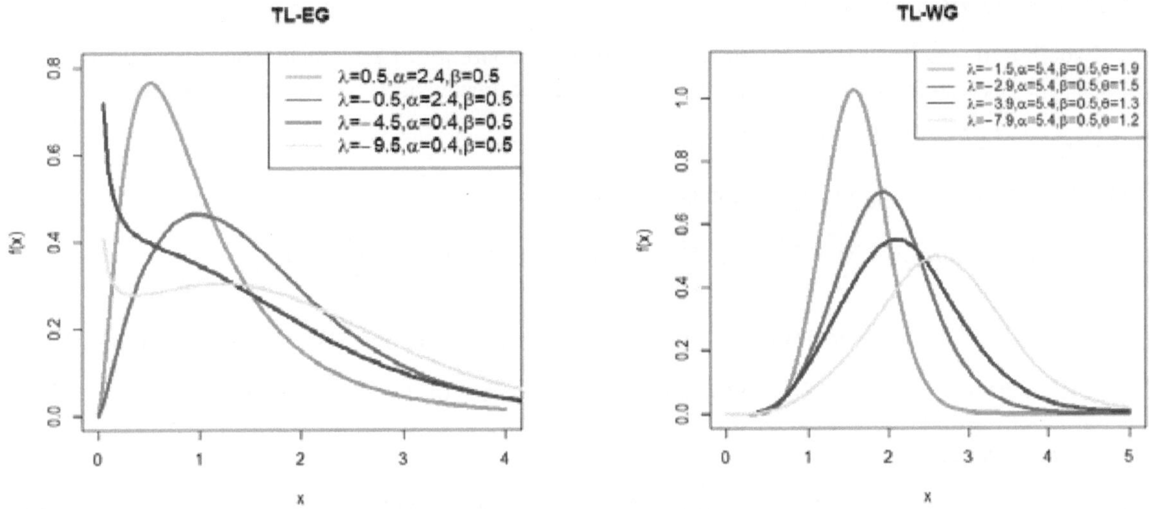

Figure 3: pdf plots of the TLEG and TLWG distribution

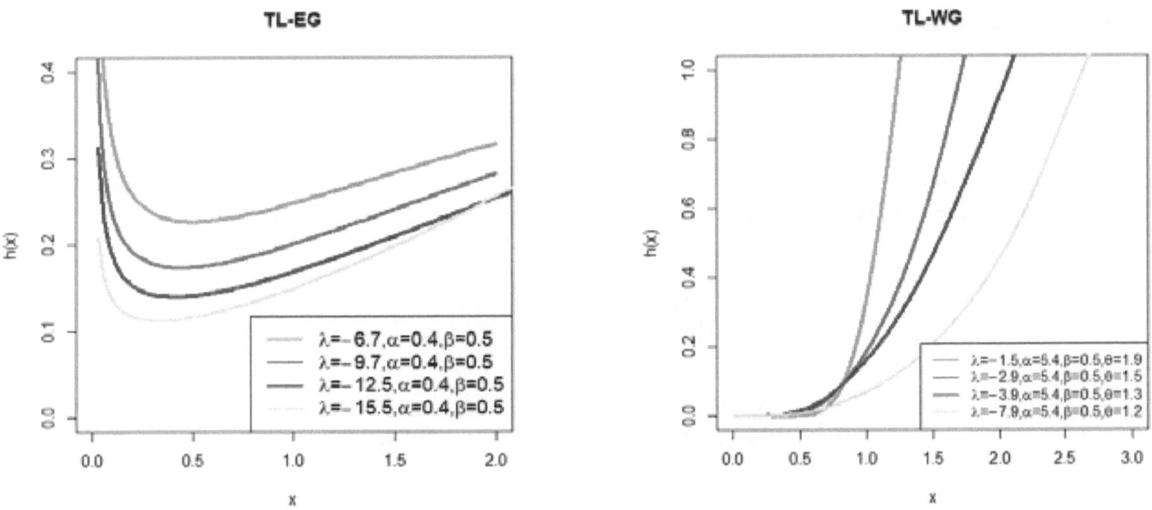

Figure 4: hrf plots of the TLEG and TLWG distribution

3. Statistical properties of TLGPS(α, λ)

3.1 Moment generating function

The moment generating function of TLGPS(α, λ) family can be easily expressed in terms of those of the exponentiated TLG(α) distribution using the results of proposition 2. For example using equation (8) it can be seen that,

$$M_X(s) = E[e^{SX}] = \int_{-\infty}^{\infty} e^{sx} f(x)\,dx$$

$$= \int_0^{\infty} e^{sx} \sum_{n=1}^{\infty} \sum_{j=0}^{n-1} \xi'_{n,j} \frac{d}{dx}[F^{\text{TLG}}(x;\alpha)]^{j+1} dx = \sum_{n=1}^{\infty} \sum_{j=0}^{n-1} \xi'_{n,j} \int_0^{\infty} e^{sx} \frac{d}{dx}[F^{\text{TLG}}(x;\alpha)]^{j+1} dx$$

$$= \sum_{n=1}^{\infty} \sum_{j=0}^{n-1} \xi'_{n,j} M_X(s)$$, where $M_X(s)$ is the mgf of a TLG (α) distribution

3.2 Power weighted moments

Greenwood et al. (1979) first proposed the idea of the probability of weighted moments (PWMs) as the expectations of certain functions of a random variable whose mean exists. The $(p, q, r)^{th}$ PWM of X is defined by $\Gamma_{p,q,r} = \int_{-\infty}^{\infty} x^p [F(x)]^q [1-F(x)]^r f(x)\,dx$.`

From equation (7) the sth moment of X can be written as,

$$E(X^s) = \int_0^{\infty} x^s \sum_{n=1}^{\infty}\sum_{j=0}^{n-1} \xi_{n,j}\, f^{\text{TLG}}(x;\alpha)\,[F^{\text{TLG}}(x;\alpha)]^j\, dx = \sum_{n=1}^{\infty}\sum_{j=0}^{n-1} \xi_{n,j}\, \Gamma_{s,\,j,\,0},$$

where, $\Gamma_{p,q,r} = \int_{-\infty}^{\infty} x^p \{F^{\text{TLG}}(x;\alpha)\}^q \{1-F^{\text{TLG}}(x;\alpha)\}^r f^{\text{TLG}}(x;\alpha)\,dx$ is the PWM of TLG(α) distribution. Thus, the moments of the TLGPS(α, λ) can be expressed in terms of the PWMs of TLG(α).

3.3 Order statistics

Consider a random sample $X_1, X_1, \cdots, X_n$, from any TLGPS(α, λ) istribution and let $X_{i:n}$ denote their i^{th} order statistic. The pdf of $X_{i:n}$ can be expressed as,

$$\begin{aligned} f_{i:n}(x) &= \frac{n!}{(i-1)!(n-i)!} f^{\text{TLGPS}}(x) F^{\text{TLGPS}}(x)^{i-1}\{1-F^{\text{TLGPS}}(x)\}^{n-i} \\ &= \frac{n!}{(i-1)!(n-i)!}\sum_{m=0}^{n-i}(-1)^m\binom{n-i}{m} f^{\text{TLGPS}}(x) F^{\text{TLGPS}}(x)^{m+i-1}. \end{aligned} \tag{9}$$

Now the cdf $F_{i:n}(x)$ of $f_{i:n}(x)$ can be expressed as $\left(\because f(x)F(x)^{a+b-1} = \frac{1}{a+b}\frac{d}{dx}[F(x)]^{a+b}\right)$

$$\begin{aligned} F_{i:n}(x) &= \frac{n!}{(i-1)!(n-i)!}\sum_{m=0}^{n-i}\binom{n-i}{m}\frac{(-1)^m}{m+i} F^{\text{TLGPS}}(x)^{m+i} \\ &= \frac{n!}{(i-1)!(n-i)!}\sum_{m=0}^{n-i}\binom{n-i}{m}\frac{(-1)^m}{m+i}\left[1-\frac{\Phi\left(\lambda[1-F^{\text{TLG}}(x;\alpha)]\right)}{\Phi(\lambda)}\right]^{m+i}. \end{aligned}$$

Taking $z = 1-\frac{\Phi\left(\lambda[1-F^{\text{TLG}}(x;\alpha)]\right)}{\Phi(\lambda)}$ we can write,

$$\begin{aligned} F_{i:n}(x) &= \frac{n!}{(i-1)!(n-i)!}\sum_{m=0}^{n-i}\binom{n-i}{m}\frac{(-1)^m}{m+i} z^{m+i} \\ &= \frac{n!\,z^i}{(i-1)!(n-i)!}\sum_{m=0}^{n-i}\frac{(i-n)_m\,\Gamma(m+i)}{\Gamma(m+i+1)}\frac{z^m}{m!} \\ &= \frac{n!\,z^i}{i(i-1)!(n-i)!}\sum_{m=0}^{n-i}\frac{(i-n)_m(i)_m}{(i+1)_m}\frac{z^m}{m!} \\ &= \frac{n!\,z^i}{i!(n-i)!}\,{}_2F_1(-n+i,\,i:i+1;z)\ \text{[Since for } m>n-i \text{ the summand is zero]} \\ &= \binom{n}{i} z^i\,{}_2F_1(-n+i,\,i:i+1;z), \text{ for all } 1\le i\le n \text{ and } x\ge 0, \text{ where,} \end{aligned}$$

$_2F_1(a, b:c;z) = \sum_{k=0}^{\infty} \frac{(a)_k\,(b)_k}{(c)_k} \frac{z^k}{k!}$ is the Gauss hypergeometric function and $(a)_i = \frac{\Gamma(a+i)}{\Gamma(i)}$ denotes the Pochhammer symbol with the convention that (0) = 1. Moreover, $F_{i:n}(x) = z^n$.

4. Estimation

In this section, estimation of the parameters of the TLGPS(α, λ; Ω) where Ω is the parameter of baseline distribution G is conducted using the maximum likelihood method. Let $x = (x_1, x_2,\ldots, x_n)$ be a random sample of size n from the TLGPS(α, λ; Ω), distribution with parameter vector $\rho = (\alpha, \lambda, \Omega)$, then the log-likelihood function for ρ is given by,

$$\ell = \ell(\rho) = 2n\log(\alpha\lambda) - n\log[\Phi(\lambda)] + \sum_{i=1}^{n}\log[g(x_i)] + \sum_{i=1}^{n}\log[\overline{G}(x_i)] + (\alpha-1)\sum_{i=1}^{n}\log[G(x_i)]$$

$$+(\alpha-1)\sum_{i=1}^{n}\log[2-G(x_i)] + \sum_{i=1}^{n}\log\Phi'\left(\lambda[1-G(x_i)^{\alpha}\{2-G(x_i)\}^{\alpha}]\right).$$

The MLEs are obtained by maximizing the log-likelihood function numerically by using an available function from R.

The asymptotic variance-covariance matrix of the MLEs of parameters can be obtained by inverting the Fisher information matrix I($\boldsymbol{\rho}$) which can be derived using the second partial derivatives of the log-likelihood function with respect to each parameter. The ij^{th} elements of $\mathrm{I}_n(\boldsymbol{\rho})$ are given by $\mathrm{I}_{ij} = -E[\partial^2 l(\boldsymbol{\rho})\,\partial\rho_i\,\partial\rho_j]$, $i,j = 1,\ldots, 2+q$, where q is the number of parameters in the G-family.

The exact evaluation of the above expectations may be cumbersome. In practice one can estimate $\mathrm{I}_n(\boldsymbol{\rho})$ by the observed Fisher's information matrix $\hat{\mathrm{I}}_n(\hat{\boldsymbol{\rho}}) = (\hat{I}_{i,j})$ defined as,

$$\hat{\mathrm{I}}_{ij} \approx \left(-\partial^2 l(\boldsymbol{\rho})/\partial\rho_i\partial\rho_j\right)_{\boldsymbol{\eta}=\hat{\boldsymbol{\eta}}}, \qquad i, j = 1,\ldots, 2+q.$$

Using the general theory of MLEs under some regularity conditions on the parameters as $n \to \infty$ the asymptotic distribution of $\sqrt{n}(\hat{\boldsymbol{\rho}} - \boldsymbol{\rho})$ is $N_k(0, V_n)$ where $V_n = (v_j) = \mathrm{I}_n^{-1}(\boldsymbol{\rho})$. The asymptotic behaviour remains valid if V_n is replaced by $\hat{V}_n = \hat{\mathrm{I}}^{-1}(\hat{\boldsymbol{\rho}})$. Using this result large sample standard errors of the jth parameter ρ_j are given by $\sqrt{\hat{v}_{jj}}$.

5. Applications

In this section, we use two real data sets to show that the TLEP distribution can be a better model than the ones based on exponential (Exp), moment exponential (ME), Marshall-Olkin exponential (MO-E) (Marshall and Olkin, 1997), generalized Marshall-Olkin exponential (GMO-E) (Jayakumar and Mathew, 2008), Kumaraswamy exponential (Kw-E) (Cordeiro and de Castro, 2011), Beta exponential (BE) (Eugene et al., 2002), Marshall-Olkin Kumaraswamy exponential (MOKw-E) (Handique et al., 2017), Kumaraswamy Marshall-Olkin exponential (KwMO-E) (Alizadeh et al., 2015), Beta Poisson exponential and Kumaraswamy Poisson exponential (KwP-E) (Chakraborty et al., 2022) distributions. The first data set represents the survival times (in days) of 72 guinea pigs infected with virulent tubercle bacilli, observed and reported by Bjerkedal (1960). The second failure time data set is about the relief times (in minutes) of patients receiving an analgesic. This data set of twenty (20) observations was reported by Gross and Clark (1975). Here some well known model selection criteria namely the AIC, BIC, CAIC and HQIC along with the Kolmogorov-Smirnov (K-S) statistics, Anderson-Darling (A) and Cramer von-mises (W) for goodness of fit are used to compare the fitted models. The findings are summarized in Tables 4, 5, 6 and 7 with visual comparison fitted density and the fitted cdf presented in Figures 6 and 7.

TTT plots and Descriptive Statistics for the data sets:

The TTT plots (see Aarset, 1987) for the data sets Figure 5 clearly indicate that both the data sets have an increasing hazard rate.

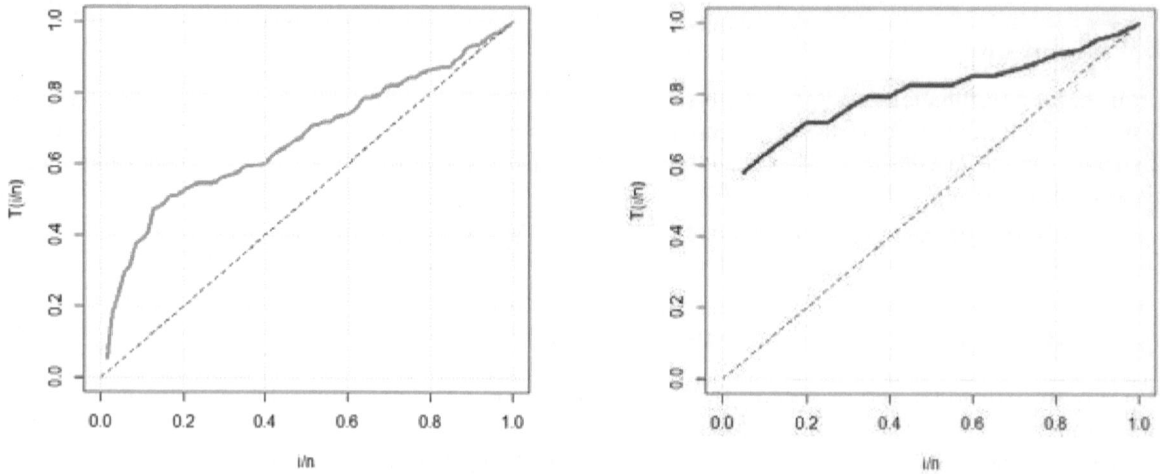

Figure 5: TTT-plots for the Data set I and Data set II.

Table 3: Descriptive Statistics for the data sets I and II.

Data sets	n	Min.	Mean	Median	s.d.	Skewness	Kurtosis	1st Qu.	3rd Qu.	Max.
I	72	0.100	1.851	1.560	1.200	1.788	4.157	1.080	2.303	7.000
II	20	1.100	1.900	1.700	0.704	1.592	2.346	1.475	2.050	4.100

Table 4: MLEs, standard errors, (in parentheses) values for the guinea pigs survival time's data set I.

Models	$\hat{\lambda}$	$\hat{\alpha}$	$\hat{a}$	$\hat{b}$	$\hat{\beta}$
Exp	---	---	---	---	0.540
(β)					(0.063)
ME	---	---	---	---	0.925
(β)					(0.077)
MO-E	---	8.778	---	---	1.379
(α,β)		(3.555)			(0.193)
GMO-E	0.179	47.635	---	---	4.465
(λ,α,β)	(0.070)	(44.901)			(1.327)
Kw-E	---	---	3.304	1.100	1.037
(a,b,β)			(1.106)	(0.764)	(0.614)
B-E	---	---	0.807	3.461	1.331
(a,b,β)			(0.696)	(1.003)	(0.855)
MOKw-E	---	0.008	2.716	1.986	0.099
(α,a,b,β)		(0.002)	(1.316)	(0.784)	(0.048)
KwMO-E	---	0.373	3.478	3.306	0.299
(α,a,b,β)		(0.136)	(0.861)	(0.779)	(1.112)
BP-E	0.014	---	3.595	0.724	1.482
(a,b,λ,β)	(0.010)		(1.031)	(1.590)	(0.516)
KwP-E	4.001	---	3.265	2.658	0.177
(a,b,λ,β)	(5.670)		(0.991)	(1.984)	(0.226)
TL-EP	-9.786	0.438	---	---	0.572
(λ,α,β)	(9.422)	(0.533)			(0.085)

Table 5: Log-likelihood, AIC, BIC, CAIC, HQIC, A, W and KS (p-value) values for the guinea pigs survival times data set I.

Models	AIC	BIC	CAIC	HQIC	A	W	KS (p-value)
Exp (β)	234.63	236.91	234.68	235.54	6.53	1.25	0.27 (0.06)
ME (β)	210.40	212.68	210.45	211.30	1.52	0.25	0.14 (0.13)
MO-E (α,β)	210.36	214.92	210.53	212.16	1.18	0.17	0.10 (0.43)
GMO-E (λ,α,β)	210.54	217.38	210.89	213.24	1.02	0.16	0.09 (0.51)
Kw-E (a,b,β)	209.42	216.24	209.77	212.12	0.74	0.11	0.08 (0.50)
B-E (a,b,β)	207.38	214.22	207.73	210.08	0.98	0.15	0.11 (0.34)
MOKw-E (α,a,b,β)	209.44	218.56	210.04	213.04	0.79	0.12	0.10 (0.44)
KwMO-E (α,a,b,β)	207.82	216.94	208.42	211.42	0.61	0.11	0.08 (0.73)
BP-E (a,b,λ,β)	205.42	214.50	206.02	209.02	0.55	0.08	0.09 (0.81)
KwP-E (a,b,λ,β)	206.63	215.74	207.23	210.26	0.48	0.07	0.09 (0.79)
TL-EP (λ,α,β)	204.51	211.34	204.86	207.23	0.44	0.05	0.07 (0.84)

Table 6: MLEs, standard errors, (in parentheses) values for failure time data set II.

Models	$\hat{\lambda}$	$\hat{\alpha}$	$\hat{a}$	$\hat{b}$	$\hat{\beta}$
Exp (β)	---	---	---	---	0.526 (0.117)
ME (β)	---	---	---	---	0.950 (0.150)
MO-E (α,β)	---	54.474 (35.582)	---	---	2.316 (0.374)
GMO-E (λ,α,β)	0.519 (0.256)	89.462 (66.278)	---	---	3.169 (0.772)
Kw-E (a,b,β)	---	---	83.756 (42.361)	0.568 (0.326)	3.330 (1.188)
B-E (a,b,β)	---	---	81.633 (120.41)	0.542 (0.327)	3.514 (1.410)
MOKw-E (α,a,b,β)	---	0.133 (0.332)	33.232 (57.837)	0.571 (0.721)	1.669 (1.814)
KwMO-E (α,a,b,β)	---	28.868 (9.146)	34.826 (22.312)	0.299 (0.239)	4.899 (3.176)
BP-E (a,b,λ,β)	1.965 (0.341)	---	13.396 (1.494)	9.600 (1.091)	0.244 (0.037)
KwP-E (a,b,λ,β)	5.983 (1.470)	---	11.837 (6.493)	3.596 (2.392)	0.225 (0.098)
TL-EP (λ,α,β)	2.096 (2.208)	26.116 (19.801)	---	---	0.866 (0.320)

Table 7: Log-likelihood, AIC, BIC, CAIC, HQIC, A, W and KS (*p*-value) values for the failure time data set II.

Models	AIC	BIC	CAIC	HQIC	A	W	KS (*p*-value)
Exp(β)	67.67	68.67	67.89	67.87	4.60	0.96	0.44 (0.004)
ME(β)	54.32	55.31	54.54	54.50	2.76	0.53	0.32 (0.07)
MO-E(α,β)	43.51	45.51	44.22	43.90	0.81	0.14	0.18 (0.55)
GMO-E(λ,α,β)	42.75	45.74	44.25	43.34	0.51	0.08	0.15 (0.78)
Kw-E(a,b,β)	41.78	44.75	43.28	42.32	0.45	0.07	0.14 (0.86)
B-E(a,b,β)	43.48	46.45	44.98	44.02	0.70	0.12	0.16 (0.80)
MOKw-E(α,a,b,β)	41.58	45.54	44.25	42.30	0.60	0.11	0.14 (0.87)
KwMO-E(α,a,b,β)	42.88	46.84	45.55	43.60	1.08	0.19	0.15 (0.86)
BP-E(a,b,λ,β)	38.07	42.02	40.73	38.78	0.39	0.06	0.14 (0.91)
KwP-E(a,b,λ,β)	38.32	42.28	40.98	39.04	0.41	0.05	0.13 (0.93)
TL-EP(λ,α,β)	37.65	40.63	39.15	38.23	0.22	0.03	0.10 (0.97)

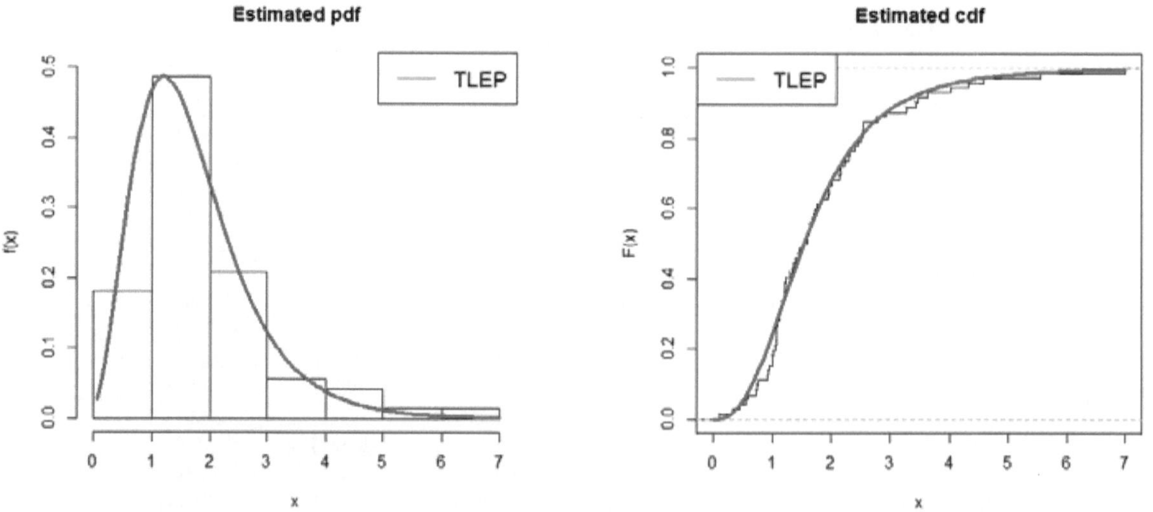

Figure 6: Plots of the observed histogram and estimated pdf on left and observed ogive and estimated cdf on right data set I.

From these findings based on the lowest values and different criteria the TLEP is found to be a better model than the models Exp, ME, MO-E, GMO-E, Kw-E, B-E, MOKw-E, KwMO-E, BP-E and KwP-E for the survival time data set. Thus the proposed distributions provide a comparatively closer fit to these data sets.

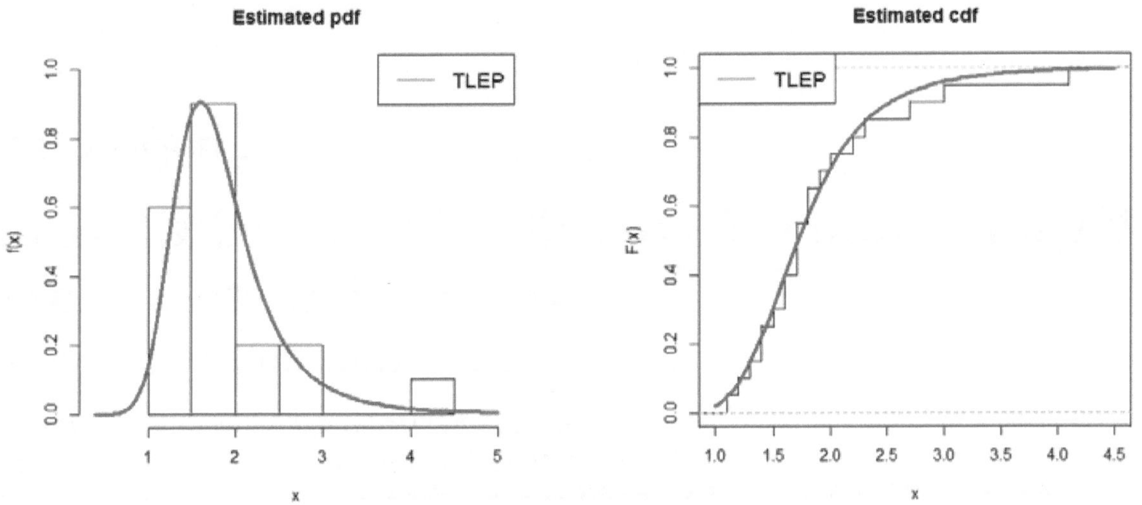

Figure 7: Plots of the observed histogram and estimated pdf on left and observed ogive and estimated cdf on right data set II.

6. Conclusion

Topp-Leone Power series distribution is extended by replacing the Topp-Leone distribution by Topp-Leone Generalized distribution. Some mathematical properties of the family of distributions are derived. Maximum likelihood estimation is discussed. Two real life applications are presented to showcase the advantage of members of the class of distributions over some competing distributions.

Reference

Aarset, M. V. (1987). How to identify a bathtub hazard rate. Ieee T Reliab., 36: 106–108.

Alizadeh, M., Tahir, M. H., Cordeiro, G. M., Zubair, M., Hamedani, G. G. et al (2015). The Kumaraswamy Marshal-Olkin family of distributions. Journal of the Egyptian Mathematical Society, 23: 546–557.

Ali, Al, Arif, O., Shawky, A., Hanif, S., Shahbaz, M. Q. et al. (2016). Topp-Leone Family of Distributions: Some Properties and Application. Pak. J. Stat. Oper. Res., 3: 443–452.

Bjerkedal, T. (1960). Acquisition of resistance in Guinea pigs infected with different doses of virulent tubercle bacilli. Am. J. Hyg., 72: 130–148.

Cordeiro, G. M. and De Castro, M. (2011). A new family of generalized distributions. J. Stat. Comput. Simul., 81: 883–893.

Chakraborty, S., Handique, L. and Jamal, F. (2022). The Kumaraswamy Poisson-G family of distribution: its properties and applications. Ann. Data. Sci., 9: 229–247.

Eugene, N., Lee, C. and Famoye, F. (2002). Beta-normal distribution and its applications. Commun. Statist. Theor. Meth., 31: 497–512.

Gross, A. J. and Clark, V. A. (1975). Survival distributions, reliability applications in the biometrical sciences. New York: Wiley.

Greenwood, J. A., Landwehr, J. M., Matalas, N. C. and Wallis, J. R. (1979). Probability weighted moments: definition and relation to parameters of several distributions expressible in inverse form. Water Resour. Res., 15: 1049–1054.

Handique, L., Chakraborty, S. and Hamedani, G. G. (2017). The Marshall-Olkin-Kumaraswamy-G family of distributions. J. Stat. Theory Appl., 16: 427–447.

Jayakumar, K. and Mathew, T (2008). On a generalization to Marshall-Olkin scheme and its application to Burr type XII distribution. Stat. Pap., 49: 421–439.

Marshall, A. and Olkin, I. (1997). A new method for adding a parameter to a family of distributions with applications to the exponential and Weibull families. Biometrika, 84: 641–652.

Rasool, R. and Nadarajah, S. (2017). A new class of topp-leone power series distributions with reliability application. J. Fail. Anal. and Preven. DOI 10.1007/s11668-017-0329-9.

Weibull, W. (1951). A statistical distribution functions of wide applicability. J. Appl. Mech., 18: 293–297.

Chapter 10

Exponentiated Generalized General Class of Inverted Distributions

Estimation and Prediction

Abeer A EL-Helbawy, Gannat R AL-Dayian, Asmaa M Abd AL-Fattah* and *Rabab E Abd EL-Kader*

1. Introduction

There are different types of lifetime data in reliability studies, lifetime testing, human mortality studies, engineering modeling, electronic sciences and biological studies. Thus, different shapes of lifetime distributions are needed to fit these types of lifetime data. Various extensions and modifications have been suggested by researchers to construct new families of distributions which provide more flexibility and applicability than the existing distributions. These new families have been used for modeling data in many applied areas, i.e., engineering, economics, biological studies, environmental sciences, lifetime analysis, finance and insurance. Different methods for constructing, extending and generalizing lifetime distributions are discussed and presented. [For more details see Lai (2013)].

Some generalized distributions are constructed by adding new parameters to the baseline model which are useful in deriving general results that could be applied to special cases to obtain new results. Marshall-Olkin generated family with a positive parameter was added to a general survival function by Marshall and Olkin (1997), the beta generating family introduced by Eugene et al. (2002). Also, Kumaraswamy generating family was presented by Jones (2009) and *Exponentiated Generalized* was provided by Cordeiro et al. (2013) as an extension of the exponentiated type distribution. Transformed-Transformer (T-X) and exponentiated (T-X) were considered by Alzaatreh et al. (2013). Yousof et al. (2015) introduced the transmuted EG-G family of distributions. The alpha power transformation distribution has recently been proposed by Mahdavi and Kundu (2017). Sindi et al. (2017) studied exponentiated general class of distributions, Oluyede et al. (2020) obtained the EG power series family of distributions and a new method for generating distributions which combines two techniques: the (T-X) and alpha power transformation approaches by Baharith and Aljuhani (2021).

The inverted distributions have a wide range of applications; in problems related to econometrics, biological sciences, survey sampling, engineering sciences, medical research, and life testing problems.

Statistics Department, Faculty of Commerce, AL-Azhar University, Girls' Branch – Cairo.
Emails: aldayian@yahoo.com; asmaa.stat@gmail.com; rbb_ibrahim@yahoo.com
* Corresponding author: aah_elhelbawy@hotmail.com

Many authors focused on the exponentiated distributions and its applications; for example, Nadarajah and Kotz (2006), Ali et al. (2007), Silva et al. (2010), Lemonte et al. (2013), Elgarhy and Shawki (2017) and Rather and Subramanian (2018).

Cordeiro et al. (2013) proposed a class of distributions as an extension of the exponentiated type distribution which has great flexibility in its tails and can be widely applied in many areas of biology and engineering. Given a non-negative continuous random variable X, the *cumulative distribution function* (cdf) of the *EG general class* (EGGC) of distribution is defined by,

$$F(x;\alpha,\beta) = \{1 - [1 - G(x)]^{\alpha}\}^{\beta}, \quad \alpha,\beta > 0, \tag{1}$$

where α and β are additional shape parameters, the corresponding *probability density function* (pdf) for (1) is given by,

$$f(x;\alpha,\beta) = \alpha\beta g(x)[1 - G(x)]^{\alpha-1}\{1 - [1 - G(x)]^{\alpha}\}^{\beta-1}, \quad \alpha,\beta > 0. \tag{2}$$

By setting $\alpha = 1$ in (1), the exponentiated type distributions is obtained, derived by Gupta et al. (1998); further the exponentiated exponential and exponentiated gamma distributions can be derived if is the exponential or gamma cdf. If $\beta = 1$ in (1) and $G(x)$ is the Gumbel or Fréchet cdf, then, one can get the exponentiated Gumbel and exponentiated Fréchet distributions, respectively, as defined by Nadarajah and Kotz (2006). Thus, the class of distributions (1) extends to both exponentiated type distributions.

The general class of distributions is important to obtain a general result that could be applied to special cases in obtaining new results. It is more flexible in dealing with statistical problems. Many authors focused on the generalized and EG distributions and its applications; for example, Oguntunde et al. (2014), Yousof et al. (2015), De Andrade et al. (2016), Mustafa et al. (2016) and Sindi et al. (2017).

Abd EL-Kader (2013) proposed a general class of inverted distributions with a positive domain which have many applications in survival analysis. Abd El-Kader general class has the pdf and cdf as given below

$$G(t \mid \underline{\theta}) = exp[-\lambda(t)], \; t > 0, \tag{3}$$

and

$$g(t \mid \underline{\theta}) = [-\grave{\lambda}(t)] \; exp[-\lambda(t)], \quad t > 0, \tag{4}$$

where, $\lambda(t) \equiv \lambda(t,\underline{\theta})$ is a non-negative continuous function of t^{-1} such that $\lambda(t) \to 0^{-}$ as $t \to 0^{+}$ and $\lambda(t) \to \infty$ as $t \to \infty$, $\grave{\lambda}(t)$ is the derivative of $\lambda(t)$ with respect to t.

Burkschat et al. (2003) studied the *dual generalized order statistics* (dgos) that enable a common approach to descending ordered random variables as reversed ordered order statistics, lower record models and lower Pfeifer records.

Let $X_{(1,n,m,k)}, X_{(2,n,m,k)}, \ldots X_{(n,n,m,k)}$ be n dgos from an absolute cdf with a corresponding pdf. Then, the joint pdf has the form,

$$f_{X(1,n,m,k)}, X_{(2,n,m,k)}, \ldots X_{(n,n,m,k)}(x_{(1)},\ldots,x_{(n)}) = k\left(\prod_{j=1}^{n-1}\gamma_j\right)\left\{\prod_{i=1}^{n-1}[F(x_{(i)})]^{m} f(x_{(i)})\right\}[F(x_{(n)})]^{k-1} f(x_{(n)}), \tag{5}$$

where $F^{-1}(1) \geq x_{(1)} \ldots \geq x_{(n)} \geq F^{-1}(0)$, $n \in N$, $k \geq 1$, $m_1,\ldots,m_{n-1} = m$, $m \in \mathbb{R}$ is the parameters such that $\gamma_r = k + (n - r)(m + 1) \geq 1$, *for all* $1 \leq r \leq n$.

The marginal pdf of the rth dgos $X(r,n,m,k)$, $1 \leq r \leq n$, is given by

$$f^{(r,n,m,k)}(x_{(r)}) = \frac{\varsigma_{r-1}}{(r-1)!}[F(x_{(r)})]^{\gamma_r - 1} f(x_{(r)}) g_m^{r-1}[F(x_{(i)})], \tag{6}$$

where, $\zeta_{r-1} = \prod_{j1=1}^{r} \gamma_{j1}$, $g_m(x) = h_m(x) - h_m(1)$, $x \in [0,1)$,

$$h_m(x) = \begin{cases} -\dfrac{1}{m+1} x^{m+1}, & m \neq -1, \\ -\ln x, & m = -1. \end{cases} \tag{7}$$

2. Exponentiated generalized general class of inverted distributions

In this section, the new EGGC of inverted distributions is introduced. Using (1)-(3), the cdf and pdf can be derived as follows:

$$F(t;\alpha,\beta) = (1 - \{1 - exp[-\lambda(t)]\}^{\alpha})^{\beta}, \qquad t > 0; \alpha,\beta > 0, \tag{8}$$

and

$$f(t;\alpha,\beta) = \alpha\beta g(t) exp[-\lambda(t)]\ \{1 - exp[-\lambda(t)]\}^{\alpha-1}\ (1 - \{1 - exp[-\lambda(t)]\}^{\alpha})^{\beta-1}, \quad t > 0;\ \alpha,\beta > 0, \tag{9}$$

where $g(t) = -\grave{\lambda}(t)$ is the first derivative of $\lambda(t)$ with respect to t.

The *reliability function* (rf), *hazard rate function* (hrf) and *reversed hazard rate function* (rhrf) are given, respectively, by,

$$R(t;\alpha,\beta) = 1 - (1 - \{1 - exp[-\lambda(t)]\}^{\alpha})^{\beta}, \qquad t > 0;\ \alpha,\beta > 0, \tag{10}$$

$$h(t;\alpha,\beta) = \frac{\alpha\beta g(t) exp[-\lambda(t)]\{1 - exp[-\lambda(t)]\}^{\alpha-1}(1 - \{1 - exp[-\lambda(t)]\}^{\alpha})^{\beta-1}}{1 - (1 - \{1 - exp[-\lambda(t)]\}^{\alpha})^{\beta-1}}, \quad t > 0; \alpha,\beta > 0. \tag{11}$$

and

$$rh(t;\alpha,\beta) = \alpha\beta g(t) exp[-\lambda(t)]\ \{1 - exp[-\lambda(t)]\}^{\alpha-1}(1 - \{1 - exp[-\lambda(t)]\}^{\alpha})^{-1}, \quad t > 0;\ \alpha,\beta > 0. \tag{12}$$

Table 1: Some resulting distributions of EGGC of inverted distributions

$\lambda(t)$	$F(t)$	**The resulting distribution**
$\left(\frac{\gamma}{t}\right)^{\theta}$	$\left(1 - \left\{1 - exp\left[-\left(\frac{\gamma}{t}\right)^{\theta}\right]\right\}^{\alpha}\right)^{\beta}$	EG-inverted Weibull $(\alpha,\beta,\gamma,\theta)$ [Elbatal and Muhammed (2014)]
$(\gamma t)^{-2}$	$(1 - \{1 - exp[-(\gamma t)^{-2})]\})^{\beta}$	EG- inverted Rayleigh (α,β,γ) [Fatima et al. (2018)]
$e^{-\frac{t-\gamma}{\theta}}$	$\left(1 - \left\{1 - exp\left[-exp - \left(\frac{t-\gamma}{\theta}\right)\right]\right\}^{\alpha}\right)^{\beta}$	EG-Gumbel $(\alpha,\beta,\gamma,\theta)$ [Cordeiro et al. (2013)]
$-ln[1 - (1 + t)^{-\gamma}]^{\theta}$	$(1 - \{1 - [1 - (1 + t)^{-\gamma}]^{\theta}\}^{\alpha})^{\beta}$	EG-IKum $(\alpha,\beta,\gamma,\theta)$

2.1 Estimation for exponentiated generalized general class of inverted distributions based on dual generalized order statistics

This subsection develops estimation of the parameters, rf and hrf from EGGC of inverted distributions based on dgos using *maximum likelihood* (ML) and Bayesian approaches, under *squared error* and *linear-exponential* (LINEX) loss functions. Also, *confidence intervals* (CIs) and credible intervals for the parameters, rf and hrf are obtained.

2.1.1 Maximum likelihood estimation for exponentiated generalized general class of inverted distributions

Suppose that $T_{(1,n,m,k)}, T_{(2,n,m,k)}, \ldots, T_{(n,n,m,k)}$ dgos n from the EGGC of an inverted distribution, the likelihood function can be obtained by substituting (8) and (9) in (5) as given below

$$L(\alpha,\beta;\underline{t}) \propto \alpha^n \beta^n exp\left[-\sum_{i=1}^{n}\lambda(t_i)\right]\prod_{i=1}^{n} g(t_i)\{1-exp[-\lambda(t_i)]\}^{\alpha-1} \times \prod_{i=1}^{n-1}(1\{1-exp[-\lambda(t_i)]\}^{\alpha})^{\beta(m+1)-1}(1-\{1-exp[-\lambda(t_n)]\}^{\alpha})^{\beta k-1}. \quad (13)$$

The natural logarithm of the likelihood function is given by,

$$\ell \equiv lnL(\alpha,\beta;\underline{t}) \propto nln\alpha + n\,ln\beta - \textstyle\sum_{i=1}^{n}\lambda(t_i))$$
$$+\sum_{i=1}^{n} ln[g(t_i)] + (\alpha-1)\sum_{i=1}^{n} ln\{1-exp[-\lambda(t_n)]\} + [\beta(m+1)]-1]\sum_{i=1}^{n} ln(1-\{1-exp[-\lambda(t_n)]\}^{\alpha}) + (\beta k-1)ln(1-\{1-exp[-\lambda(t_n)]\}^{\alpha}). \quad (14)$$

Considering that the two parameters α and β are unknown and differentiating the log likelihood function in (14) with respect to α and β, one obtains,

$$\frac{\partial\ell}{\partial\beta} = \frac{n}{\beta} + (m+1)\sum_{i=1}^{n-1} ln(1-\{1-exp[-\lambda(t_i)]\}^{\alpha}) + k\,ln(1-\{1-exp[-\lambda(t_n)]\}^{\alpha}), \quad (15)$$

and

$$\frac{\partial\ell}{\partial\alpha} = \frac{n}{\alpha} + \sum_{i=1}^{n} ln\{1-exp[-\lambda(t_n)]\} + [\beta(m+1)]-1]\sum_{i=1}^{n}\frac{\{1-exp[-\lambda(t_i)]\}^{\alpha})ln\{1-exp[-\lambda(t_i)]\}}{(1-\{1-exp[-\lambda(t_i)]\}^{\alpha})} - (\beta k-1)\frac{\{1-exp[-\lambda(t_n)]\}^{\alpha})ln\{1-exp[-\lambda(t_n)]\}}{(1-\{1-exp[-\lambda(t_n)]\}^{\alpha})}. \quad (16)$$

Equating (15) to zero, one can obtain the ML estimator of β,

$$\hat{\beta} = \frac{-n}{(m+1)\sum_{i=1}^{n-1} ln(1-\{1-exp[-\lambda(t_i)]\}^{\hat{\alpha}}) + k\,ln(1-\{1-exp[-\lambda(t_n)]\}^{\hat{\alpha}})}. \quad (17)$$

Then the ML estimator of the parameter α can be obtained numerically by substituting (17) in (16). The invariance property of the ML estimation can be used to obtain the ML estimates $\hat{R}(t)$ and $\hat{h}(t)$, by just replacing the parameters by their corresponding ML estimates, as given below:

$$\hat{R}(t) = 1 - (1 - \{1 - exp[-\lambda(t)]\}\hat{\alpha})\hat{\beta}, \quad t > 0, \quad (18)$$

and

$$\hat{h}(t) = \frac{\hat{\alpha}\hat{\beta}g(t)exp[-\lambda(t)]\{1-exp[-\lambda(t)]\}^{\hat{\alpha}-1}(1-\{1-exp[-\lambda(t)]\}^{\hat{\alpha}})^{\hat{\beta}-1}}{1-(1-\{1-exp[-\lambda(t)]\}^{\hat{\alpha}})^{\hat{\beta}-1}}, \; t > 0. \quad (19)$$

Asymptotic variance–covariance matrix of maximum likelihood estimators

The asymptotic variance-covariance matrix of the EGGC of inverted distributions for the two parameters α and β is the inverse of the Fisher information matrix as follows:

$$\tilde{I}^{-1} \approx \begin{bmatrix} \tilde{var}(\hat{\alpha}) & \tilde{cov}(\hat{\alpha},\hat{\beta}) \\ \tilde{cov}(\hat{\alpha},\hat{\beta}) & \tilde{var}(\hat{\beta}) \end{bmatrix} \approx \frac{1}{|I|}\begin{bmatrix} -\frac{\partial^2\ell}{\partial\beta^2} & \frac{\partial^2\ell}{\partial\alpha\partial\beta} \\ \frac{\partial^2\ell}{\partial\alpha\partial\beta} & -\frac{\partial^2\ell}{\partial\alpha^2} \end{bmatrix}_{\hat{\alpha},\hat{\beta}},$$

with,

$$\frac{\partial^2 \ell}{\partial \beta^2} = \frac{-n}{\beta^2}, \tag{20}$$

$$\frac{\partial^2 \ell}{\partial \alpha^2} = \frac{-n}{\alpha^2} - (\beta(m+1)-1)\sum_{i=1}^{n-1} \frac{\{1-exp[-\lambda(t_i)]\}^{\alpha}(ln\{1-exp[-\lambda(t_i)]\})^2}{(1-\{1-exp[-\lambda(t_i)]\}^{\alpha})^2}$$
$$-(\beta k-1)\frac{\{1-exp[-\lambda(t_n)]\}^{\alpha}(ln\{1-exp[-\lambda(t_n)]\})^2}{(1-\{1-exp[-\lambda(t_n)]\}^{\alpha})^2}, \tag{21}$$

and

$$\frac{\partial^2 \ell}{\partial \beta \partial \alpha} = (m+1)\sum_{i=1}^{n-1} \frac{\{1-exp[-\lambda(t_i)]\}^{\alpha} ln\{1-exp[-\lambda(t_i)]\}}{(1-\{1-exp[-\lambda(t_i)]\}^{\alpha})}$$
$$+k\frac{\{1-exp[-\lambda(t_n)]\}^{\alpha} ln\{1-exp[-\lambda(t_n)]\}}{(1-\{1-exp[-\lambda(t_n)]\}^{\alpha})}. \tag{22}$$

The asymptotic normality of the ML estimation can be used to compute the asymptotic 100 $(1-\omega)$% CIs for α and β as,

$$\hat{\alpha} \pm Z_{(1-\frac{\omega}{2})}\sqrt{\widehat{var}(\hat{\alpha})} \quad \text{and} \quad \hat{\beta} \pm Z_{(1-\frac{\omega}{2})}\sqrt{\widehat{var}(\hat{\beta})}. \tag{23}$$

Also, the asymptotic $100(1-\omega)$% CIs for are given by,

$$\hat{R}(t) \pm Z_{(1-\frac{\omega}{2})}\sqrt{\widehat{var}(\hat{R}(t))} \quad \text{and} \quad \hat{h}(t) \pm Z_{(1-\frac{\omega}{2})}\sqrt{\widehat{var}\left(\hat{h}(t)\right)}, \tag{24}$$

where $Z_{(1-\frac{\omega}{2})}$ is standard normal and $(1-\omega)$ is the confidence coefficient.

2.1.2 Bayesian estimation for exponentiated generalized general class of inverted distributions

Bayesian estimation of the parameters, rf and hrf based on dgos under the and LINEX loss functions is considered. Also, the credible intervals of the parameters are obtained.

Assuming that the parameters α and β of EGGC distributions are random variables with gamma prior distributions, then the joint prior density function of α and β is given by,

$$\pi(\alpha,\beta) \propto \alpha^{c_1-1}\, \beta^{c_2-1}\,)\, e^{-[d_1\alpha+d_2\beta]}, \tag{25}$$

where c_1, c_2, d_1, d_2 are hyper parameters.

The joint posterior of α and β can be derived by using (11) and (25) as follows:

$$\pi(\alpha,\beta|\underline{t}) \propto L(\alpha,\beta|\underline{t})\, \pi(\alpha,\beta)$$

$$= \alpha^{n+c_1-1}\beta^{n+c_2-1}e^{-\alpha(d_1+\sum_{i=1}^{n}\{1-exp[-\lambda(t_i)]\})}e^{-\beta[d_2-(m+1)\sum_{i=1}^{n-1}\ln Q_i-k\ln Q_n]}\prod_{i=1}^{n}(Q_i)^{-1},$$

where,

$$Q_i = (1-\{1-exp[-\lambda(t_i)]\}^{\alpha}), \qquad i = 1,2,\ldots,n. \tag{26}$$

Hence, the joint posterior distribution of α and β is,

$$\pi(\alpha,\beta|\underline{t}) = \frac{\alpha^{n+c_1-1}\beta^{n+c_2-1}e^{-\alpha(d_1+\sum_{i=1}^{n}\{1-exp[-\lambda(t_i)]\})}e^{-\beta[d_2-(m+1)\sum_{i=1}^{n-1}\ln Q_i-k\ln Q_n]}\prod_{i=1}^{n}(Q_i)^{-1}}{\varphi_1\Gamma(n+c_2)}, \tag{27}$$

where,

$$\varphi_1 = \int_0^\infty \frac{\alpha^{n+c_1-1} e^{-\alpha(d_1+\sum_{i=1}^{n}\{1-exp[-\lambda(t_i)]\})} \prod_{i=1}^{n}(Q_i)^{-1}}{[d_2-(m+1)\sum_{i=1}^{n-1}\ln Q_i - k\ln Q_n]^{n+c_2}} d\alpha\,. \tag{28}$$

a. Point estimation

The Bayes estimators of the parameters, rf, hrf and rhrf based on dgos under SE and LINEX loss functions of the EGGC of inverted distributions are obtained.

I. Bayesian estimation for exponentiated generalized general class of inverted distributions under squared error loss function

Under SE loss function the Bayes estimators of the parameters α and β are given by their marginal posterior expectations using (27) as given below,

$$\tilde{\alpha}_{(SE)} = E(\alpha|\underline{t}) = \int_0^\infty \frac{\alpha^{n+c_1} e^{-\alpha(d_1+\sum_{i=1}^{n}\{1-exp[-\lambda(t_i)]\})} \prod_{i=1}^{n}(Q_i)^{-1}}{\varphi_1[d_2-(m+1)\sum_{i=1}^{n-1}\ln Q_i - k\ln Q_n]^{n+c_2}} d\alpha, \tag{29}$$

and

$$\tilde{\beta}_{(SE)} = E(\beta|\underline{t}) = \int_0^\infty \frac{(n+c_2)\alpha^{n+c_1-1} e^{-\alpha(d_1+\sum_{i=1}^{n}\{1-exp[-\lambda(t_i)]\})} \prod_{i=1}^{n}(Q_i)^{-1}}{\varphi_1\left[d_2-(m+1)\sum_{i=1}^{n-1}\ln Q_i - k\ln Q_n\right]^{n+c_2+1}} d\alpha. \tag{30}$$

The Bayes estimators of the rf and hrf under the SE loss function are as follows:

$$\tilde{R}_{1(SE)}(t) = E(R(t)|\underline{t}) = 1 - \int_0^\infty \frac{\alpha^{n+c_1-1} e^{-\alpha(d_1+\sum_{i=1}^{n}\{1-exp[-\lambda(t_i)]\})} \prod_{i=1}^{n}(Q_i)^{-1}}{\varphi_1\left[d_2-(m+1)\sum_{i=1}^{n-1}\ln Q_i - k\ln Q_n - \ln Q\right]^{n+c_2}} d\alpha, \tag{31}$$

and

$$\tilde{h}_{1(SE)}(t) = E(h(t)|\underline{t}) =$$

$$\int_0^\infty\int_0^\infty \frac{\alpha^{n+c_1-1}\beta^{n+c_2-1} g(t) exp[-\lambda(t)]\{1-exp[-\lambda(t)]\}^{\alpha-1} Q^{\beta-1} e^{-\alpha(d_1+\sum_{i=1}^{n}\{1-exp[-\lambda(t_i)]\})} e^{-\beta[d_2-(m+1)\sum_{i=1}^{n-1}\ln Q_i - k\ln Q_n]} \prod_{i=1}^{n}(Q_i)^{-1}}{\varphi_1(1-Q^\beta)\Gamma(n+c_2)} d\alpha\, d\beta\,. \tag{32}$$

where,

$$Q = (1-\{1-exp[-\lambda(t)]\}^{\alpha}), \tag{33}$$

and

Q_i and Q_n are given by (26).

II. Bayesian estimation for exponentiated generalized general class of inverted distributions under linear exponential loss function

Under the LINEX loss function, the Bayes estimators for the shape parameters and are given, respectively, by,

$$\tilde{\alpha}_{(LNX)} = \frac{-1}{\nu} \ln E(e^{-\nu\alpha}|\underline{t}) = \frac{-1}{\nu} \ln\left[\int_0^\infty \frac{\alpha^{n+c_1-1} e^{-\alpha(\nu+d_1+\sum_{i=1}^{n}\{1-exp[-\lambda(t_i)]\})} \prod_{i=1}^{n}(Q_i)^{-1}}{\varphi_1[d_2-(m+1)\sum_{i=1}^{n-1}\ln Q_i - k\ln Q_n]^{n+c_2}} d\alpha\right], \tag{34}$$

and

$$\tilde{\beta}_{(LNX)} = \frac{-1}{\nu} \ln E(e^{-\nu\beta}|\underline{t}) = \frac{-1}{\nu} \ln \int_0^\infty \frac{\alpha^{n+c_1-1} e^{-\alpha(d_1+\sum_{i=1}^{n}\{1-exp[-\lambda(t_i)]\})} \prod_{i=1}^{n}(Q_i)^{-1}}{\varphi_1\left(\nu+d_2-(m+1)\sum_{i=1}^{n-1}\ln Q_i - k\ln Q_n\right)^{n+c_2}} d\alpha\,. \tag{35}$$

Also, the Bayes estimator for rf based on dgos is,

$$\tilde{R}_{1(LNX)}(t) = \frac{-1}{\nu}\ln E\left(e^{-\nu R(t)}|\underline{t}\right)$$

$$= \frac{-1}{\nu} ln\left[1 - \int_0^{\infty}\int_0^{\infty} \frac{\alpha^{n+c_1-1}\beta^{n+c_2-1}e^{-\nu(1-Q)^{\beta}}e^{-\alpha(d_1+\sum_{i=1}^{n}\{1-exp[-\lambda(t_i)]\})}}{\varphi_1\Gamma(n+c_2)} \right. \tag{36}$$

$$\left. \times e^{-\beta[d_2-(m+1)\sum_{i=1}^{n-1}\ln Q_i - k\ln Q_n]}\prod_{i=1}^{n}(Q_i)^{-1}\,d\alpha\,d\beta\right],$$

and the Bayes estimator for the hrf based on dgos is given by,

$$\tilde{h}_{1(LNX)}(t) = \frac{-1}{\nu}\ln E\left(e^{-\nu h(t)}|\underline{t}\right)$$

$$= \frac{-1}{\nu} ln\left[\int_0^{\infty}\int_0^{\infty} \frac{\alpha^{n+c_1-1}\beta^{n+c_1-1}e^{-\nu\left[\frac{\alpha\beta g(t)exp[-\lambda(t)]\{1-exp[-\lambda(t)]\}^{\alpha-1}[Q]^{\beta-1}}{1-[Q]^{\beta}}\right]}}{\varphi_1\Gamma(n+c_2)}\right. \tag{37}$$

$$\left. \times e^{-\alpha[d_1+\sum_{i=1}^{n}\{1-exp[-\lambda(t_i)]\}]}e^{-\beta[d_2-(m+1)\sum_{i=1}^{n-1}\ln Q_i - k\ln Q_n]}\prod_{i=1}^{n}(Q_i)^{-1}\,d\alpha\,d\beta\right],$$

where,

Q_i and Q_n are defined in (26), φ_1 is given in (28) and Q is defined in (33).

b. Credible intervals

In general, $(L(\underline{x}), U(\underline{x}))$ is the $100(1-\omega)\%$ credibility interval of $\underline{\theta}$ if,

$$P\left[L(\underline{t}) < \underline{\theta} < U(\underline{t})|\underline{t}\right] = \int_{L(\underline{t})}^{U(\underline{t})} \pi(\underline{\theta}|\underline{t})\,d\underline{\theta} = 1-\omega. \tag{38}$$

where, $L(\underline{x})$ and $U(\underline{x})$ are the *lower limit* (LL) and *upper limit* (UL) and ω is the level of significance. Since, the posterior distribution is given by (27), then a $100(1-\omega)\%$ credible interval for α is $(L(\underline{t}), U(\underline{t}))$, where,

$$P\left[\alpha > L(\underline{t})|\underline{t}\right] = \int_{L(\underline{t})}^{\infty} \frac{\alpha^{n+c_1-1}e^{-\alpha(d_1+\sum_{i=1}^{n}\{1-exp[-\lambda(t_i)]\})}\prod_{i=1}^{n}(Q_i)^{-1}}{\varphi_1[d_2-(m+1)\sum_{i=1}^{n-1}\ln Q_i - k\ln Q_n]^{n+c_2}}\,d\alpha = 1-\frac{\omega}{2}, \tag{39}$$

and

$$P\left[\alpha > U(\underline{t})|\underline{t}\right] = \int_{U(\underline{t})}^{\infty} \frac{\alpha^{n+c_1-1}e^{-\alpha(d_1+\sum_{i=1}^{n}\{1-exp[-\lambda(t_i)]\})}\prod_{i=1}^{n}(Q_i)^{-1}}{\varphi_1[d_2-(m+1)\sum_{i=1}^{n-1}\ln Q_i - k\ln Q_n]^{n+c_2}}\,d\alpha = \frac{\omega}{2}. \tag{40}$$

Also, $100(1-\omega)\%$ credible interval for β is$(L(\underline{t}), U(\underline{t}))$,
where,

$$P\left[\beta > L(\underline{t})|\underline{t}\right]$$

$$= \int_{L(\underline{t})}^{\infty}\int_0^{\infty} \frac{\alpha^{n+c_1-1}\beta^{n+c_2-1}e^{-\alpha(d_1+\sum_{i=1}^{n}\{1-exp[-\lambda(t_i)]\})}e^{-\beta[d_2-(m+1)\sum_{i=1}^{n-1}\ln Q_i - k\ln Q_n]}\prod_{i=1}^{n}(Q_i)^{-1}}{\varphi_1\Gamma(n+c_2)}\,d\alpha d\beta$$

$$= 1-\frac{\omega}{2}, \tag{41}$$

and

$$P[\,\beta > U(\underline{t})|\underline{t}] = \int_{U(\underline{t})}^{\infty}\int_{0}^{\infty} \frac{\alpha^{n+c_1-1}\beta^{n+c_2-1}e^{-\alpha\left(d_1+\sum_{i=1}^{n}\{1-exp[-\lambda(t_i)]\}\right)}e^{-\beta\left[d_2-(m+1)\sum_{i=1}^{n-1}\ln Q_i-k\ln Q_n\right]}\prod_{i=1}^{n}(Q_i)^{-1}}{\varphi_1\Gamma(n+c_2)}\,d\alpha d\beta$$

$$= \frac{\omega}{2}, \tag{42}$$

where,
Q_i and Q_n are defined in (26), φ_1 is given in (28) and Q is defined in (33).
All the previous equations can be solved numerically.

2.2 Prediction for exponentiated generalized general class of inverted distributions based on dual generalized order statistics

This subsection considers ML and Bayesian prediction (point and interval) for a future observation of the EGGC of inverted distributions based on dgos, under and LINEX loss functions.

Let $T(1,n,m,k),\ldots, T(r,n,m,k)$ be a dgos of size n with the pdf $f(t;\alpha,\beta)$ and suppose $Y(1,n_y, m_y, k_y),\ldots, Y(r_y,n_y,m_y,k_y)$, $k_y > 0$, $m_y \in \mathbb{R}$ is a second independent dgos of size n_y, of future observations from the same distribution. Using (6)–(9), the pdf of the dgos $Y_{(s)}$ can be obtained just replacing $t_{(r)}$ by $y_{(s)}$ as follows:

$$\begin{aligned} f(y_{(s)}|\alpha,\beta) &= \frac{\alpha\beta\,\zeta_{s-1}}{(s-1)!}[-\grave{\lambda}(y_{(s)})]exp[-\lambda(y_{(s)})]\{1-exp[-\lambda(y_{(s)})]\}^{\alpha-1} \\ &\times\left(1-\{1-exp[-\lambda(y_{(s)})]\}^{\alpha}\right)^{\beta\gamma_s-1} g_{m_y}^{s-1}[F(y_{(s)})], \end{aligned} \tag{43}$$

where, $\zeta_{s-1} = \prod_{j_1=1}^{s}\gamma_{j_1}$, $g_M(y_s) = h_M(y_s) - h_M(1)$, $\gamma_r^* = k_y + (n_y - r)(m_y + 1)$.

$$g_{m_y}^{s-1}\left(F(y_{(s)})\right) = \begin{cases} \frac{1}{(m_y+1)^{s-1}}\left[1-\left(1-\{1-exp[-\lambda(y_{(s)})]\}^{\alpha}\right)^{\beta(m_y+1)}\right]^{s-1}, & m_y \neq -1, \\ \left[-\ln\left(1-\{1-exp[-\lambda(y_{(s)})]\}^{\alpha}\right)^{\beta}\right]^{s-1}, & m_y = -1. \end{cases} \tag{44}$$

For the future sample of size n_y, let $Y_{(s)}$ denote the sth ordered lifetime, $1 \le s \le n_y$, hence the pdf of the dgos $Y_{(s)}$ from EGGC of inverted distributions is obtained by substituting (44) in (43).

Case one: for $m_y \neq -1$

$$\begin{aligned} f(y_{(s)}|\alpha,\beta) &= \frac{\alpha\beta\zeta_{s-1}}{(m_y+1)^{s-1}(s-1)!}[-\grave{\lambda}(y_{(s)})]exp[-\lambda(y_{(s)})]\{1-exp[-\lambda(y_{(s)})]\}^{\alpha-1} \\ &\times\left(1-\{1-exp[-\lambda(y_{(s)})]\}^{\alpha}\right)^{\beta\gamma_s-1}\left[1-\left(1-\{1-exp[-\lambda(y_{(s)})]\}^{\alpha}\right)^{\beta(m_y+1)}\right]^{s-1}. \end{aligned}$$

Using the binomial expansion, one obtains,

$$\begin{aligned} f(y_{(s)}|\alpha,\beta) &= \frac{\alpha\beta\zeta_{s-1}}{(m_y+1)^{s-1}(s-1)!}[-\grave{\lambda}(y_{(s)})]exp[-\lambda(y_{(s)})]\{1-exp[-\lambda(y_{(s)})]\}^{\alpha-1} \\ &\times\sum_{i_1=0}^{s-1}(-1)^{i_1}\binom{s-1}{i_1}\left(1+y_{(s)}\right)^{-(\alpha+1)}\left(1-\{1-exp[-\lambda(y_{(s)})]\}^{\alpha}\right)^{\beta\left(\gamma_s+i_1(m_y+1)\right)-1}, \end{aligned}$$

Let,

$$\xi = \frac{\zeta_{s-1}}{(m_y+1)^{s-1}(s-1)!},\ \eta_{i_1} = (-1)^{i_1}\binom{s-1}{i_1} \text{ and } \varpi_{i_1} = [\gamma_s + i_1(m_y+1)]$$

Then,

$$f(y_{(s)}|\alpha,\beta) = \xi\alpha\beta[-\grave{\lambda}(y_{(s)})]exp[-\lambda(y_{(s)})]\{1-exp[-\lambda(y_{(s)})]\}^{\alpha-1}$$
$$\times\sum_{i_1=0}^{s-1}\eta_{i_1}\left(1-\{1-exp[-\lambda(y_{(s)})]\}^{\alpha}\right)^{\beta\varpi_{i_1}-1},\ y_{(s)}>0,\ \ \alpha,\beta>0. \tag{45}$$

Case two: for $m_y = -1$

$$f(y_{(s)}|\alpha,\beta) = \frac{k_y^s\alpha\beta^s}{(s-1)!}[-\grave{\lambda}(y_{(s)})]exp[-\lambda(y_{(s)})]\{1-exp[-\lambda(y_{(s)})]\}^{\alpha-1}$$
$$\times\left(1-\{1-exp[-\lambda(y_{(s)})]\}^{\alpha}\right)^{\beta k_y-1}\left[-\ln\left(1-\{1-exp[-\lambda(y_{(s)})]\}^{\alpha}\right)\right]^{s-1}, \tag{46}$$
$$y_{(s)}>0,\ \ \alpha,\beta>0.$$

2.2.1 Maximum likelihood prediction for exponentiated generalized general class of inverted distributions

Two-sample ML point and interval predictors are obtained based on dgos. The ML prediction can be derived by using the conditional density function of the sth future dgos which are given by (45) and (46) after replacing the parameters (α,β) by their ML estimators $(\hat{\alpha},\hat{\beta})$.

a. **Point prediction**

The *ML predictors* (MLP) of the future dgos $Y_{(s)}$ can be obtained using (45) and (46) as follows:

Case one: for $M \neq -1$

$$\hat{y}_{(s)} = \int_0^{\infty} y_{(s)}\hat{\alpha}\hat{\beta}[-\grave{\lambda}(y_{(s)})]exp[-\lambda(y_{(s)})]\{1-exp[-\lambda(y_{(s)})]\}^{\hat{\alpha}-1}$$
$$\times\sum_{i_1=0}^{s-1}\eta_{i_1}\left(1-\{1-exp[-\lambda(y_{(s)})]\}^{\hat{\alpha}}\right)^{\hat{\beta}\varpi_{i_1}-1}dy_{(s)}. \tag{47}$$

Case two: for $M = -1$

$$\hat{y}_{(s)} = \int_0^{\infty} y_{(s)}\frac{k_y^s\hat{\alpha}\hat{\beta}^s}{(s-1)!}[-\grave{\lambda}(y_{(s)})]exp[-\lambda(y_{(s)})]\{1-exp[-\lambda(y_{(s)})]\}^{\hat{\alpha}-1}$$
$$\times\left(1-\{1-exp[-\lambda(y_{(s)})]\}^{\hat{\alpha}}\right)^{\hat{\beta}k_y-1}\left[-\ln\left(1-\{1-exp[-\lambda(y_{(s)})]\}^{\hat{\alpha}}\right)\right]^{s-1}. \tag{48}$$

b. **Interval prediction**

In general, the *ML predictive bounds* (MLPB) for the future dgos $Y_{(s)}$, $1 \le s \le N$, can be derived as follows:

$$P[L(\underline{t}) < Y_{(s)} < U(\underline{t})|\underline{t}] = \int_{L(\underline{t})}^{U(\underline{t})} f(y_{(s)}|\hat{\alpha},\hat{\beta};\underline{t})dy_{(s)} = 1-\omega,$$

Case one: for $M \neq -1$

$$P[Y_{(s)} > L(\underline{t})|\underline{t}]$$
$$= \int_{L(\underline{x})}^{\infty}\xi\hat{\alpha}\hat{\beta}[-\grave{\lambda}(y_{(s)})]exp[-\lambda(y_{(s)})]\{1-exp[-\lambda(y_{(s)})]\}^{\hat{\alpha}-1}$$
$$\times\sum_{i_1=0}^{s-1}\eta_{i_1}\left(1-\{1-exp[-\lambda(y_{(s)})]\}^{\hat{\alpha}}\right)^{\hat{\beta}\varpi_{i_1}-1}dy_{(s)} = 1-\frac{\omega}{2}, \tag{49}$$

and

$$P[Y_{(s)} > U(\underline{t})|\underline{t}] = \int_{U(\underline{x})}^{\infty} \xi\hat{\alpha}\hat{\beta}[-\grave{\lambda}(y_{(s)})]exp[-\lambda(y_{(s)})]\{1-exp[-\lambda(y_{(s)})]\}^{\hat{\alpha}-1}$$
$$\times \sum_{i_1=0}^{s-1} \eta_{i_1}\left(1-\{1-exp[-\lambda(y_{(s)})]\}^{\hat{\alpha}}\right)^{\hat{\beta}\varpi_{i_1}-1} dy_{(s)} = \frac{\omega}{2}. \quad (50)$$

Case two: for $M = -1$

$$P[Y_{(s)} > L(\underline{t})|\underline{t}]$$
$$= \int_{L(\underline{x})}^{\infty} \frac{k_y^s\hat{\alpha}\hat{\beta}^s}{(s-1)!}[-\grave{\lambda}(y_{(s)})]exp[-\lambda(y_{(s)})]\{1-exp[-\lambda(y_{(s)})]\}^{\hat{\alpha}-1}$$
$$\times\left(1-\{1-exp[-\lambda(y_{(s)})]\}^{\hat{\alpha}}\right)^{\hat{\beta}k_y-1}\left[-\ln\left(1-\{1-exp[-\lambda(y_{(s)})]\}^{\hat{\alpha}}\right)\right]^{\hat{s}-1} dy_s \quad (51)$$
$$= 1-\frac{\omega}{2},$$

$$P[Y_{(s)} > U(\underline{t})|\underline{t}]$$
$$= \int_{U(\underline{x})}^{\infty} \frac{k_y^s\hat{\alpha}\hat{\beta}^s}{(s-1)!}[-\grave{\lambda}(y_{(s)})]exp[-\lambda(y_{(s)})]\{1-exp[-\lambda(y_{(s)})]\}^{\hat{\alpha}-1}$$
$$\times\left(1-\{1-exp[-\lambda(y_{(s)})]\}^{\hat{\alpha}}\right)^{\hat{\beta}k_y-1}\left[-\ln\left(1-\{1-exp[-\lambda(y_{(s)})]\}^{\hat{\alpha}}\right)\right]^{s-1} dy_s \quad (52)$$
$$= \frac{\omega}{2},$$

2.1.2 Bayesian prediction for exponentiated generalized general class of inverted distributions

The Bayesian two-sample prediction is considered based on dgos for the future observation $Y_{(s)}$, $1 \le s \le n_y$. The *Bayesian predictive density* (BPD) function can be derived as follows:

The BPD function can be derived as follows:

$$f(y_{(s)}|\underline{t}) = \int_0^{\infty}\int_0^{\infty} f(y_{(s)}|\alpha,\beta)\,\pi(\alpha,\beta|\underline{t})\,d\alpha d\beta, \quad (53)$$

where, $\pi(\alpha,\beta \mid \underline{t})$ is the joint posterior distribution of α,β and $f(y(s) \mid \alpha,\beta)$ is the pdf of $Y_{(s)}$.

Case one: for $m_y \neq -1$

The BPD of $Y_{(s)}$ given is obtained by substituting (27) and (45) in (53) as given below

$$f(y_{(s)}|\underline{t}) = \int_0^{\infty} \frac{\xi(n+c_2)\alpha^{n+c_1}[-\grave{\lambda}(y_{(s)})]exp[-\lambda(y_{(s)})]}{\varphi_1\left(1-\{1-exp[-\lambda(y_{(s)})]\}^{\alpha}\right)\prod_{i=1}^{n} Q_i\ \tau_3^{(n+c_2+1)}}$$
$$\times\{1-exp[-\lambda(y_{(s)})]\}^{\alpha-1}e^{-\alpha(d_1+\sum_{i=1}^{n}\{1-exp[-\lambda(t_i)]\})}d\alpha, \quad (54)$$

where, $\tau_3 = [(d_2-(m+1)\sum_{i=1}^{n-1}\ln Q_i - k\ln Q_n) - \sum_{i_1=0}^{s-1}\varpi_{i_1}\ln\left[\eta_{i_1}\left(1-\{1-exp[-\lambda(y_{(s)})]\}^{\alpha}\right)\right]]$,

$\xi = \frac{\zeta_{s-1}}{(m_y+1)^{s-1}(s-1)!}$, $\eta_{i_1} = (-1)^{i_1}\binom{s-1}{i_1}$, $\varpi_{i_1} = [\gamma_s + i_1(m_y+1)]$, Q_i and Q_n are defined in (26)

Case two: for $m_y \neq -1$

Substituting (27) and (46) in (53), the BPD of $Y_{(s)}$ given $\underline{t}$ can be obtained as follows:

$$f(y_{(s)}|\underline{t}) = \int_0^\infty \frac{k_y^s \alpha^{n+c_1}[-\grave{\lambda}(y_{(s)})]exp[-\lambda(y_{(s)})]\{1-exp[-\lambda(y_{(s)})]\}^{\alpha-1}\Gamma(n+c_2+s)}{\varphi_1(s-1)!\left(1-\{1-exp[-\lambda(y_{(s)})]\}^{\alpha}\right)\prod_{i=1}^{n} Q_i\Gamma(n+c_2)\,\tau_4^{(n+c_2+s)}} \times \left[-\ln\left(1-\{1-exp[-\lambda(y_{(s)})]\}^{\alpha}\right)\right]^{s-1} e^{-\alpha(d_1+\sum_{i=1}^{n}\{1-exp[-\lambda(t_i)]\})}\, d\alpha, \quad (55)$$

where,
$\tau_4 = \left[(d_2-(m+1)\sum_{i=1}^{n-1}\ln Q_i - k\ln Q_n) - k_y\, ln\left(1-\{1-exp[-\lambda(y_{(s)})]\}^{\alpha}\right)\right]$ Q_i and Q_n are defined in (26).

a. Point prediction

The *Bayes predictors* (BP) of the future dgos $Y_{(s)}$ can be derived under SE and LINEX loss functions as given below:

Case one: for $m_y \neq -1$

The BP of the future dgos $Y_{(s1)}$ can be obtained under the SE loss function using (54) as follows:

$$y^*_{(s1)(SE)} = E(y_{(s1)}|\underline{t}) = \int_0^\infty y_{(s1)}\, f_2(y_{(s1)}|\underline{t})\; dy_{(s1)},$$

$$= \int_0^\infty\int_0^\infty \frac{y_{(s1)}\xi(n+c_2)\alpha^{n+c_1}[-\grave{\lambda}(y_{(s1)})]exp[-\lambda(y_{(s1)})]\{1-exp[-\lambda(y_{(s1)})]\}^{\alpha-1}e^{-\alpha(d_1+\sum_{i=1}^{n}\{1-exp[-\lambda(t_i)]\})}}{\varphi_1\left(1-\{1-exp[-\lambda(y_{(s1)})]\}^{\alpha}\right)\prod_{i=1}^{n} Q_i\,\tau_3^{(n+c_2+1)}}\, d\alpha dy_{(s1)} \quad (56)$$

The BP of the future dgos $Y_{(s1)}$ can be obtained under LINEX loss function using (54) as given below:

$$y^*_{(s1)(LNX)} = \frac{-1}{\nu}\ln E(e^{-\nu y_{(s1)}}|\underline{t}),$$

$$E(e^{-\nu y_{(s1)}}|\underline{t}) = \int_0^\infty e^{-\nu y_{(s1)}}\, f_2(y_{(s1)}|\underline{t})\; dy_{(s1)} \quad (57)$$

$$= \int_0^\infty\int_0^\infty \frac{e^{-\nu y_{(s1)}}\xi(n+c_2)\alpha^{n+c_1}[-\grave{\lambda}(y_{(s1)})]exp[-\lambda(y_{(s1)})]\{1-exp[-\lambda(y_{(s1)})]\}^{\alpha-1}e^{-\alpha(d_1+\sum_{i=1}^{n}\{1-exp[-\lambda(t_i)]\})}}{\varphi_1(1-\{1-exp[-\lambda(y_{(s1)})]\}^{\alpha})\prod_{i=1}^{n} Q_i\,\tau_3^{(n+c_2+1)}}\, d\alpha\, dy_{(s1)}.$$

Case two: for $m_y = -1$

The BP of dgos $Y_{(s2)}$ can be obtained under SE loss function using (55) as follows:

$$y^*_{(s2)(SE)} = E\left(y_{(s2)}\middle|\underline{t}\right) =$$

$$\int_0^\infty\int_0^\infty \frac{y_{(s2)}k_y^{s2}\alpha^{n+c_1}\left[-\grave{\lambda}\left(y_{(s2)}\right)\right]exp\left[-\lambda\left(y_{(s2)}\right)\right]\left\{1-exp\left[-\lambda\left(y_{(s2)}\right)\right]\right\}^{\alpha-1}\left[-\ln\left(1-\left\{1-exp\left[-\lambda(y_{(s2)})\right]\right\}^{\alpha}\right)\right]^{s2-1}e^{-\alpha(d_1+\sum_{i=1}^{n}\{1-exp[-\lambda(t_i)]\})}\Gamma(n+c_2+s2)}{\varphi_1(s2-1)!\left(1-\left\{1-exp\left[-\lambda(y_{(s2)})\right]\right\}^{\alpha}\right)\prod_{i=1}^{n} Q_i\Gamma(n+c_2)\,\tau_4^{(n+c_2+s2)}}\, d\alpha dy_{(s2)}. \quad (58)$$

The BP of the future dgos can be obtained under LINEX loss function using (55) as follows:

$$y^*_{(s2)(LNX)} = \frac{-1}{\nu}\ln E(e^{-\nu y_{(s2)}}|\underline{t}), \quad (59)$$

where,

$$E\left(e^{-\nu y_{(s2)}}|\underline{t}\right) =$$

$$\int_0^\infty \int_0^\infty \frac{\Gamma(n+c_2+s2)e^{-\nu y_{(s)}} k_y^{s2} \alpha^{n+c_1}\left[-\grave{\lambda}\left(y_{(s2)}\right)\right] exp\left[-\lambda\left(y_{(s2)}\right)\right]\left\{1-exp\left[-\lambda\left(y_{(s2)}\right)\right]\right\}^{\alpha-1}}{\varphi_1(s2-1)!\left(1-\left\{1-exp\left[-\lambda(y_{(s2)})\right]\right\}^{\alpha}\right)\prod_{i=1}^{n} Q_i \Gamma(n+c_2)\ \tau_4^{(n+c_2+s2)}}$$
$$\times \left[-\ln\left(1-\left\{1-exp\left[-\lambda(y_{(s2)})\right]\right\}^{\alpha}\right)\right]^{s2-1} e^{-\alpha\left[d_1+\sum_{i=1}^{n}(1-exp[-\lambda(t_i)])\right]} d\alpha dy_{(s2)} \ .$$

b. Interval prediction

The *Bayes predictive bounds* (BPB) of the future dogs $Y(s)$ can be obtained using (54) and (55) as given below:

Case one: for $m_y \neq -1$

$$P\left[Y_{(s1)} > L(\underline{t})|\underline{t}\right]$$
$$= \int_{L(\underline{t})}^\infty \int_0^\infty \frac{\xi(n+c_2)\alpha^{n+c_1}\left[-\grave{\lambda}\left(y_{(s1)}\right)\right] exp\left[-\lambda\left(y_{(s1)}\right)\right]}{\varphi_1\left(1-\left\{1-exp\left[-\lambda(y_{(s1)})\right]\right\}^{\alpha}\right)\prod_{i=1}^{n} Q_i\ \tau_3^{(n+c_2+1)}}$$
$$\times \left\{1-exp\left[-\lambda\left(y_{(s1)}\right)\right]\right\}^{\alpha-1} e^{-\alpha\left(d_1+\sum_{i=1}^{n}\{1-exp[-\lambda(t_i)]\}\right)} d\alpha\, dy_{(s1)} = 1-\frac{\omega}{2}, \tag{60}$$

and

$$P\left[Y_{(s1)} > U(\underline{t})|\underline{t}\right]$$
$$= \int_{U(\underline{t})}^\infty \int_0^\infty \frac{\xi(n+c_2)\alpha^{n+c_1}\left[-\grave{\lambda}\left(y_{(s1)}\right)\right] exp\left[-\lambda\left(y_{(s1)}\right)\right]}{\varphi_1\left(1-\left\{1-exp\left[-\lambda(y_{(s1)})\right]\right\}^{\alpha}\right)\prod_{i=1}^{n} Q_i\ \tau_3^{(n+c_2+1)}}$$
$$\times \left\{1-exp\left[-\lambda\left(y_{(s1)}\right)\right]\right\}^{\alpha-1} e^{-\alpha\left(d_1+\sum_{i=1}^{n}\{1-exp[-\lambda(t_i)]\}\right)} d\alpha\, dy_{(s1)} = \frac{\omega}{2}. \tag{61}$$

Case two: for $m_y = -1$

$P[Y_{(s2)} > L(\underline{t})|\underline{t}] =$

$$\int_{L(\underline{t})}^\infty \int_0^\infty \frac{k_y^{s2}\alpha^{n+c_1}\left[-\grave{\lambda}\left(y_{(s2)}\right)\right] exp\left[-\lambda\left(y_{(s2)}\right)\right]\left\{1-exp\left[-\lambda\left(y_{(s2)}\right)\right]\right\}^{\alpha-1}\left[-\ln\left(1-\left\{1-exp\left[-\lambda(y_{(s2)})\right]\right\}^{\alpha}\right)\right]^{s2-1} e^{-\alpha\left(d_1+\sum_{i=1}^{n}\{1-exp[-\lambda(t_i)]\}\right)}\Gamma(n+c_2+s2)}{\varphi_1(s2-1)!\left(1-\left\{1-exp\left[-\lambda(y_{(s2)})\right]\right\}^{\alpha}\right)\prod_{i=1}^{n} Q_i \Gamma(n+c_2)\ \tau_4^{(n+c_2+s2)}} d\alpha\, dy_{(s2)}$$
$$= 1-\frac{\omega}{2}, \tag{62}$$

and $P[Y_{(s2)} > U(\underline{t})|\underline{t}] =$

$$\int_{L(\underline{t})}^\infty \int_0^\infty \frac{k_y^{s2}\alpha^{n+c_1}\left[-\grave{\lambda}\left(y_{(s2)}\right)\right] exp\left[-\lambda\left(y_{(s2)}\right)\right]\left\{1-exp\left[-\lambda\left(y_{(s2)}\right)\right]\right\}^{\alpha-1}\left[-\ln\left(1-\left\{1-exp\left[-\lambda(y_{(s2)})\right]\right\}^{\alpha}\right)\right]^{s2-1} e^{-\alpha\left(d_1+\sum_{i=1}^{n}\{1-exp[-\lambda(t_i)]\}\right)}\Gamma(n+c_2+s2)}{\varphi_1(s2-1)!\left(1-\left\{1-exp\left[-\lambda(y_{(s2)})\right]\right\}^{\alpha}\right)\prod_{i=1}^{n} Q_i \Gamma(n+c_2)\ \tau_4^{(n+c_2+s2)}} d\alpha\, ay_{(s2)}$$
$$= \frac{\omega}{2}. \tag{63}$$

All the previous equations can be solved numerically.

3. The exponentiated generalized inverted Kumaraswamy distribution

This section is devoted to introducing the *EG inverted Kumaraswamy* (EG-IKum) distribution as a new distribution which is a special model from EGGC of inverted distributions. Some of its properties are studied through, rf, hrf and rhrf, graphical description, moments, and related measures, mean residual life and mean past lifetime, the ML estimators, CIs for the parameters, rf and hrf of the EG-IKum distribution based on dgos are obtained. Also, the shape parameters, rf and hrf of the EG-IKum distribution are estimated using the Bayesian approach. The Bayes estimators are derived under the SE and the LINEX loss functions based on dgos. Credible intervals for the parameters, rf and hrf are derived. The Bayesian prediction (point and interval) for a future observation of the EG-IKum distribution are obtained based on dgos. All results are specialized to lower record values and a numerical study is presented. Moreover, three real data sets are used to illustrate the flexibility of the distribution.

Assuming that T is a random variable distributed as EG-IKum distribution which is a special model from EGGC of inverted distributions in (8) when $\lambda(t) = [-ln(1-(1+t)^{-\gamma})^{\theta}]$ with shape parameters, $\theta = (\theta_1, \theta_2, \theta_3, \theta_4)' > \underline{0}$, denoted by $T \sim$ EG-IKum ($\underline{\theta}$), the cdf and pdf are given by,

$$F_{EGIK}(t;\underline{\theta}) = \left(1-\left\{1-\left[1-(1+t)^{-\theta_3}\right]^{\theta_4}\right\}^{\theta_2}\right)^{\theta_1}, \; t>0,\; \underline{\theta}>\underline{0},\; j=1,2,3,4, \tag{64}$$

and

$$f_{EGIK}(t;\underline{\theta}) = \prod_{j=1}^{4}\theta_j\,(1+t)^{-(\theta_3+1)}\left[1-(1+t)^{-\theta_3}\right]^{\theta_4-1}\left\{1-\left[1-(1+t)^{-\theta_3}\right]^{\theta_4}\right\}^{\theta_2-1}$$
$$\times\left(1-\left\{1-\left[1-(1+t)^{-\theta_3}\right]^{\theta_4}\right\}^{\theta_2}\right)^{\theta_1-1}, \quad t>0,\; \underline{\theta}>\underline{0}. \tag{65}$$

i. Reliability function, hazard rate and reversed hazard rate functions

The rf, hrf and rhrf of EG-IKum are given, respectively, by,

$$R_{EGIK}(t;\underline{\theta}) = 1-\left(1-\left\{1-\left[1-(1+t)^{-\theta_3}\right]^{\theta_4}\right\}^{\theta_2}\right)^{\theta_1}, \qquad t>0,\; \underline{\theta}>\underline{0}, \tag{66}$$

$$h_{EGIK}(t;\underline{\theta}) = \frac{\prod_{j=1}^{4}\theta_j\,(1+t)^{-(\theta_3+1)}\left[1-(1+t)^{-\theta_3}\right]^{\theta_4-1}}{1-(1-\{1-[1-(1+t)^{-\theta_3}]^{\theta_4}\}^{\theta_2})^{\theta_1}}\left(1-\left(1-(1+t)^{-\theta_3}\right)^{\theta_4}\right)^{\theta_2-1}$$
$$\times\left(1-\left\{1-\left[1-(1+t)^{-\theta_3}\right]^{\theta_4}\right\}^{\theta_2}\right)^{\theta_1-1}, t>0,\; \underline{\theta}>\underline{0}, \tag{67}$$

and

$$rh_{EGIK}(t;\underline{\theta}) = \prod_{j=1}^{4}\theta_j\,(1+t)^{-(\theta_3+1)}\left[1-(1+t)^{-\theta_3}\right]^{\theta_4-1}\left\{1-\left[1-(1+t)^{-\theta_3}\right]^{\theta_4}\right\}^{\theta_2-1}$$
$$\times\left(1-\left\{1-\left[1-(1+t)^{-\theta_3}\right]^{\theta_4}\right\}^{\theta_2}\right)^{-1}, \; t>0,\; \underline{\theta}>\underline{0}. \tag{68}$$

Plots of the pdf, hrf and rhrf of EG-IKum are given, respectively in Figures 1–3. The plots, in Figure 1 and Figure 2 show the behavior of the pdf and hrf which are positively skewed and uni-modal with different values of the shape parameters. Also, the behavior of the hrf indicates that the model possesses the non-monotone property, which is useful for application in reliability studies, clinical trial studies and analyzing different data sets. From Figure 3 one can observe that the curves of the rhrf for all values are decreasing and then constant.

ii. Quantiles

The quantile function of the EG-IKum is given by

$$t_q = \left(\left\{1-\left[1-(1-q^{\frac{1}{\theta_1}})^{\frac{1}{\theta_2}}\right]^{\frac{1}{\theta_4}}\right\}^{-\frac{1}{\theta_3}}-1\right), \qquad 0<q<1. \tag{69}$$

Special cases can be obtained using (69) such as the second quartile (median), when $q = 0.5$, first quartile when $q = 0.25$ and third quartile when $q = 0.75$.

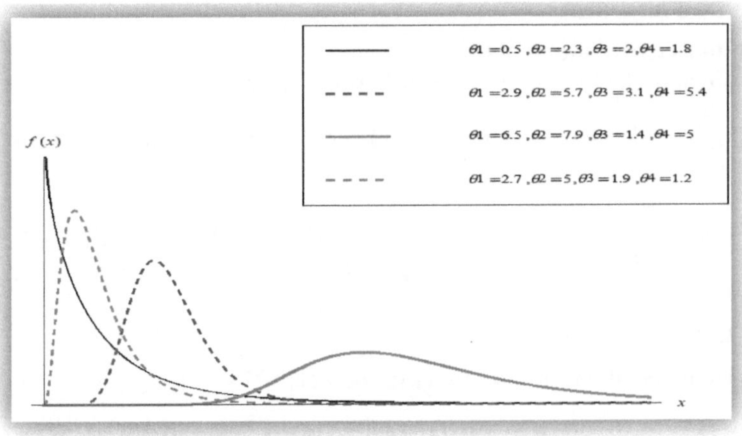

Figure 1: Plots of the pdf of the EG-IKum distribution.

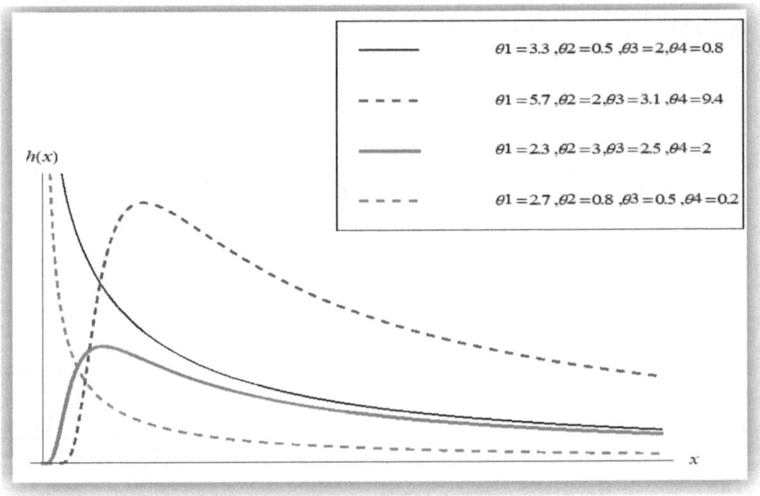

Figure 2: Plots of the hrf of the EG-IKum distribution.

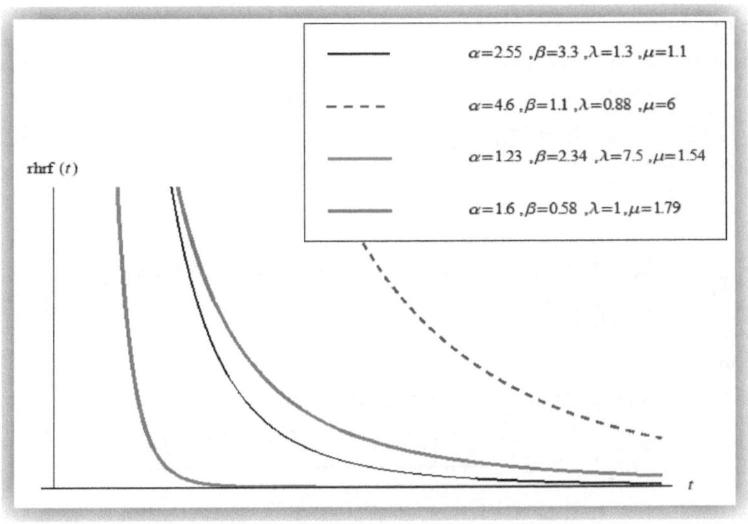

Figure 3: Plots of the rhrf of the EG-IKum distribution.

iii. Moments and related measures

a. The central and non-central moments

The rth non central moment of the EG-IKum ($\underline{\theta}$) distribution is given by,

$$\acute{\mu}_r = \prod_{j=1}^{4} \theta_j \sum_{\kappa^*} \Omega \ \mathbf{B}\{(r+1), [\,\theta_3(\kappa_3+1)-r]\}, \quad r = 1,2,\dots, \quad \theta_3(\kappa_3+1) > r, \tag{70}$$

where,

$$\sum_{\kappa^*} \Omega = \sum_{\kappa_1=0}^{\infty} \sum_{\kappa_2=0}^{\infty} \sum_{\kappa_3=0}^{\infty} (-1)^{\kappa_1+\kappa_2+\kappa_3} \binom{\theta_1-1}{\kappa_1} \binom{\theta_2(\kappa_1+1)-1}{\kappa_2} \binom{\theta_4(\kappa_2+1)-1}{\kappa_3}, \tag{71}$$

and **B** (.,.) is the beta Type II function.

Thus, the mean and variance of EG-IKum distribution are,

$$\acute{\mu}_1 = E(X) = \prod_{j=1}^{4} \theta_j \sum_{\kappa^*} \Omega \ \mathbf{B}\{2, [\,\theta_3(\kappa_3+1)-1]\} = \prod_{j=1}^{4} \theta_j \sum_{\kappa^*} \Omega \ \frac{\Gamma[\theta_3(\kappa_3+1)-1]}{\Gamma[\theta_3(\kappa_3+1)+1]}, \tag{72}$$

$$\acute{\mu}_2 = E(X^2) = \prod_{j=1}^{4} \theta_j \sum_{\kappa^*} \Omega \ \mathbf{B}\{3, [\,\theta_3(\kappa_3+1)-2]\} = \prod_{j=1}^{4} \theta_j \sum_{\kappa^*} \Omega \ \frac{2\Gamma[\theta_3(\kappa_3+1)-2]}{\Gamma[\theta_3(\kappa_3+1)+1]}, \tag{73}$$

and,

$$V(X) = \acute{\mu}_2 - (\acute{\mu}_1)^2$$
$$= \prod_{j=1}^{4} \theta_j \sum_{\kappa^*} \Omega \ \frac{2\Gamma[\theta_3(\kappa_3+1)-2]}{\Gamma[\theta_3(\kappa_3+1)+1]} - \left(\prod_{j=1}^{4} \theta_j \sum_{\kappa^*} \Omega \ \frac{\Gamma[\theta_3(\kappa_3+1)-1]}{\Gamma[\theta_3(\kappa_3+1)+1]}\right)^2. \tag{74}$$

The coefficient of variation is given by

$$\gamma = \frac{\sqrt{V(X)}}{E(X)}$$
$$= \frac{\sqrt{\left(\prod_{j=1}^{4} \theta_j \sum_{\kappa^*} \Omega \ \frac{2\Gamma[\theta_3(\kappa_3+1)-2]}{\Gamma[\theta_3(\kappa_3+1)+1]} - \left\{\prod_{j=1}^{4} \theta_j \sum_{\kappa^*} \Omega \ \frac{\Gamma[\theta_3(\kappa_3+1)-1]}{\Gamma[\theta_3(\kappa_3+1)+1]}\right\}^2\right)}}{\prod_{j=1}^{4} \theta_j \sum_{\kappa^*} \Omega \ \frac{\Gamma[\theta_3(\kappa_3+1)-1]}{\Gamma[\theta_3(\kappa_3+1)+1]}}, \tag{75}$$

where $\sum_{\kappa^*}\Omega$ is given in (71)

iv. Mean residual life and mean past lifetime

The mean residual life of EG-IKum (distribution can be obtained as given below:

$$m_1(t_0) = \frac{\mu - \left(t_0 - \sum_{I^*} \tau_{I^*} \left\{\left[(1+t_0)^{-(\theta_3 I_3 - 1)}\right] - 1\right\}\right)}{R_{EGIK}(t_0; \underline{\theta})}, \tag{76}$$

where,

$$\sum_{I^*} \tau_{I^*} = \sum_{I_1=0}^{\infty} \sum_{I_2=0}^{\infty} \sum_{I_3=0}^{\infty} (-1)^{I_1+I_2+I_3} \binom{\theta_1}{I_1} \binom{\theta_2 I_1}{I_2} \binom{\theta_4 I_2}{I_3} \frac{1}{1-\theta_3 I_3}, \tag{77}$$

and $R_{EGIK}(t_0;\underline{\theta})$ is given by (66).

The mean past lifetime of the component can be obtained as follows:

$$m_2(t_0) = \frac{\sum_{I^*} \tau_{I^*} \left\{\left[(1+t_0)^{-(\theta_3 I_3 - 1)}\right] - 1\right\}}{F_{EGIK}(t_0; \underline{\theta})}, \tag{78}$$

where, $\sum_{I^*} \tau_{I^*}$ is defined in (77) and $F_{EGIK}(t_0;\underline{\theta})$ is given by (64).

3.1 Estimation for exponentiated generalized inverted Kumaraswamy distribution based on dual generalized order statistics

This subsection proposed the ML estimators, CIs for the parameters, rf and hrf of EG-IKum distribution based on dgos. Also, the shape parameters, rf and hrf of the EG-IKum distribution are estimated using Bayesian method. The Bayes estimators are derived under SE and LINEX loss functions based on dgos. Credible intervals for the parameters, rf and hrf are derived.

3.1.1 Maximum likelihood estimation

The ML method is used to estimate the parameters, rf and hrf of the EG-IKum distribution based on dgos. Also, the asymptotic $100(1-\omega)\%$ CIs for $\underline{\theta}$ are obtained.

Suppose that $T_{(1,n,m,k)}, T_{(2,n,m,k)}, \ldots, T_{(n,n,m,k)}$ are n dgos from the EG-IKum distribution, then the likelihood function can be derived by substituting (64) and (65) in (5) as follows:

$$L_{EGIK}(\underline{\theta};\underline{t}) \propto \prod_{j=1}^{4}\theta_j^n \prod_{i=1}^{n}(1+t_i)^{-(\theta_3+1)}\left[1-(1+t_i)^{-\theta_3}\right]^{\theta_4-1}\left\{1-\left[1-(1+t_i)^{-\theta_3}\right]^{\theta_4}\right\}^{\theta_2-1}$$
$$\times \prod_{i=1}^{n-1}\left(1-\left\{1-\left[1-(1+t_i)^{-\theta_3}\right]^{\theta_4}\right\}^{\theta_2}\right)^{\theta_1(m+1)-1} \tag{79}$$
$$\times\left(1-\left\{1-\left[1-(1+t_n)^{-\theta_3}\right]^{\theta_4}\right\}^{\theta_2}\right)^{\theta_1 k-1}.$$

The likelihood function can be rewritten as,

$$L_{EGIK}(\underline{\theta};\underline{t}) \propto \prod_{i=1}^{n}\delta_i^*\,(Q_i^*)^{-1}\prod_{j=1}^{4}\theta_j^n\, e^{\theta_1[(m+1)\sum_{i=1}^{n-1}ln(Q_i^*)+k ln(Q_n^*)]}, \tag{80}$$

where,

$$\delta_i^* = (1+t_i)^{-(\theta_3+1)}\left[1-(1+t_i)^{-\theta_3}\right]^{\theta_4-1}\left\{1-\left[1-(1+t_i)^{-\theta_3}\right]^{\theta_4}\right\}^{\theta_2-1}, \tag{81}$$

$$Q_i^* = \left(1-\left\{1-\left[1-(1+t_i)^{-\theta_3}\right]^{\theta_4}\right\}^{\theta_2}\right), \qquad i=1,2,\ldots,n. \tag{82}$$

The natural logarithm of the likelihood function is given by,

$$\ell_1 \equiv ln\, L_{EGIK}(\underline{\theta};\underline{t}) \propto n\,ln\,\theta_1 + n\,ln\,\theta_2 + n ln\theta_3 + n ln\,\theta_4 - (\theta_3+1)\sum_{i=1}^{n}ln(1+t_i)$$
$$+(\theta_4-1)\sum_{i=1}^{n}ln\left[1-(1+t_i)^{-\theta_3}\right] + (\theta_2-1)\sum_{i=1}^{n}ln\left\{1-\left[1-(1+t_i)^{-\theta_3}\right]^{\theta_4}\right\} \tag{83}$$
$$+[\theta_1(m+1)-1]\sum_{i=1}^{n-1}ln(Q_i^*) + (\theta_1 k-1)\,ln(Q_n^*).$$

Considering the parameters $\theta_j, j=1,2,3,4$ are unknown and differentiating the log likelihood function in (83) with respect to θ_j one gets,

$$\frac{\partial\ell_1}{\partial\theta_1} = \frac{n}{\theta_1} + (m+1)\sum_{i=1}^{n-1}ln(Q_i^*) + k\,ln(Q_n^*)\,, \tag{84}$$

$$\frac{\partial \ell_1}{\partial \theta_2} = \frac{n}{\theta_2} - \sum_{i=1}^{n} ln\left\{1-\left[1-(1+t_i)^{-\theta_3}\right]^{\theta_4}\right\} - (\theta_1(m+1)-1)\sum_{i=1}^{n-1} \frac{\left\{1-\left[1-(1+t_i)^{-\theta_3}\right]^{\theta_4}\right\}^{\theta_2} \ln\left\{1-\left[1-(1+t_i)^{-\theta_3}\right]^{\theta_4}\right\}}{Q_i^*} - \frac{(\theta_1 k-1)\left\{1-\left[1-(1+t_n)^{-\theta_3}\right]^{\theta_4}\right\}^{\theta_2} ln\left\{1-\left[1-(1+t_n)^{-\theta_3}\right]^{\theta_4}\right\}}{Q_n^*}, \quad (85)$$

$$\frac{\partial \ell_1}{\partial \theta_3} = \frac{n}{\theta_3} - \sum_{i=1}^{n} ln(1+t_{(i)}) + (\theta_4-1)\sum_{i=1}^{n} \frac{(1+t_i)^{-\theta_3} ln(1+t_i)}{[1-(1+t_i)^{-\theta_3}]} - (\theta_2-1)\sum_{i=1}^{n} \frac{\theta_4\left[1-(1+t_i)^{-\theta_3}\right]^{\theta_4-1}(1+t_i)^{-\theta_3} ln(1+t_i)}{\{1-[1-(1+t_i)^{-\theta_3}]^{\theta_4}\}} + [\theta_1(m+1)-1] \times \sum_{i=1}^{n-1} \frac{\theta_2\theta_4\left\{1-\left[1-(1+t_i)^{-\theta_3}\right]^{\theta_4}\right\}^{\theta_2-1}\left[1-(1+t_i)^{-\theta_3}\right]^{\theta_4-1}(1+t_i)^{-\theta_3} ln(1+t_i)}{Q_i^*} + \frac{\theta_2\theta_4(\theta_1 k-1)\left\{1-\left[1-(1+t_n)^{-\theta_3}\right]^{\theta_4}\right\}^{\theta_2-1}\left[1-(1+t_n)^{-\theta_3}\right]^{\theta_4-1}(1+t_n)^{-\theta_3} ln(1+t_n)}{Q_n^*} \quad (86)$$

and

$$\frac{\partial \ell_1}{\partial \theta_4} = \frac{n}{\theta_4} + \sum_{i=1}^{n} ln\left[1-(1+t_i)^{-\theta_3}\right] - (\theta_2-1)\sum_{i=1}^{n} \frac{\left[1-(1+t_i)^{-\theta_3}\right]^{\theta_4} ln\left[1-(1+t_i)^{-\theta_3}\right]}{\{1-[1-(1+t_i)^{-\theta_3}]^{\theta_4}\}} + [\theta_1(m+1)-1] \times \sum_{i=1}^{n-1} \frac{\theta_2\left\{1-\left[1-(1+t_i)^{-\theta_3}\right]^{\theta_4}\right\}^{\theta_2-1}\left[1-(1+t_i)^{-\theta_3}\right]^{\theta_4} ln\left[1-(1+t_i)^{-\theta_3}\right]}{(1-\{1-[1-(1+t_i)^{-\theta_3}]^{\theta_4}\}^{\theta_2})} + \frac{\theta_2(\theta_1 k-1)\left\{1-\left[1-(1+t_n)^{-\theta_3}\right]^{\theta_4}\right\}^{\theta_2-1}\left[1-(1+t_n)^{-\theta_3}\right]^{\theta_4} ln\left[1-(1+t_n)^{-\theta_3}\right]}{\left(1-\left\{1-\left[1-(1+t_n)^{-\theta_3}\right]^{\theta_4}\right\}^{\theta_2}\right)}, \quad (87)$$

where Q_i^* and Q_n^* are given in (82).

Equating (84) to zero, then solving it numerically, one can obtain the ML estimator of θ_1,

$$\hat{\theta}_1 = \frac{-n}{(m+1)\sum_{i=1}^{n-1} \ln(\hat{Q}_i^*) + k\, ln(\hat{Q}_n^*)}, \quad (88)$$

where,

$$\hat{Q}_i^* = \left(1-\left\{1-\left[1-(1+t_i)^{-\hat{\theta}_3}\right]^{\hat{\theta}_4}\right\}^{\hat{\theta}_2}\right), \qquad i = 1,2,\dots,n. \quad (89)$$

The system of non-linear Equations (84)-(87) can be solved numerically using Newton-Raphson method, to obtain the ML estimates $\hat{\theta}_1$, $\hat{\theta}_1$, $\hat{\theta}_3$ and $\hat{\theta}_4$.

The invariance property can be used to obtain the ML estimates $\hat{R}_{EGIK}(t)$ and $\hat{h}_{EGIK}(t)$, just replacing the parameters $\theta_j, j = 1,2,3,4$ by their corresponding ML estimates.

Asymptotic variance-covariance matrix of the maximum likelihood estimators

For large sample sizes, the ML estimators under appropriate regularity conditions are consistent and asymptotically unbiased as well as asymptotically normally distributed. Therefore, the asymptotic normality of ML estimation can be used to compute the two sided approximate 100 $(1-\omega)$% CIs for $\underline{\theta}$ as follows:

$$L_{\theta_j} = \hat{\theta}_j - Z_{\left(1-\frac{\omega}{2}\right)}\sqrt{\hat{var}(\hat{\theta}_j)} \quad \text{and} \quad U_{\theta_j} = \hat{\theta}_j + Z_{\left(1-\frac{\omega}{2}\right)}\sqrt{\hat{var}(\hat{\theta}_j)}\ . \tag{90}$$

Also, the asymptotic 100 $(1-\omega)$% confidence intervals for rf and hrf are,

$$\hat{R}_{EGIK}(t) \pm Z_{\left(1-\frac{\omega}{2}\right)}\sqrt{\hat{var}(\hat{R}_{EGIK}(t))} \quad \text{and} \quad \hat{h}_{EGIK}(t) \pm Z_{\left(1-\frac{\omega}{2}\right)}\sqrt{\hat{var}\left(\hat{h}_{EGIK}(t)\right)}, \tag{91}$$

where $Z_{\left(1-\frac{\omega}{2}\right)}$ is the standard normal percentile and $(1-\omega)$ is the confidence coefficient.

3.1.2 Bayesian estimation

The parameters, rf and hrf are estimated based on dgos using the Bayesian method, under and LINEX loss functions. Also, the credible intervals are obtained.

a. Point estimation

Let θ_1, θ_2, θ_3 and θ_4 be independent random variables with a gamma prior distribution. Hence, a joint prior density function of $\underline{\theta} = (\theta_1, \theta_2, \theta_3, \theta_2)'$ is given by,

$$\pi(\underline{\theta}) \propto \prod_{j=1}^{4} \theta_j^{c_j-1} e^{-d_j\theta_j} .\ \theta_j, d_j, c_j > 0, \qquad j = 1,2,3,4, \tag{92}$$

where c_j, d_j are the hyper parameters.

The joint posterior density can be derived by using (80) and (92) as follows:

$$\pi^*_{EGIK}(\underline{\theta}|\underline{t}) = \Psi \prod_{j=2}^{4} \theta_j^{c_j+n-1} e^{-d_j\theta_j} \prod_{i=1}^{n} \delta_i^*\ (Q_i^*)^{-1} \theta_1^{c_1+n-1} e^{-\theta_1[d_1-(m+1)\sum_{i=1}^{n-1} ln(Q_i^*) - k ln(Q_n^*)]}\ , \tag{93}$$

where Ψ is the normalizing constant and

$$\Psi^{-1} = \Gamma(c_1+n) \int_0^\infty \int_0^\infty \int_0^\infty \left(\prod_{j=2}^{4} \theta_j^{c_j+n-1} e^{-d_j\theta_j} \prod_{i=1}^{n} \delta_i^*\ (Q_i^*)^{-1} \right) \times \left[\frac{1}{\left[d_1 - (m+1)\sum_{i=1}^{n-1} ln(Q_i^*) - k ln(Q_n^*)\right]^{c_1+n}} \right] d\theta_2 d\theta_3 d\theta_4.$$

Let,

$$\varphi_1^* = \int_0^\infty \int_0^\infty \int_0^\infty \left(\prod_{j=2}^{4} \theta_j^{c_j+n-1} e^{-d_j\theta_j} \prod_{i=1}^{n} \delta_i^*\ (Q_i^*)^{-1} \right) \times \left\{ \frac{1}{\left[d_1 - (m+1)\sum_{i=1}^{n-1} ln(Q_i^*) - k ln(Q_n^*)\right]^{c_1+n}} \right\} d\theta_2 d\theta_3 d\theta_4. \tag{94}$$

Hence, the joint posterior distribution of $\underline{\theta}$ given $\underline{t}$ can be written as follows:

$$\pi^*_{EGIK}(\underline{\theta}|\underline{t})$$

$$= \frac{1}{\varphi_1^*\Gamma(c_1+n)}\left[\prod_{j=2}^{4}\theta_j^{c_j+n-1}e^{-d_j\theta_j}\prod_{i=1}^{n}\delta_i^*\,(Q_i^*)^{-1}\theta_1^{c_1+n-1}e^{-\theta_1[d_1-(m+1)\sum_{i=1}^{n-1}ln(Q_i^*)-kln(Q_n^*)]}\right], \tag{95}$$

where Q_i^* and Q_n^* are given in (82).

I. Bayesian estimation of exponentiated generalized inverted Kumaraswamy distribution under squared error loss function based on dual generalized order statistics

Under the SE loss function, the Bayes estimators of the parameters $\underline{\theta}$ are given by their marginal posterior expectations using (93) as shown below:

$$\tilde{\theta}_{j(SE)EGIK} = E(\theta_j|\underline{t}) = \int_{\underline{\theta}}^{\infty}\theta_j\,\pi^*_{EGIK}(\underline{\theta}|\underline{t})\,d\underline{\theta}\,, \qquad j=1,2,3,4, \tag{96}$$

where,

$$\int_{\underline{\theta}}^{\infty} = \int_0^{\infty}\int_0^{\infty}\int_0^{\infty}\int_0^{\infty} \quad \text{and } d\underline{\theta} = d\theta_1 d\theta_2 d\theta_3 d\theta_4$$

The Bayes estimators of the rf and hrf under the SE loss function can be obtained using (66), (67) and (95) as follows:

$$\tilde{R}_{(SE)EGIK}(t) = E(R_{EGIK}(t)|\underline{t}) = 1 - \int_{\underline{\theta}}(Q^*)^{\theta_1}\,\pi^*_{EGIK}(\underline{\theta}|\underline{t})\,d\underline{\theta}$$

$$= 1 - \int_0^{\infty}\int_0^{\infty}\int_0^{\infty}\frac{\left(\prod_{j=2}^{4}\theta_j^{c_j+n-1}e^{-d_j\theta_j}\prod_{i=1}^{n}\delta_i^*(Q_i^*)^{-1}\right)}{\varphi_1^*[d_1-(m+1)\sum_{i=1}^{n-1}ln(Q_i^*)-kln(Q_n^*)-ln(Q^*)]^{c_1+n}}d\theta_2\,d\theta_3 d\theta_4\,, \tag{97}$$

and

$$\tilde{h}_{(SE)EGIK}(t) = E(h_{EGIK}(t)|\underline{t})$$

$$= \int_{\underline{\theta}} h_{EGIK}(t)\frac{1}{\varphi_1^*\Gamma(c_1+n)}\left[\prod_{j=2}^{4}\theta_j^{c_j+n-1}e^{-d_j\theta_j}\prod_{i=1}^{n}\delta_i^*\,(Q_i^*)^{-1}\theta_1^{c_1+n-1}e^{-\theta_1[d_1-(m+1)\sum_{i=1}^{n-1}ln(Q_i^*)-kln(Q_n^*)]}\right]d\underline{\theta}, \tag{98}$$

where, $Q^* = [1-(1-(1-(1+t)^{-\theta_3})^{-\theta_4})^{-\theta_2}]$, Q_i^* and Q_n^* are given in (82) and φ_1^* is defined in (94).
To obtain the Bayes estimates of the parameters, rf and hrf, (96) - (98) should be solved numerically.

II. Bayesian estimation of exponentiated generalized inverted Kumaraswamy distribution under LINEX loss function based on dual generalized order statistics

Under the LINEX loss function, the Bayes estimators for the shape parameters $\underline{\theta}$ are given, respectively, by

$$\tilde{\theta}_{j(LNX)EGIK} = \frac{-1}{v}\ln E(e^{-v\theta_j}|\underline{t})$$

$$= \frac{-1}{v}ln\left\{\int_{\underline{\theta}}^{\infty}e^{-v\theta_j}\frac{1}{\varphi_1^*\Gamma(c_1+n)}\left[\prod_{j=2}^{4}\theta_j^{c_j+n-1}e^{-d_j\theta_j}\prod_{i=1}^{n}\delta_i^*\,(Q_i^*)^{-1}\theta_1^{c_1+n-1}e^{-\theta_1[d_1-(m+1)\sum_{i=1}^{n-1}ln(Q_i^*)-kln(Q_n^*)]}\right]d\underline{\theta}\right\},$$

$$j=1,2,3,4, \tag{99}$$

where δ_i^* is given in (81), Q_i^* and Q_n^* are given in (82) and φ_1^* is defined in (94).

Also, the Bayes estimator for rf and hrf based on dgos can be obtained as follows:

$$\begin{aligned}\tilde{R}_{(LNX)EGIK}(t) &= \frac{-1}{v}\ln E\left(e^{-vR_{(EGIK)}(t)}|\underline{t}\right) \\ &= \frac{-1}{v} ln\left[\int_{\underline{\theta}} e^{-vR_{(EGIK)}(t)}\pi(\underline{\theta}|\underline{t})\, d\underline{\theta}\right],\end{aligned} \tag{100}$$

and

$$\begin{aligned}\tilde{h}_{(LNX)EGIK}(t) &= \frac{-1}{v}\ln E\left(e^{-vh_{(EGIK)}(t)}|\underline{t}\right) \\ &= \frac{-1}{v} ln\left[\int_{\underline{\theta}} e^{-vh_{(EGIK)}(t)}\pi(\underline{\theta}|\underline{t})\, d\underline{\theta}\right].\end{aligned} \tag{101}$$

To obtain the Bayes estimates of the parameters, rf and hrf, Equations (99)–(101) should be solved numerically.

b. Credible interval

Since, the posterior distribution is given by (95), then a $100(1 - \omega)$ % credible interval for is $\left(L(\underline{t}), U(\underline{t})\right)$, where $P\left[L(\underline{t}) < \underline{\theta} < U(\underline{t})|\underline{t}\right] = \int_{L(\underline{t})}^{U(\underline{t})} \pi^*_{EGIK}(\underline{\theta}|\underline{t})\, d\underline{\theta} = 1 - \omega$.

Then a $100(1 - \omega)$% credibility interval for θ_j based on dgos is $(L_j(\underline{t}), U_j(\underline{t}))$ where,

$$P\left[\theta_j > L_j(\underline{t})|\underline{t}\right] = \int_{L_j(\underline{t})}^{\infty} \pi^*_{EGIK}(\underline{\theta}|\underline{t})\, d\theta_j = 1 - \frac{\omega}{2}, \qquad j = 1,2,\ldots,4, \tag{102}$$

and

$$P\left[\theta_j > U_j(\underline{t})|\underline{t}\right] = \int_{U_j(\underline{t})}^{\infty} \pi^*_{EGIK}(\underline{\theta}|\underline{t})\, d\theta_j = \frac{\omega}{2}, \qquad j = 1,2,\ldots,4. \tag{103}$$

3.2 Prediction for exponentiated generalized inverted Kumaraswamy distribution based on dual generalized order statistics

In this subsection, the ML and Bayesian prediction (point and interval) for a future observation of the EG-IKum of distributions based on dgos, are considered under and LINEX loss functions.

Let $T(1,n,m,k),\ldots,T(r,n,m,k)$ be a dgos of size n with the pdf; $f(t;\ \underline{\theta})$, and suppose $Y(1,n_y,m_y,k_y),\ldots,Y(r_y,n_y,m_y,k_y)$, $k_y > 0$, $m_y \in \mathbb{R}$ is a second independent dgos of size n_y, of future observations from the same distribution. Using (6), (7), (64) and (65), the pdf of the dgos $Y_{(s)}$ can be obtained by replacing $t_{(r)}$ by $y_{(s)}$ as follows:

$$\begin{aligned}f_{EGIK}(y_{(s)}|\underline{\theta}) &= \frac{\zeta_{s-1}\prod_{j=1}^{4}\theta_j}{(s-1)!}\left(1+y_{(s)}\right)^{-(\theta_3+1)}\left[1-\left(1+y_{(s)}\right)^{-\theta_3}\right]^{\theta_4-1} \\ &\times\left\{1-\left[1-\left(1+y_{(s)}\right)^{-\theta_3}\right]^{\theta_4}\right\}^{\theta_2-1}\left(Q^*_{y_{(s)}}\right)^{\theta_1\gamma_s-1} g^{s-1}_{1m_y}\left[F(y_{(s)})\right],\end{aligned} \tag{104}$$

where,

$$\begin{aligned}&Q^*_{y_{(s)}} = \left(1-\left\{1-\left[1-\left(1+y_{(s)}\right)^{-\theta_3}\right]^{\theta_4}\right\}^{\theta_2}\right), \\ &\zeta_{s-1} = \prod_{j_1=1}^{s}\gamma_{j_1},\ g_M(y_s) = h_M(y_s) - h_M(1),\ \gamma_s = k_y + (n_y - s)(m_y+1),\ for\ all\ 1 \le s \le n_y,\end{aligned} \tag{105}$$

and

$$g^{s-1}_{1m_y}\left[F(y_{(s)})\right] = \begin{cases}\frac{1}{(m_y+1)^{s-1}}\left[1-\left(Q^*_{y_{(s)}}\right)^{\theta_1(m_y+1)}\right]^{s-1}, & m_y \neq -1, \\ \left[-ln\left(Q^*_{y_{(s)}}\right)^{\theta_1}\right]^{s-1}, & m_y = -1.\end{cases} \tag{106}$$

For the future sample of size n_y, let $Y_{(s)}$ denotes the sth ordered lifetime, $1 \le s \le n_y$. The pdf of the dgos $y_{(s)}$ from EG-IKum distribution is obtained by substituting (106) in (104).

Case one: for $m_y \neq -1$

$$f_{EGIK}(y_{(s)}|\underline{\theta}) = \frac{\xi \prod_{j=1}^{4} \theta_j \left(1+y_{(s)}\right)^{-(\theta_3+1)} \left[1-\left(1+y_{(s)}\right)^{-\theta_3}\right]^{\theta_4-1}}{Q^*_{y_{(s)}}}$$
$$\times \left\{1-\left[1-\left(1+y_{(s)}\right)^{-\theta_3}\right]^{\theta_4}\right\}^{\theta_2-1} e^{\theta_1 \sum_{i_1=0}^{s-1} \varpi_{i_1} ln\left(\eta_{i_1} Q^*_{y_{(s)}}\right)}, \quad y_{(s)} > 0;\ \theta_j > 0, \tag{107}$$

where,

$\xi = \frac{\zeta_{s-1}}{(m_y+1)^{s-1}(s-1)!}$, $\eta_{i_1} = (-1)^{i_1}\binom{s-1}{i_1}$ and $\varpi_{i_1} = [\gamma_s + i_1(m_y+1)]$.

Case two: for for $m_y = -1$

$$f_{EGIK}(y_{(s)}|\underline{\theta}) = \frac{\prod_{j=1}^{4} \theta_j k_y^s \theta_1^{s-1}}{(s-1)!} \left(1+y_{(s)}\right)^{-(\theta_3+1)} \left[1-\left(1+y_{(s)}\right)^{-\theta_3}\right]^{\theta_4-1}$$
$$\times \left\{1-\left[1-\left(1+y_{(s)}\right)^{-\theta_3}\right]^{\theta_4}\right\}^{\theta_2-1} \left(Q^*_{y_{(s)}}\right)^{\theta_1 k_y - 1} \left[-ln\left(Q^*_{y_{(s)}}\right)\right]^{s-1}, \tag{108}$$
$$y_{(s)} > 0;\ \theta_j > 0,$$

where $Q^*_{y(s)}$ is given in (105).

3.2.1 Maximum likelihood prediction for exponentiated generalized inverted Kumaraswamy distribution

The ML point and interval prediction are obtained considering a two-sample prediction based on dgos. The ML prediction can be derived by using the conditional pdf of the sth future dgos which are given by (107) and (116) after replacing the parameters (θ_j) by their ML estimators ($\hat{\theta}_j$).

a. **Point prediction**

The MLP of the future dgos $Y_{(s)}$ can be obtained using (107) and (108) as follows:

$$\hat{y}_{(s)(ML)} = E\left(y_{(s)}|\hat{\underline{\theta}}\right) = \int_0^\infty y_{(s)} f\left(y_{(s)}|\hat{\underline{\theta}}\right) dy_{(s)}. \tag{109}$$

Case one: for $M \neq -1$

$$\hat{y}_{(s)(ML)} = \int_0^\infty y_{(s)}\, \xi \prod_{j=1}^{4} \hat{\theta}_j \left(1+y_{(s)}\right)^{-(\hat{\theta}_3+1)} \left[1-\left(1+y_{(s)}\right)^{-\hat{\theta}_3}\right]^{\hat{\theta}_4-1}$$
$$\times \left\{1-\left[1-\left(1+y_{(s)}\right)^{-\hat{\theta}_3}\right]^{\hat{\theta}_4}\right\}^{\hat{\theta}_2-1} \sum_{\varsigma=0}^{s-1} \eta_\varsigma \left[u_{y_{(s)}}\right]^{\hat{\theta}_1 \varpi_i - 1} dy_{(s)}. \tag{110}$$

Case two: for $M = -1$

$$\hat{y}_{(s)(ML)} = \int_0^\infty y_{(s)} \frac{\prod_{j=1}^{4} \hat{\theta}_j k_y^s \hat{\theta}_1^{s-1}}{(s-1)!} \left(1+y_{(s)}\right)^{-(\hat{\theta}_3+1)} \left[1-\left(1+y_{(s)}\right)^{-\hat{\theta}_3}\right]^{\hat{\theta}_4-1}$$
$$\times \left\{1-\left[1-\left(1+y_{(s)}\right)^{-\hat{\theta}_3}\right]^{\hat{\theta}_4}\right\}^{\hat{\theta}_2-1} \left[u_{y_{(s)}}\right]^{\hat{\theta}_1 k_y - 1} \left[-\ln\left(u_{y_{(s)}}\right)\right]^{s-1} dy_{(s)}. \tag{111}$$

b. Interval prediction

The MLPB for the future dgos $Y_{(s)}$, $1 \leq s \leq N$ can be derived from the following probabilities

$$P\left[Y_{(s)} > L(\underline{t})|\underline{t}\right] = \int_{L(\underline{t})}^{\infty} f\left(y_{(s)}|\underline{\hat{\theta}};\underline{t}\right)dy_{(s)} = 1 - \frac{\omega}{2}, \tag{112}$$

and

$$P\left[Y_{(s)} > U(\underline{t})|\underline{t}\right] = \int_{U(\underline{t})}^{\infty} f\left(y_{(s)}|\underline{\hat{\theta}};\underline{t}\right)dy_{(s)} = \frac{\omega}{2}. \tag{113}$$

Substituting (107) and (108) in (112) and (113), then the lower and upper bounds are obtained as given below:

Case one: for $M \neq -1$

$$P\left[Y_{(s)} > L(\underline{t})|\underline{t}\right]$$
$$= \int_{L(\underline{t})}^{\infty} \xi \prod_{j=1}^{4} \hat{\theta}_j \left(1 + y_{(s)}\right)^{-(\hat{\theta}_3+1)} \left[1 - \left(1 + y_{(s)}\right)^{-\hat{\theta}_3}\right]^{\hat{\theta}_4-1}$$
$$\times \left\{1 - \left[1 - \left(1 + y_{(s)}\right)^{-\hat{\theta}_3}\right]^{\hat{\theta}_4}\right\}^{\hat{\theta}_2-1} \sum_{\varsigma=0}^{s-1} \eta_\varsigma \left(u_{y_{(s)}}\right)^{\hat{\theta}_1\varpi_i-1} dy_{(s)} = 1 - \frac{\omega}{2}, \tag{114}$$

and

$$P\left[Y_{(s)} > U(\underline{t})|\underline{t}\right]$$
$$= \int_{U(\underline{t})}^{\infty} \xi \prod_{j=1}^{4} \hat{\theta}_j \left(1 + y_{(s)}\right)^{-(\hat{\theta}_3+1)} \left[1 - \left(1 + y_{(s)}\right)^{-\hat{\theta}_3}\right]^{\hat{\theta}_4-1}$$
$$\times \left\{1 - \left[1 - \left(1 + y_{(s)}\right)^{-\hat{\theta}_3}\right]^{\hat{\theta}_4}\right\}^{\hat{\theta}_2-1} \sum_{\varsigma=0}^{s-1} \eta_\varsigma \left(u_{y_{(s)}}\right)^{\hat{\theta}_1\varpi_i-1} dy_{(s)} = \frac{\omega}{2}. \tag{115}$$

Case two: for $M = -1$

$$P\left[Y_{(s)} > L(\underline{t})|\underline{t}\right]$$
$$= \int_{L(\underline{t})}^{\infty} \frac{\prod_{j=1}^{4} \hat{\theta}_j \, k_y^s \, \hat{\theta}_1^{\,s-1}}{(s-1)!} \left(1 + y_{(s)}\right)^{-(\hat{\theta}_3+1)} \left[1 - \left(1 + y_{(s)}\right)^{-\hat{\theta}_3}\right]^{\hat{\theta}_4-1}$$
$$\times \left\{1 - \left[1 - \left(1 + y_{(s)}\right)^{-\hat{\theta}_3}\right]^{\hat{\theta}_4}\right\}^{\hat{\theta}_2-1} \left(u_{y_{(s)}}\right)^{\hat{\theta}_1 k_y-1} \left[-\ln\left(u_{y_{(s)}}\right)\right]^{s-1} dy_s = 1 - \frac{\omega}{2}, \tag{116}$$

and

$$P\left[Y_{(s)} > U(\underline{t})|\underline{t}\right]$$
$$= \int_{U(\underline{t})}^{\infty} \frac{\prod_{j=1}^{4} \hat{\theta}_j \, k_y^s \, \hat{\theta}_1^{\,s-1}}{(s-1)!} \left(1 + y_{(s)}\right)^{-(\hat{\theta}_3+1)} \left[1 - \left(1 + y_{(s)}\right)^{-\hat{\theta}_3}\right]^{\hat{\theta}_4-1}$$
$$\times \left\{1 - \left[1 - \left(1 + y_{(s)}\right)^{-\hat{\theta}_3}\right]^{\hat{\theta}_4}\right\}^{\hat{\theta}_2-1} \left(u_{y_{(s)}}\right)^{\hat{\theta}_1 k_y-1} \left[-\ln\left(u_{y_{(s)}}\right)\right]^{s-1} dy_s = \frac{\omega}{2}. \tag{117}$$

3.2.2 Bayesian prediction for exponentiated generalized inverted Kumaraswamy distribution

Considering a two-sample Bayesian prediction based on dgos for the future observation $Y_{(s)}$, $1 \leq s \leq n_y$, the BPD function can be derived as follows:
The BPD function is,

$$f_{EGIK}(y_{(s)}|\underline{t}) = \int_{\underline{\theta}} f_{EGIK}(y_{(s)}|\underline{\theta})\pi^*_{EGIK}(\underline{\theta}|\underline{t})d\underline{\theta}\,, \tag{118}$$

where $\pi^*_{EGIK}(\underline{\theta}\,|\,\underline{t})$ is the joint posterior distribution of $\underline{\theta}$ and $f_{EGIK}(y_{(s)}\,|\,\underline{\theta})$ is the pdf of $y_{(s)}$.

Case one: for $m_y \neq -1$

The BPD of $y(s)$ given $\underline{t}$ is obtained by substituting (95) and (107) in (118), hence,

$$f_{EGIK}(y_{(s)}|\underline{t}) = \int_0^\infty\int_0^\infty\int_0^\infty \frac{\xi(c_1+n)(1+y_{(s)})^{-(\theta_3+1)}\left[1-(1+y_{(s)})^{-\theta_3}\right]^{\theta_4-1}}{\varphi_1^*\prod_{i=1}^n \delta_i^*\,(Q_i^*)^{-1}\,\tau_5^{(c_1+n+1)}} \times\left\{1-\left[1-(1+y_{(s)})^{-\theta_3}\right]^{\theta_4}\right\}^{\theta_2-1}\prod_{j=2}^{4}\theta_j^{c_j+n}e^{-d_j\theta_j}\,d\theta_2 d\theta_3 d\theta_4, \tag{119}$$

where,
$\tau_5 = \left[d_1-(m+1)\sum_{i=1}^{n-1} ln(Q_i^*) - k ln(Q_n^*) - \sum_{i_1=0}^{s-1}\varpi_{i_1}\, ln\left(\eta_{i_1} Q^*_{y_{(s)}}\right)\right]$, $Q^*_{y_{(s)}}$ is given in (105), δ_i^* is defined in (81), Q_i^* and Q_n^* are given in (82) and φ_1^* is defined in (94).

Case two: for $m_y = -1$

Substituting (95) and (108) in (118), the BP density of $Y_{(s)}$ given $\underline{t}$ one can obtain,

$$f_{EGIK}(y_{(s)}|\underline{t}) = \int_0^\infty\int_0^\infty\int_0^\infty \frac{k_y^s\prod_{j=2}^4\theta_j^{c_j+n}e^{-d_j\theta_j}(1+y_{(s)})^{-(\theta_3+1)}\left[1-(1+y_{(s)})^{-\theta_3}\right]^{\theta_4-1}}{\varphi_1^*(s-1)!\,Q^*_{y_{(s)}}\prod_{i=1}^n\delta_i^*\,(Q_i^*)^{-1}\tau_2^{(c_1+n+s)}\Gamma(c_1+n)} \times\left\{1-\left[1-(1+y_{(s)})^{-\theta_3}\right]^{\theta_4}\right\}^{\theta_2-1}\left[-ln\left(Q^*_{y_{(s)}}\right)\right]^{s-1}\Gamma(c_1+n+s)\,d\theta_2 d\theta_3 d\theta_4, \tag{120}$$

where,
$\tau_6 = \left[d_1-(m+1)\sum_{i=1}^{n-1} ln(Q_i^*) - k ln(Q_n^*) - k_y ln\left(Q^*_{y_{(s)}}\right)\right]$, $Q^*_{y_{(s)}}$ is given in (105), δ_i^* is defined in (81), Q_i^* and Q_n^* are given in (82) and φ_1^* is defined in (94).

a. Point prediction

The BP of the future dgos $Y_{(s)}$ can be derived under SE and LINEX loss functions as follows:

Case one: for $m_y \neq -1$

The BP of the future dgos $Y_{(s)}$ can be obtained under SE loss function using (119) as follows:

$$y^*_{EGIK(S)(SE)} = E(y_{(s)}|\underline{t}) = \int_0^\infty y_{(s)}\, f_{EGIK}(y_{(s)}|\underline{t})\; dy_{(s)}$$

$$= \int_0^\infty\int_0^\infty\int_0^\infty\int_0^\infty \frac{\xi y_{(s)}(c_1+n)(1+y_{(s)})^{-(\theta_3+1)}\left[1-(1+y_{(s)})^{-\theta_3}\right]^{\theta_4-1}}{\varphi_1^*\prod_{i=1}^n\delta_i^*\,(Q_i^*)^{-1}\,\tau_5^{(c_1+n+1)}} \times\left\{1-\left[1-(1+y_{(s)})^{-\theta_3}\right]^{\theta_4}\right\}^{\theta_2-1}\prod_{j=2}^{4}\theta_j^{c_j+n}e^{-d_j\theta_j}\,d\theta_2 d\theta_3 d\theta_4 dy_{(s)}. \tag{121}$$

The BP of the future dgos $Y_{(s)}$ can be obtained under LINEX loss function using (119) as given below:

$$y^*_{EGIK(S)(LNX)} = \frac{-1}{v} ln\left[\int_0^\infty\int_0^\infty\int_0^\infty\int_0^\infty \frac{\xi e^{-vy_{(s)}}(c_1+n)\left(1+y_{(s)}\right)^{-(\theta_3+1)}\left[1-\left(1+y_{(s)}\right)^{-\theta_3}\right]^{\theta_4-1}}{\varphi_1^*\prod_{i=1}^n \delta_i^*\left(Q_i^*\right)^{-1}\tau_5^{(c_1+n+1)}}\right. \tag{122}$$

$$\left.\times\left\{1-\left[1-\left(1+y_{(s)}\right)^{-\theta_3}\right]^{\theta_4}\right\}^{\theta_2-1}\prod_{j=2}^4 \theta_j^{c_j+n}e^{-d_j\theta_j}\,d\theta_2 d\theta_3 d\theta_4 dy_{(s)}\right],$$

where,

$\tau_5 = \left[d_1-(m+1)\sum_{i=1}^{n-1} ln(Q_i^*) - kln(Q_n^*) - \sum_{i_1=0}^{s-1}\varpi_{i_1} ln\left(\eta_{i_1}Q^*_{y_{(s)}}\right)\right]$, $Q^*_{y_{(s)}}$ is given in (105), δ_i^* is defined in (81), Q_i^* and Q_n^* are given in (82) and φ_1^* is defined in (94).

Case one: for $m_y = -1$

The BP of dgos $Y_{(s)}$ can be obtained under SE loss function using (120) as follows:

$$y^*_{EGIK(S)(SE)} = E\left(y_{(s)}|\underline{t}\right)$$

$$= \int_0^\infty\int_0^\infty\int_0^\infty\int_0^\infty \frac{y_{(s)}\prod_{j=2}^4 \theta_j^{c_j+n}e^{-d_j\theta_j}k_y^s\left[-ln\left(Q^*_{y_{(s)}}\right)\right]^{s-1}\Gamma(c_1+n+s)}{\varphi_1^*(s-1)!\,Q^*_{y_{(s)}}\prod_{i=1}^n\delta_i^*\left(Q_i^*\right)^{-1}\tau_6^{(c_1+n+s)}\Gamma(c_1+n)} \tag{123}$$

$$\times\left(1+y_{(s)}\right)^{-(\theta_3+1)}\left[1-\left(1+y_{(s)}\right)^{-\theta_3}\right]^{\theta_4-1}$$

$$\times\left\{1-\left[1-\left(1+y_{(s)}\right)^{-\theta_3}\right]^{\theta_4}\right\}^{\theta_2-1} d\theta_2 d\theta_3 d\theta_4 dy_{(s)}.$$

The BP of the future dgos Y_s can be obtained under LINEX loss function using (120) as given below:

$$y^*_{EGIK(S)(LNX)} = \frac{-1}{v} ln\left[\int_0^\infty\int_0^\infty\int_0^\infty\int_0^\infty \frac{\prod_{j=2}^4 \theta_j^{c_j+n}e^{-d_j\theta_j}k_y^s e^{-vy_{(s)}}\left[-ln\left(Q^*_{y_{(s)}}\right)\right]^{s-1}\Gamma(c_1+n+s)}{\varphi_1^*(s-1)!\,Q^*_{y_{(s)}}\prod_{i=1}^n\delta_i^*\left(Q_i^*\right)^{-1}\tau_6^{(c_1+n+s)}\Gamma(c_1+n)}\right.$$

$$\times\left(1+y_{(s)}\right)^{-(\theta_3+1)}\left[1-\left(1+y_{(s)}\right)^{-\theta_3}\right]^{\theta_4-1} \tag{124}$$

$$\left.\times\left\{1-\left[1-\left(1+y_{(s)}\right)^{-\theta_3}\right]^{\theta_4}\right\}^{\theta_2-1} d\theta_2 d\theta_3 d\theta_4 dy_{(s)}\right],$$

where, $\tau_6 = \left[d_1-(m+1)\sum_{i=1}^{n-1} ln(Q_i^*) - kln(Q_n^*) - k_y ln\left(Q^*_{y_{(s)}}\right)\right]$, $Q^*_{y_{(s)}}$ is given in (105), δ_i^* is defined in (81), Q_1^* and Q_n^* are given in (84) and φ_1^* is defined in (94).

b. Interval prediction

The BPB of the future dogs $Y_{(s)}$ can be obtained using (119) and (120) as follows:

Case one: for $m_y \neq -1$

$$P\left[Y_{EGIK(s)} > L(\underline{t})|\underline{t}\right] = \int_{L(\underline{t})}^{\infty}\int_0^{\infty}\int_0^{\infty}\int_0^{\infty} \frac{\xi(c_1+n)\left(1+y_{(s)}\right)^{-(\theta_3+1)}\left[1-\left(1+y_{(s)}\right)^{-\theta_3}\right]^{\theta_4-1}}{\varphi_1^* \prod_{i=1}^{n}\delta_i^*\left(Q_i^*\right)^{-1}\tau_5^{(c_1+n+1)}} \times \left\{1-\left[1-\left(1+y_{(s)}\right)^{-\theta_3}\right]^{\theta_4}\right\}^{\theta_2-1}\prod_{j=2}^{4}\theta_j^{c_j+n}e^{-d_j\theta_j}\,d\theta_2 d\theta_3 d\theta_4\,dy_{(s)} = 1-\frac{\omega}{2}, \quad (125)$$

and

$$P\left[Y_{EGIK(s)} > U(\underline{t})|\underline{t}\right] = \int_{U(\underline{t})}^{\infty}\int_0^{\infty}\int_0^{\infty}\int_0^{\infty} \frac{\xi(c_1+n)\left(1+y_{(s)}\right)^{-(\theta_3+1)}\left[1-\left(1+y_{(s)}\right)^{-\theta_3}\right]^{\theta_4-1}\left\{1-\left[1-\left(1+y_{(s)}\right)^{-\theta_3}\right]^{\theta_4}\right\}^{\theta_2-1}\prod_{j=2}^{4}\theta_j^{c_j+n}e^{-d_j\theta_j}}{\varphi_1^* \prod_{i=1}^{n}\delta_i^*\left(Q_i^*\right)^{-1}\tau_5^{(c_1+n+1)}} d\theta_2 d\theta_3 d\theta_4\,dy_{(s)} = \frac{\omega}{2}. \quad (126)$$

Case two: for $m_y = -1$

$$P\left[Y_{EGIK(s)} > L(\underline{t})|\underline{t}\right] = \int_{L(\underline{t})}^{\infty}\int_0^{\infty}\int_0^{\infty}\int_0^{\infty} \frac{k_y^s \prod_{j=2}^{4}\theta_j^{c_j+n}e^{-d_j\theta_j}\left(1+y_{(s)}\right)^{-(\theta_3+1)}\left[1-\left(1+y_{(s)}\right)^{-\theta_3}\right]^{\theta_4-1}}{\varphi_1^*(s-1)!\,Q_{y_{(s)}}^* \prod_{i=1}^{n}\delta_i^*\left(Q_i^*\right)^{-1}\tau_6^{(c_1+n+s)}\Gamma(c_1+n)} \times \left\{1-\left[1-\left(1+y_{(s)}\right)^{-\theta_3}\right]^{\theta_4}\right\}^{\theta_2-1}\left[-\ln\left(Q_{y_{(s)}}^*\right)\right]^{s-1}\Gamma(c_1+n+s)\,d\theta_2 d\theta_3 d\theta_4\,dy_{(s)} = 1-\frac{\omega}{2}, \quad (127)$$

and

$$P\left[Y_{EGIK(s)} > U(\underline{t})|\underline{t}\right] = \int_{U(\underline{t})}^{\infty}\int_0^{\infty}\int_0^{\infty}\int_0^{\infty} \frac{k_y^s \prod_{j=2}^{4}\theta_j^{c_j+n}e^{-d_j\theta_j}\left(1+y_{(s)}\right)^{-(\theta_3+1)}\left[1-\left(1+y_{(s)}\right)^{-\theta_3}\right]^{\theta_4-1}}{\varphi_1^*(s-1)!\,Q_{y_{(s)}}^* \prod_{i=1}^{n}\delta_i^*\left(Q_i^*\right)^{-1}\tau_6^{(c_1+n+s)}\Gamma(c_1+n)} \times \left\{1-\left[1-\left(1+y_{(s)}\right)^{-\theta_3}\right]^{\theta_4}\right\}^{\theta_2-1}\left[-\ln\left(Q_{y_{(s)}}^*\right)\right]^{s-1}\Gamma(c_1+n+s)\,d\theta_2 d\theta_3 d\theta_4\,dy_{(s)} = \frac{\omega}{2}. \quad (128)$$

4. Numerical results

This section aims to illustrate the theoretical results of the Bayes estimates and BP under SE and LINEX loss functions. Numerical results are presented for the EG-IKum distribution based on lower record values through a simulation study and three applications.

4.1 Simulation study

A simulation study is introduced to examine the performance of the Bayes estimates and BP for different sample sizes of lower record values and for different parameter values for EG-IKum distribution.

4.1.1 Bayesian estimation

The lower record values can be obtained as a special case from dgos by setting $m = -1$ and $k = 1$, therefore the estimation results obtained in Subsection 3.1 can be specialized to lower records. The Bayes estimates of θ_j, where, $j = 1,2,3,4$, are evaluated. Also, rf and hrf and their average estimates, *estimated risks* (ERs) are computed based on lower record values according to the following steps:

a) For given values of $\theta_j, j = 1,2,3,4$, random samples of size n are generated from the EG-IKum distribution observing that if U is uniform distribution (0,1), then,

$$t_{ij} = \left[(1-(1-(1-(U_{ij})^{\frac{1}{\theta_1}})^{\frac{1}{\theta_2}})^{\frac{1}{\theta_4}})^{-\frac{1}{\theta_s}} - 1 \right], \text{ is EG-IKum } (\underline{\theta}) \text{ distribution.}$$

b) For each sample size n, consider that the first observation is the first lower record value t_1, then denote it by R_1 and the second observation t_2, denote it by R_2; $(t_1 > t_2)$ record and $t_1 \leq t_2$ if ignore it and repeat until you get a sample of records Rv.

c) The Bayes estimates of the parameters, rf and hrf under SE and LINEX loss functions are computed; at a specified number of surviving units with population parameter values θ_j and hyper parameters of the prior distribution. The computations are performed using R programming language.

d) Tables 2 and 3 present the Bayes averages under SE and LINEX loss functions of the parameters and their ERs and credible intervals based on lower record values for different population parameter values for $\theta_1 = (0.8,0.2)$, $\theta_2 = (0.6,0.3)$, $\theta_3 = (1.2,0.4)$ and $\theta_4 = (1.5, 0.7)$ based on records of size $Rv = 3, 5, 7$ and *number of replication* (NR) = 10000.

e) Table 4 displays the Bayes averages, ERs and 95% credible intervals of the rf and hrf at $t_0 = 0.5,1,2$ from EG-IKum distribution based on lower record values for different samples of records of size $Rv = 3,7$ and NR = 10000.

4.1.2 Bayesian prediction

The predictors of the future lower record values can be obtained from the above results of dgos when $m = -1$, $k = 1$, $m_y = -1$ and $k_y = 1$.

a. Determine the value of s, $1 \leq s \leq n_y$, which is the index of the future unobserved lower record value from the second sample.

b. The BP for the future lower records is calculated under SE and LINEX loss functions.

c. Table 5 displays the point predictors and 95% credible intervals for the future lower record values of $Y_{(s)}$ from EG-IKum distribution, where $Rv = 6$, $\theta_1 = 0.8$, $\theta_2 = 0.5$, $\theta_3 = 1.2$ and $\theta_4 = 0.7$.

Table 2: Bayes averages, estimated risks and credible intervals from EG-IKum distribution based on lower records ($\theta_1 = 0.8$, $\theta_2 = 0.6$, $\theta_3 = 1.2$, $\theta_4 = 1.5$ and NR = 10000)

Rv	Loss functions	Parameters	Averages	ER	LL	UL	Length
3	SE	θ_1	0.8013	0.0021	0.7993	0.8994	0.0032
		θ_2	0.6020	0.0020	0.5999	0.6034	0.0036
		θ_3	1.1983	0.0016	1.1968	1.2001	0.0033
		θ_4	1.4987	0.0013	1.4957	1.5004	0.0047
	LINEX	θ_1	0.7979	0.0030	0.7960	0.7999	0.0039
		θ_2	0.5983	0.0019	0.5958	0.6007	0.0048
		θ_3	1.2007	0.0006	1.1997	1.2013	0.0015
		θ_4	1.4994	0.0007	1.4971	1.5006	0.0035
5	SE	θ_1	0.7991	0.0014	0.7973	0.7994	0.0028
		θ_2	0.6005	0.0009	0.5986	0.6018	0.0032
		θ_3	1.1991	0.0009	1.1975	1.2000	0.0025
		θ_4	1.4992	0.0009	1.4969	1.5005	0.0037
	LINEX	θ_1	0.7978	0.0029	0.7963	0.7993	0.0030
		θ_2	0.5994	0.0008	0.5974	0.6003	0.0029
		θ_3	1.1994	0.0006	1.1978	1.2000	0.0022
		θ_4	1.4993	0.0005	1.4980	1.4959	0.0021
7	SE	θ_1	0.8000	0.0004	0.7991	0.8002	0.0013
		θ_2	0.5994	0.0008	0.5975	0.6005	0.0019
		θ_3	1.1992	0.0007	1.1980	1.1998	0.0023
		θ_4	1.4994	0.0006	1.4977	1.5005	0.0018
	LINEX	θ_1	0.8003	0.0005	0.7994	0.8007	0.0013
		θ_2	0.5990	0.0008	0.5981	0.5998	0.0017
		θ_3	1.1998	0.0003	1.1988	1.2006	0.0018
		θ_4	1.4994	0.0005	1.4981	1.5003	0.0018

Table 3: Bayes averages, estimated risks and credible intervals from EG-IKum distribution based on lower records ($\theta_1 = 0.2$, $\theta_2 = 0.3$, $\theta_3 = 0.4$, $\theta_4 = 0.7$ and NR = 10000)

Rv	Loss functions	Parameters	Averages	ER	LL	UL	Length
3	SE	θ_1	0.1990	0.0060	0.1973	0.2002	0.0029
		θ_2	0.2985	0.0046	0.2969	0.3008	0.0040
		θ_3	0.3989	0.0035	0.3972	0.4003	0.0031
		θ_4	0.6980	0.0031	0.6964	0.6997	0.0033
	LINEX	θ_1	0.2027	0.0160	0.1995	0.2046	0.0051
		θ_2	0.3005	0.0024	0.2987	0.3015	0.0028
		θ_3	0.4018	0.0050	0.3999	0.4027	0.0028
		θ_4	0.6996	0.0013	0.6977	0.7004	0.0027
5	SE	θ_1	0.2007	0.0050	0.1993	0.2002	0.0023
		θ_2	0.2991	0.0028	0.2977	0.3000	0.0022
		θ_3	0.3990	0.0027	0.3979	0.4002	0.0023
		θ_4	0.7017	0.0028	0.6996	0.7030	0.0033
	LINEX	θ_1	0.1993	0.0056	0.1975	0.2007	0.0031
		θ_2	0.3007	0.0024	0.2991	0.3015	0.0024
		θ_3	0.3998	0.0017	0.3984	0.4009	0.0025
		θ_4	0.6988	0.0019	0.6975	0.6997	0.0022
7	SE	θ_1	0.1995	0.0031	0.1986	0.1997	0.0017
		θ_2	0.2991	0.0022	0.2983	0.2998	0.0014
		θ_3	0.3997	0.0013	0.3987	0.4005	0.0018
		θ_4	0.7001	0.0007	0.6990	0.7011	0.0021
	LINEX	θ_1	0.1999	0.0023	0.1987	0.2007	0.0020
		θ_2	0.2992	0.0022	0.2979	0.3000	0.0020
		θ_3	0.4006	0.0024	0.3995	0.4015	0.0021
		θ_4	0.7011	0.0019	0.6990	0.7022	0.0022

Table 4: Bayes averages, estimated risks and credible intervals for the rf and hrf at $t_0 = 0.5,1,2$, from EG-IKum distribution based on SE and LINEX loss functions for different sample size of records Rv and NR = 10000

Rv		Loss functions	rf and hrf	Averages	ER	LL	UL	Length
3	0.5	SE	$R_{EGIK}(t_0)$ $h_{EGIK}(t_0)$	0.7802 0.4581	0.0014 0.0027	0.7782 0.4557	0.7812 0.4596	0.0030 0.0039
		LINEX	$R_{EGIK}(t_0)$ $h_{EGIK}(t_0)$	0.7801 0.4594	0.0012 0.0013	0.7785 0.4586	0.7810 0.4601	0.0025 0.0015
	1	SE	$R_{EGIK}(t_0)$ $h_{EGIK}(t_0)$	0.6368 0.3541	0.0006 0.0082	0.6356 0.3507	0.6373 0.3562	0.0017 0.0054
		LINEX	$R_{EGIK}(t_0)$ $h_{EGIK}(t_0)$	0.6378 0.3580	0.0020 0.0046	0.6361 0.3560	0.6388 0.3588	0.0026 0.0027
	2	SE	$R_{EGIK}(t_0)$ $h_{EGIK}(t_0)$	0.4751 0.2398	0.0015 0.0092	0.4738 0.2382	0.4764 0.2418	0.0025 0.0035
		LINEX	$R_{EGIK}(t_0)$ $h_{EGIK}(t_0)$	0.4770 0.2401	0.0038 0.0071	0.4751 0.2391	0.4780 0.2410	0.0029 0.0019
7	0.5	SE	$R_{EGIK}(t_0)$ $h_{EGIK}(t_0)$	0.7797 0.4584	0.0007 0.0019	0.7788 0.4569	0.7804 0.4597	0.0016 0.0027
		LINEX	$R_{EGIK}(t_0)$ $h_{EGIK}(t_0)$	0.7799 0.4589	0.0009 0.0007	0.7790 0.4581	0.7805 0.4596	0.0014 0.0015
	1	SE	$R_{EGIK}(t_0)$ $h_{EGIK}(t_0)$	0.6365 0.3577	0.0005 0.0041	0.6359 0.3556	0.6371 0.3587	0.0012 0.0031
		LINEX	$R_{EGIK}(t_0)$ $h_{EGIK}(t_0)$	0.6360 0.3574	0.0012 0.0033	0.6351 0.3562	0.6368 0.3587	0.0017 0.0025
	2	SE	$R_{EGIK}(t_0)$ $h_{EGIK}(t_0)$	0.4886 0.2053	0.0010 0.0023	0.4874 0.2045	0.4894 0.2057	0.0020 0.0019
		LINEX	$R_{EGIK}(t_0)$ $h_{EGIK}(t_0)$	0.4891 0.2056	0.0037 0.0014	0.4878 0.2048	0.4899 0.2064	0.0037 0.0014

Table 5: Point predictors and 95% credible intervals for the future lower record values $y^*_{(s)}$ from EG-IKum distribution ($Rv = 6, \theta_1 = 0.8$, $\theta_2 = 0.5$, $\theta_3 = 1.2$ and $\theta_4 = 0.7$)

s	Loss functions	$y^*_{(s)}$	LL	UL	Length
1	SE	0.8995	0.8977	0.9006	0.0028
	LINEX	0.9004	0.8997	0.9009	0.0012
3	SE	0.9011	0.8990	0.9018	0.0029
	LINEX	0.9009	0.8998	0.9015	0.0017
6	SE	0.9131	0.9097	0.9156	0.0059
	LINEX	0.9091	0.9074	0.9100	0.0025

4.2 Applications

In this subsection, three applications to real data sets are provided to illustrate the importance, applicability, and flexibility of the EG-IKum distribution based on lower records. The three applications demonstrate the superiority of the EG-IKum distribution over some known distributions namely IKum and G-IKum distributions.

Table 6: ML estimates of the parameters and standard errors for the three applications based on lower records

Application	*Rv*	Parameters	Estimates	Standard Errors
I	3	θ_1	0.8890	0.0007
		θ_2	0.6578	0.0293
		θ_3	0.8255	0.0003
		θ_4	1.6405	0.0018
II	4	θ_1	0.9164	0.0001
		θ_2	1.8020	0.0304
		θ_3	0.8763	0.0002
		θ_4	6.1045	0.0036
III	7	θ_1	0.4745	0.0026
		θ_2	0.9516	0.0507
		θ_3	0.5345	0.0030
		θ_4	0.7715	0.0008

Table 7: ML estimates of rf, hrf and standard errors from EG-IKum distribution for the three applications based on lower records

Application	*Rv*	rf and hrf	Estimates	Standard Errors
I	3	$R_{EGIK}(t_0)$	0.8875	0.0001
		$h_{EGIK}(t_0)$	0.2618	0.0014
II	4	$R_{EGIK}(t_0)$	0.7961	0.0033
		$h_{EGIK}(t_0)$	0.2042	0.0024
III	7	$R_{EGIK}(t_0)$	0.8396	0.0011
		$h_{EGIK}(t_0)$	0.2261	0.0006

Table 8: Bayes estimates for the parameters and standard errors from EG-IKum distribution for the three applications based on lower records

Application	*Rv*	Loss functions	Parameters	Estimates	Standard Errors
I	3	SE	θ_1	0.9009	8.88e-05
			θ_2	0.9001	1.01e-04
			θ_3	0.8990	1.04e-04
			θ_4	1.7008	1.12e-04
		LINEX	θ_1	0.9007	9.11e-05
			θ_2	0.8978	1.84e-04
			θ_3	0.9014	1.53e-04
			θ_4	1.7012	2.27e-04
II	4	SE	θ_1	0.9009	3.94e-05
			θ_2	1.5994	6.25e-05
			θ_3	0.8999	5.94e-05
			θ_4	6.0491	4.04e-05
		LINEX	θ_1	0.8996	4.54e-05
			θ_2	1.5994	4.33e-05
			θ_3	0.9009	6.57e-05
			θ_4	6.0501	4.97e-05
III	7	SE	θ_1	0.5996	5.26e-05
			θ_2	0.8992	8.39e-05
			θ_3	0.4000	6.54e-05
			θ_4	0.6982	2.17e-04
		LINEX	θ_1	0.6031	2.88e-04
			θ_2	0.9006	8.60e-05
			θ_3	0.4015	1.83e-04
			θ_4	0.7004	1.06e-04

Table 9: Bayes estimates for rf, hrf and standard error from EG-IKum distribution for the three applications based on lower records

Application	*Rv*	Loss functions	rf and hrf	Estimates	Standard Errors
I	3	SE	$R_{EGIK}(t_0)$ $h_{EGIK}(t_0)$	0.9501 0.3421	8.68e-05 1.49e-04
		LINEX	$R_{EGIK}(t_0)$ $h_{EGIK}(t_0)$	0.9486 0.3426	8.06e-05 8.46e-05
II	4	SE	$R_{EGIK}(t_0)$ $h_{EGIK}(t_0)$	0.7871 0.0304	7.10e-05 2.99e-05
		LINEX	$R_{EGIK}(t_0)$ $h_{EGIK}(t_0)$	0.7867 0.0320	1.09e-04 3.62e-05
III	7	SE	$R_{EGIK}(t_0)$ $h_{EGIK}(t_0)$	0.7983 0.2558	0.0002 0.0002
		LINEX	$R_{EGIK}(t_0)$ $h_{EGIK}(t_0)$	0.7981 0.2587	0.0001 0.0002

Table 10: Point predictors and 95% credible intervals for the future lower record values $y^*_{(s)}$ from the three applications

Application	*s*	SE				LINEX			
		$y^*_{(s)}$	Credible interval			$y^*_{(s)}$	Credible interval		
			LL	UL	Length		LL	UL	Length
I	1	1.2007	1.1989	1.2015	0.0026	1.1996	1.1986	1.2013	0.0026
	3	1.2012	1.1994	1.2027	0.0033	1.1999	1.1974	1.2014	0.0040
II	1	1.2020	1.1999	1.2038	0.0039	1.1986	1.1973	1.1995	0.0022
	3	1.2019	1.1993	1.2033	0.0040	1.2017	1.1999	1.2026	0.0027
III	1	0.9008	0.8987	0.9018	0.0030	0.8994	0.8982	0.9002	0.0020
	3	0.90213	0.8999	0.9048	0.0048	0.9013	0.8995	0.9032	0.0036

Tables 6 and 7 display the ML estimates of the parameters, rf, hrf and standard error for the three real data sets based on lower records. Tables 8 and 9 present the Bayes estimates of the parameters, rf, hrf and standard errors for the real data sets based on lower records. Point predictors and 95% credible intervals for the future lower record values $Y_{(s)}$ from the three real data sets are shown in Table 10.

To check the validity of the fitted model, Kolmogorov-Smirnov goodness of fit test is performed for each data set and the p values in each case indicate that the model fits the data very well. Figure 4: displays the fitted pdf, PP-Plot and Q-Q plot of the EG-IKum distribution for the first real data. Figure 5: presents the fitted pdf, PP-Plot and Q-Q plot of the EG-IKum distribution for the second real data. Also, Figure 6: gives the fitted pdf, PP-Plot and Q-Q plot of the EG-IKum distribution for the third real data, which indicates that the EG-IKum distribution provides better fits to these data sets.

I. The first application is given by Hinkley (1977). It consists of thirty successive values of March precipitation (in inches) in Minneapolis/St Paul. The data is 0.77, 1.74, 0.81, 1.20, 1.95, 1.20, 0.47, 1.43, 3.37, 2.20, 3.00, 3.09, 1.51, 2.10, 0.52, 1.62, 1.31, 0.32, 0.59, 0.81, 2.81, 1.87, 1.18, 1.35, 4.75, 2.48, 0.96, 1.89, 0.90, 2.05.

From the data, one can observe that the following are the lower record values: 0.77, 0.47, 0.32, with p-value = 0.1866.

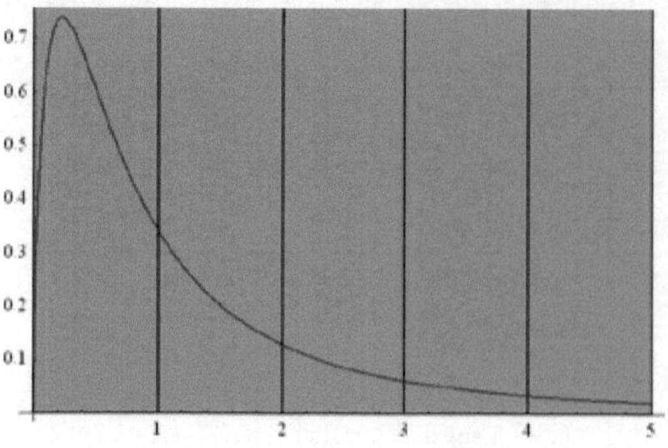

Fitted pdf of the first application

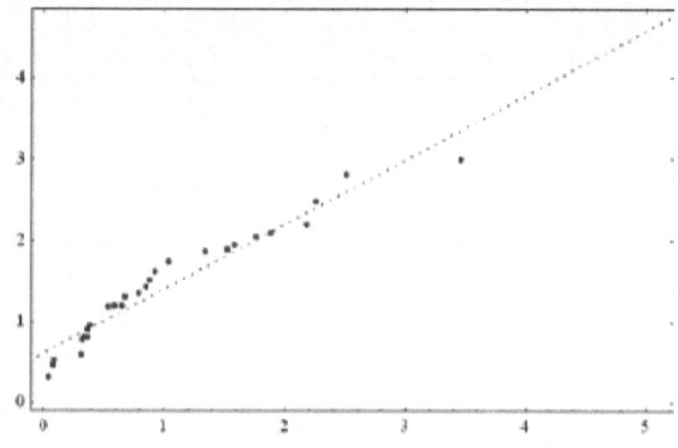

Q-Q Plot of the first application

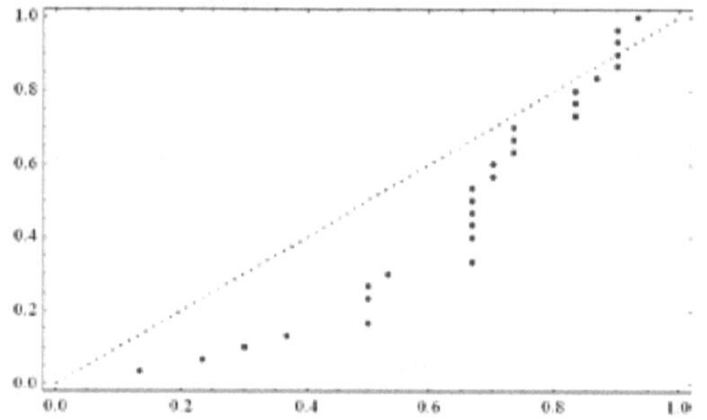

PP- Plot of the first application

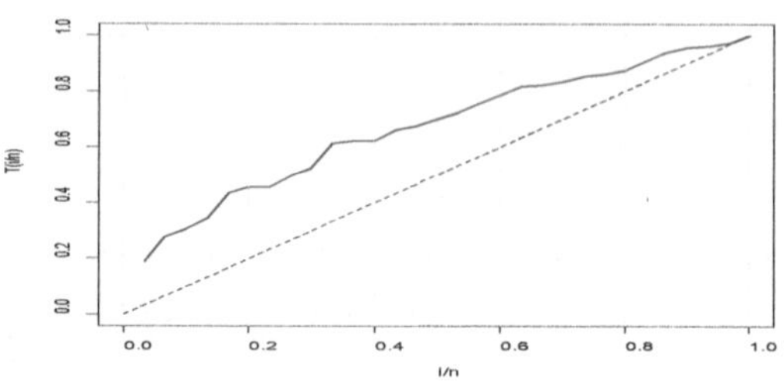

The empirical scaled TTT-transform plot

Boxplot

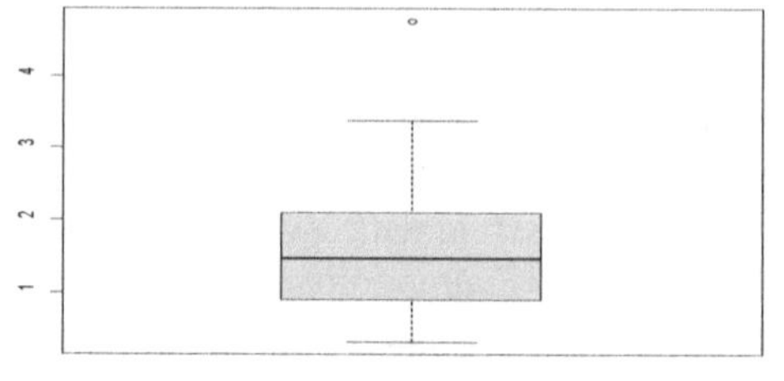

Figure 4: The fitted pdf, Q-Q Plot, PP- Plot, the empirical scaled TTT Plot and boxplot of the EG-IKum distribution for the first application

II. The second application is a real data set obtained from Lee and Wang (2003). It represents the remission times (in months) of a random sample of 128 bladder cancer patients. The data is

0.08, 2.09, 3.48, 4.87, 6.94, 8.66, 13.11, 23.63, 0.2, 2.23, 0.26, 0.31, 0.73, 0.52, 4.98, 6.97, 9.02, 13.29, 0.4, 2.26, 3.57, 5.06, 7.09, 11.98, 4.51, 2.07, 0.22, 13.8, 25.74, 0.5, 2.46, 3.64, 5.09, 7.26, 9.47, 14.24, 19.13, 6.54, 3.36, 0.82, 0.51, 2.54, 3.7, 5.17, 7.28, 9.74, 14.76, 26.31, 0.81, 1.76, 8.53, 6.93, 0.62, 3.82, 5.32, 7.32, 10.06, 14.77, 32.15, 2.64, 3.88, 5.32, 3.25, 12.03, 8.65, 0.39, 10.34, 14.83, 34.26, 0.9, 2.69, 4.18, 5.34, 7.59, 10.66, 4.5, 20.28, 12.63, 0.96, 36.66, 1.05, 2.69, 4.23, 5.41, 7.62, 10.75, 16.62, 43.01, 6.25, 2.02, 22.69, 0.19, 2.75, 4.26, 5.41, 7.63, 17.12, 46.12, 1.26, 2.83, 4.33, 8.37, 3.36, 5.49, 0.66, 11.25, 17.14, 79.05, 1.35, 2.87, 5.62, 7.87, 11.64, 17.36, 12.02, 6.76, 0.4, 3.02, 4.34, 5.71, 7.93, 11.79, 18.1, 1.46, 4.4, 5.85, 2.02, 12.07.

From the data, the following are the lower record values : 8, 2.09, 0.2, 0.19, with p-value = 0.4922.

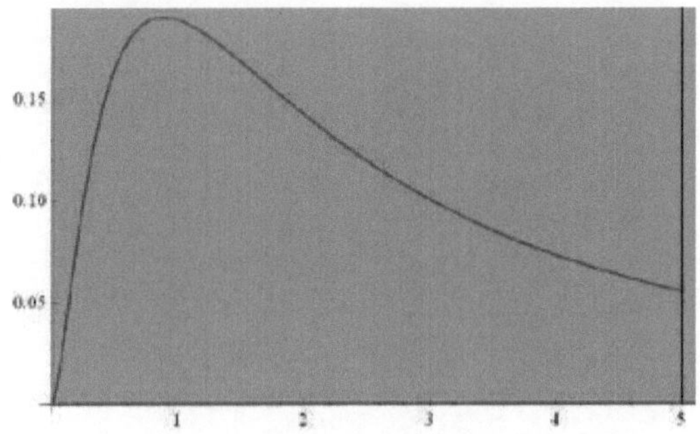

Fitted pdf of the second application

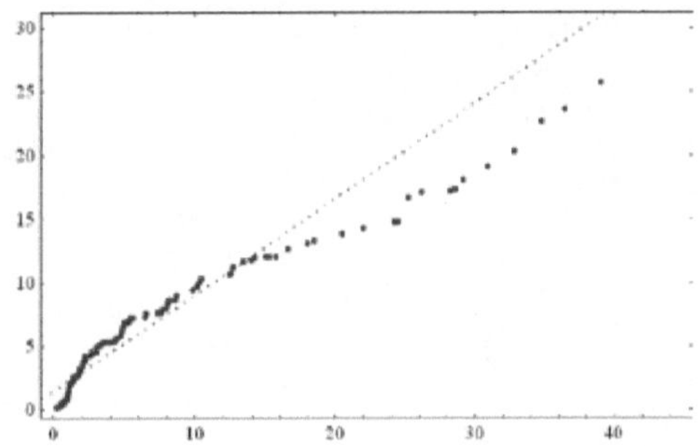

Q-Q Plot of the second application

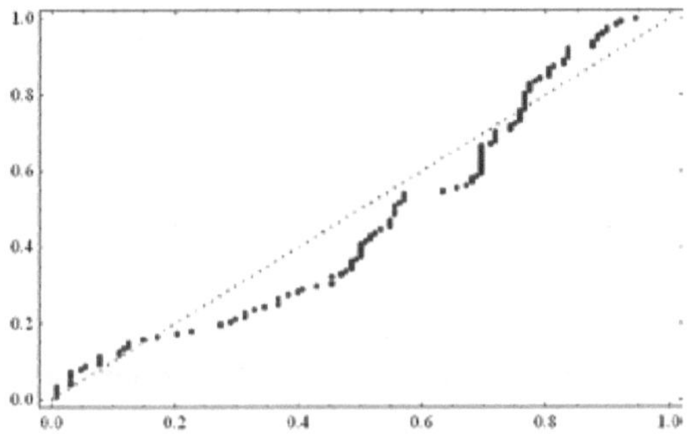

PP- Plot of the second application

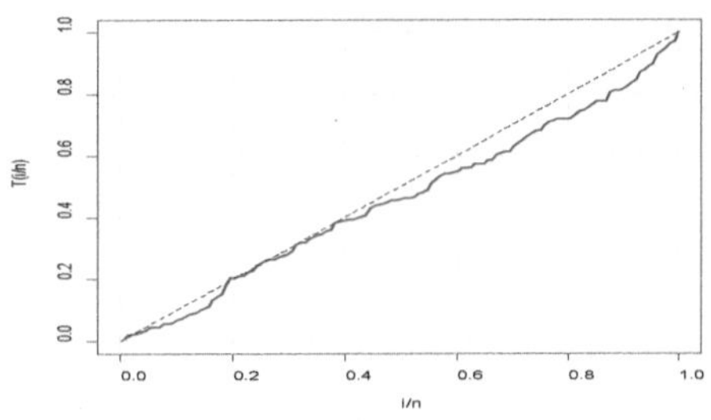

The empirical scaled TTT-transform plot

Boxplot

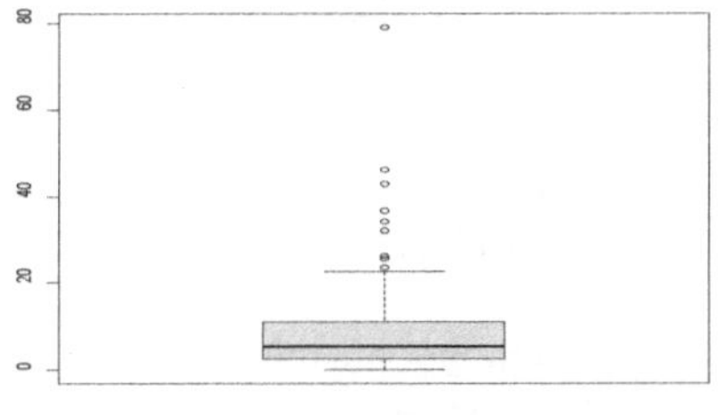

Figure 5: The fitted pdf, Q-Q Plot, PP-Plot, the empirical scaled TTT-transform plot and boxplot of the EG-IKum distribution for the second application

III. The third application is the vinyl chloride data obtained from clean upgrading, monitoring wells in mg/L; this data set was used for Bhaumik et al. (2009). The data is: 5.1, 1.2, 1.3, 0.6, 0.5, 2.4, 0.5, 1.1, 8.0, 0.8, 0.4, 0.6, 0.9, 0.4, 2.0, 0.5, 5.3, 3.2, 2.7, 2.9, 2.5, 2.3, 1.0, 0.2, 0.1, 0.1, 1.8, 0.9, 2.0, 4.0, 6.8, 1.2, 0.4, 0.2.

From the original data, one can observe the following lower record values: 5.1, 1.2, 0.6, 0.5, 0.4, 0.2, 0.1, where the p-value = 0.185.

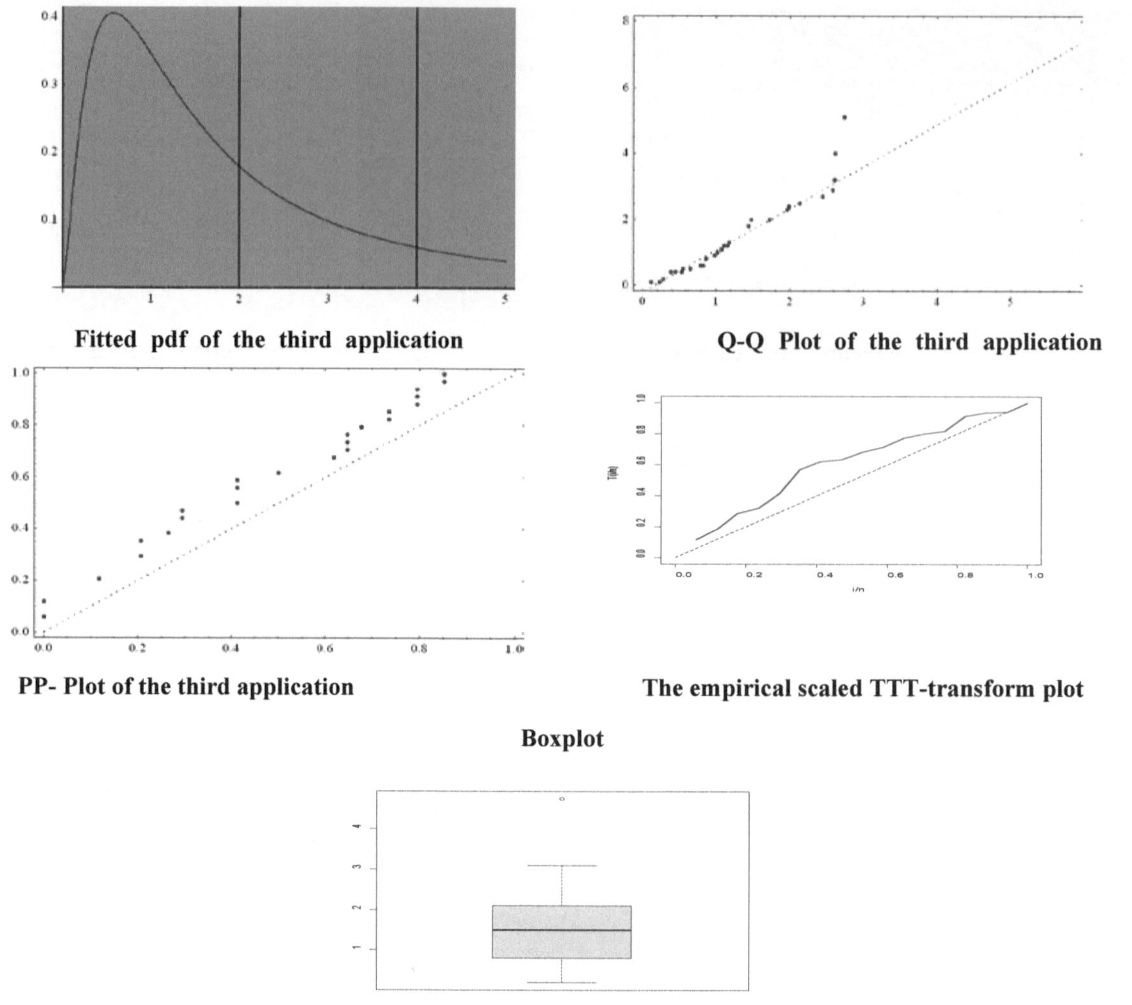

Fitted pdf of the third application

Q-Q Plot of the third application

PP- Plot of the third application

The empirical scaled TTT-transform plot

Boxplot

Figure 6: The fitted pdf, Q-Q Plot, PP-Plot, the empirical scaled TTT-transform plot and boxplot of the EG-IKum distribution for the third application

- The proposed distribution is compared to demonstrate the superiority of EG-IKum distribution over some known distributions namely IKum and G-IKum distributions. To verify which distribution fits better to the real data set, the Kolmogorov-Smirnov test is employed. Other criteria including (maximized log-likelihood), *Akaike information criterion* (AIC), *Akaike information criterion corrected* (AICC) and *Bayesian information criterion* (BIC) are used to compare the fit of the competitor distributions, where,

$$\mathrm{AIC} = 2k - 2\log(\mathrm{L}), \qquad AICC = AIC + \frac{2k(k+1)}{n-k-1} \quad \text{and} \quad BIC = k\log(n) - 2\log(L),$$

where k is the number of the parameters in the statistical model, n is the sample size and L is the maximized value of the LF for the estimated model. The best distribution corresponds to the highest p-value and the lowest values of $-2\log(\mathrm{L})$, AIC, AICC and BIC.

- Tables 11, 12 and 13 show the p-values of the Kolmogorov-Smirnov test, $-2\log(\mathrm{L})$, AIC, AICC and BIC for the three real data sets.

Table 11: The goodness of fit measures for fitted models of Application 1

Model	P-value	–2LL	AIC	AICC	BIC
EG-IKum	0.184	7.004	0.995	2.595	6.6001
IKum	0.071	8.646	16.646	18.246	22.25
G-IKum	0.134	11.256	17.256	18.179	21.459

Table 12: The goodness of fit measures for fitted models of Application 2

Model	P-value	-2LL	AIC	AICC	BIC
EG-IKum	0.4922	7.371	15.371	16.971	20.975
IKum	0.1127	14.195	22.195	23.795	27.800
G-IKum	0.0639	21.115	27.115	28.038	31.318

Table 13: The goodness of fit measures for fitted models of Application 3

Model	P-value	-2LL	AIC	AICC	BIC
EG-IKum	0.185	6.5805	14.581	16.181	20.185
IKum	0.083	10.899	18.899	20.499	24.504
G-IKum	0.098	21.366	27.366	28.289	31.569

4.3 Concluding remarks

- From Tables 2 and 3 one can notice that the ERs of the Bayes estimates of the shape parameters decrease when the sample size increases. Also, the lengths of the credible intervals become narrower as the sample size of records increases.
- It is clear from Table 4 that the ERs of rf and hrf perform better when the sample size increases, and the lengths of the credible intervals get shorter when the sample size increases.
- One can observe that the ERs for the estimates of the parameters, rf and hrf under the LINEX loss function have lesser values than the corresponding ERs of the estimates under the SE loss function.
- From Table 5, one can observe that the lengths of the BPB increase when s increases.
- Also, the lengths of the BPB under the LINEX loss function perform better than the corresponding lengths under the SE loss function.
- Regarding the standard errors for the Bayes estimates of the parameters, rf and hrf in Tables 8 and 9, the LINEX loss function seems to perform better than the SE loss function.
- From Table 10, one can notice that the predictive intervals include the predictive values (between the LL and UL). Also, the BPs of the future observations are very close to the actual observations.
- From Tables 11, 12 and 13 one can observe that the EG-IKum distribution has the lowest K-S values and the highest p-values for the three applications. Thus, it provides the best fit for the data compared to the other competitors of distributions. Moreover, the EG-IKum distribution has the smallest values of the –2log (L), AIC, AICC and BIC which implies that the proposed model is the best among the other competitors of distributions (IKum and G-IKum).

Some suggestions for future research

1. Considering other methods of estimation for the parameters, rf and hrf such as modified maximum likelihood or modified moments.
2. One sample Bayesian prediction from EG-IKum distribution based on dgos can be obtained.
3. Considering different types of loss functions such as balanced loss functions i.e., balanced absolute, balanced binary, balanced modified LINEX and balanced general entropy loss functions.
4. Transmuted EG-IKum distribution may be studied.
5. Recurrence relations for single and product moments of EG-IKum distribution may be derived.
6. E-Bayesian estimation for EG-IKum distribution may be studied.
7. ML and Bayesian estimation for EG-IKum distribution based on Type I and Type II censored samples can be obtained.
8. Empirical Bayesian for EG-IKum distribution may be studied.

Acknowledgements

The authors would like to thank the referees and the editor for their comments which led to the improvement of the earlier version of this article.

Abbreviations

AIC	Akaike information criterion
AICC	Akaike information criterion corrected
BIC	Bayesian information criterion
BPD	Bayesian predictive density
BP	Bayes predictors
BPB	Bayes predictive bounds
cdf	Cumulative distribution function
CIs	Confidence intervals
dgos	Dual generalized order statistics
EG	Exponentiated generalized
EGGC	EG general class
EG-IKum	EG inverted Kumaraswamy
hrf	Hazard rate function
LINEX	Linear exponential
LL	Lower limit
pdf	Probability density function
ML	Maximum likelihood
MLP	ML predictors
MLPB	ML predictive bounds
rhrf	Reversed hazard rate function
rf	Reliability function
SE	Squared error
UL	Upper limit

References

Abd EL-Kader, R. E. (2013). A general class of some inverted distributions. Ph.D. Thesis, Statistical Department, Faculty of commerce, AL-Azhar University, Girls' Branch, Cairo.

Ali, M. M., Pal, M. and Woo, J. (2007). Some exponentiated distributions. The Korean Communications in Statistics. 14: 93–109.

Alzaatreh, A., Lee, C. and Famoye, F. (2013). A new method for generating families of continuous distributions. Metron, 71(1): 63–79.

Alzaghal, A., Famoye, F. and Lee, C. (2013). Exponentiated T-X family of distributions with some applications. International Journal of Statistics and Probability, 2(3): 31.

Baharith, L. A. and Aljuhani, W. H. (2021). New method for generating new families of distribution. Symmetry. 13(4): 726.

Bhaumik, D. K, Kapur, K. and Gibbons, R. D. (2009). Testing parameters of a gamma distribution for small samples. Technometrics. 51: 326–334.

Burkschat, M., Cramer, E. and Kamps, U. (2003). Dual generalized order statistics. International Journal of Statistics. LX1(1): 13–26.

Cordeiro, G. M., Ortega, E. M. and da Cunha D. C. (2013). The exponentiated generalized class of distributions. Journal of Data Science, 11: 1–27.

De Andrade, T. A. N., Bourguignon, M. and Cordeiro, G. M. (2016). The exponentiated generalized extended exponential distribution. Journal of Data Science 14: 393–414.

Elbatal, I. and Muhammed, H.Z. (2014). Exponentiated generalized inverse Weibull distribution. Applied Mathematical Sciences, 8(81): 3997–4012.

Elgarhy, M. and Shawki, A. W. (2017). Exponentiated SUSHIL A distribution. International Journal of Scientific Engineering and Science, 1(7): 9–-2.

Eugene, N., Lee, C. and Famoye, F. (2002). Beta-normal distribution and its applications. Communications in Statistics Theory and Methods, 31(4): 497–512.

Fatima, K., Naqash, S. and Ahmad, S.P. (2018). Exponentiated generalized inverse Rayleigh distribution with applications in medical sciences. Pakistan Journal of Statistics, 34(5): 425–439.

Gupta, R. C., Gupta, P. L. and Gupta, R. D. (1998). Modeling failure time data by Lehman alternatives. Communication in Statistics-Theory and Methods, 27: 887–904.

Hinkley, D. (1977). On quick choice of power transformations. Journal of the American Statistician, 26: 67–69. https://doi.org/10.3390/sym13040726.

Jones, M. C. (2009). Kumaraswamy's distribution: A beta-type distribution with some tractability advantages. Statistical Methodology, 6(1): 70–81.

Lai, C. D. (2013). Constructions and applications of lifetime distributions. Applied Stochastic Models in Business and Industry, 29(2): 127–129.

Lee, E. T. and Wang, J. (2003). Statistical methods for survival data analysis, John Wiley & Sons, Hoboken, NJ, USA. 476.

Lemonte, A. J., Barreto-Souza, W. and Cordeiro G. M. (2013). The exponentiated Kumaraswamy distribution and its log-transform. Brazilian Journal of Probability and Statistics, 27(1): 31–53.

Mahdavi, A. and Kundu, D. (2017). A new method for generating distributions with an application to exponential distribution. Communications in Statistics Theory and Methods, 46: 6543–6557.

Marshall, A. W. and Olkin, I. (1997). A new method for adding a parameter to a family of distributions with application to the exponential and Weibull families. Biometrika, 84(3): 641–652.

Mustafa, A., EL-Desouky, B. S. and AL-Garash, S. (2016). The exponentiated generalized flexible Weibull extension distribution. Fundamental Journal of Mathematics and Mathematical Sciences, 6(2): 75-98. http://www.frdint.com/.

Nadarajah, S. and Kotz, S. (2006). The exponentiated type distributions. Acta Applicandae Mathematicae, 92: 97–111.

Oguntunde, P. E., A. dejumo, A.O. and Balogun, O. S. (2014). Statistical properties of the exponentiated generalized inverted exponential distribution. Applied Mathematics, 4(2): 47–55. http://journal.sapub.org/am.

Oluyede, B. O., Mashabe, B. Fagbamigbe, A., Makubate, B., Wanduku, D. et al. (2020). The exponentiated generalized power series family of distributions: theory, properties and applications. Heliyon. 6(8).

Rather, A. A. and Subramanian, C. (2018). Exponentiated Mukherjee-Islam distribution. Journal of Statistics Applications and Probability, 7(2): 357–361.

Silva, R. B., Barreto-Souza, W. and Cordeiro, G. M. (2010). A new distribution with decreasing, increasing and upside-down bathtub failure rate, Computational Statistics and Data Analysis, 54: 935–944.

Sindi, S. N., AL-Dayian, G. R. and Shahbaz, S. H. (2017). Inference for exponentiated general class of distributions based on record values. Pakistan Journal of Statistics and Operation Research. XIII. (3): 575–587.

Yousof, H. M, Afify. A. Z., Alizadeh, M, Butt. N. S., Hamedan, G. G. et al. (2015). The transmuted exponentiated generalized-G family of distributions. Pakistan Journal of Statistics and Operation Research. 11(4): 441–464. [http://dx.doi.org/10.18187/pjsor.v11i4.1164].

Chapter 11

A New Class of Discrete Distribution Arising as an Analogue of Gamma-Lomax Distribution:

Properties and Applications

Indranil Ghosh,[1,*] *Ayman Alzaatreh*[2] and *GG Hamedani*[3]

1. Introduction

Since the beginning of the first world war (and may be even prior to that) there have been many industrial disputes ranging from low labor wages to uncomfortable work conditions that led to several strikes in the UK. According to the leading daily newspaper in Britain, the Guardian, "the number of workers who went on strike in Britain last year fell to the lowest level since the 1890's". Furthermore, data from the Office for National Statistics in the UK show 33,000 workers were involved in labor disputes in 2017, down from 154,000 a year earlier. This is the lowest number since records began in 1893, the year of Britain's first national coal strike, when the figure was 634,000. The major seminal event that is worth mentioning here is the miners' strike of 1984−1985. Indeed, this may be considered as an example of a large scale industrial reaction to stop the operation of the British coal industry. For further details on this event, an interested reader is suggested to look in the Wikipedia. It is to be noted that many other industries in the UK have received major set-back in terms of productivity and loss in revenue because of strikes due to various reasons. Needless to say, it has been a matter of great concern to the industries as to how one can analyze the quantum of these strikes and take appropriate measures. There exists a sizable number of research articles in the literature where quantitative insights into strikes are discussed. Among them some noteworthy models are those described in Velden (2000), Skeels and McGrath (1991), Leigh (1984), Buck (1984), Mauleon and Vannetelbosch (1998) and the references cited therein. While a majority of these earlier works focus on establishing a regression type modeling in search for a causal relation, a few of the proposed models search for an appropriate discrete probability model that might discuss the behaviors and patterns arising due to such strikes. This serves as a major motivation to carry out this research work.

In recent years there has been a growing interest in exploring several discrete distributions in univariate, bivariate, and in the multivariate domain albeit computational complexity and the absence of tractable probability mass functions. For a non-exhaustive list of references on such developments of discrete

[1] Department of Mathematics and Statistics, University of North Carolina, Wilmington, NC 28403, USA.
[2] Department of Mathematics and Statistics, American University of Sharjah, PO Box 26666, Sharjah, United Arab Emirates.
[3] Department of Mathematical and Statistical Sciences, Marquette University, Milwaukee WI 53233.
Emails: aalzaatreh@aus.edu; gholamhoss.hamedani@marquette.edu
* Corresponding author: ghoshi@uncw.edu

probability models in univariate and in higher domains, the readers are encouraged to see the book by Johnson et al. (2005) on discrete multivariate distributions; characterizations of recently developed twenty discrete distributions by Hamedani et al. (2021); a new versatile discrete distribution by Turner (2021); the new discrete distribution with application to COVID-19 data by Almetwally et al. (2022); a one-parameter discrete distribution for over-dispersed data by Eliwa and El-Morshedy (2021); a new discrete analog of the continuous Lindley distribution by Al-Batain et al. (2020); a discrete Pareto (type-IV) distribution by Ghosh (2020), and the references cited therein. While several of them are developed as a continuous analog of certain known distributions, such as Pareto (type-IV) and Lindley, several of them have been developed using some other techniques of generating a new class of discrete distributions. However, none of the above-mentioned probability models have been applied to model strike data which we aim to discuss in this paper. The main objective of this article is to establish that the industrial strikes data, specifically strikes data sets arising out of several industries in the UK, can be described by a discrete probability model, namely the discrete gamma-Lomax model. We begin our discussion by providing a general framework which leads to our specific discrete gamma-Lomax distribution from its continuous analogue model.

Suppose that is the cumulative distribution function of any random variable X, and is the probability density function of a random variable R defined on $[0,\infty)$. The probability density function of the gamma-X family of distributions defined by Alzaatreh, et al. (2013, 2014) is given by,

$$f_X(x) = \frac{1}{\Gamma(\alpha)\beta^{\alpha}} f_R(x)(-\log(1-F_R(x)))^{\alpha-1}(1-F_R(x)))^{\frac{1}{\beta}-1}, x \in \mathbb{R}. \quad (1)$$

If the random variable R follows the Lomax distribution with the density function $f_R(x) = k\theta^{-1}(1 + x/\theta)^{-(k+1)}$, $k > 0$; $x > 0$, then (1) reduces to the gamma-Lomax distribution (GLD) as,

$$f_X(x) = \frac{1}{\theta\Gamma(\alpha)c^{\alpha}}\left\{1+\frac{x}{\theta}\right\}^{-1-\frac{1}{c}}\left(\log\left[1+\frac{x}{\theta}\right]\right)^{\alpha-1}, \quad (2)$$

$x > 0$, where $c = \beta/k$, α and θ are positive parameters.

Note that if $X + \theta$ is replaced by X, then (2) reduces to the gamma-Pareto distribution which was proposed and studied in Alzaatreh et al. (2012a). From (2), the cumulative distribution function of the gamma-Lomax distribution is given by,

$$F_X(x) = \frac{1}{\Gamma(\alpha)}\gamma\left\{\alpha, c^{-1}\log\left[1+\frac{x}{\theta}\right]\right\}, x \geq 0, \quad (3)$$

where $\gamma(\alpha,t) = \int_0^t u^{\alpha-1} e^{-u} du$ is the incomplete gamma function.

There is a variety of works available in the literature that extends the Lomax distribution under the continuous paradigm. For example, (1) was studied by Cordeiro et al. (2015) as a particular case of Zografos- Balakrishnan (G) family of distributions, where G is any baseline continuous distribution (from the perspective of a gamma generated model). Lemonte and Cordeiro (2013) proposed the McDonald Lomax distribution, which is an extension of the classical Lomax distribution. Ghitany et al. (2007) studied properties of a continuous probability model derived from the Lomax model and utilized the Marshall and Olkin type generator. However, not much work has been done towards discrete Lomax mixture type models. In the work of Prieto et al. (2014), the authors considered the discrete generalized Pareto model (mixing with zero-inflated Poisson distribution) in modeling road accident black spots data. Ghosh (2020) discussed and studied a discrete version of Pareto (Type IV) model and established that the associated p.m.f. can be approximately symmetric, left- or right-skewed with applications in modeling several types of count data. Dzidzornu et al. (2021) studied the performance of the discrete generalized Pareto models in modeling non-insurance claims. Amponash et al. (2021) discussed various estimation strategies regarding several variants of discrete Pareto models. These are some references from which one could find a strong motivation to carry out this work which is stated as follows: (a) As pointed out in Hitz et al. (2018), there are several situations in which modeling extreme value events [in the univariate case, an approach that often works quite well

in practice is to model observations above a large threshold with a parametric family of distributions] for count data, the limiting distribution behaves like a generalized Pareto distribution. However, not much work has been done to address the issue with discrete data, and, in particular, if the data can be modeled with a discrete Lomax distribution, how such extremes for discrete values will behave asymptotically; (b) In income modeling if the data (cross-sectional over time) is both discrete and continuous, for the discrete portion, the DGLD developed in this article can play as a natural candidate distribution. Therefore, the development of this discrete analog of a continuous Lomax distribution is of paramount importance. It is observed that the discrete model is unimodal and the p.m.f. (probability mass function) is always a decreasing function. So, the model is somehow restricted in nature. Now, the proposed discrete gamma-Lomax distribution (henceforth, for short, DGLD) is not always decreasing. Consequently, it has greater flexibility.
The discrete gamma-Lomax distribution can be defined as follows:

$$\begin{aligned} g(x) &= P(x \le X < x+1) \\ &= S(x) - S(x+1) \qquad (4) \\ &= \frac{1}{\Gamma(\alpha)}\left[\gamma\left\{\alpha, c^{-1}\log\left[1+\frac{x+1}{\theta}\right]\right\} - \gamma\left\{\alpha, c^{-1}\log\left[1+\frac{x}{\theta}\right]\right\}\right], \end{aligned}$$

$x \in \mathbb{N}^*$; where $\mathbb{N}^* = \mathbb{N} \cup \{0\}$, and $\mathbb{N}$ is the set of all positive integers.
From (4), the cumulative distribution and the survival functions of the DGLD are, respectively, given by,

$$G(x) = \frac{1}{\Gamma(\alpha)}\gamma\left\{\alpha, c^{-1}\log\left[1+\frac{\lfloor x \rfloor + 1}{\theta}\right]\right\}, \quad x \ge 0, \qquad (5)$$

$$S(x) = 1 - \frac{1}{\Gamma(\alpha)}\gamma\left\{\alpha, c^{-1}\log\left[1+\frac{\lfloor x \rfloor + 1}{\theta}\right]\right\}, \quad x \ge 0, \qquad (6)$$

where $\lfloor x \rfloor = max\{m \in \mathbb{Z} | m \le x\}$ is the floor function. The probability mass function in (4) will be utilized for the maximum likelihood estimation of the parameters, see Section 4. The survival function in (6) is useful for censored maximum likelihood estimation of the parameters, see Subsection 4.2. Figure 1 shows various plots of the DGLD, where the scale parameter $\theta = 1$ for various values of the shape parameters c and α. The plots indicate that DGLD exhibits various shapes including reversed J and right skewed unimodal shapes.

The rest of the paper is organized as follows. In Section 2, we discuss some structural properties of the proposed DGLD including shapes and transformations. Moments and order statistics are discussed in Section 3. Section 4 deals with certain characterizations of the DGLD. In Section 5, maximum likelihood estimation under regular and censored data setup are discussed, also the performance of the ML method is investigated through a simulation study. Three different real life data sets are considered to illustrate the applicability of the DGLD in Section 6. Finally, some concluding remarks are provided in Section 7.

2. Structural properties

In this Section, we discuss some useful structural properties of the DGLD distribution. The following Lemma is useful for simulating random samples from the DGLD.

Lemma 1.

(a) If the random variable Y has the gamma-Lomax distribution with parameters c, α and θ, then the random variable $X = \lfloor Y \rfloor$ follows the DGLD (c, α, θ).

(b) If the random variable Y has the gamma distribution with parameters α and c, then the random variable $X = \lfloor \theta(e^Y - 1) \rfloor$ follows the DGLD (c,α,θ).

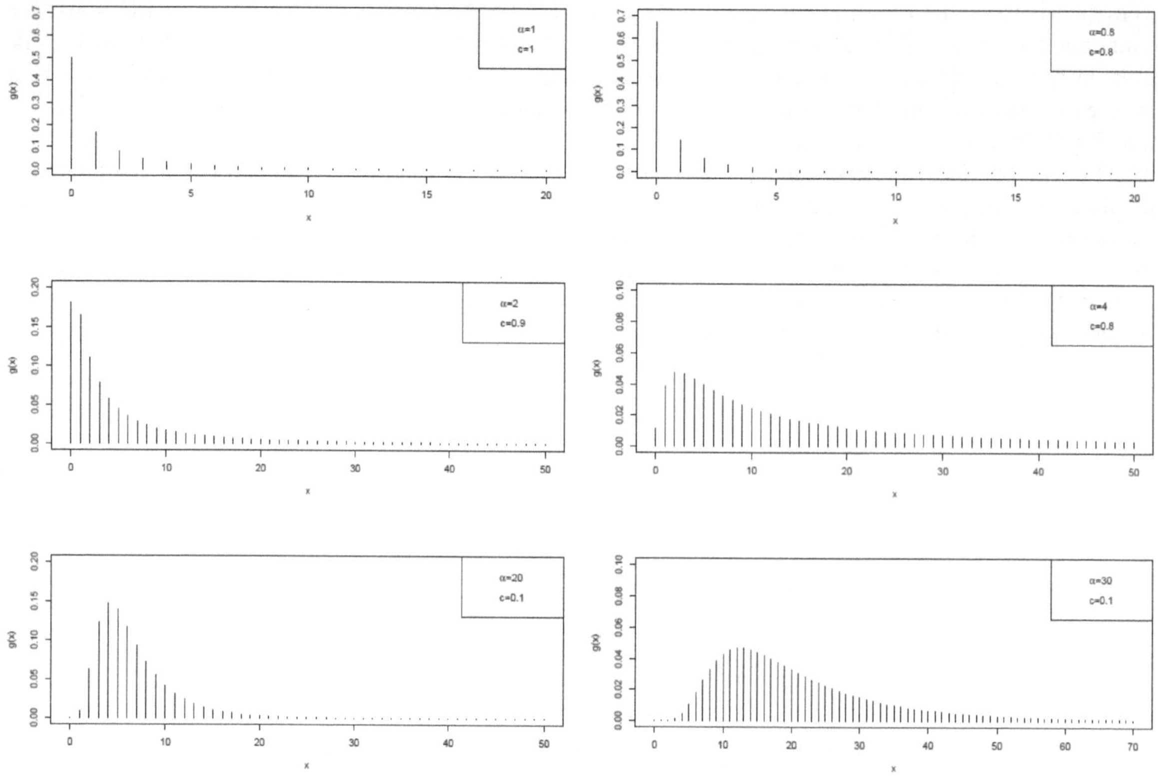

Figure 1: Plots of DGLD for various values of c and α.

Proof. The proof follows immediately from (3) and (5).

The hazard function associated with the DGLD is given by,

$$h(x)=\frac{\left[\gamma\left\{\alpha,c^{-1}\log\left[1+\frac{x+1}{\theta}\right]\right\}-\gamma\left\{\alpha,c^{-1}\log\left[1+\frac{x}{\theta}\right]\right\}\right]}{\Gamma(\alpha)-\gamma\left\{\alpha,c^{-1}\log\left[1+\frac{x}{\theta}\right]\right\}},x\in\mathbb{N}^*. \tag{7}$$

Discrete hazard rates originate in various common situations in reliability theory in which clock time is not the best scale on which to describe lifetime. For example, in ammunition reliability, the number of rounds fired until the first failure (say) is more important than the failure age. A similar scenario can be envisioned when a piece of equipment operates in cycles and the observation is the number of cycles successfully completed prior to the failure. In other situations, a device is monitored only once per time period and the observation then is the number of time periods successfully completed prior to the failure of the device, for details, see Shaked et al. (1995) and the references cited therein.

Lemma 2. The discrete gamma-Lomax distribution has a decreasing probability mass function and hence a decreasing failure rate for $\alpha \leq 1$.

Proof. By differentiating (2), it is easy to show that the DGLD has a decreasing density for $\alpha \leq 1$. Now let $x_1 \leq x_2$, then $g(x_1) \leq g(x_2)$. Also, by the definition of the survival function we have $S(x_2) \leq S(x_1)$, and therefore $h(x_1) \leq h(x_2)$. Hence the proof.

Theorem 1. The DGLD is unimodal, and the mode is $x = m$, where $m \in\{\lfloor x_0\rfloor - 1, \lfloor x_0\rfloor, \lfloor x_0\rfloor + 1\}$ and $x_0 = max\{0,\theta\{exp(c(\alpha - 1)/(c + 1))-1\}\}$ Furthermore, if $\lfloor x_0\rfloor = 0$, then the mode m is 0 or 1.

Proof. By differentiating (2), we can see that the gamma-Lomax distribution is unimodal with mode at $x_0 = 0$ for $\alpha \le 1$ and at $x_0 = \theta \exp(c(\alpha - 1)/(c + 1))$ for $\alpha > 1$. Consequently, the DGLD is also unimodal with mode at m as given in Theorem 2 of Alzaatreh et al. (2012b). This completes the proof.

Remark. The discrete gamma-Lomax p.m.f. can be written as linear combinations of discrete Pareto p.m.f's.

Proof of the Remark. We consider the following series expansion,

$$\gamma\{\alpha,x\} = \sum_{m=0}^{\infty}(-1)^m \frac{x^{m+\alpha}}{m!(m+\alpha)}, \tag{8}$$

from Nadarajah and Pal (2008). Also, for any $c \in \mathbb{R}$, we have

$$\left\{\log\left[1+\frac{x}{\theta}\right]\right\}^c = c\sum_{k=0}^{\infty}\binom{k-c}{k}\sum_{j=0}^{\infty}\frac{(-1)^j}{(c-j)}\binom{k}{j}P_{k,j}\left(\frac{x}{\theta}\right)^{k+c} = \sum_{k=0}^{\infty}\sum_{j=0}^{\infty}\phi_{(c,k,j)}\left(\frac{x}{\theta}\right)^{k+c}, \tag{9}$$

where $\phi_{(c,k,j)} = c\binom{k-c}{k}\binom{k}{j}\frac{(-1)^j}{(c-j)}P_{k,j}$, $P_{j,0} = 1$ and $P_{k,j}\,\frac{1}{k}\sum_{m=1}^{k}\frac{(-1)^{m+1}(jm-k+m)}{m+1}P_{j,k-m}$, $k = 1,2,\cdots$

The p.m.f. in (4) can be rewritten as,

$$g(x) = \sum_{m,k,j=0}^{\infty}(-1)^m \frac{c^{-(m+\alpha)}}{m!(m+\alpha)}\left\{\phi_{(m+\alpha,k,j)}\left[\left(\frac{x+1}{\theta}\right)^{k+m+\alpha} - \left(\frac{x}{\theta}\right)^{k+m+\alpha}\right]\right\}, \quad x \in \mathbb{N}^*. \tag{10}$$

Equation (10) shows that the DGLD can be written as linear combinations of discrete Pareto distributions with parameters $k + m + \alpha$ and θ. For more information about the discrete Pareto distribution and some of its properties, the reader is referred to Buddana and Kozubowski (2014), Krishna and Pundir (2009) and Alzaatreh et al. (2012b).

3. Moments

The rth moment of the DGLD is given by,

$$E(X^r) = \frac{1}{\Gamma(\alpha)}\sum_{x=0}^{\infty}x^r\left[\gamma\left\{\alpha, c^{-1}\log\left[1+\frac{x+1}{\theta}\right]\right\} - \gamma\left\{\alpha, c^{-1}\log\left[1+\frac{x}{\theta}\right]\right\}\right]. \tag{11}$$

Theorem 2. If $c < 1/r$, then the rth moment of the DGLD (c, α, θ) exists.

Proof. Assume that X follows the GLD. Alzaatreh et al. (2012a) showed that if $c < 1/r$, then $E(X^r)$ exists for all r. Then, the proof immediately follows due to the fact that $0 \le \lfloor X\rfloor \le X$.

Table 1 provides the mean, variance, skewness, kurtosis and the mode of the DGLD with scale parameter $\theta = 1$, and for various values of the shape parameters α and c. For a fixed α, the mean, the variance and the mode of the DGLD are nondecreasing functions of c. Also, for a fixed c, the mean, the variance and the mode of the DGLD are nondecreasing functions of α. The skewness of the DGLD is always positive, and the kurtosis value shows that the DGLD possesses a heavy tail characteristic.

Table 1: Mean, variance, skewness, kurtosis and mode of the DGLD for various values of α and c.

α	c	mean	variance	skewness	kurtosis	mode
0.5	0.1	0.0002	0.0002	74.3055	5926.7829	0
	0.15	0.0025	0.0029	24.9902	873.6643	0
	0.2	0.0097	0.0131	17.3414	672.0767	0
1	0.1	0.0010	0.0010	33.6432	1236.0707	0
	0.15	0.0106	0.0125	12.7101	241.0927	0
	0.2	0.0369	0.0525	9.8004	245.2655	0
5	0.1	0.1971	0.2001	2.4807	11.3941	0
	0.15	0.7369	0.9638	2.7579	24.7541	0
	0.2	1.5422	3.6585	5.2599	160.4486	1
10	0.1	1.3662	1.1875	1.8591	11.7731	1
	0.15	3.5800	9.6806	3.8991	53.5608	2
	0.2	7.8133	78.7267	9.3104	906.2119	3
15	0.1	3.3571	4.9135	2.3565	16.1693	2
	0.15	9.9476	79.6694	5.0778	106.7949	5
	0.2	26.9217	1319.0287	16.1348	3555.6579	9

3.1 Order statistics

Let $X_1, X_2, \cdots, X_n$ be a random sample drawn from the p.m.f. in (4). Then, the p.m.f. of the ith order statistic, $X_{i:n}$, is given by,

$$P(X_{i:n}=x)=\frac{n!}{(i-1)!(n-i)!}\int_{F(x-1)}^{F(x)}u^{i-1}(1-u)^{n-i}\,du$$
$$=\frac{n!}{(i-1)!(n-i)!}\sum_{j=0}^{n-i}(-1)^j\binom{n-i}{j}\int_{F(x-1)}^{F(x)}u^{i+j-1}\,du$$
$$=\frac{n!}{(i-1)!(n-i)!}\sum_{j=0}^{n-i}(-1)^j\binom{n-i}{j}\frac{1}{(i+j)(\Gamma(\alpha))^{i+j}}\times\{A(x)-A(x-1)\},$$

where,

$$A(x)=\sum_{k_1=0}^{\infty}\sum_{k_2=0}^{\infty}\cdots\sum_{k_{i+j}=0}^{\infty}\frac{(-1)^{s_{i+j}}c^{-s_{i+j}-(i+j)\alpha}}{p_{i+j}}\left[\log\left(1+\frac{x+1}{\theta}\right)\right]^{s_{i+j}+(i+j)\alpha}$$

and $s_{i+j}=\sum_{\ell=1}^{i+j}k_\ell$ and $p_{k,j}=\prod_{\ell=1}^{i+j}k_\ell!(k_\ell+\alpha)$. Now, one can use (11) to obtain a general r[th] order moment of $X_{i:n}$.

The distribution of maximum and minimum order statistics may be obtained as follows. Let X_i, $i = 1,2,\cdots, n$ be independent DGLD random variables with parameters c_i, α_i and θ. Define $U = min(X_1, X_2,\cdots, X_n)$ U = $\max(X_1, X_2,\cdots, X_n)$ and $W = \max(X_1, X_2,\cdots, X_n)$. Then the cumulative distribution function of U, on using (8), can be written as,

$$P(U\le u)=1-\prod_{i=1}^{n}\left\{1-\frac{1}{\Gamma(\alpha_i)}\sum_{k=0}^{\infty}\frac{(-1)^k c_i^{-k-\alpha_i}}{k!(k+\alpha_i)}\left[\log\left(1+\frac{u+1}{\theta}\right)\right]^{k+\alpha_i}\right\}.$$

Hence, the p.m.f. of U can be obtained using P(U = u) = P(U ≤ u)−P(U ≤ u−1). Next, by a similar approach, the cumulative distribution function of W is given by,

$$P(W \le u) = \prod_{i=1}^{n} \left\{ \frac{1}{\Gamma(\alpha_i)} \sum_{k=0}^{\infty} \frac{(-1)^k c_i^{-k-\alpha_i}}{k!(k+\alpha_i)} \left[\log\left(1+\frac{w+1}{\theta}\right) \right]^{k+\alpha_i} \right\}.$$

For the distribution of the range (R), one can use equations (10) and (11) from Kabe et al. (1969) to find the joint distribution of the maximum and the minimum order statistics $P(X_{1:n} = x_{1:n}, X_{n:n} = x_{n:n})$ as well as $P(X_{1:n} = X_{n:n})$.

4. Characterizations of the DGLD

To understand the behavior of the data obtained through a given process, we need to be able to describe this behavior via its approximate probability law. This, however, requires establishing conditions which govern the required probability law. In other words we need to have certain conditions under which we may be able to recover the probability law of the data. So, characterization of a distribution is important in applied sciences, where an investigator is vitally interested to find out if the proposed model follows the selected distribution. Therefore, the investigator relies on conditions under which the model follows a specified distribution. A probability distribution can be characterized in different directions, one being based on the truncated moments. This type of characterization initiated by Galambos and Kotz (1978) and followed by other authors such as Kotz and Shanbhag (1980), Gla¨nzel et al. (1984), Gla¨nzel (1987), Gla¨nzel and Hamedani (2001) and Kim and Jeon (2013), to name a few. For example, Kim and Jeon (2013) proposed a credibility theory based on the truncation of the loss data to estimate conditional mean loss for a given risk function. It should also be mentioned that characterization results are mathematically challenging and elegant. In this section, we present two characterizations of the DGLD based on: (i) conditional expectation (truncated moment) of a certain function of a random variable and (ii) the reverse hazard function. Next, we provide characterizations of the DGLD in terms of the reverse hazard function and conditional expectations of a certain function of a random variable. The following lemma is useful for this purpose.

Lemma 3. Let Z be a discrete random variable taking values of natural numbers 0,1,2,..., with probability mass and distribution functions g and G, respectively. Then the following hold:

(1) $E(G(Z) + G(Z-1)|Z \le k) = G(k),\ k = 0,1,2,\ldots,$

(2) $\dfrac{g(k+2)}{G(k+1)} - \dfrac{g(k+1)}{G(k)} = \dfrac{G(k+2)G(k) - G^2(k+1)}{G(k)G(k+1)},\ k = 0,1,2,\ldots$

Proof. Straightforward computation with $g(k) = G(k) - G(k-1)$.

4.1 Characterization of the DGLD in terms of the conditional expectation of a certain function of a random variable

Proposition 1. Let X : Ω → $N^* = N \cup \{0\}$ be a random variable. The p.m.f. of X is (4) if and only if,

$$E\left\{\left[\gamma\left\{\alpha, c^{-1}\log\left[1+\frac{X+1}{\theta}\right]\right\} + \gamma\left\{\alpha, c^{-1}\log\left[1+\frac{X}{\theta}\right]\right\}\right] \mid X \le k\right\} = \gamma\left\{\alpha, c^{-1}\log\left[1+\frac{x+1}{\theta}\right]\right\}. \quad (12)$$

Proof. If X has p.m.f. in (4), then (12) holds by Lemma 3. Conversely, if (12) holds, then,

$$\sum_{x=0}^{k}\left\{\left[\gamma\left\{\alpha,c^{-1}\log\left[1+\frac{x+1}{\theta}\right]\right\}+\gamma\left\{\alpha,c^{-1}\log\left[1+\frac{x}{\theta}\right]\right\}\right]g(x)\right\}$$
$$=G(k)\gamma\left\{\alpha,c^{-1}\log\left[1+\frac{x+1}{\theta}\right]\right\}. \tag{13}$$

From (13), we also have,

$$\sum_{x=0}^{k-1}\left\{\left[\gamma\left\{\alpha,c^{-1}\log\left[1+\frac{x+1}{\theta}\right]\right\}+\gamma\left\{\alpha,c^{-1}\log\left[1+\frac{x}{\theta}\right]\right\}\right]g(x)\right\}$$
$$=\{G(k)-g(x)\}\gamma\left\{\alpha,c^{-1}\log\left[1+\frac{k}{\theta}\right]\right\}, \tag{14}$$

where we used $G(k-1) = G(k) - g(k)$.
Now, subtracting (14) from (13), yields,

$$\gamma\left\{\alpha,c^{-1}\log\left[1+\frac{k+1}{\theta}\right]\right\}g(x)$$
$$=G(k)\left\{\left[\gamma\left\{\alpha,c^{-1}\log\left[1+\frac{k+1}{\theta}\right]\right\}-\gamma\left\{\alpha,c^{-1}\log\left[1+\frac{k}{\theta}\right]\right\}\right]\right\}.$$

From the above equality, we have,

$$rG(k)=\frac{g(x)}{G(k)}=-1\left(\frac{\gamma\left\{\alpha,c^{-1}\log\left[1+\frac{k}{\theta}\right]\right\}}{\gamma\left\{\alpha,c^{-1}\log\left[1+\frac{k+1}{\theta}\right]\right\}}\right).$$

which is the reverse hazard function of the random variable X with the p.m.f. in (4).

4.2 Characterization of the DGLD in terms of the reverse hazard function

Proposition 2. Let $X : \Omega \rightarrow N^* = N \cup \{0\}$ be a random variable. The p.m.f. of X is (4) if and only if its reverse hazard function, $r_G(k)$, satisfies the difference equation,

$$r_G(k+1)-r_G(k)=\frac{\gamma\left\{\alpha,c^{-1}\log\left[1+\frac{k}{\theta}\right]\right\}\gamma\left\{\alpha,c^{-1}\log\left[1+\frac{k+2}{\theta}\right]\right\}-\gamma^2\left\{\alpha,c^{-1}\log\left[1+\frac{k+1}{\theta}\right]\right\}}{\gamma\left\{\alpha,c^{-1}\log\left[1+\frac{k+1}{\theta}\right]\right\}\gamma\left\{\alpha,c^{-1}\log\left[1+\frac{k+2}{\theta}\right]\right\}}, \tag{15}$$

with initial condition $r_G(0) = 1$.

Proof. If X has a p.m.f. in (4), then (15) holds by Lemma 3. Now, if (15) holds, then for x ∈ N, we have,

$$\sum_{k=0}^{x-1}\{r_G(k+1)-r_G(k)\}$$

$$=\sum_{k=0}^{x-1}\left\{\frac{\gamma\left\{\alpha,c^{-1}\log\left[1+\frac{k}{\theta}\right]\right\}}{\gamma\left\{\alpha,c^{-1}\log\left[1+\frac{k+1}{\theta}\right]\right\}}-\frac{\gamma\left\{\alpha,c^{-1}\log\left[1+\frac{k+1}{\theta}\right]\right\}}{\gamma\left\{\alpha,c^{-1}\log\left[1+\frac{k+2}{\theta}\right]\right\}}\right\},$$

$$=-\frac{\gamma\left\{\alpha,c^{-1}\log\left[1+\frac{x}{\theta}\right]\right\}}{\gamma\left\{\alpha,c^{-1}\log\left[1+\frac{x+1}{\theta}\right]\right\}},$$

or

$$r_G(x)-r_G(0)=-\frac{\gamma\left\{\alpha,c^{-1}log\left[1+\frac{x}{\theta}\right]\right\}}{\gamma\left\{\alpha,c^{-1}log\left[1+\frac{k+1}{\theta}\right]\right\}},$$

or, in view of $r_G(0) = 1$, we have,

$$r_G(x)=1-\left(\frac{\gamma\left\{\alpha,c^{-1}log\left[1+\frac{k}{\theta}\right]\right\}}{\gamma\left\{\alpha,c^{-1}log\left[1+\frac{k+1}{\theta}\right]\right\}}\right).$$

5. Estimation

5.1 Regular maximum likelihood estimation

In this subsection, we consider the maximum likelihood estimation method in order to estimate the model parameters of the DGLD. The maximum likelihood estimators (MLEs) enjoy desirable properties and can be used to construct confidence intervals and regions, and also test statistics. The large sample asymptotics for the estimators obtained in this approach, under mild regularity conditions, can be easily handled numerically. To apply the method of maximum likelihood for estimating the parameter vector $\Delta = (c,\alpha,\theta)^T$ of DGLD, we assume that $\vec{x} = (x_1, x_2, \cdots, x_n)_T$ is a random sample of size n from a X ~ DGLD (c,α,θ). The log-likelihood function becomes,

$$\ell=-n\log\Gamma(\alpha)+n\sum\left[\left[\gamma\left\{\alpha,c^{-1}\log\left[1+\frac{x_i+1}{\theta}\right]\right\}-\gamma\left\{\alpha,c^{-1}\log\left[1+\frac{x_i}{\theta}\right]\right\}\right]\right].$$

The above equation can be maximized using available statistical software such as R (optim function) and SAS (PROC NLMIXED), or by solving the nonlinear likelihood equations obtained by differentiating the likelihood function. Regarding interval estimation and hypothesis tests, one may use standard likelihood techniques based on the observed Fisher information matrix, since in this case it is difficult to obtain expected values of the elements of the Fisher Information matrix. For example, the asymptotic covariance matrix of Δ can be approximated by the inverse of the observed Fisher Information matrix evaluated at Δ. One may consider appropriate Likelihood ratio (LR) test(s) to test for the model parameters of the DGLD.

5.2 Censored maximum likelihood estimation

One may also consider the estimation under censoring. Censoring is common in lifetime data sets. There are many types of censoring: type I censoring, type II censoring, and others. A general form known as multi-censoring can be described as follows: there are m lifetimes of which

- m_0 have failed at times $T_1,\cdots,T_{m0}$;
- m_1 have failed at times belonging to $(S_{i-1},S_i]$, $i = 1,\cdots,m_1$;
- m_2 have survived the times R_i, $i = 1,\cdots,m_2$ but have no longer been observed. It is obvious that, $m = m_0 + m_1 + m_2$.

For the multi-censoring data, the associated log likelihood function will be,

$$\log L(c,\alpha,\theta)$$

$$= -(m_0+m_1)\log(\Gamma(\alpha)) + \sum_{j=1}^{m_0}\log\left[\gamma\left\{\alpha, c^{-1}\log\left[1+\frac{T_j+1}{\theta}\right]\right\} - \gamma\left\{\alpha, c^{-1}\log\left[1+\frac{T_j}{\theta}\right]\right\}\right]$$

$$+\sum_{j=1}^{m_1}\log\left[\gamma\left\{\alpha, c^{-1}\log\left[1+\frac{S_j+1}{\theta}\right]\right\} - \gamma\left\{\alpha, c^{-1}\log\left[1+\frac{S_j}{\theta}\right]\right\}\right]$$

$$+\sum_{j=1}^{m_2}\log\left[1-\frac{1}{\Gamma(\alpha)}\gamma\left\{\alpha, c^{-1}\log\left[1+\frac{\lfloor R_j\rfloor+1}{\theta}\right]\right\}\right].$$

The maximum likelihood estimators of c, α and θ can be obtained by maximizing the above function.

5.3 Simulation study

To evaluate the performance of the MLE method, a simulation study is conducted for a total of eighteen parameter combinations and the process is repeated 3000 times. Three different sample sizes n = 100, 200 and 500 are considered. The bias (estimate-actual) and the root mean square errors (RMSE) of the parameter estimates for the MLE are presented in Tables 2, 3 and 4 respectively. It is noted from the tables that the bias and RMSE for the parameter c are in most cases higher than the bias and RMSE for α and θ. In general, the ML method performs well in estimating the DGLD parameters. As expected, reduction in the bias and the RMSE values are observed for all parameter combinations with an increase in the sample size.

6. Application

In this Section, the discrete gamma-Lomax distribution is applied to several data sets. These data sets are taken from Consul (1989). The three data sets represent the observed frequencies of the number of outbreaks of strike in three leading industries in the U.K. during 1948–1959. These industries are Coal-mining, Vehicle manufacturing and Transportation. The data are depicted in Tables 5, 7 and 9. Consul (1989) fitted the data for the three industries to the generalized Poisson distribution (GPD) with p.m.f. given by,'

$$P(\theta,\lambda) = \begin{cases} \theta(\theta+\lambda x)^{x-1}e^{-(\theta+\lambda x)}/x! & x = 0,1,2,\ldots \\ 0 & \text{for } x > m \text{ if } \lambda < 0, \end{cases}$$

where, $\theta > 0$, $\max(-1,-\theta/m) \le \lambda \le 1$ and $m \ge 4$ is the largest positive integer for which $\theta + m\lambda \ge 0$ when $\lambda < 0$.

The results showed that the GPD does not provide an adequate fit to the coal-mining industry data sets. From the goodness of fit summary values, it appears that the discrete gamma-Lomax distribution provides a

Table 2: Bias and RMSE for the parameter estimates using MLE method for n = 100.

Actual values			Bias			RMSE		
c	α	θ	$\hat{c}$	$\hat{\alpha}$	$\hat{\theta}$	$\hat{c}$	$\hat{\alpha}$	$\hat{\theta}$
0.1	1	1	0.1423	–0.1238	0.0877	0.8850	0.1614	0.2210
		2	0.1268	0.0492	0.1529	0.8566	0.1271	0.4352
	5	1	0.1311	0.0568	0.0098	1.3633	0.2229	0.1150
		2	0.1573	0.0904	0.0538	1.1027	0.2826	0.2184
	10	1	–0.1407	0.2693	0.0253	0.9337	0.6942	0.0408
		2	–0.1528	0.2781	0.0161	0.8668	0.7487	0.0652
0.5	1	1	–1.1769	0.1524	0.2709	1.8470	0.2067	0.4075
		2	–1.3217	0.1660	0.3520	1.8728	0.2259	0.6470
	5	1	–0.4169	0.2123	0.0616	2.3357	0.4445	0.1552
		2	–1.0375	0.3312	0.2132	2.3711	0.5265	0.3633
	10	1	1.5882	1.2261	0.6632	1.9993	1.5346	0.7292
		2	–1.0357	1.0231	0.0587	1.7280	1.4654	0.3731
1	1	1	–1.2633	0.1524	0.2709	1.8949	0.1957	0.4302
		2	–1.3518	0.1660	0.3802	2.0138	0.2340	0.5467
	5	1	0.4145	0.2123	0.1533	2.1962	0.4556	0.2321
		2	1.1132	0.3351	0.5591	2.4516	0.5378	0.9881
	10	1	–1.8528	1.2261	0.1463	2.2154	1.5356	0.3470
		2	1.2936	1.0231	0.0578	1.9173	1.5127	0.7002

Table 3: Bias and RMSE for the parameter estimates using MLE method for n = 200.

Actual values			Bias			RMSE		
c	α	θ	$\hat{c}$	$\hat{\alpha}$	$\hat{\theta}$	$\hat{c}$	$\hat{\alpha}$	$\hat{\theta}$
0.1	1	1	0.0357	0.0064	0.0119	0.4013	0.0436	0.0806
		2	–0.0937	0.0136	0.0217	0.4466	0.0522	0.1959
	5	1	0.0214	0.0169	0.0052	0.4376	0.0936	0.0431
		2	0.1022	0.0141	0.0326	0.4025	0.0939	0.0923
	10	1	–0.0523	0.0738	0.0152	0.4155	0.2356	0.0211
		2	–0.1205	0.1017	0.0138	0.3390	0.2687	0.0297
0.5	1	1	–0.2371	0.0315	0.0579	0.5312	0.0893	0.1696
		2	–0.1534	0.0265	0.1020	1.0196	0.0788	0.3047
	5	1	–0.2108	0.0607	0.0214	1.0525	0.1881	0.0849
		2	–0.0999	0.0448	0.0352	1.2412	0.1231	0.1849
	10	1	–0.4528	0.0952	0.2354	0.9858	0.1830	0.4751
		2	–0.6322	0.3218	0.0172	0.9756	0.4576	0.0250
1	1	1	0.4482	–0.1293	–0.0346	0.9925	0.1877	0.2403
		2	0.4051	–0.1104	–0.0836	0.6944	0.1652	0.4956
	5	1	0.3167	–0.0348	–0.0574	0.8260	0.4455	0.1292
		2	0.8452	–0.0698	–0.1348	1.0160	0.3666	0.4882
	10	1	0.9452	0.1726	–0.1152	1.1006	0.3991	0.6132
		2	0.5367	–0.1628	–0.0243	0.5967	0.5884	0.0282

Table 4: Bias and RMSE for the parameter estimates using MLE method for n = 500.

Actual values			Bias			RMSE		
c	α	θ	$\hat{c}$	$\hat{\alpha}$	$\hat{\theta}$	$\hat{c}$	$\hat{\alpha}$	$\hat{\theta}$
0.1	1	1	0.0237	−0.0031	0.0219	0.0238	0.0116	0.0228
		2	0.0241	−0.0085	−0.0135	0.0242	0.0128	0.0136
	5	1	0.0187	−0.0104	−0.0028	0.0188	0.0417	0.0030
		2	0.0153	−0.0102	−0.0157	0.0154	0.0412	0.0157
	10	1	0.0204	−0.0201	−0.0079	0.0205	0.3461	0.0090
		2	0.0197	0.0845	−0.0102	0.0200	0.3736	0.0102
0.5	1	1	0.0416	−0.0129	−0.0110	0.0424	0.0131	0.0111
		2	0.0340	−0.0083	−0.0068	0.0349	0.0086	0.0084
	5	1	0.0470	0.0260	−0.0151	0.0479	0.0269	0.0157
		2	0.0470	0.0260	−0.0227	0.0477	0.0263	0.0227
	10	1	0.0504	0.0882	−0.0324	0.0514	0.1115	0.0327
		2	0.0577	0.0595	−0.0911	0.0592	0.0913	0.0911
1	1	1	0.3426	−0.0513	−0.0210	0.3752	0.0520	0.1336
		2	0.5548	−0.0131	−0.0453	0.5708	0.0147	0.2322
	5	1	0.2893	−0.0123	0.0289	0.3457	0.0356	0.0845
		2	0.1749	0.0321	0.0578	0.1993	0.0386	0.0660
	10	1	0.3584	0.1494	0.0279	0.3820	0.1514	0.0962
		2	0.3325	0.1494	0.0095	0.3758	0.1690	0.0381

Table 5: The number of outbreaks of strike in the coal-mining industry in UK.

x-value	Observed	Three-parameter DGLD	Two-parameter DGLD	GPD
0	46	45.99	45.21	50.01
1	76	75.59	78.30	65.77
2	24	26.23	23.74	32.23
3	9	6.32	6.11	7.23
>=4	1	1.87	2.64	0.76
Total	156	156	156	156

Table 6: The estimated parameters and goodness of fit for the outbreaks of strike in the coal-mining industry in UK data.

Model	Parameters	K-S	W	χ^2	χ^2 df	χ^2 p-value
Three-parameter DGLD	$\hat{\alpha}=4.5109$ $\hat{c}=0.0468$ $\hat{\theta}=6.2260$	0.0116	0.0027	1.7279	1	0.1887
Two-parameter DGLD	$\hat{\alpha}=8.8492$ $\hat{c}=0.0.0989$	0.0105	0.0078	2.4755	2	0.2900
GPD	$\hat{\lambda}=-0.1450$ $\hat{\theta}=1.1377$	0.0400	0.1194	4.5234	2	0.0334

Table 7: The number of outbreaks of strike in the vehicle-manufacture industry in UK.

x-value	Observed	Three-parameter DGLD	Two-parameter DGLD	GPD
0	110	109.90	119.64	109.82
1	33	33.43	34.73	33.36
2	9	8.98	7.77	9.24
3	3	2.54	3.86 }	3.58 }
>= 4	1	1.15		
Total	156	156	156	156

Table 8: The estimated parameters and goodness of fit for the outbreaks of strike in the vehicle-manufacture industry in UK data.

Model	Parameters	K-S	W	χ^2	χ^2 df	χ^2 p-value
Three-parameter DGLD	$\hat{\alpha} = 1.2332$ $\hat{c} = 0.0546$ $\hat{\theta} = 11.5823$	0.0341	0.0774	0.1082	1	0.7422
Two-parameter DGLD	$\hat{\alpha} = 2.9637$ $\hat{c} = 0.1931$	0.0088	0.0021	0.2882	1	0.5914
GPD	$\hat{\lambda} = -0.144$ $\hat{\theta} = 0.351$	0.0727	0.1225	0.0600	1	0.8065

Table 9: The number of outbreaks of strike in the transport industry in UK.

x-value	Observed	Three-parameter DGLD	Two-parameter DGLD	GPD*
0	114	114.20	114.15	114.41
1	35	34.13	33.97	26.01
2	4	5.58	5.85	4.83
3	2	1.35	1.36	0.85
>= 4	1	0.74	0.67	9.88
Total	156	156	156	156

Table 10: The estimated parameters and goodness of fit for the outbreaks of strike in the transport industry in UK data.

Model	Parameters	K-S	W	χ^2	χ^2 df	p-value
Three-parameter DGLD	$\hat{\alpha} = 7.6182$ $\hat{c} = 0.1569$ $\hat{\theta} = 0.3166$	0.0058	0.0006	0.8707	1	0.3508
Two-parameter DGLD	$\hat{\alpha} = 3.6368$ $\hat{c} = 0.1522$	0.0062	0.0008	1.0772	2	0.5836
GPD*	$\hat{\lambda} = 0.098$ $\hat{\theta} = 0.31$	0.0155	0.0055	12.788	2	0.0017

good fit to all data sets. Since the GPD has two parameters, we fit the data sets to the two-parameter DGLD (the scale parameter $\theta = 1$) and the three-parameter DGLD. Tables 6, 8 and 10 show the results of fitting these data sets to the two-parameter discrete gamma-Lomax, three-parameter discrete gamma-Lomax and the generalized Poisson distribution. The method of maximum likelihood is applied to estimate the parameters for the assumed discrete probability models.

Arnold and Emerson (2011) proposed a discrete analogue of Cramer-von Mises and Kolmogorov Smirnov goodness of fit statistics. In order to assess the goodness of fit for the fitted models, the discrete Kolmogorov-Smirnov (k-S), discrete Cramer-von Mises (W) and the Chi-square (χ^2) goodness of fit statistics for the fitted distributions are depicted in Tables 6, 8 and 10.

Notice that (*) values in Tables 9 and 10 are different from the computed values in Consul (1989, p.120). From Tables 6, 8 and 10, it appears (based on χ^2 p-value) that the GPD provides an adequate fit to the automobile manufacturing industry but does not provide an adequate fit to coal-mining and the transport industries. The two-parameter and three-parameter DGLD provide an adequate fit to all data sets. For all data sets, the values of discrete K-S and W statistics for DGLDs are smaller than the ones obtained from GPD. In addition, it can be observed that the DGL distribution fits the left and right tails of the three data sets well. These reaffirm the fact that the DGLD can provide an adequate fit to industrial strike data sets. Consequently, the DGLD can be used as a baseline distribution for modeling the number of strikes in industries.

7. Conclusion

In this paper, we have proposed and derived some distributional properties of a new discrete analogue of the continuous gamma-Lomax distribution (DGLD). From Figure 1, it appears that the DGLD offers great flexibilities in terms of shapes for the probability mass functions. The real data application section shows that DGLD can be useful in fitting various strike data sets and provides a better alternative to the existing GPD probability model. We have also discussed the estimation of the parameters in the standard situation (with all the data available), and also under the multi-censoring setup. An extension of the proposed DGLD to two or multi-dimensional setups can be a possible future research topic.

References

Al-Babtain, A. A., Ahmed, A. H. N. and Afify, A. Z. (2020). A new discrete analog of the continuous Lindley distribution, with reliability applications. Entropy, 22(6): 603.

Almetwally, E. M., Abdo, D. A., Hafez, E. H., Jawa, T. M., Sayed-Ahmed et al. (2022). The new discrete distribution with application to COVID-19 Data. Results in Physics, 32: 104987.

Alzaatreh, A., Famoye, F. and Lee, C. (2012a). Gamma-Pareto distribution and its applications. Journal of Modern Applied Statistical Methods, 11: 78–94.

Alzaatreh, A., Lee, C. and Famoye, F. (2012b). On the discrete analogues of continuous distributions. Statistical Methodology, 9: 589–603.

Alzaatreh, A., Lee, C. and Famoye, F. (2013). A new method for generating families of continuous distributions. Metron, 71: 63–79.

Amponsah, C. K. and Kozubowski, T. J. (2021). A computational approach to estimation of discrete Pareto parameters. Communications in Statistics-Simulation and Computation, 1–20.

Arnold, T. B. and Emerson, J. W. (2011). Nonparametric goodness-of-fit tests for discrete null distributions. R. Journal, 3: 34–39.

Buck, A. (1984). Modeling a Poisson Process: Strike Frequency in Great Britain. Atlantic Economic Journal, 12: 60–64.

Buddana, A. and Kozubowski, T. J. (2014). Discrete pareto distributions. Stochastics and Quality Control, 29(2): 143–156.

Consul, P. C. (1989). Generalized Poisson Distributions: Properties and Applications, Marcel Dekker, Inc., New York, NY.

Cordeiro, G. M., Ortega, M. M. E. and Popovic, V. B. (2015). The gamma-Lomax distribution. Journal of Statistical Computation and Simulation, 85: 305–319.

Dzidzornu, S. B. and Minkah, R. (2021). Assessing the performance of the discrete generalised pareto distribution in modelling non-life insurance claims. Journal of Probability and Statistics.

Eliwa, M. S. and El-Morshedy, M. (2021). A one-parameter discrete distribution for over-dispersed data: Statistical and reliability properties with applications. Journal of Applied Statistics, 1–21.

Ghitany, M. E., AL-Awadhi, F. A. and Alkhalfan, L. A. (2007). Marshall-Olkin extended Lomax distribution and its applications to censored data. Communications in statistics-Theory and Methods, 36: 1855–1866.

Ghosh, I. (2020). A new discrete Pareto type (IV) model: theory, properties, and applications. Journal of Statistical Distributions and Applications, 7(1): 1–17.

Glänzel, W. A. and Schubert, A. (1984). Characterization by truncated moments and its application to Pearson-type distributions. Z. Wahrsch. Verw. Gebiete, 66: 173–182.

Glänzel, W. A. (1987). Characterization theorem based on truncated moments and its application to some distribution families. In Mathematical Statistics and Probability Theory (pp. 75–84). Springer, Dordrecht.

Glänzel, W. A. and Hamedani, G. G. (2001). Characterizations of the univariate distributions. Studia Scien. Math. Hung., 37: 83–118.

Hamedani, G. G., Najaf, M., Roshani, A. and Butt, N. S. (2021). Characterizations of twenty (2020–2021) proposed discrete distributions. Pakistan Journal of Statistics and Operation Research, 847–884.

Hitz, A., Davis, R. and Samorodnitsky, G. (2017). Discrete extremes. arXiv preprint arXiv:1707.05033.

Johnson, N. L., Kotz, S. and Kemp, A. W. (2005). Univariate Discrete Distributions. John Wiley & Sons.

Kim, J. H. and Jeon, Y. (2013). Credibility theory based on trimming. Insur. Math. Econ. 53: 36–47.

Kotz, S. and Shanbhag, D. N. (1980). Some new approach to probability distributions. Adv. Appl. Probab. 12: 903–921.

Krishna, H. and Pundir, P. S. (2009). Discrete Burr and discrete Pareto distributions. Statistical Methodology, 6(2): 177–188.

Leigh, J. (1984). A Bargaining Model and Empirical Analysis of Strike Activity Across Industries. Journal of Labor Research, 5: 127–137.

Lemonte, A. J. and Cordeiro, G. M. (2013). An extended Lomax distribution. Statistics: A Journal of Theoretical and Applied Statistics, 47: 800–816.

Mauleon, A. and Vannetelbosch, V. (1998). Profit Sharing and Strike Activity in Cournot Oligopoly. Journal of Economics, 69: 19–40.

Prieto, F., G-Dz, E. and Sarabia, J. M. (2014). Modeling rodad accident blackspots data with the discrete generalized pareto distribution. Accident Analysis and Prevention, 71: 38–49.

Shaked, M., Shanthikumar, J. G. and Valdez-Torres, J. B. (1995). Discrete hazard rate functions. Computers and Operations Research, 22: 391–402.

Skeels, J. and McGrath, P. (1991). A test of Uncertainty, Expectations, and Error Response in Two Strike Models. Journal of Labor Research, 7: 205–222.

Turner, R. (2021). A New Versatile Discrete Distribution. The R Journal.

Velden, S. V. (2000). Stakingen in Nederland. Amsterdam: Stichting beheer IISG/NIWI.

Chapter 12

New Compounding Lifetime Distributions with Application to Hard Drive Reliability

A Asgharzadeh,[1] *Hassan S Bakouch,*[2,3] *L Esmaeili*[1] and *S Nadarajah*[4,*]

1. Introduction

1.1 Main problem

The following are failure times in days of hard drives reported in https://www.backblaze.com/hard-drive-test-data.html:

```
0.9017917 0.8358333 0.9012917 0.8121250 0.9626667
0.9628333 0.7802917 0.8363750 0.9565000 0.8223750
0.9895417 0.4602500 0.9893333 0.3791667 0.9894167
0.9892917 0.8224167 0.6422917 0.9893750 0.6539583
0.1529583 0.8223750 0.9893750 0.7193333 0.8875000
0.6462500 0.9580417 0.8009583 0.9577917 0.8655000
0.9515417 0.1469583 0.8091250 0.6420417 0.9412917
0.9413750 0.9415833 0.9413750 0.9415000 0.4512500
0.9511667 0.6892500 0.9444167 0.9445000 0.7495833
0.7857917 0.8794583 0.8796250 0.8795417 0.8795000
0.8794583 0.8796250 0.8805417 0.9395417 0.9508333
0.9495833 0.7594167 0.9507917 0.9507083 0.9451667
0.7592500 0.9446667 0.9508333 0.7592917 0.9507083
0.7593750 0.9201667 0.8556250 0.7337500 0.7337500
0.7332083 0.9687500 0.9686667 0.7337917 0.9705417
0.8881250 0.7337083 0.9707500 0.8881250 0.7337083
```

[1] Department of Statistics, University of Mazandaran, Babolsar, Iran.
[2] Department of Mathematics, College of Science, Qassim University, Buraydah, Saudi Arabia.
[3] Department of Mathematics, Faculty of Science,Tanta University, Tanta, Egypt.
[4] Department of Mathematics, University of Manchester, Manchester, UK.
Emails: a.asgharzadeh@umz.ac.ir; hassan.bakouch@science.tanta.edu.eg; esmaily.laila@gmail.com
* Corresponding author: mbbsssn2@manchester.ac.uk

```
0.9690833 0.8882083 0.6025417 0.6353333 0.7610833
0.7613750 0.9068333 0.7614583 0.9068333 0.6356667
0.9365833 0.7615417 0.9365417 0.7490000 0.7292500
0.7292083 0.7440000 0.7489583 0.7293333 0.7429583
```

For computational stability, we have divided each observation by 1000. Some summary statistics of the data are: minimum $= 0.147$, first quartile $= 0.748$, median $= 0.880$, mean $= 0.825$, third quartile $= 0.825$, maximum $= 0.990$, inter quartile range $= 0.197$, range $= 0.843$, skewness $= -1.914$, kurtosis $= 8.036$, 10th percentile $= 0.646$, 5th percentile $= 0.595$, 1st percentile $= 0.153$, 90th percentile $= 0.969$, 95th percentile $= 0.989$ and 99th percentile $= 0.989$. A histogram of the data is shown later in Figure 5.

The problem studied in this paper is to find a simple yet accurate model for the distribution of these failure times. This is an important problem. An accurate modeling of failure times can lead to the production of more robust hard drives. We are not aware of any paper modeling failure times of hard drives – at least physically motivated models.

1.2 Proposed model

We can suppose that a hard drive consists of components working in series (i.e., the hard drive will fail if and only if any one of its components fails) or that a hard drive consists of components working in parallel (i.e., the hard drive will fail if and only if all of its components fail). We can also suppose a mixture of series and parallel systems for the components of the hard drive, but that would complicate things. Of the two scenarios stated, the former is the more reasonable one. So, we shall suppose from now on that a hard drive consists of say N independent components working in series, the assumption of independence is made for simplicity. The value of N may vary depending on the type of hard drive, manufacturer and other factors like length and weight. We can take N to follow a discrete uniform, geometric, binomial, negative binomial or a Poisson distribution. The simplest model for N is a discrete uniform distribution with probability mass function $\Pr(N = n) = \frac{1}{k}$, $n = 1, 2, \ldots, k$, where $k = 1, 2, \ldots$ is an unknown parameter. For study of systems similar to that of a hard drive, we refer the readers to Kumar and Saini (2014), Saini and Kumar (2020) and Saini et al. (2020).

Let $X_1, X_2, \ldots, X_N$ denote the failure times of the N components. Then the failure time of hard drive is $X = \min(X_1, X_2, \ldots, X_N)$. Suppose $X_1, X_2, \ldots$ are independent and identical random variables with cumulative distribution function (cdf) and probability density function (pdf) specified by $G(\cdot; \boldsymbol{\alpha})$ and $g(\cdot; \boldsymbol{\alpha})$, respectively, where $\boldsymbol{\alpha}$ are some unknown parameters. Suppose $X_1, X_2, \ldots$ are independent of N. Assume throughout that $G(\cdot; \boldsymbol{\alpha})$ and $g(\cdot; \boldsymbol{\alpha})$ are absolutely continuous functions. It is easy to show that the cdf of X is,

$$F(x) = 1 - \frac{\overline{G}(x)\left[1 - \left(\overline{G}(x)\right)^k\right]}{k\left[1 - \overline{G}(x)\right]} = 1 - \omega\left(\overline{G}(x)\right) \tag{1}$$

for $x > 0$ and $k = 1, 2, \ldots$, where $\overline{G}(x) = 1 - G(x)$ and

$$\omega(t) = \frac{t\left(1 - t^k\right)}{k(1 - t)}, \tag{2}$$

which is also the probability generating function of N. The pdf of X is,

$$f(x) = g(x)\frac{d\omega\left(\overline{G}(x)\right)}{dx} = g(x)\omega'\left(\overline{G}(x)\right),$$

where,

$$\omega'(t) = \frac{d\omega(t)}{dt} = \frac{1 - (k+1)t^k + kt^{k+1}}{k(t-1)^2}. \tag{3}$$

The pdf of X can be reexpressed as,

$$f(x) = g(x)\frac{1-(k+1)\overline{G}^k(x)+k\overline{G}^{k+1}(x)}{k\left[\overline{G}(x)-1\right]^2}. \tag{4}$$

The hazard rate function of X can be expressed in the forms,

$$h(x) = \frac{g(x)\omega'\left(\overline{G}(x)\right)}{\omega\left(\overline{G}(x)\right)} = -\frac{d}{dx}\log\left[\omega\left(\overline{G}(x)\right)\right]$$

or

$$h(x) = h_g(x)\frac{1-(k+1)\overline{G}^k(x)+k\overline{G}^{k+1}(x)}{\left[1-\overline{G}(x)\right]\left[1-\overline{G}^k(x)\right]}, \tag{5}$$

where $h_g(x) = g(x)/\overline{G}(x)$. Hereafter, a random variable X with cdf given by (1) shall be denoted by $X \sim \text{GU}(\boldsymbol{\alpha}, k)$. We shall suppose throughout that k is a positive real parameter, equivalent to approximating the discrete uniform random variable by a continuous uniform random variable. Approximating discrete random variables by continuous ones is a common practice. Also certain components of a hard drive not working to their full capacity can correspond to non-integer values of k. It is obvious that $F(x) = G(x)$ when $k = 1$.

Variates of $\text{GU}(\boldsymbol{\alpha}, k)$ can be simulated as follows: if u is a variate of the uniform $[0,1]$ distribution then,

$$x = G^{-1}\left(1-\omega^{-1}(u)\right)$$

or the root of,

$$\overline{G}^{k+1}(x) - \left[1+(1-u)k\right]\overline{G}(x) + (1-u)k = 0$$

is a variate from $\text{GU}(\boldsymbol{\alpha}, k)$.

A simpler method for simulation when k is an integer is as follows Ristic et al. (2007): simulate n from a discrete uniform distribution; simulate a random sample $X_1, X_2, \ldots, X_n$ independently from the distribution specified by the cdf G; then $\min\left(X_1, X_2, \ldots, X_n\right)$ is a variate of $\text{GU}(\boldsymbol{\alpha}, k)$.

Distributions constructed by means of $X = \min\left(X_1, X_2, \ldots, X_N\right)$ are known as compound distributions of a minimum of a random number of random variables. Such distributions have received considerable attention in recent years. Twenty prominent distributions introduced recently are the exponential geometric distribution (Adamidis and Loukas 1998), the exponential Poisson distribution (Kus 2007), the exponential logarithmic distribution (Tahmasbi and Rezaei 2008), the Weibull geometric distribution (Barreto-Souza et al. 2011, Morais and Barreto-Souza 2011), the Weibull logarithmic distribution (Morais and Barreto-Souza 2011), the Weibull negative binomial distribution (Morais and Barreto-Souza 2011), the exponentiated exponential binomial distribution (Bakouch et al. 2012), the exponential negative binomial distribution (Hajebi et al. 2013), the modified Weibull geometric distribution (Silva et al. 2013), the exponentiated exponential Poisson distribution (Ristic and Nadarajah 2014), the extended generalized gamma geometric distribution (Bortolini et al. 2017), the exponentiated inverse Weibull geometric distribution (Chung et al. 2017), the reflected generalized Topp-Leone power series distribution (Condino and Domma 2017), Topp–Leone power series distributions (Roozegar and Nadarajah 2017), the power Lomax Poisson distribution (Hassan and Nassr 2018), compounded inverse Weibull distributions (Chakrabarty and Chowdhury 2019), an extended generalized half logistic distribution (Muhammad and Liu 2019), generalized linear exponential geometric distributions (Okasha and Al-Shomrani 2019), an extended Poisson family of life distributions (Ramos et al. 2020) and the exponential-discrete Lindley distribution (Kemaloglu and Yilmaz 2020). Families of distributions obtained by compounding a minimum of a random number of random variables include the Marshall-Olkin family of distributions (Marshall and Olkin 1997) and the generalized Marshall-Olkin family of distributions (Nadarajah et al. 2013).

We shall show later that the GU distribution provides better fits than ten of these distributions for the hard drive data set. We shall also show that the GU distribution provides better fits than the gamma, Weibull, exponentiated exponential (originally due to Gupta and Kundu 1999) and exponentiated Rayleigh (originally due to Surles and Padgett 2001) distributions.

We have motivated the GU distribution by the hard drive data set. It could have been motivated by other real examples too. Two such examples are:

- Wu and Chang (2003) and Wu et al. (2007) considered a scheme where a random number of units is removed every time a failure occurs. This random number was supposed to follow a discrete uniform distribution. Let N denote the number of units removed and $X_1, X_2, \ldots, X_N$ some characteristic (for example, age) of the units.

- Chang and Huang (2014) supposed that job processing times in a reentrant flow shop follow a discrete uniform distribution. Let N denote the job processing time and $X_1, X_2, \ldots, X_N$ the costs of waiting while the job is processed.

In both examples, a variable of interest is $X = \min\left(X_1, X_2, \ldots, X_N\right)$.

The calculations in this paper involve the gamma function defined by,

$$\Gamma(a) = \int_0^\infty t^{a-1} \exp(-t) dt,$$

and the beta function defined by,

$$B(a, b) = \int_0^1 t^{a-1} (1-t)^{b-1} dt.$$

The properties of these special functions can be found in Gradshteyn and Ryzhik (2000).

1.3 Purposes

The purposes of this paper are to:

- derive various mathematical properties of $X \sim \mathrm{GU}(\alpha, k)$, see Section 2. These include expansions for the pdf and the cdf, shape of the pdf, shape of the hazard rate function, moments, moment generating function, reversed residual life moments, order statistic properties, and the Rényi entropy.

- estimate parameters of the GU distribution by the method of maximum likelihood and discuss the asymptotic properties of the estimators, see Section 3.

- show that the GU distribution can be a good model for the failure times data, see Section 4.

Some of the mathematical properties reported involve single and double infinite sums, see Section 2.6 and the appendix. Numerical computations not reported here show that each of these infinite sums can be truncated at 20 to yield a relative error less than 10^{-25} for a wide range of parameter values and for a wide range of choices for g and G. This shows that the mathematical properties can be computed for most practical uses with their infinite sums truncated at twenty. The computations were performed using Maple. Maple took only a fraction of a second to compute the truncated versions. The computational times for the truncated versions were significantly smaller than those for the untruncated versions.

2. Mathematical properties

2.1 Expansions

Some of the mathematical properties of the GU distribution cannot be expressed in closed form. In these cases, it is useful to have expansions for the pdf and the cdf. As mentioned in Section 1, the truncated versions of these expansions can be of practical use.

One can demonstrate that the cdf in (1) and the pdf in (4) can be expressed as

$$F(x) = \frac{1}{k}\sum_{j=1}^{k} Q_j(x) \tag{6}$$

and

$$f(x) = \frac{1}{k}\sum_{j=1}^{k} q_j(x), \tag{7}$$

respectively, where $Q_a(x) = 1 - \overline{G}^a(x)$ and $q_a(x) = ag(x)\overline{G}^{a-1}(x)$. Note that q_a and Q_a are the pdf and cdf of the exponentiated-G distribution. Structural properties of several exponentiated-G distributions have been studied by many authors: see Marshall and Olkin (2007) for an excellent account.

It follows from (6) that (1) can be expressed as a finite linear combination of cdfs of exponentiated-G distributions. It follows from (7) that (4) can be expressed as a finite linear combination of pdfs of exponentiated-G distributions. Hence, any mathematical property of the GU distribution can be expressed as a finite linear combination based on exponentiated-G distributions.

2.2 Shape

In this section, we investigate the shapes of (1), (4) and (5). Shape properties are important because they allow the practitioner to see if the distribution can be fitted to a given data set (this can be seen by comparing the shape of the histogram of the data with possible shapes of the pdf). Shape properties are also useful to see if the distribution can model increasing failure rates, decreasing failure rates or bathtub shaped failure rates.

The shapes of the pdf, (4), and the hazard rate function, (5), can be described analytically. The critical points of the pdf are the roots of the equation:

$$g'(x)\omega'\left(\overline{G}(x)\right) = g'(x)g(x)\omega''\left(\overline{G}(x)\right). \tag{8}$$

There may be more than one root to (8). If $x = x_0$ is a root of (8) then it corresponds to a local maximum, a local minimum or a point of inflexion depending on whether $\lambda(x_0) < 0$, $\lambda(x_0) > 0$ or $\lambda(x_0) = 0$, where,

$$\lambda(x) = g^3(x)\omega'''\left(\overline{G}(x)\right) - 2g(x)g'(x)\omega''\left(\overline{G}(x)\right) + g''(x)\omega'\left(\overline{G}(x)\right) - \left(g'(x)\right)^2\omega''\left(\overline{G}(x)\right).$$

The critical points of the hazard rate function, (5), are the roots of the equation:

$$g^2(x)\frac{\omega''\left(\overline{G}(x)\right)}{\omega^2\left(\overline{G}(x)\right)} = g^2(x)\left[\frac{\omega'\left(\overline{G}(x)\right)}{\omega\left(\overline{G}(x)\right)}\right]^2 + g'(x)\frac{\omega'\left(\overline{G}(x)\right)}{\omega\left(\overline{G}(x)\right)}. \tag{9}$$

There may be more than one root to (9). If $x = x_0$ is a root of (9) then it corresponds to a local maximum, a local minimum or a point of inflexion depending on whether $\lambda(x_0) < 0$, $\lambda(x_0) > 0$ or $\lambda(x_0) = 0$, where,

$$\begin{aligned}\lambda(x) &= g^3(x)\frac{\omega'''\left(\overline{G}(x)\right)}{\omega^2\left(\overline{G}(x)\right)} - 2g^3(x)\frac{\omega'\left(\overline{G}(x)\right)\omega''\left(\overline{G}(x)\right)}{\omega^3\left(\overline{G}(x)\right)} - 2g(x)g'(x)\frac{\omega''\left(\overline{G}(x)\right)}{\omega^2\left(\overline{G}(x)\right)} \\ &\quad -2g^3(x)\frac{\omega'\left(\overline{G}(x)\right)\omega''\left(\overline{G}(x)\right)}{\omega^2\left(\overline{G}(x)\right)} + 2g^3(x)\left[\frac{\omega'\left(\overline{G}(x)\right)}{\omega\left(\overline{G}(x)\right)}\right]^3 + 3g(x)g'(x)\left[\frac{\omega'\left(\overline{G}(x)\right)}{\omega\left(\overline{G}(x)\right)}\right]^2 \\ &\quad -g(x)g'(x)\frac{\omega''\left(\overline{G}(x)\right)}{\omega\left(\overline{G}(x)\right)} + g''(x)\frac{\omega'\left(\overline{G}(x)\right)}{\omega\left(\overline{G}(x)\right)}.\end{aligned}$$

Here, $\omega(\cdot)$ and $\omega'(\cdot)$ are given by (2) and (3), respectively. The second and third order derivatives are,

$$\omega''(t) = \frac{t^{k+1}k^2 - 2t^k k^2 - t^{k+1}k + t^{k-1}k^2 + t^{k-1}k + 2t^k - 2}{k\,(t-1)^3}$$

and

$$\omega'''(t) = -\frac{-t^{k+1}k^3 + 3t^k k^3 - 3t^{k-1}k^3 - 2t^{k+1}k + 3t^{k+1}k^2 + t^{k-2}k^3 - 3t^k k - 6t^k k^2 + 6t^{k-1}k + 3t^{k-1}k^2 - t^{k-2}k + 6t^k - 6}{k\,(t-1)^4},$$

respectively.

The asymptotes of (1), (4) and (5) as $x \to 0, \infty$ are,

$$\begin{aligned}
f(x) &\sim \frac{(k+2)}{2}g(x) \quad as\ x \to 0, \\
f(x) &\sim \frac{g(x)}{k} \quad as\ x \to \infty, \\
F(x) &\sim 1 \quad as\ x \to \infty, \\
F(x) &\sim \frac{1-\overline{G}^k(x)}{k(1-\overline{G}(x))} \quad as\ x \to 0, \\
h(x) &\sim \frac{(k+2)}{2}h_g(x) \quad as\ x \to 0, \\
h(x) &\sim h_g(x) \quad as\ x \to \infty.
\end{aligned}$$

2.3 Special cases

For $G(x) = 1 - e^{-\lambda x^\beta}$, $G(x) = 1 - \left(1 + x^\lambda\right)^{-\beta}$, $G(x) = e^{-\lambda x^{-\beta}}$ and $G(x) = \left(1 + x^{-\lambda}\right)^{-\beta}$, the GU distribution reduces to the Weibull uniform (WU), Burr uniform (BU), inverse Weibull uniform (IWU) and inverse Burr uniform (IBU) distributions, respectively. Table 1 lists the probability density and survival functions of these distributions. Table 2 lists the corresponding hazard rate functions.

We have chosen the special cases to correspond to the Weibull and Burr distributions because: i) Weibull distribution is the most popular model for lifetime data; ii) Burr distribution is one of the most versatile distributions in statistics. As shown by Rodriguez (1977) and Tadikamalla (1980), the Burr distribution contains the shape characteristics of the normal, lognormal, gamma, logistic and exponential distributions as well as a significant portion of the Pearson type I, II, V, VII, IX and XII families.

Figures 1 and 2 plot the pdfs of the WU, IWU, BU and IBU distributions. Figures 3 and 4 plot the hazard rate functions of the WU, IWU, BU and IBU distributions. We see that monotonically decreasing and unimodal shapes are possible for the pdfs. Monotonically decreasing, monotonically increasing, unimodal and upside down bathtub shapes are possible for the hazard rate functions. We have chosen $k = 5$ in Figures 1 to 4. Other values of k did not exhibit different shapes.

Table 1: Probability density and survival functions of some special GU distributions.

	$f(x)$	$1-F(x)$
WU	$\frac{\beta\lambda x^{\beta-1}e^{-\lambda x^{\beta}}\left[1-(k+1)e^{-k\lambda x^{\beta}}+ke^{-(k+1)\lambda x^{\beta}}\right]}{k\left(1-e^{-\lambda x^{\beta}}\right)^{2}}$	$\frac{e^{-\lambda x^{\beta}}\left(1-e^{-k\lambda x^{\beta}}\right)}{k\left(1-e^{-\lambda x^{\beta}}\right)}$
BU	$\frac{\beta\lambda x^{\lambda-1}\left(1+x^{\lambda}\right)^{-(\beta+1)}\left[1-(k+1)\left(1+x^{\lambda}\right)^{-k\beta}+k\left(1+x^{\lambda}\right)^{-(k+1)\beta}\right]}{k\left[1-\left(1+x^{\lambda}\right)^{-\beta}\right]^{2}}$	$\frac{\left(1+x^{\lambda}\right)^{-\beta}\left[1-\left(1+x^{\lambda}\right)^{-k\beta}\right]}{k\left[1-\left(1+x^{\lambda}\right)^{-\beta}\right]}$
IWU	$\frac{\beta\lambda x^{-(\beta+1)}\left[1-(k+1)\left(1-e^{-\lambda x^{-\beta}}\right)^{k}+k\left(1-e^{-\lambda x^{-\beta}}\right)^{k+1}\right]}{ke^{-\lambda x^{-\beta}}}$	$\frac{\left(1-e^{-\lambda x^{-\beta}}\right)\left[1-\left(1-e^{-\lambda x^{-\beta}}\right)^{k}\right]}{ke^{-\lambda x^{-\beta}}}$
IBU	$\frac{\beta\lambda x^{-(\lambda+1)}\left\{1-(k+1)\left[1-\left(1+x^{-\lambda}\right)^{-\beta}\right]^{k}+k\left[1-\left(1+x^{-\lambda}\right)^{-\beta}\right]^{k+1}\right\}}{k\left(1+x^{-\lambda}\right)^{-(\beta-1)}}$	$\frac{\left[1-\left(1+x^{-\lambda}\right)^{-\beta}\right]\left\{1-\left[1-\left(1+x^{-\lambda}\right)^{-\beta}\right]^{k}\right\}}{k\left(1+x^{-\lambda}\right)^{-\beta}}$

Table 2: Hazard rate functions of some special GU distributions.

	$h(x)$
WU	$\frac{\beta\lambda x^{\beta-1}\left[1-(k+1)e^{-k\lambda x^{\beta}}+ke^{-(k+1)\lambda x^{\beta}}\right]}{\left(1-e^{-k\lambda x^{\beta}}\right)\left(1-e^{-\lambda x^{\beta}}\right)}$
BU	$\frac{\beta\lambda x^{\lambda-1}\left[1-(k+1)\left(1+x^{\lambda}\right)^{-k\beta}+k\left(1+x^{\lambda}\right)^{-(k+1)\beta}\right]}{(1+x^{\lambda})\left[1-(1+x^{\lambda})^{-\beta}\right]\left[1-(1+x^{\lambda})^{-k\beta}\right]}$
IWU	$\frac{\beta\lambda x^{-(\beta+1)}\left[1-(k+1)\left(1-e^{-\lambda x^{-\beta}}\right)^{k}+k\left(1-e^{-\lambda x^{-\beta}}\right)^{k+1}\right]}{\left(1-e^{-\lambda x^{-\beta}}\right)\left[1-\left(1-e^{-\lambda x^{-\beta}}\right)^{k}\right]}$
IBU	$\frac{\beta\lambda x^{-(\lambda+1)}\left(1+x^{-\lambda}\right)^{2\beta-1}\left\{1-(k+1)\left[1-\left(1+x^{-\lambda}\right)^{-\beta}\right]^{k}+k\left[1-\left(1+x^{-\lambda}\right)^{-\beta}\right]^{k+1}\right\}}{\left[1-(1+x^{-\lambda})^{-\beta}\right]\left\{1-\left[1-(1+x^{-\lambda})^{-\beta}\right]^{k}\right\}}$

2.4 Moments and moment generating function

Moments properties are fundamental for any distribution. For example, the first four moments can be used to describe any data fairly well. Moments are also useful for estimation.

We derive two representations for the nth moment of $X \sim \mathrm{GU}(\boldsymbol{\alpha}, k)$. The first is immediate from (7):

$$E\left(X^{n}\right)=\frac{1}{k}\sum_{j=1}^{k}E\left(Y_{j}^{n}\right), \tag{10}$$

where Y_a is a random variable with the cdf and pdf specified by $Q_a(x) = 1 - \overline{G}^{a}(x)$ and $q_a(x) = ag(x)\overline{G}^{a-1}(x)$, respectively. If N is a discrete uniform random variable and $X_1, X_2, \ldots$ are independent random variables distributed according to the cdf G and are independent of N then,

$$E\left(X^{n}\right)=\frac{1}{k}\sum_{j=1}^{k}E\left(\left[\min\left(X_{1},X_{2},\ldots,X_{j}\right)\right]^{n}\right).$$

The ordinary moments of several GU distributions can be calculated directly from (10), see Table 3.

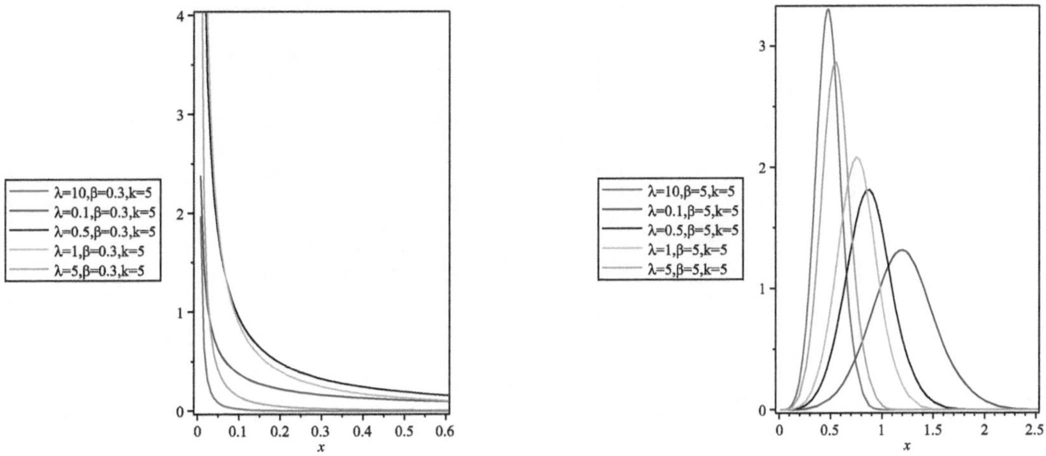

a) WU distribution

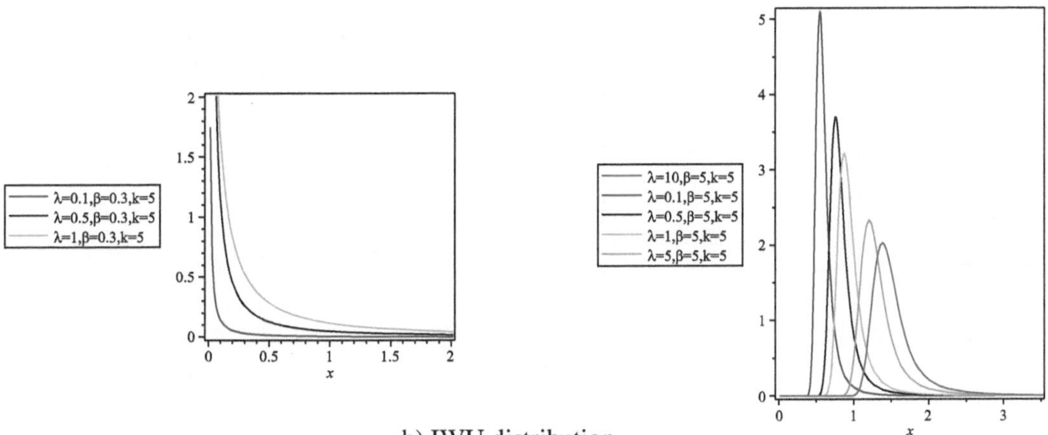

b) IWU distribution

Figure 1: Plots of the probability density functions of the WU and IWU distributions.

Table 3: Moments of some special GU distributions.

GU distribution	$Q_a(x)$	$E\left(Y_a^n\right)$
WU	$1-\left(e^{-\lambda x^{\beta}}\right)^{a}$	$\frac{\Gamma\left(\frac{n}{\beta}+1\right)}{(a\lambda)^{\frac{n}{\beta}}}$
BU	$1-\left(1+x^{\lambda}\right)^{-a\beta}$	$a\sum_{l=0}^{\infty}\binom{\frac{n}{\lambda}}{l}(-1)^{l}B\left(a,\frac{\lambda l-n}{\beta\lambda}+1\right), n<\lambda\beta$
IWU	$1-\left(1-e^{-\lambda x^{-\beta}}\right)^{a}$	does not exist
IBU	$1-\left[1-\left(1+x^{-\lambda}\right)^{-\beta}\right]^{a}$	$a\sum_{l=0}^{\infty}\binom{\frac{-n}{\lambda}}{l}(-1)^{l}B\left(\frac{n}{\lambda}+\frac{l}{\beta}+1,a\right)$

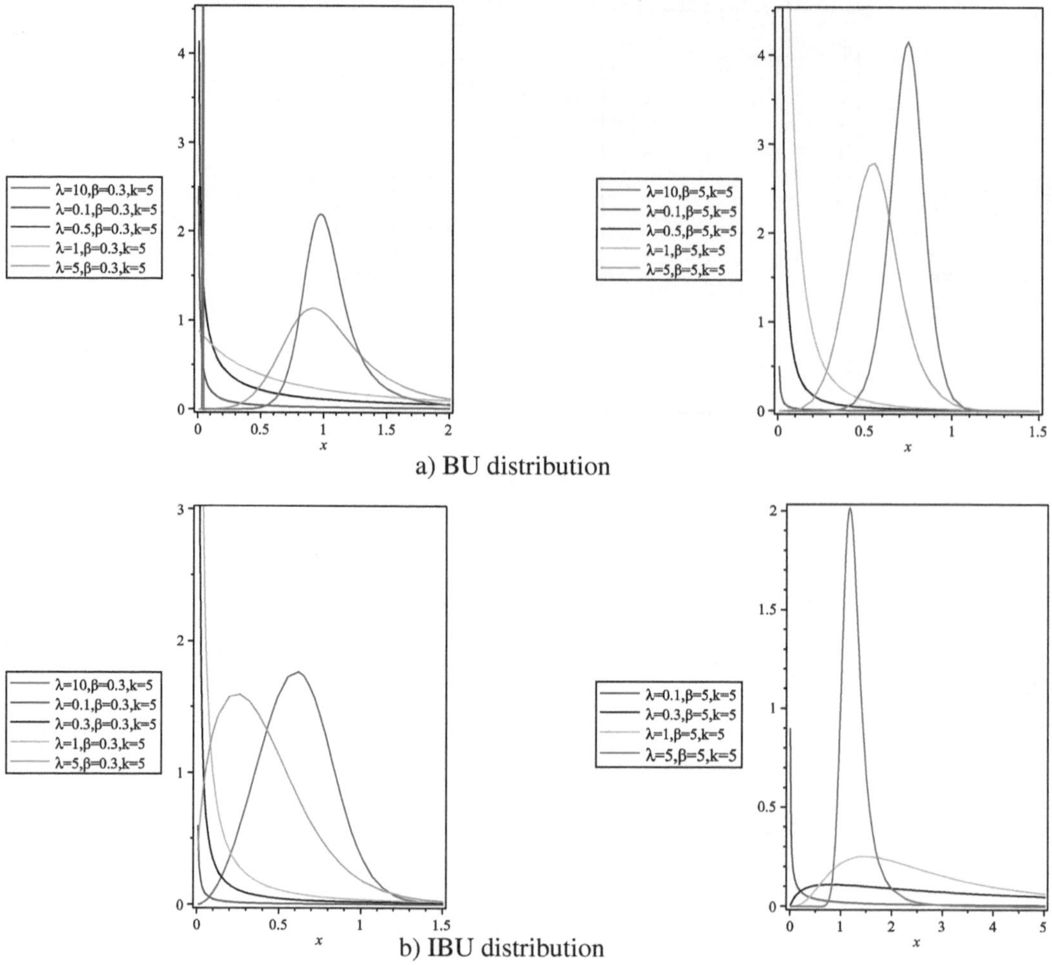

Figure 2: Plots of the probability density functions of the BU and IBU distributions.

We derive two representations for the moment generating function $M(t) = E\left[\exp(tX)\right]$ of $X \sim \mathrm{GU}(\alpha, k)$. The first one is,

$$M(t) = \sum_{j=0}^{\infty} \frac{\mu'_j}{j!} t^j,$$

where $\mu'_j = E\left(X^j\right)$ is given by (10). The other one comes from (7):

$$M(t) = \frac{1}{k} \sum_{j=1}^{k} M_j(t),$$

where $M_a(t)$ denotes the moment generating function of Y_a. So, the moment generating function of several GU distributions can be determined from the moment generating function of Y_a.

Moments of residual lifetime and reversed residual lifetime random variables are extensively used in actuarial sciences in the analysis of risks. Given survival to time t, the residual life is the period from time t

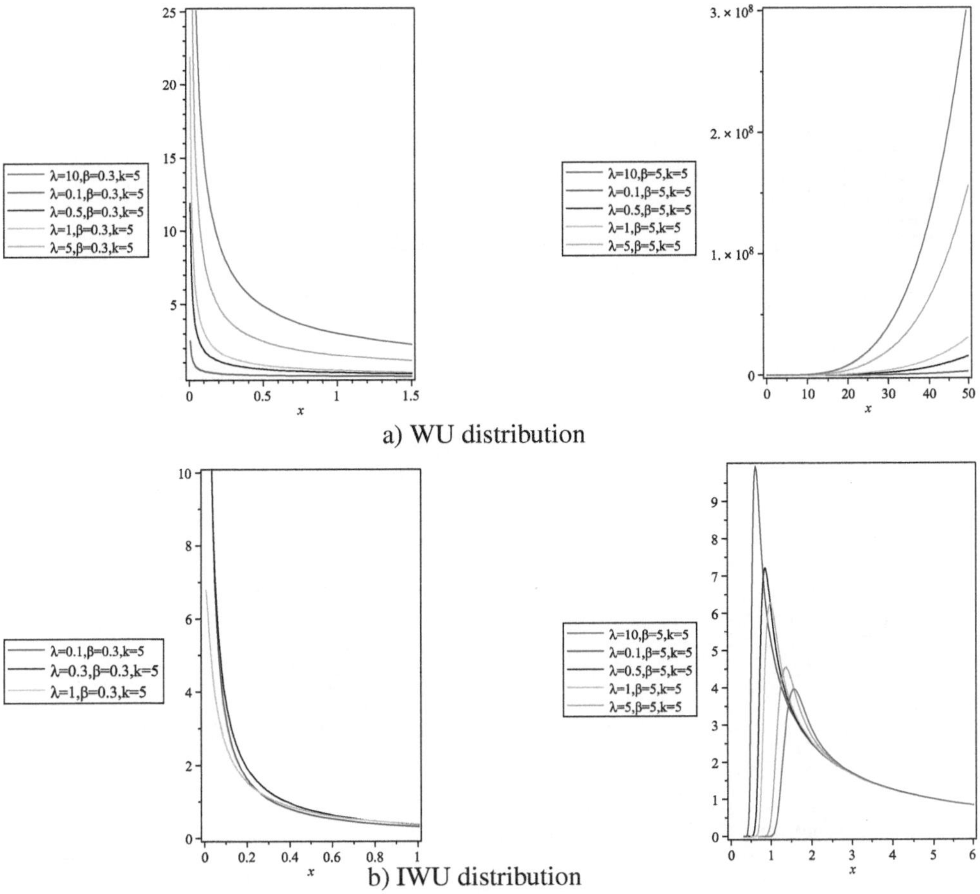

Figure 3: Plots of the hazard rate functions of the WU and IWU distributions.

until the time of failure. The rth moment of residual lifetime and reversed residual lifetime random variables for $X \sim \text{GU}(\boldsymbol{\alpha}, k)$ can be expressed as,

$$\begin{aligned} \mu_r(t) &= E\left[(X-t)^r | X>t\right] \\ &= \frac{1}{1-F(t)} \int_t^{\infty} (x-t)^r f(x) dx \\ &= \frac{1}{1-F(t)} \sum_{i=0}^{r} \binom{r}{i} (-1)^{r-i} t^{r-i} L(\alpha, k, i, t) \end{aligned}$$

and

$$\begin{aligned} m_r(t) &= E\left[(X-t)^r | X<t\right] \\ &= \frac{1}{F(t)} \int_0^{t} (x-t)^r f(x) dx \\ &= \frac{1}{F(t)} \sum_{i=0}^{r} \binom{r}{i} (-1)^{r-i} t^{r-i} \left[E\left(X^i\right) - L(\alpha, k, i, t)\right], \end{aligned}$$

respectively, where $L(\cdots)$ is defined in the appendix.

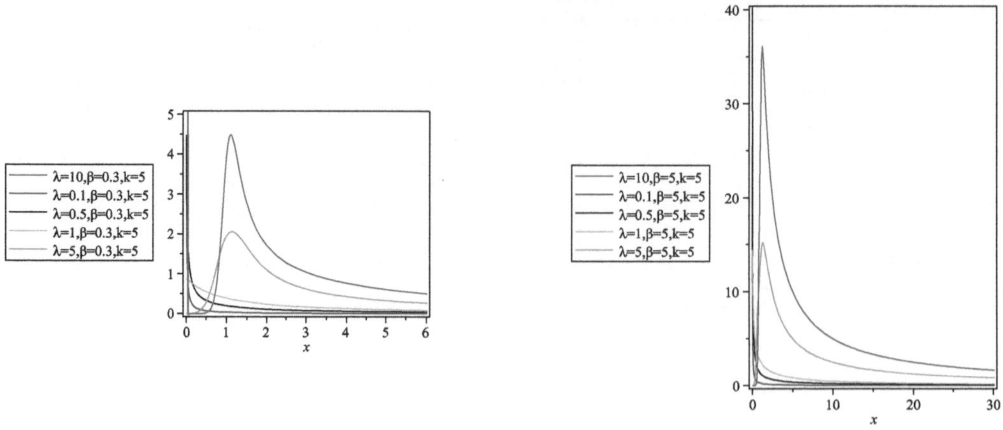

a) BU distribution

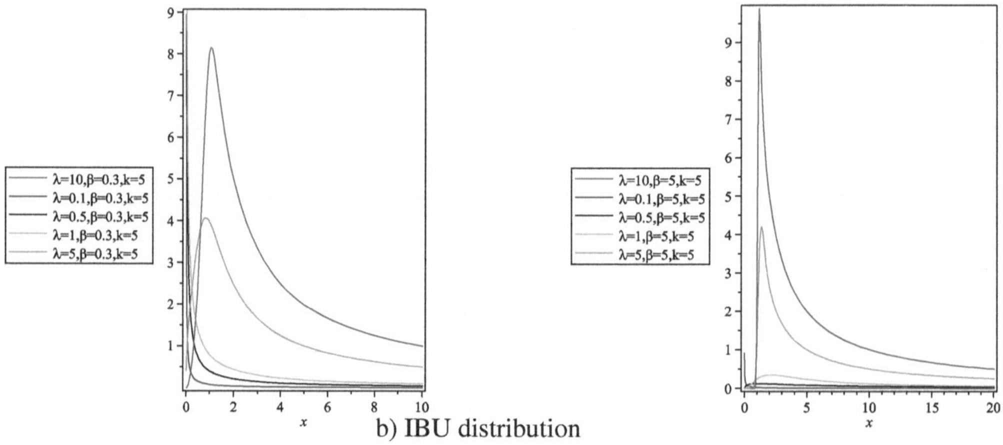

b) IBU distribution

Figure 4: Plots of the hazard rate functions of the BU and IBU distributions.

2.5 Order statistics

Order statistics have been used in a wide range of problems, including robust statistical estimation, detection of outliers, estimation using L moments, characterization of probability distributions, goodness-of-fit tests, entropy estimation, analysis of censored samples, reliability analysis, quality control and strength of materials.

Suppose $X_1, X_2, \ldots, X_n$ is a random sample from the GU distribution. Let $X_{i:n}$ denote the ith order statistic when $X_1, X_2, \ldots, X_n$ are arranged in the increasing order. The pdf of $X_{i:n}$ can be expressed as,

$$\begin{aligned} f_{i:n}(x) &= \frac{f(x)}{B(i, n-i+1)} \sum_{j=0}^{n-i} \binom{n-i}{j} (-1)^j F^{i+j-1}(x) \\ &= \frac{1}{B(i, n-i+1)} \sum_{j=0}^{n-i} \binom{n-i}{j} (-1)^j k^{-i-j} \\ &\cdot \sum_{p=1}^{k} \sum_{r_1=1}^{k} \sum_{r_2=1}^{k} \cdots \sum_{r_{i+j-1}=1}^{k} q_p(x) Q_{r_1}(x) Q_{r_2}(x) \cdots Q_{r_{i+j-1}}(x). \end{aligned}$$

The sth moment of $X_{i:n}$ can be expressed as,

$$E\left(X_{i:n}^{s}\right) = \frac{1}{B(i,n-i+1)}\sum_{j=0}^{n-i}\binom{n-i}{j}(-1)^{j}k^{-i-j}$$
$$\cdot\sum_{p=1}^{k}\sum_{r_1=1}^{k}\sum_{r_2=1}^{k}\cdots\sum_{r_{i+j-1}=1}^{k}\int_{0}^{\infty}x^{s}q_{p}(x)Q_{r_1}(x)Q_{r_2}(x)\cdots Q_{r_{i+j-1}}(x)dx.$$

The moment generating function of $X_{i:n}$ can be expressed as

$$M_{i:n}(t) = \frac{1}{B(i,n-i+1)}\sum_{j=0}^{n-i}\binom{n-i}{j}(-1)^{j}k^{-i-j}$$
$$\cdot\sum_{p=1}^{k}\sum_{r_1=1}^{k}\sum_{r_2=1}^{k}\cdots\sum_{r_{i+j-1}=1}^{k}\int_{0}^{\infty}\exp(tx)q_{p}(x)Q_{r_1}(x)Q_{r_2}(x)\cdots Q_{r_{i+j-1}}(x)dx.$$

2.6 Rényi entropy

The entropy of a random variable, X, a measure of its uncertainty, has many applications in various fields of science and engineering. The most popular entropy measure is the Rényi entropy (Rényi 1961). For a random variable X with pdf, $f(x)$, it is defined as

$$I_R(\gamma) = \frac{1}{1-\gamma}\log\left(\int_{\Re}f^{\gamma}(x)dx\right)$$

for $\gamma > 0$ and $\gamma \neq 1$.

Recent applications of the Rényi entropy include: ultrasonic molecular imaging (Hughes et al. 2009); molecular imaging of tumors using a clinically relevant protocol (Marsh et al. 2010); sparse kernel density estimation and its application in variable selection (Han et al. 2011).

For $X \sim \mathrm{GU}(\boldsymbol{\alpha}, k)$,

$$\int_{\Re}f^{\gamma}(x)dx = \sum_{j=0}^{\infty}\sum_{m=0}^{j}\binom{\gamma}{j}\binom{j}{m}(-1)^{j}k^{m-\gamma}\int_{\Re}G(x)^{m-2\gamma}\overline{G}(x)^{kj}g(x)dx$$
$$= \sum_{j=0}^{\infty}\sum_{m=0}^{j}\binom{\gamma}{j}\binom{j}{m}(-1)^{j}k^{m-\gamma}I(\gamma,j,m),$$

where,

$$I(\gamma,j,m) = \int_{0}^{1}u^{m-2\gamma}(1-u)^{kj}g^{\gamma-1}\left(G^{-1}(u)\right)du.$$

Therefore, the Rényi entropy for $X \sim \mathrm{GU}(\boldsymbol{\alpha}, k)$ can be expressed as

$$I_R(\gamma) = \frac{1}{1-\gamma}\log\left\{\sum_{j=0}^{\infty}\sum_{m=0}^{j}\binom{\gamma}{j}\binom{j}{m}(-1)^{j}k^{m-\gamma}I(\gamma,j,m)\right\}.$$

Rényi entropy for the WU, IWU, BU and IBU distributions are given in the appendix.

3. Estimation

Suppose $x_1, \ldots, x_n$ is a random sample from the GU distribution with unknown parameters $\Theta = (\alpha, k)^\top$, a $r \times 1$ parameter vector say. We determine the maximum likelihood estimators of the parameters.

The log-likelihood function of Θ is,

$$\ell_n(\Theta) = -n \log k + \sum_{i=1}^{n} \log g(x_i) - 2 \sum_{i=1}^{n} \log G(x_i) + \sum_{i=1}^{n} \log \left[1 - kG(x_i)\overline{G}^k(x_i) - \overline{G}^k(x_i)\right]. \quad (11)$$

The log-likelihood function can be maximized either directly by using SAS (PROC NLMIXED) or the Ox program (subroutine MaxBFGS) (see Doornik 2007) or by solving the nonlinear normal equations obtained by differentiating (11). The latter are,

$$\begin{aligned}
\frac{\partial \ell_n}{\partial \alpha} &= \sum_{i=1}^{n} \frac{\frac{\partial g(x_i)}{\partial \alpha}}{g(x_i)} - 2 \sum_{i=1}^{n} \frac{\frac{\partial G(x_i)}{\partial \alpha}}{G(x_i)} \\
&\quad + \sum_{i=1}^{n} \frac{\frac{\partial G(x_i)}{\partial \alpha}\overline{G}^{k-1}(x_i)\left[k^2 G(x_i) + k\overline{G}(x_i) + k\right]}{1 - \overline{G}^k(x_i)\left[kG(x_i) + 1\right]}, \\
\frac{\partial \ell_n}{\partial k} &= -\frac{n}{k} + \sum_{i=1}^{n} \frac{G(x_i) + \log \overline{G}(x_i)\left[kG(x_i) + 1\right]}{kG(x_i) + 1 - \overline{G}^k(x_i)}.
\end{aligned}$$

In the practical data application presented in Section 4, the SAS procedure was used to obtain the maximum likelihood estimates. Numerical computations not reported here showed that the surface of (11) was smooth for given smooth functions $g(\cdot)$ and $G(\cdot)$. The SAS procedure was able to locate the maximum of the likelihood surface for a wide range of smooth functions and for a wide range of starting values, including starting values determined by the method of moments, i.e., the simultaneous solutions of $E\left(X^i\right) = \frac{1}{n}\sum_{j=1}^{n} x_j^i$ for $i = 1, 2, \ldots, r$, where the left hand side is given by (10). These equations were solved numerically using SOLVE in SAS as (10) is not in closed form. The solutions for the maximum likelihood and estimates of moments were unique for all starting values.

Let $\widehat{\Theta} = \left(\widehat{\alpha}, \widehat{k}\right)^\top$ denote the maximum likelihood estimator of $\Theta = (\alpha, k)^\top$. Under certain regularity conditions (see, for example, Ferguson (1996) and Lehmann and Casella (1998), pages 461-463), the distribution of $\widehat{\Theta}$ as $n \to \infty$ is the r-variate normal with mean Θ and covariance given by the inverse of,

$$\mathbf{I} = \begin{pmatrix} \mathbf{I}_{11} & \mathbf{I}_{12} \\ \mathbf{I}_{21} & I_{22} \end{pmatrix} = \begin{pmatrix} E\left(-\frac{\partial^2 \ell_n}{\partial \alpha^2}\right) & E\left(-\frac{\partial^2 \ell_n}{\partial \alpha \partial k}\right) \\ E\left(-\frac{\partial^2 \ell_n}{\partial k \partial \alpha}\right) & E\left(-\frac{\partial^2 \ell_n}{\partial k^2}\right) \end{pmatrix}.$$

Here, $\mathbf{I}$ is the expected information matrix.

In practice, n is finite. The recommended approximation (see, for example, Cox and Hinkley 1979) for the distribution of $\widehat{\Theta}$ is the r-variate normal distribution with mean Θ and covariance taken to be the inverse of

$$\mathbf{J} = \begin{pmatrix} \mathbf{J}_{11} & \mathbf{J}_{12} \\ \mathbf{J}_{21} & J_{22} \end{pmatrix} = \left.\begin{pmatrix} -\frac{\partial^2 \ell_n}{\partial \alpha^2} & -\frac{\partial^2 \ell_n}{\partial \alpha \partial k} \\ -\frac{\partial^2 \ell_n}{\partial k \partial \alpha} & -\frac{\partial^2 \ell_n}{\partial k^2} \end{pmatrix}\right|_{\alpha=\widehat{\alpha}, k=\widehat{k}}.$$

Here, **J** is the observed information matrix. This is known to be a better approximation than the one based on the expected information matrix, see Cox and Hinkley (1979).

4. Practical data analysis

In this section, we return to the hard drive data set discussed in Section 1. As explained there, the failure time of a hard drive can be modeled as the failure of a system having components working in series. We fitted the following distributions to the data: the two-parameter gamma distribution specified by the pdf,

$$f(x) = \frac{b^a x^{a-1} e^{-bx}}{\Gamma(a)}$$

for $x > 0$, $a > 0$ and $b > 0$; the two-parameter Weibull distribution specified by the pdf,

$$f(x) = abx^{a-1} e^{-bx^a}$$

for $x > 0$, $a > 0$ and $b > 0$; the two-parameter exponentiated exponential (Gupta and Kundu 1999) distribution specified by the pdf,

$$f(x) = abe^{-bx} \left[1 - e^{-bx}\right]^{a-1}$$

for $x > 0$, $a > 0$ and $b > 0$; the two-parameter exponentiated Rayleigh (Surles and Padgett 2001) distribution specified by the pdf,

$$f(x) = 2ab^2 xe^{-b^2 x^2} \left[1 - e^{-b^2 x^2}\right]^{a-1}$$

for $x > 0$, $a > 0$ and $b > 0$; the three-parameter WU distribution; the three-parameter IWU distribution; the four-parameter BU distribution; the four-parameter IBU distribution; the two-parameter exponential geometric (Adamidis and Loukas 1998) distribution specified by the pdf,

$$f(x) = \frac{b(1-p)e^{-bx}}{\left[1 - pe^{-bx}\right]^2}$$

for $x > 0$, $b > 0$ and $0 < p < 1$; the two-parameter exponential Poisson distribution (Kus 2007) specified by the pdf,

$$f(x) = \frac{\lambda b e^{-\lambda - bx + \lambda e^{-bx}}}{1 - e^{-\lambda}}$$

for $x > 0$, $b > 0$ and $\lambda > 0$; the two-parameter exponential logarithmic distribution (Tahmasbi and Rezaei 2008) specified by the pdf,

$$f(x) = \frac{b(1-p)e^{-bx}}{(-\log p)\left[1 - (1-p)e^{-bx}\right]}$$

for $x > 0$, $b > 0$ and $0 < p < 1$; the three-parameter Weibull geometric (Barreto-Souza et al. 2011) distribution specified by the pdf,

$$f(x) = \frac{ab^a (1-p) x^{a-1} e^{-(bx)^a}}{\left[1 - pe^{-(bx)^a}\right]^2}$$

for $x > 0$, $a > 0$, $b > 0$ and $0 < p < 1$; the three-parameter Weibull logarithmic distribution (Morais and Barreto-Souza 2011) specified by the pdf,

$$f(x) = \frac{ab^a(1-p)x^{a-1}e^{-(bx)^a}}{(-\log p)\left[1-(1-p)e^{-(bx)^a}\right]}$$

for $x > 0$, $a > 0$, $b > 0$ and $0 < p < 1$; the four-parameter Weibull negative binomial distribution (Morais and Barreto-Souza 2011) specified by the pdf,

$$f(x) = \frac{ab^a spe^{-(bx)^a}\left[1-pe^{-(bx)^a}\right]^{-s-1}}{(1-p)^{-s}-1}$$

for $x > 0$, $a > 0$, $b > 0$, $0 < p < 1$ and $s > 0$; the four-parameter exponentiated exponential binomial distribution (Bakouch et al. 2012) specified by the pdf,

$$f(x) = \frac{abspe^{-bx}\left(1-e^{-bx}\right)^{a-1}\left[1-p\left(1-e^{-bx}\right)^a\right]^{s-1}}{1-(1-\theta)^s}$$

for $x > 0$, $a > 0$, $b > 0$, $0 < p < 1$ and $s > 0$; the three-parameter exponential negative binomial distribution (Hajebi et al. 2013) specified by the pdf,

$$f(x) = \frac{kb(1-p)^k e^{-kbx}}{\left[1-pe^{-bx}\right]^{k+1}}$$

for $x > 0$, $b > 0$, $0 < p < 1$ and $k > 0$; the four-parameter modified Weibull geometric (Silva et al. 2013) distribution specified by the pdf,

$$f(x) = \frac{b^a(1-p)x^{a-1}(a+\lambda x)e^{\lambda x-(bx)^a e^{\lambda x}}}{\left[1-pe^{-(bx)^a e^{\lambda x}}\right]^2}$$

for $x > 0$, $a > 0$, $b > 0$, $\lambda \geq 0$ and $0 < p < 1$; the three-parameter exponentiated exponential Poisson distribution (Ristic and Nadarajah 2014) specified by the pdf,

$$f(x) = \frac{ab\lambda e^{-bx}\left(1-e^{-bx}\right)^{a-1}e^{-\lambda\left(1-e^{-bx}\right)^a}}{1-e^{-\lambda}}$$

for $x > 0$, $a > 0$, $b > 0$ and $\lambda > 0$. In total, eighteen distributions were fitted to the hard drive data set. Four of these distributions have two parameters each. Six of these distributions have three parameters each. The remaining five distributions have four parameters each. The distributions were fitted by the methods of maximum likelihood and moments. The WU, IWU, BU and IBU distributions were fitted by following the procedures in Section 3.

Many of the fitted distributions are not tested. Discrimination among them was performed using various criteria:

- the Akaike information criterion due to Akaike (1974) defined by,

$$\mathrm{AIC} = 2q - 2\log L\left(\widehat{\boldsymbol{\Theta}}\right),$$

where $\boldsymbol{\Theta}$ is the vector of unknown parameters, $\widehat{\boldsymbol{\Theta}}$ is the maximum likelihood estimate of $\boldsymbol{\Theta}$ and q is the number of unknown parameters;

- the Bayesian information criterion due to Schwarz (1978) defined by,

$$\mathrm{BIC} = q \log n - 2 \log L\left(\widehat{\boldsymbol{\Theta}}\right);$$

- the consistent Akaike information criterion (CAIC) due to Bozdogan (1987) defined by,

$$\mathrm{CAIC} = -2 \log L\left(\widehat{\boldsymbol{\Theta}}\right) + q\left(\log n + 1\right);$$

- the corrected Akaike information criterion (AICc) due to Hurvich and Tsai (1989) defined by,

$$\mathrm{AICc} = \mathrm{AIC} + \frac{2q(q+1)}{n-q-1};$$

- the Hannan-Quinn criterion due to Hannan and Quinn (1979) defined by,

$$\mathrm{HQC} = -2 \log L\left(\widehat{\boldsymbol{\Theta}}\right) + 2q \log \log n;$$

- the p-value of the Kolmogorov-Smirnov statistic (Kolmogorov 1933, Smirnov 1948) defined by,

$$\sup_{x} \left| \frac{1}{n} \sum_{i=1}^{n} I\left\{x_i \leq x\right\} - \widehat{F}(x) \right|,$$

  where, $I\left\{\cdot\right\}$ denotes the indicator function and $\widehat{F}(\cdot)$ the maximum likelihood estimate of $F(x)$;

- the p-value of the Kolmogorov-Smirnov statistic (Kolmogorov 1933, Smirnov 1948) is defined by,

$$\sup_{x} \left| \frac{1}{n} \sum_{i=1}^{n} I\left\{x_i \leq x\right\} - \widetilde{F}(x) \right|,$$

  where, $\widetilde{F}(\cdot)$ is the method of moments estimate of $F(x)$;

- the p-value of the Anderson-Darling statistic (Anderson and Darling 1954) is defined by,

$$-n - \sum_{i=1}^{n} \left\{ \log \widehat{F}\left(x_{(i)}\right) + \log\left[1 - \widehat{F}\left(x_{(n+1-i)}\right)\right] \right\},$$

  where, $x_{(1)} \leq x_{(2)} \leq \cdots \leq x_{(n)}$ is the observed data arranged in increasing order;

- the p-value of the Anderson-Darling statistic (Anderson and Darling 1954) is defined by,

$$-n - \sum_{i=1}^{n} \left\{ \log \widetilde{F}\left(x_{(i)}\right) + \log\left[1 - \widetilde{F}\left(x_{(n+1-i)}\right)\right] \right\};$$

- the p-value of the Cramér-von Mises statistic (Cramér 1928, von Mises 1931) is defined by,

$$\frac{1}{12n} + \sum_{i=1}^{n} \left[\frac{2i-1}{2n} - \widehat{F}\left(x_{(i)}\right) \right]^2;$$

- the p-value of the Cramér-von Mises statistic (Cramér 1928, von Mises 1931) defined by,

$$\frac{1}{12n}+\sum_{i=1}^{n}\left[\frac{2i-1}{2n}-\widetilde{F}\left(x_{(i)}\right)\right]^{2}.$$

The smaller the values of AIC, BIC, CAIC, AICc, and HQC the better the fit. For more discussion on these criteria, see Burnham and Anderson (2004) and Fang (2011).

Since the Kolmogorov-Smirnov, Anderson-Darling and Cramér-von Mises tests assume that the fitted distribution gives the "true" parameter values, the p-values were computed by simulation as follows:

(i) fit the distribution to the data and compute the corresponding Kolmogorov-Smirnov/Anderson-Darling/Cramér-von Mises statistic;

(ii) generate 10000 samples each of the same size as the data from the fitted model in step (i);

(iii) refit the model to each of the 10000 samples;

(iv) compute the Kolmogorov-Smirnov/Anderson-Darling/Cramér-von Mises statistic for the 10000 fits in step (iii);

(v) construct an empirical cdf of the 10000 values of the Kolmogorov-Smirnov/Anderson-Darling/Cramér-von Mises statistic obtained in step (iv);

(vi) compare the Kolmogorov-Smirnov/Anderson-Darling/Cramér-von Mises statistic obtained in step (i) with the empirical cdf in step (v) to get the p-value.

The values of $-\log L$, AIC, BIC, CAIC, AICc and HQC for the eighteen fitted distributions are given in Table 4. The p-values of the Kolmogorov-Smirnov, Anderson-Darling and Cramér-von Mises statistics for the eighteen fitted distributions are given in Table 5. In these tables, EE, ER, EG, EP, EL, WG, WL, WNB, EEB, ENB, MWG and EEP denote the exponentiated exponential, exponentiated Rayleigh, exponential geometric, exponential Poisson, exponential logarithmic, Weibull geometric, Weibull logarithmic, Weibull negative binomial, exponentiated exponential binomial, exponential negative binomial, modified Weibull geometric and the exponentiated exponential Poisson distributions, respectively.

We can see that the smallest AIC, the smallest BIC, the smallest CAIC, the smallest AICc, the smallest HQC and the largest p-values are for the WU distribution. The second smallest AIC, the second smallest BIC, the second smallest CAIC, the second smallest AICc, the second smallest HQC and the second largest p-values are for the Weibull distribution. The largest AIC, the largest BIC, the largest CAIC, the largest AICc, the largest HQC and the smallest p-values are for the ENB distribution.

There is not much difference between the p-values obtained by the methods of maximum likelihood and moments. The relative performances of the eighteen distributions with the respect to the p-values appear the same for both methods. At the five percent level, all of the fitted distributions appear acceptable except for the EP and ENB distributions. However, the best fitting distribution in terms of the twelve criteria is the WU distribution. It is pleasing that the WU distribution gives the best fit in spite of having one parameter less than the five distributions each having four parameters.

The parameter estimates for the best fitting WU distribution are $\widehat{k}=0.077(0.016)$, $\widehat{\lambda^{-1/\beta}}=0.820(0.009)$ and $\widehat{\beta}=6.881(0.092)$. The standard errors given in brackets were computed by inverting the observed information matrix, see Section 3. The parameter estimates imply that failure time of a hard drive can be modeled as the failure time of a discrete uniform number of components working in series, the average number being 0.538. The average number being less than one means that hard drives do not work to their full capacity on average. The failure time of each component has a Weibull distribution with mean equal to 0.767 and variance equal to 0.017.

Table 4: Log-likelihood values and information criteria for the distributions fitted to the hard drive failure data.

Distribution	$-\log L$	AIC	BIC	CAIC	AICc	HQC
Gamma	-18.430	-32.860	-27.650	-25.650	-32.736	-30.751
Weibull	-50.759	-97.517	-92.307	-90.307	-97.393	-95.408
EE	-0.289	3.422	8.632	10.632	3.545	5.530
ER	-21.323	-38.646	-33.435	-31.435	-38.522	-36.537
EEP	21.321	48.641	56.457	59.457	48.891	51.805
EP	81.252	166.503	171.714	173.714	166.627	168.612
EG	82.333	168.666	173.876	175.876	168.790	170.775
ENB	82.333	170.666	178.481	181.481	170.916	173.829
EL	11.975	27.950	33.160	35.160	28.074	30.059
WG	-8.919	-11.839	-4.023	-1.023	-11.589	-8.675
WNB	-8.919	-9.839	0.582	4.582	-9.418	-5.621
WL	-3.686	-1.373	6.443	9.443	-1.123	1.790
EEB	3.939	15.878	26.298	30.298	16.299	20.095
MWG	-8.919	-9.839	0.582	4.582	-9.418	-5.621
WU	-53.893	-101.787	-93.971	-90.971	-101.537	-98.624
IWU	49.310	104.619	112.435	115.435	104.869	107.783
BU	-44.827	-81.654	-71.234	-67.234	-81.233	-77.437
IBU	-12.226	-16.453	-6.032	-2.032	-16.032	-12.235

Table 5: p-values of goodness of fit statistics for the distributions fitted to the hard drive failure data.

	KS p-value		AD p-value		CV p-value	
Distribution	MLE	MME	MLE	MME	MLE	MME
Gamma	0.196	0.233	0.269	0.199	0.196	0.173
Weibull	0.267	0.256	0.276	0.249	0.294	0.236
EE	0.136	0.102	0.066	0.108	0.094	0.108
ER	0.211	0.246	0.274	0.243	0.232	0.208
EEP	0.105	0.061	0.046	0.036	0.074	0.060
EP	0.064	0.058	0.027	0.024	0.053	0.026
EG	0.051	0.039	0.026	0.018	0.041	0.025
ENB	0.044	0.012	0.006	0.011	0.017	0.019
EL	0.108	0.068	0.054	0.042	0.075	0.060
WG	0.147	0.183	0.083	0.174	0.143	0.149
WNB	0.140	0.114	0.071	0.125	0.113	0.124
WL	0.147	0.183	0.083	0.174	0.143	0.149
EEB	0.127	0.086	0.054	0.066	0.080	0.077
MWG	0.146	0.158	0.077	0.148	0.123	0.136
WU	0.289	0.279	0.278	0.296	0.296	0.265
IWU	0.072	0.059	0.040	0.027	0.057	0.053
BU	0.260	0.248	0.275	0.245	0.259	0.214
IBU	0.163	0.224	0.268	0.198	0.191	0.173

The probability-probability, quantile-quantile and density plots for the best fitting WU distribution are shown in Figures 5, 6 and 7. We see that its fit is reasonable except possibly in the lower tail. A future work is required to find distributions providing better fits to the lower tail.

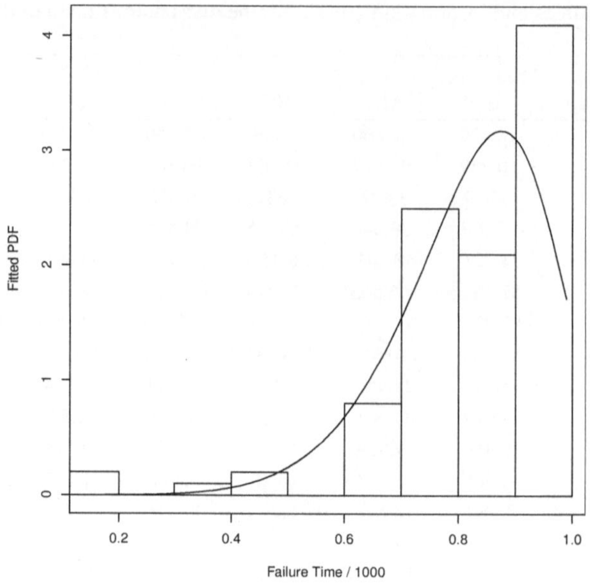

Figure 5: Fitted WU pdf and the histogram for the hard drive failure data.

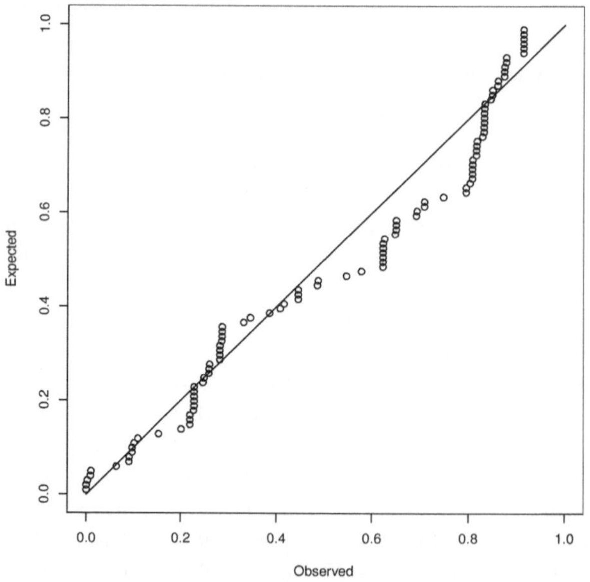

Figure 6: Probability plot for the fitted WU distribution for the hard drive failure data.

5. Conclusions

Motivated by a failure times data set of hard drives, we have proposed a class of distributions. We have studied various mathematical properties of the distributions, derived estimators by the method of maximum likelihood and discussed the asymptotic properties of the estimators. In particular, we have shown that the hazard rate of the failure time of a hard drive can be monotonically decreasing, monotonically increasing, unimodal or upside down bathtub shaped.

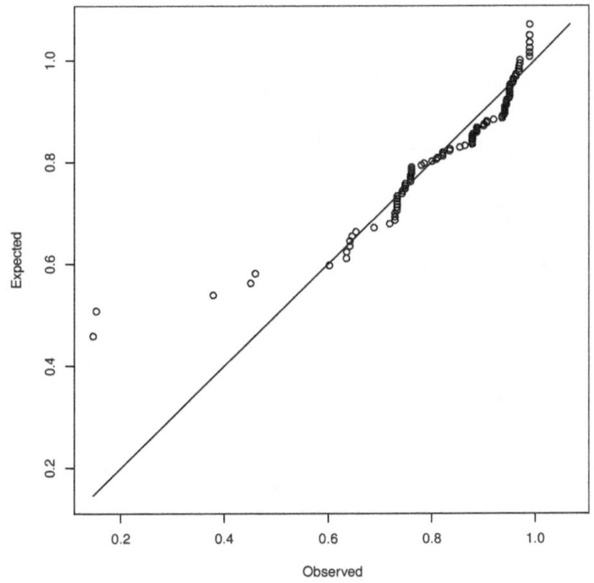

Figure 7: Quantile plot for the fitted WU distribution for the hard drive failure data.

We have shown that the proposed distributions fit the failure times data well. They provide better fits than at least fourteen other known distributions, including the gamma, Weibull, exponentiated exponential and exponentiated Rayleigh distributions. None of these distributions have been previously used (neither have they been physically motivated) to model failure time of a hard drive.

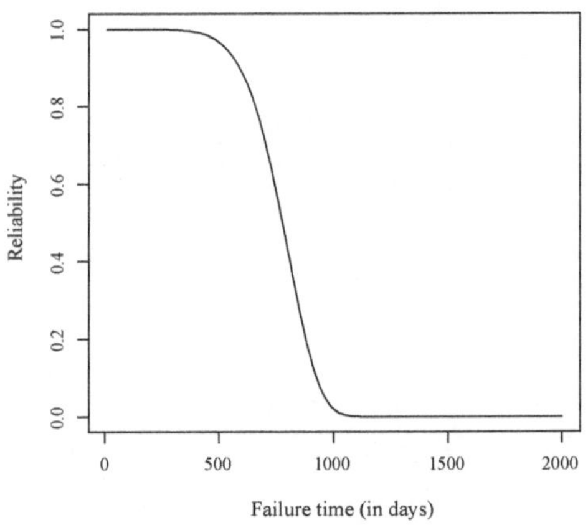

Figure 8: Reliability of the hard drive.

The reliability associated with the failure time of a hard drive is shown in Figure 8. As expected, the reliability is a decreasing function of the failure time. For example, the probabilities that the hard drive will continue to operate without failures after 10, 20, 30, 40, 50, 60, 70, 80, 90 and 100 days are 0.9999995, 0.9999393, 0.9990117, 0.9928671, 0.9673065, 0.8899872, 0.7141655, 0.4300986, 0.1499379 and 0.01988396, respectively. The probability that the hard drive will continue to operate without failures after 1000 days is almost zero. In particular, the probability that the hard drive will continue to operate without failures after 1300 days is less than 10^{-10}.

Appendix

The function L in Section 2.4 can be expressed as,

$$L(\boldsymbol{\alpha}, k, c, t) = \int_t^{\infty} x^c f(x)dx = \sum_{j=0}^{\infty} \left\{ \frac{L_1(j+1, \boldsymbol{\alpha}, c)}{k} + \frac{(j+1)L_1(j+k+2, \boldsymbol{\alpha}, c)}{j+k+2} \right.$$
$$\left. - \frac{(j+1)(k+1)L_1(j+k+1, \boldsymbol{\alpha}, c)}{k(j+k+1)} \right\},$$

where,

$$L_1(a, \boldsymbol{\alpha}, c, t) = \int_t^{\infty} ax^c g(x)\overline{G}(x)dx = a\int_{G(t)}^{1} \left[G^{-1}(u)\right]^n (1-u)^{a-1}du.$$

Rényi entropy in Section 2.6 for the WU, IWU, BU and IBU distributions can be expressed as,

$$I_R(\gamma) = -\log\beta - \frac{1}{\beta}\log\lambda + \frac{\log\Gamma\left(\frac{(\gamma-1)(\beta-1)}{\beta}+1\right)}{1-\gamma} + \frac{1}{1-\gamma}\left\{\log\left[\sum_{j=0}^{\infty}\sum_{m=0}^{j}\sum_{l=0}^{\infty}\binom{\gamma}{j}\right.\right.$$
$$\left.\left.\cdot\binom{j}{m}\binom{m-2\gamma}{l}(-1)^{j+l}\frac{k^{m-\gamma}}{(kj+\gamma+l)^{\frac{(\gamma-1)(\beta-1)}{\beta}+1}}\right]\right\},$$

$$I_R(\gamma) = -\log\beta - \frac{1}{\beta}\log\lambda + \frac{\log\Gamma\left(\frac{(\gamma-1)(\beta+1)}{\beta}+1\right)}{1-\gamma} + \frac{1}{1-\gamma}\left\{\log\left[\sum_{j=0}^{\infty}\sum_{m=0}^{j}\sum_{l=0}^{k}j\binom{\gamma}{j}\right.\right.$$
$$\left.\left.\cdot\binom{j}{m}\binom{kj}{l}(-1)^{j+l}\frac{k^{m-\gamma}}{(m-\gamma+l)^{\frac{(\gamma-1)(\beta+1)}{\beta}+1}}\right]\right\},$$

$$I_R(\gamma) = -\log\beta\lambda + \frac{1}{1-\gamma}\left\{\log\left[\sum_{j=0}^{\infty}\sum_{m=0}^{j}\sum_{l=0}^{\infty}\binom{\gamma}{j}\binom{j}{m}\binom{\frac{(\lambda-1)(\gamma-1)}{\lambda}}{l}\right.\right.$$
$$\left.\left.\cdot(-1)^{j+l}k^{m-\gamma}B\left(m-2\gamma+1, kj+\frac{l+(\gamma-1)\left(\beta+\frac{1}{\lambda}\right)}{\beta}+1\right)\right]\right\}$$

and

$$I_R(\gamma) = -\log\beta\lambda + \frac{1}{1-\gamma}\left\{\log\left[\sum_{j=0}^{\infty}\sum_{m=0}^{j}\sum_{l=0}^{\infty}\binom{\gamma}{j}\binom{j}{m}\binom{\frac{(\lambda+1)(\gamma-1)}{\lambda}}{l}\right.\right.$$
$$\left.\left.\cdot(-1)^{j+l}k^{m-\gamma}B\left(m-2\gamma+\frac{l+(\gamma-1)\left(\beta-\frac{1}{\lambda}\right)}{\beta}+1, kj+1\right)\right]\right\},$$

respectively.

Funding Not applicable.
Data Availability The data can be obtained from the corresponding author.
Code Availability The code can be obtained by contacting the corresponding author.
Ethical Approval All authors kept the 'Ethical Responsibilities of Authors'.
Consent to Publish All authors gave explicit consent to publish this manuscript.
Competing Interests Authors declare no conflicts of interest.

References

Adamidis, K. and Loukas, S. (1998). A lifetime distribution with decreasing failure rate. Statistics and Probability Letters, 39: 35–42.

Akaike, H. (1974). A new look at the statistical model identification. IEEE Transactions on Automatic Control, 19: 716–723.

Anderson, T. W. and Darling, D. A. (1954). A test of goodness of fit. Journal of the American Statistical Association, 49: 765–769.

Bakouch, H. S., Ristic, M. M., Asgharzadeh, A., Esmaily, L., Al-Zahrani, B. M. et al. (2012). An exponentiated exponential binomial distribution with application. Statistics and Probability Letters, 82: 1067–1081.

Barreto-Souza, W., de Morais, A. L. and Cordeiro, G. M. (2011). The Weibull-geometric distribution. Journal of Statistical Computation and Simulation, 81: 645–657.

Bortolini, J., Pascoa, M. A. R., de Lima, R. R. and Oliveira, A. C. S. (2017). The extended generalized gamma geometric distribution. International Journal of Statistics and Probability, 6: 48–69.

Bozdogan, H. (1987). Model selection and Akaike's Information Criterion (AIC): The general theory and its analytical extensions. Psychometrika, 52: 345–370.

Burnham, K. P. and Anderson, D. R. (2004). Multimodel inference: Understanding AIC and BIC in model selection. Sociological Methods and Research, 33: 261–304.

Chakrabarty, J. B. and Chowdhury, S. (2019). Compounded inverse Weibull distributions: Properties, inference and applications. Communications in Statistics—Simulation and Computation, 48: 2012–2033.

Chang, Y. C. and Huang, W. T. (2014). An enhanced model for SDBR in a random reentrant flow shop environment. International Journal of Production Research, 52: 1808–1826.

Chung, Y., Dey, D. K. and Jung, M. (2017). The exponentiated inverse Weibull geometric distribution. Pakistan Journal of Statistics, 33: 161–178.

Condino, F. and Domma, F. (2017). A new distribution function with bounded support: The reflected generalized Topp-Leone power series distribution. Metron, 75: 51–68.

Cox, D. R. and Hinkley, D. V. (1979). Theoretical Statistics. Chapman and Hall, London.

Cramér, H. (1928). On the composition of elementary errors: II. Statistical applications. Skandinavisk Aktuarietidskrift, 11: 141–180.

Doornik, J. A. O. M. (2007). Introduction to Ox: An Object-Oriented Matrix Programming Language. Timberlake Consultants, London.

Fang, Y. (2011). Asymptotic equivalence between cross-validations and Akaike Information Criteria in mixed-effects models. Journal of Data Science, 9: 15–21.

Ferguson, T. S. (1996). A Course in Large Sample Theory. Chapman and Hall, London.

Gradshteyn, I. and Ryzhik, I. (2000). Table of Integrals, Series and Products, Sixth Edition. Academic Press, San Diego.

Gupta, R. D. and Kundu, D. (1999). Generalized exponential distributions. Australian and New Zealand Journal of Statistics, 41: 173–188.

Hajebi, M., Rezaei, S. and Nadarajah, S. (2013). An exponential-negative binomial distribution. REVSTAT, 11: 191–210.

Han, M., Liang, Z. and Li, D. (2011). Sparse kernel density estimations and its application in variable selection based on quadratic Rényi entropy. Neurocomputing, 74: 1664–1672.

Hannan, E. J. and Quinn, B. G. (1979). The determination of the order of an autoregression. Journal of the Royal Statistical Society, B, 41: 190–195.

Hassan, A. S. and Nassr, S. G. (2018). Power Lomax Poisson distribution: Properties and estimation. Journal of Data Science, 18: 105–128.

Hughes, M. S., Marsh, J. N., Arbeit, J. M., Neumann, R. G., Fuhrhop, R. W. et al. (2009). Application of Rényi entropy for ultrasonic molecular imaging. Journal of the Acoustical Society of America, 125: 3141–3145.

Hurvich, C. M. and Tsai, C. -L. (1989). Regression and time series model selection in small samples. Biometrika, 76: 297–307.

Kemaloglu, S. A. and Yilmaz, M. (2020). Exponential-discrete Lindley distribution: Properties and applications. Mathematical Sciences and Applications E-Notes, 8: 21–31.

Kolmogorov, A. (1933). Sulla determinazione empirica di una legge di distribuzione. Giornale dell'Istituto Italiano degli Attuari, 4: 83–91.

Kumar, A. and Saini, M. (2014). Cost-benefit analysis of a single-unit system with preventive maintenance and Weibull distribution for failure and repair activities. Journal of Applied Mathematics, Statistics and Informatics, 10: 5–19.

Kus, C. (2007). A new lifetime distribution. Computational Statistics and Data Analysis, 51: 4497–4509.

Lehmann, E. L. and Casella, G. (1998). Theory of Point Estimation, Second Edition. Springer Verlag, New York.

Marsh, J. N., Wallace, K. D., Lanza, G. M., Wickline, S. A., Hughes, M. S. et al. (2010). Application of a limiting form of the Rényi entropy for molecular imaging of tumors using a clinically relevant protocol. In: Proceedings of the 2010 IEEE Ultrasonics Symposium, pp. 53–56.

Marshall, A. W. and Olkin, I. (1997). A new method for adding a parameter to a family of distributions with application to the exponential and Weibull families. Biometrika, 84: 641–652.

Marshall, A. W. and Olkin, I. (2007). Life Distributions: Structure of Nonparametric, Semiparametric and Parametric Families. Springer Verlag, New York.

Morais, A. L. and Barreto-Souza, W. (2011). A compound class of Weibull and power series distributions. Computational Statistics and Data Analysis, 55: 1410–1425.

Muhammad, M. and Liu, L. (2019). A new extension of the generalized half logistic distribution with applications to real data. Entropy, 21. Doi: 10.3390/e21040339.

Nadarajah, S., Jayakumar, K. and Ristic, M. M. (2013). A new family of lifetime models. Journal of Statistical Computation and Simulation, 83: 1389–1404.

Okasha, H. M. and Al-Shomrani, A. A. (2019). Generalized linear exponential geometric distributions and its applications. Journal of Computational and Applied Mathematics, 351: 198–211.

Ramos, P. L., Dey, D. K., Louzada, F. and Lachos, V. H. (2020). An extended Poisson family of life distribution: A unified approach in competitive and complementary risks. Journal of Applied Statistics, 47: 306–322.

Rényi, A. (1961). On measures of entropy and information. Paper Presented at the Fourth Berkeley Symposium on Mathematical Statistics and Probability, University of California Press, Berkeley.

Ristic, M. M. and Nadarajah, S. (2014). A new lifetime distribution. Journal of Statistical Computation and Simulation, 84: 135–150.

Ristic, M. M., Jose, K. K. and Ancy, J. (2007). A Marshall-Olkin gamma distribution and minification process. STARS, 11: 107–117.

Rodriguez, R. N. (1977). A guide to the Burr type XII distributions. Biometrics, 64: 129–134.

Roozegar, R. and Nadarajah, S. (2017). A new class of Topp-Leone power series distributions with reliability application. Journal of Failure Analysis and Prevention, 17: 955–970.

Saini, M. and Kumar, A. (2020). Stochastic modeling of a single-unit system operating under different environmental conditions subject to inspection and degradation. Proceedings of the National Academy of Sciences, India Section A: Physical Sciences, 90: 319–326.

Saini, M., Kumar, A. and Shankar, V. G. (2020). A study of microprocessor systems using RAMD approach. Life Cycle Reliability and Safety Engineering, 9: 181–194.

Schwarz, G. E. (1978). Estimating the dimension of a model. Annals of Statistics, 6: 461–464.

Silva, R. B., Bourguignon, M., Dias, C. R. B., Cordeiro, G. M. et al. (2013). The compound class of extended Weibull power series distributions. Computational Statistics and Data Analysis, 58: 352–367.

Smirnov, N. (1948). Table for estimating the goodness of fit of empirical distributions. Annals of Mathematical Statistics, 19: 279–281.

Surles, J. G. and Padgett, W. J. (2001). Inference for reliability and stress-strength for a scaled Burr type X distribution. Lifetime Data Analysis, 7: 187–200.

Tadikamalla, P. R. (1980). A look at the Burr and related distributions. International Statistical Review, 48: 337–344.

Tahmasbi, R. and Rezaei, S. (2008). A two-parameter lifetime distribution with decreasing failure rate. Computational Statistics and Data Analysis, 52: 3889–3901.

von Mises, R. (1931). Wahrscheinlichkeitsrechnung und ihre Anwendung in der Statistik und Theoretischen Physik. Deuticke, Leipzig.

Wu, S. J. and Chang, C. T. (2003). Inference in the Pareto distribution based on progressive type II censoring with random removals. Journal of Applied Statistics, 30: 163–172.

Wu, S. J., Chen, Y. J. and Chang, C. T. (2007). Statistical inference based on progressively censored samples with random removals from the Burr type XII distribution. Journal of Statistical Computation and Simulation, 77: 19–27.

Chapter 13

Comparing the Performance of G-family Probability Distribution for Modeling Rainfall Data

Md Mostafizur Rahman, Md Abdul Khalek* and *M Sayedur Rahman*

1. Introduction

Rain is an important natural resource which occurs from the interaction between several complex atmospheric processes. The quantity of rainfall confirms the amount of water availability of a particular area which is essential for agriculture production, industrial development and other human activities. All kinds of plants need atleast some amount of water to survive. Sufficient rain is a blessing for agriculture but it becomes dangerous if its scarce or in excess. So, the distribution of rainfall in time and space is necessary for the development of a particular economy. Statistical probability distributions are a proven tool to describe many natural and social problems by providing suitable models and methods. Simple summary statistics give some idea about the rainfall status but prior knowledge about rainfall is enhanced by different statistical distributions. Maliva and Missimer (2012) showed that Normal, Gamma, Gumbel and Weibull probability distributions gave better results for fitting rainfall data from arid and semi-arid regions. Sen and Eljadid (1999) investigated the performance of statistical distributions in the case of monthly Libyan rainfall data over 20 years and found the Gamma distribution provided the best performing results. Al-Mansory (2005) compared the performance of different statistical distributions such as Normal, Log-Normal, Log-Normal type III, Pearson type III, Log-Person type III, and Gumbel for maximum monthly rainfall data of Basrah station, Iraq and found that Person type III and Gumbel distributions performed better than other distributions. Olumide et al. (2013) fitted Gumbel, log Gumbel, Normal and Log-Normal probability distribution models to various rainfalls and runoffs for the Tagwai dam site in Minna, Nigeria and found Normal and log-Normal distributions were most appropriate for the prediction of yearly maximum daily-rainfall and yearly maximum daily-runoff respectively for this study area. Alghazali and Alawadi (2014) fitted three statistical distributions Normal, Gamma and Weibull in the case of thirteen Iraqi weather stations and found that the Gamma distribution was suitable for fitting five stations by the Chi-square test where Normal and Weibull distributions were not appropriate and the Kolmogorov-Smirnov test indicated that none of these three distributions were suitable for either of these five stations.

Although classical statistical distributions show better fitting results in many areas of real world situations, they have some restrictions and limitations, which led researchers to build new distributions which are more flexible and can overcome them. This new distribution is well known as a member of the G-family

Environment and Data Mining Research Group, Department of Statistics, University of Rajshahi, Rajshahi-6205, Bangladesh.
* Corresponding author: mostafiz_bd21@yahoo.com

distributions which are defined by adding one or more parameters to the cumulative distribution function of a classical statistical distribution. Several authors proposed the G-family distribution and found that in most of the cases these extended distribution perform better than classical distributions. For example, Marshall and Olkin (1997) proposed the Marshall-Olkin G family distribution, Kumaraswamy-G family was proposed by Cordeiro and Castro (2011), Alexander et al. (2012) proposed McDonald G family distribution, Bourguignon et al. (2014) proposed the Weibull-G family distribution, the exponential half-logistic G family distribution was proposed by Cordeiro et al. (2013), Tahir et al. (2016) proposed the Logistic-X G family distribution, Kumaraswamy Marshal-Olkin G family was proposed by Alizadeh et al. (2015), Generalized Transmuted G family was proposed by Nofal et al. (2017), Exponentiated Transmuted-G family was proposed by Merovcia et al. (2017) and Yousof et al. (2018) proposed the Marshall-Olkin generalized-G family distribution and so on.

From the above discussion it is clear that the Normal, log-Normal, Gamma, Gumbel and Weibull distributions are well suited for fitting rainfall data. Recently researchers have been trying to find distributions performing better by adding extra parameters to the existing classical distribution. This family of distributions is also known as G-family distributions which perform better in fitting and predicting a variety real data. The application of the G-family distribution in environmental science especially to rainfall data is rare and quite interesting. So, the aim of this paper is to find the best performing distribution from a set of different G-family distributions such as Gamma uniform G family, Kumaraswamy G family, Marshall-Olkin G family and Weibull G family distribution in case of rainfall data of the Rajshahi division, Bangladesh.

2. G-family distribution

Rainfall data analysis depends on different distribution patterns . It is always interesting for researchers to find the best performing distribution for modeling rainfall data of certain areas. In this methodology we present mathematical forms of different G-family distributions and also present some measures of goodness of fit tests. Teimouri and Nadarajah (2019) developed a Maximum Product Spacing (MPS) package for computing probability density functions, cumulative distribution functions, parameter estimation and drawing q-q plots from different G-family distributions. In our study we used Teimouri and Nadarajah's (2019) MPS package for data analysis.

2.1 Gamma uniform G distribution

The general form of the probability density function of the Gamma Uniform G distribution proposed by Torabi and Montazeri (2012) is given by,

$$f(x,\Theta)=\frac{h(x-\mu,\theta)}{\Gamma(\alpha)(1-H(x-\mu,\theta))^2}\left(\frac{H(x-\mu,\theta)}{1-H(x-\mu,\theta)}\right)^{\alpha-1} exp\left(\frac{H(x-\mu,\theta)}{1-H(x-\mu,\theta)}\right)$$

where, θ is the baseline family parameter vector, $\alpha > 0$ and μ are extra parameters induced to the baseline cumulative distribution function H whose probability density function is h. The general form for the cumulative distribution function of this distribution can be written as:

$$F(x,\Theta)=\int_0^{\frac{H(x-\mu,\theta)}{1-H(x-\mu,\theta)}}\frac{y^{\alpha-1}\exp(-y)}{\Gamma(\alpha)}$$

The baseline H refers to the cumulative distribution function of different families such as Chen, Frechet, Log-Normal and Weibull distributions. The parameter vector is $\Theta = (\alpha,\theta,\mu)$ where θ denotes the baseline G family parameters which contain the shape and scale parameters. The parameter μ is the location parameter.

2.2 Kumaraswamy G distribution

Cordeiro and Castro (2011) proposed Kumaraswamy G distribution. The probability density function of Kumaraswamy G distribution is given below:

$$f(x,\Theta) = \alpha\beta h(x-\mu,\theta)\,(H(x-\mu,\theta))^{\alpha-1}\,[1-(H(x-\mu,\theta))^{\alpha}]^{\beta-1}$$

where, θ is the baseline family parameter vector $\alpha > 0$, $\beta > 0$ and μ are extra parameters induced to the baseline cumulative distribution function H whose probability density function is h. The general form for the cumulative distribution function of the Kumaraswamy G distribution is given by,

$$F(x,\Theta) = 1-[1-(H(x-\mu,\theta))^{\alpha}]^{\beta}$$

The baseline H refers to the cumulative distribution function of different families such as Chen, Frechet, Log-normal and Weibull distributions. The parameter vector is $\Theta = (\alpha,\beta,\theta,\mu)$ where θ denotes the baseline G family parameters which contain the shape and scale parameters. In this model α and β are the first and second scale parameters respectively and μ is the location parameter.

2.3 Marshall-Olkin G distribution

Marshall and Olkin (1997) proposed the one G family distribution which is known as the Marshall and Olkin G distribution. The probability density function of this distribution is given below:

$$f(x,\Theta) = \frac{\alpha h(x-\mu,\theta)}{[1-(1-\alpha)(1-H(x-\mu,\theta))]^2}$$

where, θ is the baseline family parameter vector. $\alpha > 0$ and μ are extra parameters induced to the baseline cumulative distribution function H whose probability density function is h. The cumulative distribution function of the Marshall-Olkin G distribution is given by,

$$F(x,\Theta) = 1-\frac{\alpha(1-G(x-\mu,\theta))}{[1-(1-\alpha)(1-G(x-\mu,\theta))]}$$

The baseline H refers to the cumulative distribution function of different families such as Chen, Frechet, Log-normal and Weibull distributions. The parameter vector is $\Theta = (\alpha,\beta,\theta,\mu)$ where θ denotes the baseline G family parameters which contain the shape and scale parameters and μ is the location parameter.

2.4 Weibull G distribution

The Weibull G distribution was proposed by Alzaatreh et al. (2013). The general form for the probability density function of this distribution can be written as:

$$f(x,\Theta) = \frac{\alpha}{\beta^{\alpha}}\frac{H(x-\mu,\theta)}{1-H(x-\mu,\theta)}[-\log(-H(x-\mu,\theta))]^{\alpha-1}\exp\left(-\frac{-\log(1-H(x-\mu,\theta))}{\beta}\right)^{\alpha}$$

where, θ is the baseline family parameter vector, $\alpha > 0$, $\beta > 0$ and μ are extra parameters induced to the baseline cumulative distribution function H whose probability density function is h. The general form for the cumulative distribution function of the Weibull G distribution is given by,

$$F(x,\Theta) = 1-\exp\left(-\frac{-\log(1-H(x-\mu,\theta))}{\beta}\right)^{\alpha}$$

The Weibull G distribution is the special case of Alzaatreh et al.'s (2013), Weibull-X family distribution . The baseline H refers to the cumulative distribution functions of different families such as Chen, Frechet, Log-normal and Weibull distributions. The parameter vector is $\Theta = (\alpha,\beta,\theta,\mu)$ where θ denotes the baseline G family parameters which contain the shape and scale parameters and μ is the location parameter.

2.5 Model evaluation statistics

Selecting the appropriate model is a challenging task for researchers. Researchers confirm that the following test statistics are well known for checking goodness-of-fit tests. These are Akaike Information Criterion (AIC), Bayesian Information Criterion (BIC), Anderson Darling (AD) statistic, Log-Likelihood (LL) statistic. The Kolmogorov-Smirnov (KS) test statistic and corresponding p-values. Smirnov (1948) introduced the table for estimating the goodness of fit for different empirical distributions.

2.5.1 Akaike Information Criterion (AIC)

The AIC (Akaike Information Criterion) is a method for scoring and selecting a model. It may be shown based on information theory and frequentist inference (The AIC statistic is defined as follows:

$$AIC = -\left(\frac{2}{N}\right)LL + 2\left(\frac{k}{N}\right)$$

where, N is the sample size, LL is the log-likelihood value of the model and k is the number of parameters. In case of model selection the lowest value of AIC gives a better result (Brownlee, 2019).

2.5.2 Bayesian Information Criterion (BIC)

The BIC (Bayesian Information Criterion) was derived from the Bayesian probability and inference technique and it is appropriate for fitting models by the maximum likelihood estimation framework. The estimation formula for the BIC statistic is given below:

$$BIC = -2LL + \ln(N)k$$

In this equation LL represents the log-likelihood value of the model, N and k are the sample size and number of parameters respectively and represents the natural logarithm with base e. The minimum value of BIC confirms a better model (Brownlee 2019).

2.5.3 Log-likelihood (LL)

The likelihood method is a measure which confirms how well a parameter explains the observed data. The logarithm of the likelihood function in maximum likelihood estimation is computationally a very simple method (Robinson 2016). Let $x_1, x_2, \ldots, x_n$ be independently identically distributed random variables with probability density function $f(x)$. The their joint density is,

$$f_{x_1,x_2,\ldots x_n},(x_1,x_2,\ldots x_n) = f(x_1) * f(x_2) * \cdots f(x_n) = \prod_{i=1}^{n} f_x(x_i)$$

The log of any product is the sum of the logs of the multiplied terms, so the equation can be written as: $\sum^{n}_{i=1} ln(f_x(x_i))$. This relationship can be shown as $l(\Theta) = \ln[L(\Theta)]$. The highest value of Log-likelihood statistics confirms a better result.

2.5.4 Anderson-Darling (AD) Test

In 1952 Anderson and Darling introduced the Anderson-Darling (AD) test (Anderson and Darling, 1952, 1954). This test is treated as an alternative test for detecting sample distribution departure from normality. The mathematical formula for one sample AD test is defined as:

$$AD = -n - \frac{1}{n}\sum_{i=1}^{n}(2i-1)(\ln(x_i)) + \ln(1-(x_{(n+1-i)}))$$

where, $\{x_{(1)}, x_{(2)}, \ldots, x_{(n)}\}$ is the ordered sample with sample size n and $F(x)$ is the cumulative distribution to which the sample is compared. The null hypothesis that $\{x_{(1)} < x_{(2)} \ldots < x_{(n)}\}$comes from the underlying distribution $F(x)$ is rejected if AD is larger than the critical value AD_{α}. The critical value for different sample sizes is given by D'Agostino and Stephens (1986). Engmann and Cousineau (2011) introduced a two sample Anderson-Darling (AD) test statistic for a goodness of fit test for comparing distributions.

2.5.5 Kolmogorov-Simrnoff (KS) test

Kolmogorov (1933, 1941) and Smirnoff (1939) proposed the Kolmogorov-Smirnoff (KS) test statistics as a test of the distance between the empirical distribution and the postulated theoretical distribution. The KS test statistic for a given theoretical cumulative distribution function $F(x)$ is,

$$KS = \sqrt{n}\ sup_x|F_n(x) - F(x)|$$

where, $F(x)$ is the theoretical cumulative distribution function and $F_n(x)$ is the empirical cumulative distribution function for a sample size n. The null hypothesis is rejected if the empirical value of the KS test statistic is larger than the theoretical value of the KS test statistics (Massey, 1951).

3. Case study

Rajshahi Division is one of the oldest administrative divisions of Bangladesh covering an area 18,174.4 square kilometers and consists of eight districts namely, Rajshshi, Natore, Pabna, Bogura (Bogra), Chapainawabganj, Joypurhat, Naogaon and Sirajganj district. According to the 2011 census the total population of this division is 18,484,858. The geographical location indicates that Rajshahi is located in the western part of Bangladesh. This division is surrounding by the Khulna division in the South, Dhaka and Mymensingh in the East, Rangpur in the North and West Bengal state of India in the West. The two main rivers Padma and Jamuna are crossing over this area and besides these two rivers, they have numerous tributaries , Atrai, Karatoya and Mahananda. The land of this study area consists mainly of flat plains which produce a large variety of crops and vegetables like rice, wheat, pulses, potatos, carrots, onions and sugarcane . The administrative map of Rajshahi division is given in Figure 1.

The data used in this work was collected from the Bangladesh Meteorological Department (BMD). The daily data covered the period from January, 1971 to December, 2015 which creates a total number of 16016 observations from Rajshahi, Bogra and Pabna district of the Rajshahi division. We convert this daily rainfall data in to monthly rainfall data using Microsoft Excel. In environmental studies missing data is common. These missing values were random, and continuous missing data for one month to several months was also found in some years. We estimate the missing data by the smoothing SPSS software technique and then finally prepared the data for analysis.

4. Results and discussion

In this study we consider rainfall data from Rajshahi, Bogura and Pabna districts from Rajshahi division, Bangladesh. For rainfall data Papalexious (2012) proposed three steps for choosing a probability distribution which are (1) choose a priori some parametric families of distributions, (2) estimate the parameters with an appropriate fitting method, (3) find the best performing model based on some model evaluation criteria. Recently proposed G-family distributions are performing better than traditional probability distribution in many cases. So, in this study Gamma Uniform G, Kumaraswamy G, Marshall-Olkin G and Weibull G distribution with Chen, Frechet, Log-normal and Weibull distributions are considerd for investigation. The summary statistics of rainfall data for three districts, Rajshahi, Bogura and Pabna is given in Table 1 below.

Table 1 showed that the average monthly rainfall for Rajshahi, Bogura and Pabna districts are 12.71, 6.24 and 6.96 mm respectively. The rainfall amount is maximum in Rajshahi and minimum in Bogura district. All these rainfall series produce positive skewness and kurtosis is less than 3. The parameter estimation result with standard error and goodness of fit test for Rajshahi district is given in Table 2.

The estimated result from Table 2 indicates that all the parameters from Gamma Uniform Chen, Gamma Uniform Frechet, Gamma Uniform Log-Normal, Gamma Uniform Weibull, Kumaraswamy Chen, Kumaraswamy Frechet, Kumaraswamy Log-Normal, Kumaraswamy Weibull, Marshall-Olkin Chen,

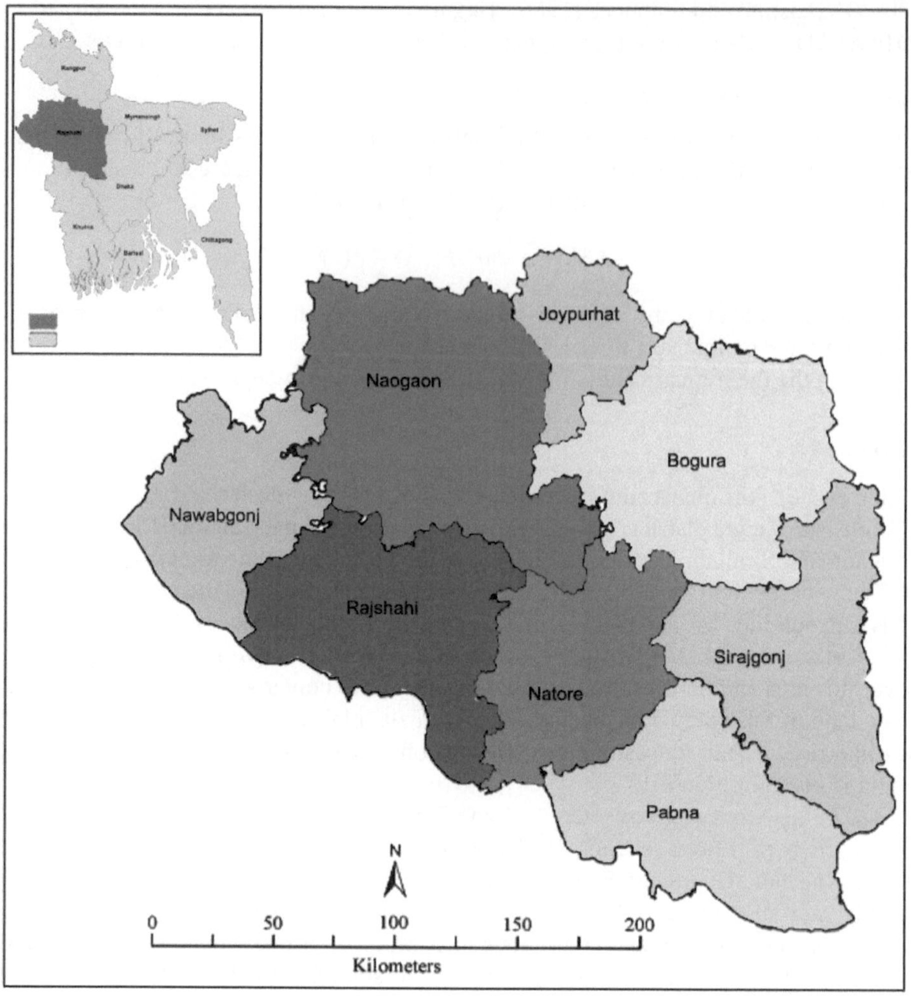

Figure 1: Map of Rajshahi division in Bangladesh (Source: author's modified).

Table 1: Summary statistics for monthly rainfall data.

	Min	First quartile	Median	Mean	Third quartile	Max.	Skewness	Kurtosis
Rajshahi	0.30	2.29	6.98	12.71	20.45	76.23	1.3346	1.8765
Bogura	0.02	1.43	5.25	6.24	9.80	19.59	0.6345	0.7543
Pabna	0.12	0.81	5.11	6.96	11.30	31.82	0.5828	0.6590

(Source: authors' own calculation)

Marshall-Olkin Frechet, Marshall-Olkin Log-Normal, Marshall-Olkin Weibull, Weibull Chen, Weibull Frechet, Weibull Log-Normal and Weibull weibull distribution meet the criteria for significant parameters. The standard error from Marshall-Olkin chen produce the lowest standard error whereas the Gamma uniform Frechet distribution produces a higher Standard Error (SE). The model evaluation criteria AIC and BIC indicate that the Marshall-Olkin Chen distribution provides a better result for fitting the distribution to the rainfall data of Rajshahi district. The log likelihood value also shows a similar conclusion. The goodness of fit test statistic Anderson Darling statistic (AD) and Kolmogorov-Smirnov (KS) test statistic show that all

Table 2: Parameter estimation and model evaluation statistic for Rajshahi district.

					Model evaluation statistic				
Model	**Parameter estimation**				**Model evaluation**			**Test statistic**	
	$\hat{\alpha}$ (SE)	$\hat{\beta}$ (SE)	$\hat{\beta}$ (SE)	$\hat{\mu}$ (SE)	AIC	BIC	-LL	AD	KS (p-value)
Gamma uniform chen	4.339 (1.341)		0.0923 (0.528)	0.653 (0.731)	6975	6993	3483	5.35	0.772 (0.001)
Gamma uniform frechet	5.432 (1.651)		0.659 (0.987)	0.701 (1.653)	7096	7136	3501	7.64	0.812 (0.010)
Gamma uniform log normal	0.218 (2.446)		0.927 (1.421)	0.365 (0.564)	6944	6962	3468	4.03	0.067 (0.008)
Gamma uniform weibull	1.648 (1.564)		0.398 (0.557)	0.956 (0.876)	6821	6838	3406	6.07	0.082 (0.000)
Kumaraswamy chen	2.381 (1.112)	0.150 (0.652)	0.422 (0.453)	0.887 (0.765)	6843	6865	3416	5.0	0.058 (0.033)
Kumaraswamy frechet	7.622 (1.313)	24.65 (0.870)	2.691 (0.798)	0.549 (0.654)	7013	7035	3501	6.95	0.804 (0.000)
Kumaraswamy log normal	6.353 (1.776)	52.34 (2.321)	6.209 (0.897)	0.576 (0.743)	6990	7012	3490	5.26	0.074 (0.002)
Kumaraswamy weibull	0.794 (1.202)	0.899 (0.781)	120.19 (0.387)	0.965 (0.657)	6876	6898	3433	5.69	0.602 (0.019)
Marshall-Olkin chen	1.769 (1.023)		0.051 (0.045)	0.786 (0.121)	6739	6756	3365	3.86	0.012 (0.001)
Marshall-Olkin frechet	22.82 (1.453)		0.590 (0.450)	0.415 (0.540)	7051	7068	3521	7.35	0.079 (0.001)
Marshall-Olkin log normal	3.101 (2.541)		1.460 (1.242)	0.173 (0.909)	7005	7023	3499	5.88	0.071 (0.004)
Marshall-Olkin weibull	0.969 (1.876)		115.18 (0.896)	0.866 (0.887)	6883	6900	3437	6.07	0.061 (0.021)
Weibull chen	2.405 (1.432)	0.559 (0.566)	0.082 (0.098)	0.424 (0.766)	6993	7015	3492	5.92	0.087 (0.000)
Weibull frechet	5.262 (1.856)	10.71 (1.234)	0.134 (0.754)	0.131 (0.562)	6992	7014	3491	6.81	0.085 (0.002)
Weibull log normal	16.99 (1.760)	18.65 (0.765)	6.54 (0.543)	0.913 (0.656)	6969	6991	3479	5.29	0.078 (0.002)
Weibull weibull	0.901 (1.558)	0.739 (0.665)	15.21 (0.543)	0.994 (0.356)	6909	6931	3450	6.14	0.065 (0,009)

(Source: authors' own calculation)

these models are significant. For comparing models the Marshall-Olkin Chen model produces the lowest value for the Anderson and Darling test statistic and the Gamma uniform Frechet model produces the highest value. For baseline distributions different distributions such as Chen, Frechet, Log-normal and Weibull show better fitting results in the case of different models. The Q-Q plot of these models for the Rajshahi district is shown in Figure 2. The Q-Q plot also confirms that the Marshall-Olkin Chen model provides a better result for fitting the monthly rainfall data of the Rajshahi district.

The estimated result from Table 2 indicates that all the parameters from Gamma uniform Chen, Gamma uniform Frechet, Gamma uniform Log-normal, Gamma uniform Weibull, Kumaraswamy Chen, Kumaraswamy Frechet, Kumaraswamy Log-normal, Kumaraswamy Weibull, Marshall-Olkin Chen, Marshall-Olkin Frechet, Marshall-Olkin Log-normal, Marshall-Olkin Weibull, Weibull Chen, Weibull Frechet, Weibull Log-normal and Weibull distributions meet the criteria for significant parameters . The

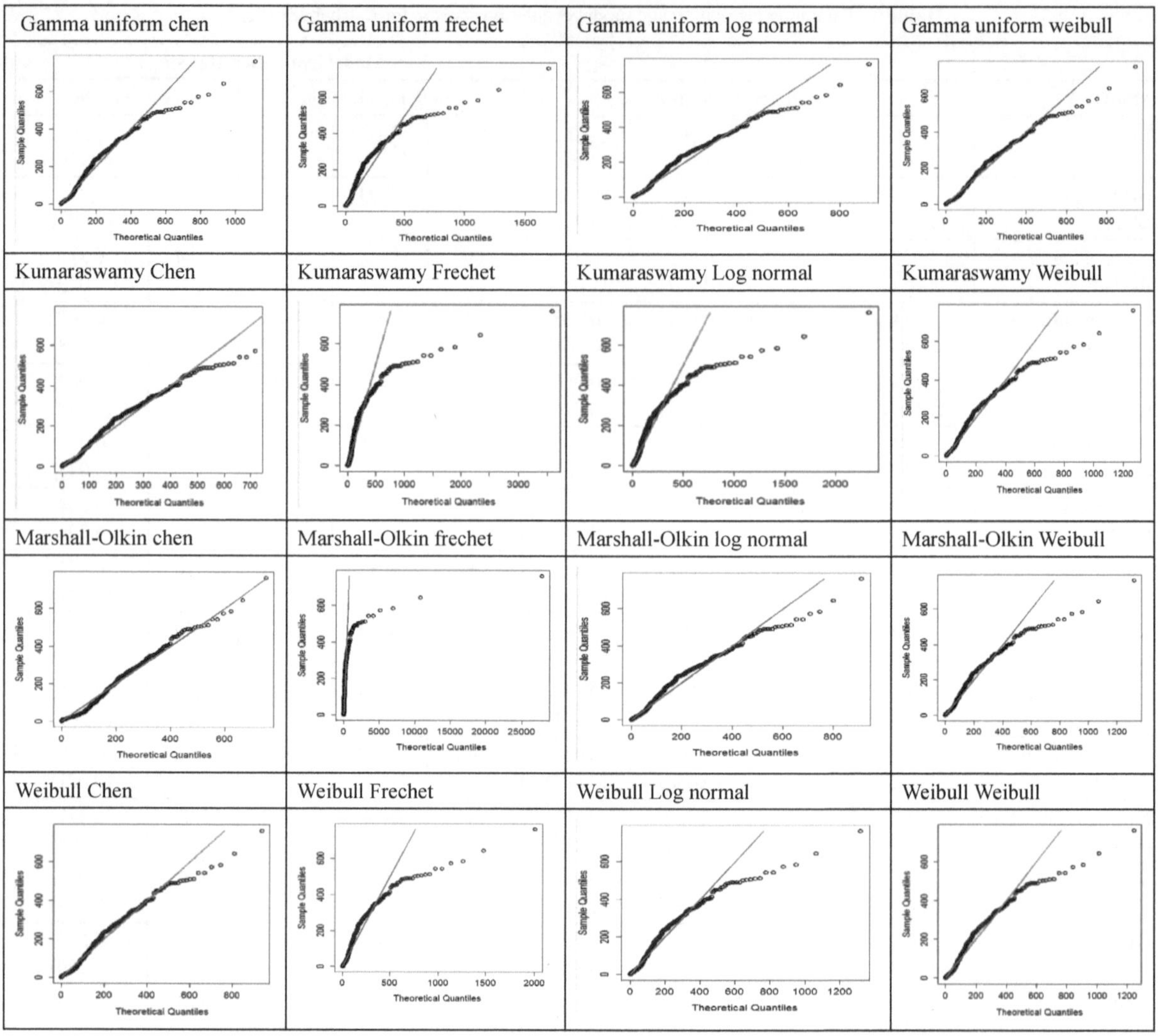

Figure 2: The Q-Q plot for the Rajshahi district (Source: created by authors)

standard error from the Marshall-Olkin Chen is the lowest whereas the Gamma uniform Frechet distribution produces a higher standard error. The model evaluation criteria AIC and BIC indicate that the Marshall-Olkin Chen distribution provides a better result for fitting distributions to the rainfall data of the Rajshahi district. The log likelihood value also gives a similar conclusion. The goodness of fit test statistic Anderson Darling statistic (AD) and Kolmogorov-Smirnov (KS) test statistic show that all these models are significant. For comparing models the Marshall-Olkin Chen model produced the lowest value of the Anderson and Darling test statistic and the Gamma uniform Frechet model produces the highest value. For baseline distributions different distributions such as Chen, Frechet, Log-normal and Weibull show better fitting results in the case of different models. The Q-Q plots of these models for the Rajshahi district are shown in Figure 2. The Q-Q plots also confirm that the Marshall-Olkin Chen model provides better results for fitting the monthly Rajshahi district rainfall data.

All sixteen models have been used in the case of monthly rainfall data of the Bogura district of Bangladesh and the estimated results are listed in Table 3.

The estimated result from Table 3 indicates that most of the parameters meet the criteria for the parameters of the original distribution. Among these models, the estimated parameters from the Gamma uniform Weibull

Table 3: Parameter estimation and model evaluation statistic for Bogura district.

					Model evaluation statistic				
Model	**Parameter estimation**				**Model evaluation**			**Test statistic**	
	$\hat{\alpha}$ (SE)	$\hat{\beta}$ (SE)	$\hat{\beta}$ (SE)	$\hat{\mu}$ (SE)	AIC	BIC	-LL	AD	KS (p-value)
Gamma uniform chen	1.791 (0.987)		0.248 (0.564)	0.017 (0.865)	3337	3354	1664	4.55	0.076 (0.001)
Gamma uniform frechet	1.153 (1.121)		3.692 (0.754)	0.734 (0.877)	3452	3470	1722	12.2	0.126 (0.000)
Gamma uniform log normal	0.506 (0.980)		1.052 (0.643)	0.145 (0.789)	3387	3404	1689	10.5	0.124 (0.000)
Gamma uniform weibull	0.284 (0.325)		16.67 (0.456)	0.016 (0.665)	3305	3312	1649	2.05	0.053 (0.004)
Kumaraswamy chen	1.740 (1.110)	0.117 (0.769)	0.946 (0.632)	0.008 (0.056)	3347	3369	1668	5.51	0.821 (0.000)
Kumaraswamy frechet	27.51 (1.716)	17.04 (0.879)	0.144 (0.709)	2.234 (0.234)	3509	3531	1749	12.2	0.12 (0.000)
Kumaraswamy log normal	0.155 (0.087)	0.891 (0.311)	0.107 (0.587)	0.441 (0.212)	3401	3423	1695	13.1	0.136 (0.000)
Kumaraswamy weibull	0.289 (0.843)	0.454 (0.301)	7.041 (0.476)	0.016 (0.543)	3327	3349	1659	3.23	0.063 (0.016)
Marshall-Olkin chen	2.213 (0.856)		0.181 (0.451)	0.018 (0.122)	3348	3366	1670	4.33	0.070 (0.005)
Marshall-Olkin frechet	7.26 (1.980)		2.24 (0.881)	1.58 (0.786)	3529	3546	1760	13.3	0.11 (0.000)
Marshall-Olkin log normal	8.87 (1.102)		1.27 (0.421)	0.059 (0.431)	3452	3470	1722	11	0.103 (0.000)
Marshall-Olkin weibull	2.511 (0.999)		3.767 (0.571)	0.011 (0.010)	3376	3394	1684	7.34	0.095 (0.000)
Weibull chen	1.252 (1.210)	2.381 (0.656)	0.288 (0.657)	0.016 (0.332)	3345	3367	1668	5.35	0.084 (0.000)
Weibull frechet	9.504 (0.897)	0.648 (0.660)	0.381 (0.613)	0.053 (0.229)	3414	3436	1702	10.9	0.120 (0.000)
Weibull log normal	24.52 (1.111)	1.382 (0.543)	19.62 (0.587)	0.014 (0.213)	3394	3416	1692	10.1	0.105 (0.000)
Weibull weibull	1.035 (0.867)	0.837 (0.430)	7.595 (0.555)	0.071 (0.056)	3391	3413	1690	8.85	0.111 (0.000)

(Source: authors' own calculation)

distribution produce the lowest standard error whereas Marshall Olkin Frechet shows the higher standard error. The model evaluation criteria AIC and BIC indicate that lowest value of these two statistics is obtained from the Gamma uniform Weibull distribution which confirms the best performance. The log-like likelihood value shows that the same model provides a higher log likelihood value whereas Marshall-Olkin Frechet produces higher AIC and BIC values and also a lower log likelihood value. The goodness of fit test statistics AD and KS show that most of the models are statistically significant. For comparing the performance among these models both test statistics show that the Gamma uniform Chen distribution is the most suitable distribution and gives better fitting results whereas the Marshall-Olkin Frechet distribution provides worse fitting results. A similar result is obtained from the Q-Q plot of these distributions in Figure 3.

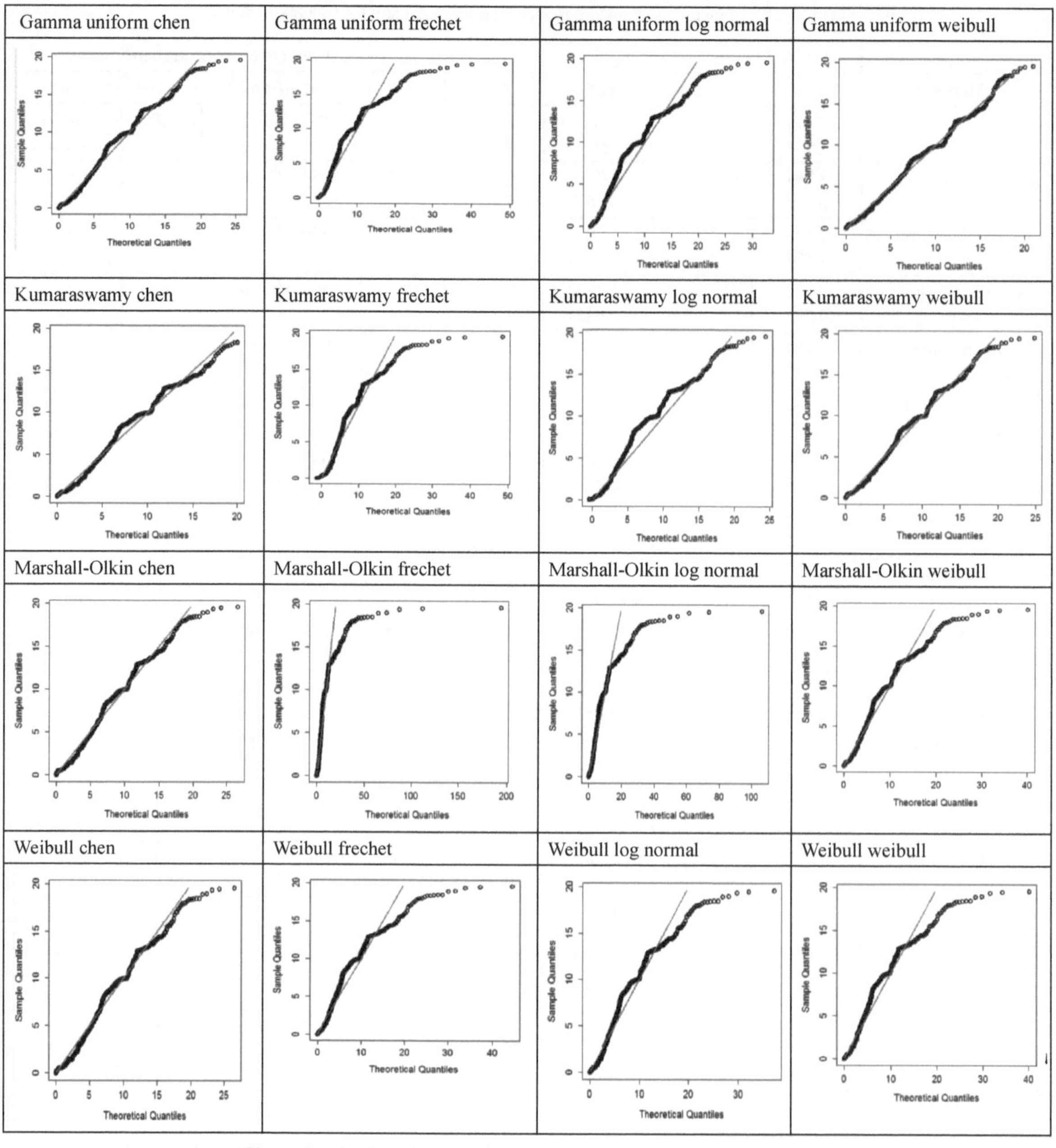

Figure 3: The Q-Q plot for Bogura district (Source: created by authors)

For Pabna, the following Gamma uniform G family, Kumaraswamy G family, Marshall-Olkin G family and Weibull G family distribution with four baseline distributions, Chen, Frechet, Log-normal and Weibull distributions fitted the monthly rainfall data. The estimated parameter values with their standard error and model evaluation criteria are given in Table 4.

The estimated result from Table 4 indicates that all the parameters satisfy the condition for the parameters of G family distribution. The Marshall-Olkin Chen model give a lower standard error and the Weibull Chen model gives a higher standard error. The AIC, BIC criteria may be used as relative goodness of fit measures which indicate that the lowest values indicate the best fitted models. These two criteria show the lowest

Table 4: Parameter estimation and model evaluation statistic for Pabna district.

					Model evaluation statistic				
Model	**Parameter estimation**				**Model evaluation**			**Test statistic**	
	$\hat{\alpha}$ (SE)	$\hat{\beta}$ (SE)	$\hat{\beta}$ (SE)	$\hat{\mu}$ (SE)	AIC	BIC	-LL	AD	KS (p-value)
Gamma uniform chen	4.423 (1.434)		0.767 (0.556)	0.034 (0.121)	3441	3458	1716	8.62	0.105 (0.000)
Gamma uniform frechet	1.821 (1.501)		0.586 (0.631)	0.055 (0.212)	3529	3546	1760	14.3	0.131 (0.000
Gamma uniform log normal	0.558 (1.398)		1.618 (0.480)	0.011 (0.187)	3452	3470	1722	10.9	0.123 (0.001)
Gamma uniform weibull	1.049 (1.222)		11.610 (0.497)	0.654 (0.212)	3398	3416	1695	5.55	0.098 (0.001)
Kumaraswamy chen	1.374 (1.029)	0.254 (0.565)	0.691 (0.641)	0.014 (0.165)	3362	3384	1676	5.61	0.078 (0.001)
Kumaraswamy frechet	11.34 (1.616)	4.849 (0.632)	0.130 (0.417)	0.155 (0.172)	3618	3640	1804	18.3	0.14 (0.000)
Kumaraswamy log normal	0.223 (1.566)	1.299 (0.457)	1.141 (0.390)	0.004 (0.012)	3552	3574	1771	12	0.108 (0.002)
Kumaraswamy weibull	0.262 (0.989)	0.656 (0.563)	10.211 (0.453)	0.005 (0.012)	3383	3405	1687	4.59	0.073 (0.003)
Marshall-Olkin chen	1.616 (1.087)		0.204 (0.267)	0.114 (0.011)	3326	3343	1659	5.01	0.049 (0.000)
Marshall-Olkin frechet	5.223 (1.451)		0.065 (0.765)	0.030 (0.211)	3603	3621	1798	15.2	0.117 (0.000)
Marshall-Olkin log normal	10.85 (1.232)		1.861 (0.546)	0.032 (0.145)	3514	3532	1753	11.1	0.098 (0.001)
Marshall-Olkin weibull	2.326 (1.243)		3.121 (0.551)	0.130 (0.125)	3437	3455	1715	7.78	0.095 (0.000)
Weibull chen	0.503 (1.996)	1.952 (0.765)	0.778 (0.889)	0.124 (0.332)	3763	3785	1877	16.6	0.333 (0.000)
Weibull frechet	11.49 (1.324)	0.507 (0.667)	1.298 (0.687)	0.136 (0.432)	3437	3496	1732	8.74	0.11 (0.001)
Weibull log normal	15.40 (1.245)	2.27 (0.556)	1.57 (0.530)	3.84 (0.232)	3451	3473	1721	8.48	0.105 (0.001)
Weibull weibull	1.123 (1.123)	0.794 (0.332)	8.470 (0.352)	0.056 (0.356)	3450	3472	1720	8.95	0.106 (0.000)

(Source: authors' own calculation)

value in the case of he Marshall-Olkin Chen model and the same model gives a higher log-likelihood value. The AD and KS test statistics indicate that the Marshall-Olkin Chen model outperforms the others. So, the Marshall-Olkin Chen model is the most suitable model for fitting monthly Pabna district rainfall data. The Q-Q plot from Figure 4 also shows a similar performance.

5. Conclusion

The study of the distribution of rainfall is important for the development of the economy. Although statistical distributions perform better in many cases they have some restrictions or limitations. To overcome these restrictions the G-family distribution is introduced. In this study Gamma uniform Chen, Gamma uniform Frechet, Gamma uniform Log-normal, Gamma uniform Weibull, Kumaraswamy Chen, Kumaraswamy

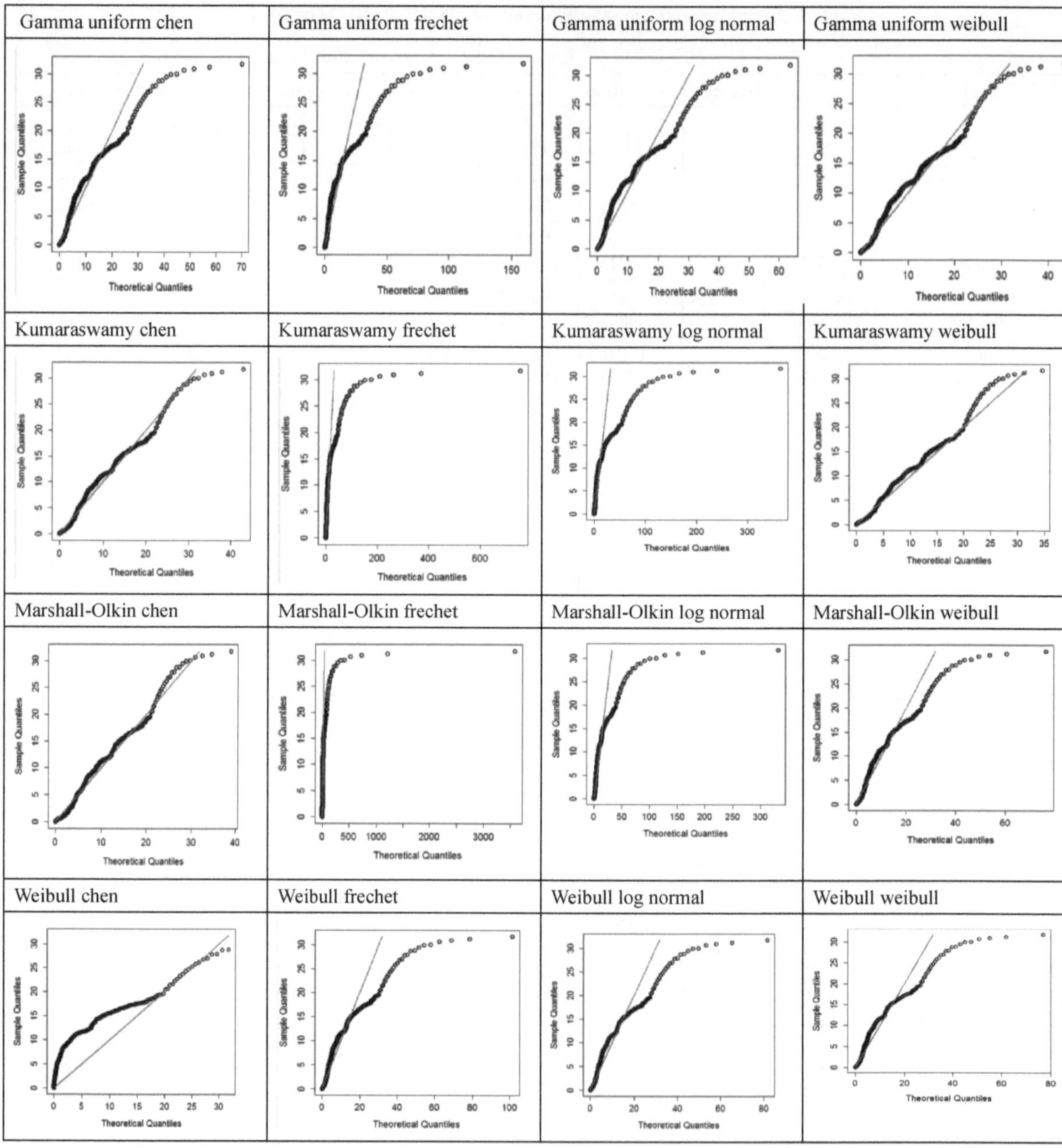

Figure 4: The Q-Q plot for Pabna district (Source: created by authors).

Frechet, Kumaraswamy Log-normal, Kumaraswamy Weibull, Marshall-Olkin Chen, Marshall-Olkin Frechet, Marshall-Olkin Log-normal, Marshall-Olkin Weibull, Weibull Chen, Weibull Frechet, Weibull Log-normal and Weibull Weibull distributions have been used for modeling monthly rainfall data from the time period January, 1971 to December, 2015 in the case of Rajshahi, Bogura and Pabna districts. The model evaluation criteria indicate that the Marshall-Olkin Chen distribution gives the best fitting results in the case of Rajshahi and Pabna districts. The Gamma uniform Weibull distribution shows the best fitting result among all sixteen G-family distributions in the case of Bogura district. As a base line distribution Chen and Weibull distribution provides better results than the Frechet and Log-normal distributions for most of the cases.

In our study we only consider three districts Rajshahi, Bogura and Pabna from Rajshahi division and could not find unique a G-family distribution which gives the best fitting result for all three districts. So, we need to investigate more G-family distributions and base line distributions and also consider rainfall data from other districts. This is put off for the future .

References

Al-Mansory, H. B. (2005). Statistical Analysis of Extreme Monthly Rainfall in Basrah City, South of Iraq. Marina Mesopotamica, Basrah, Iraq, 20(2): 283–296.

Alexander, C., Cordeiro, G. M. and Ortega, E. M. M. (2012). Generalized beta-generated distributions, Computational Statistics and Data Analysis, 56: 1880–1897.

Alghazali, N. O. S. and Alawadi, D. A. H. (2014). Fitting statistical distributions of monthly rainfall for some iraqi stations. Civil and Environmental Research. 6(6): 40–47.

Alzaatreh, A., Lee, C. and Famoye, F. (2013). A new method for generating families of continuous distributions, Metron, 71: 63–79.

Alizadeh, M., Cordeiro, G. M., Brito, E. and Demetrio, C. G. (2015). The beta Marshall-Olkin family of distributions. Journal of Statistical Distributions and Applications, 4(2): 1–18.

Anderson, T. W., Darling, D. A. (1952). Asymptotic theory of certain "goodness-of-fit" criteria based on stochastic processes. Ann. Math. Stat., 23: 193–212, doi:10.1214/aoms/1177729437.

Anderson, T. W. and Darling, D. A. (1954). A Test of Goodness-of-Fit. J. Am. Stat. Assoc., 49: 765–769, doi:10.2307/2281537.

Bourguignon, M., Silva, R. B. and Gauss, M. (2014). The Weibull-G family of probability distributions. Journal of Data Science, 12: 53–68.

Brownlee. (2019). Probabilistic model selection with AIC, BIC and MDL. Machine Learning Mastery, Making Developers Awesome at Machine Learning. https://machinelearningmastery.com/probabilistic-model-selection-measures/.

Cordeiro, G. M. and Castro, M. (2011). A new family of generalized distributions, Journal of Statistical Computation and Simulation, 81: 883–898.

Cordeiro, G. M., Ortega, E. M. M. and da Cunha, D. C. C. (2013). The exponentiated generalized class of distributions. Journal of Data Science, 11: 1–27.

D'Agostino, R. B. and Stephens, M. A. (1986). Goodness-of-Fit Techniques. New York: Marcel Dekker.

Engmann, S. and Cousineau, D. (2011). Comparing distributions: the two-sample Anderson-Darling test as an alternative to the Kolmogorov-Smirnov test. Journal of Applied Quantitative Methods, 6(3): 1–17.

Kolmogorov, A. (1933). Sulla determinazione empirica di una legge di distribuzione. Giornale dell'Istituto Italiano degli Attuari, 4: 83–91.

Kolmogorov, A. N. (1941). Confidence limits for an unknown distribution function. Annals of Mathematical Statistics, 12: 461–463.

Maliva, R. and Missimer, T. (2012). Arid Lands and Water Evaluation and Management, Springer-Verlag Berlin Heidelberg, ISBN 978-3-642-29104-3.

Marshall, A. W. and Olkin, I. (1997). A new method for adding a parameter to a family of distributions with application to the exponential and Weibull families, Biometrika, 84: 641–652.

Massey, F. J. (1951). The Kolmogorov-Smirnov test of goodness of fit. Journal of the American Statistical Association, 46: 68–78.

Merovcia, F., Alizadeh, M., Yousof, H. M. and Hamedani, G. G. (2017). The exponentiated transmuted-G family of distributions: Theory and applications, Communications in Statistics-Theory and Methods, 46(21): 10800–10822.

Nofal, Z. M., Afify, A. Z., Yousuf, H. M. and Corderio, G. M. (2017). The generalized transmuted-G family of distributions. Communications in Statistics–Theory and Methods, 46: 4119–4136.

Olumide, B. A., Saidu, M. and Oluwasesan, A. (2013). Evaluation of best fit probability distribution models for the prediction of rainfall and runoff volume (Case Study Tagwai Dam, Minna-Nigeria). International Journal of Engineering and Technology, U. K., 3(2): 94–98.

Papalexiou, S. M., Koutsoyiannis, D. and Makropoulos, C. (2012). How Extreme is Extreme? An Assessment of Daily rainfall Distribution Tails, Discussion Paper. Journal of Hydrology and Earth System Sciences, 9: 5757–5778.

Robinson, E. (2016). Introduction to Likelihood Statistics. Retrieved April 16, 2021 from: https://hea-www.harvard.edu/AstroStat/aas227_2016/lecture1_Robinson.pdf.

Sen, Z. and Eljadid, A. G. (1999). Rainfall Distribution Function for Libya and Rainfall Prediction, Hydrological Sciences Journal, Istanbul, Turkey, 44(5): 665–680.

Smirnoff, H. (1939). Sur les Écarts de la Courbe de la Distribution Empirique. Receuil Mathémathique (Matematiceskii Sbornik), 6: 3–26.

Smirnov, N. (1948). Table for estimating the goodness of fit of empirical distributions. Ann. Math. Stat., 19: 279–281.

Tahir, M. H., Cordeiro, G. M., Alzaatreh, A. and Mansoor, M. (2014). The Logistic-X family of distributions and its applications. Communication in Statistics-Theory and Methods, 45(24), DOI:10.1080/03610926.2014.980516.

Teimouri, M. and Nadarajah, S. (2019). Estimating Through the Maximum Product Spacing Approach. https://cran.r-project.org/web/packages/MPS/MPS.pdf.

Torabi, H. and Montazeri, N. H. (2012). The gamma uniform distribution and its applications, Kybernetika, 48: 16–30.

Yousof, H. M., Afify, A. Z., Nadarajah, S. Hamedani, G., Aryal, G. R. et al. (2018). The Marshall-Olkin Generalized-G family of Distributions with Applications. STATISTICA, anno LXXVIII, no. 3. 273–295.

Chapter 14

Record-Based Transmuted Kumaraswamy Generalized Family of Distributions

Properties and Application

Qazi J Azhad,[1] *Mohd Arshad,*[2,*] *Bhagwati Devi,*[1] *Nancy Khandelwal*[3] and *Irfan Ali*[3]

1. Introduction

Kumaraswamy probability distribution is continuous and, is defined on double-bounded support. It is the most popular alternative of the beta distribution and has many of its properties. One of the main striking differences between the two is the availability of Kumaraswamy's cumulative distribution function in a closed form. So, the quantile function is much easier to calculate and work on. This property of Kumaraswamy distribution stands out and makes it more suitable for use in computational aspects using simulations. As the computational aspect in research has gained much grown rapidly during the past two decades, the use of the Kumaraswamy distribution has gained momentum and found publicity in the research fraternity due to its easier and efficient working. As computer programming using a complex system of distributional forms is difficult, the availability of Kumaraswamy in place of the Beta distribution has helped researchers to work easily for modeling data. The initial use of the Kumaraswamy distribution was seen in modeling hydrological phenomena (Kumaraswamy (1980)), but later on, it has been found to have extensive uses. Readers can go through Courard-Hauri (2007), Ganji et al. (2006), Sanchez et al. (2007) among others to have more insights.

The focus of the researchers nowadays is to model complex data using more flexible forms of distributions. To obtain such distributions, several generalizations of Kumaraswamy, Weibull, Rayleigh, Gamma, Exponential and many other distributions, have been proposed. Despite having a plethora of models, we still need more models to cater to the demand of today's world which is having big data in a complex form. Some of the most popular techniques of construction of a new probability model were studied by Balakrishnan and Risti´c (2016), He et al. (2016), Tahir et al. (2020), Ghosh et al. (2021) and others.

In this chapter, the authors have adopted the recently proposed record-based transmuted map to generate new probability models by Balakrishnan and He (2021). Let us have a sequence of independent and identically distributed (iid) random variables $X_1, X_2, X_3, \ldots,$ with distribution function (DF) G(·). Let $X_{U(1)}$ and $X_{U(2)}$ be

[1] Department of Mathematics and Statistics, Banasthali Vidyapith, Rajasthan India.
[2] Department of Mathematics, Indian Institute of Technology Indore, Simrol, Indore, India.
[3] Department of Statistics and Operations Research, Aligarh Muslim University, Aligarh, India.
* Corresponding author: arshad.iitk@gmail.com

the first two upper records from this sequence of iid random variables. Then, define a new random variable as,

$$Y = \begin{cases} X_{U(1)}, & \text{with probability } 1-p \\ X_{U(2)}, & \text{with probability } p, \end{cases}$$

where, $0 \le p \le 1$. Then the DF of the random variable Y can be easily obtained as,

$$\begin{aligned} F_Y(x) &= (1-p)P(X_{U(1)} \le x) + p\,P(X_{U(1)} \le x) \\ &= (1-p)G(x) + p\left[1 - \overline{G}(x)\sum_{k=0}^{1}\frac{(-\log\overline{G}(x))^k}{k!}\right] \\ &= (1-p)G(x) + p[1 - \overline{G}(x)(1 - \log\overline{G}(x))] \\ &= G(x) + p\overline{G}(x)\log\overline{G}(x), x \in \mathbb{R}, \end{aligned}$$

where, $\bar{G}(x) = 1 - G(x)$ denotes the survival function of the baseline distribution.
The probability density function and the failure density are given respectively as,

$$f_Y(x) = g(x)[1 - p - p\,\log\bar{G}(x)], x \in \mathbb{R}, \tag{1}$$

and

$$h_Y(x) = h_X(x)\left[\frac{1 - p - p\log\overline{G}(x)}{1 - p\log\overline{G}(x)}\right], \quad x \in \mathbb{R} \tag{2}$$

where $g(x)$ is the pdf of the baseline distribution and $h_x(x) = \frac{g(x)}{\overline{G}(x)}$ is the hazard function of the baseline distribution. We will utilize the above class of distributions and introduce Record-Based Transmuted Kumaraswamy Generalized Family of Distributions, which will be discussed in the next section.

2. Record-based transmuted kumaraswamy generalized family of distributions

Let us consider the Kumaraswamy distribution with parameters (α,θ) as the baseline distribution. The distribution function and the density function of Kumaraswamy(α,θ) are defined below, respectively, as

$$G(x) = 1 - [1 - h(x)^\alpha]^\theta, x \in (0,1). \tag{3}$$

The baseline distribution $G(x)$, defined in (3), is a generalized form of the Kumaraswamy distribution which also has another baseline distribution $h(x)$. Here, it can be easily observed that by considering different baseline $h(x)$, we get new forms of probability distributions based on record transmuted maps. In this chapter, we are considering $h(x) = x$ which reduces $h(x)$ to the Kumaraswamy distribution. So, using the transformation, we get

$$G(x) = 1 - [1 - x^\alpha]^\theta, x \in (0,1) \tag{4}$$

$$g(x) = \alpha\theta x^{\alpha-1}\,(1 - [1 - x^\alpha]^{\theta-1}, x \in (0,1). \tag{5}$$

After utilizing the density and DF of the baseline distribution defined in (4) and (5) in (1), we get

$$f_Y(x) = \alpha\theta x^{\alpha-1}\,[1 - x^\alpha]^{\theta-1}\,\{1 - p[1 + \theta\log[1 - x^\alpha]\}, \quad x \in (0,1), \alpha > 0, \theta > 0, p \in [0,1]. \tag{6}$$

The corresponding DF is given as,

$$F_Y(x) = 1 - [1 - x^\alpha]^{\theta-1} + p\,[1 - x^\alpha]^\theta\,\log[1 - x^\alpha]^\theta, \quad x \in (0,1), \alpha > 0, \theta > 0, p \in [0,1]. \tag{7}$$

The probability distribution defined in (6) and (7) is called Record-Based Transmuted Kumaraswamy (RTGK) distribution. Now we visualize the shape of the probability density and distribution functions of the RTGK distribution.

For different setups of parameters, we have plotted the DF. In Figure 1, we have fixed $\alpha = 2$ and $\theta = 5$ and taken $p = (0.1, 0.3, 0.5, 0.7, 0.9)$ and we have fixed $\theta = 4$ and $p = 0.3$ and plotted the function for varying values of α. In Figure 2, the 2D and 3D plots of the probability density function for various configurations of parameters are depicted.

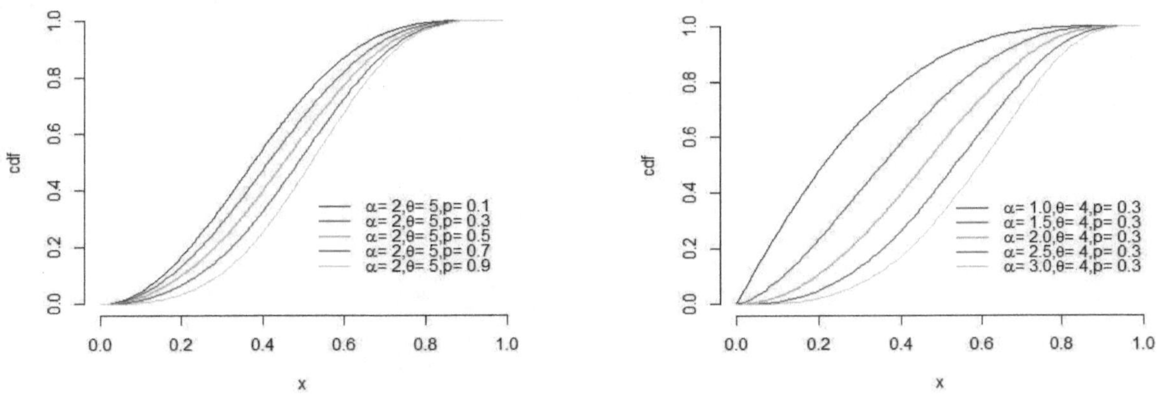

Figure 1: Distribution function plots of RTGK distribution.

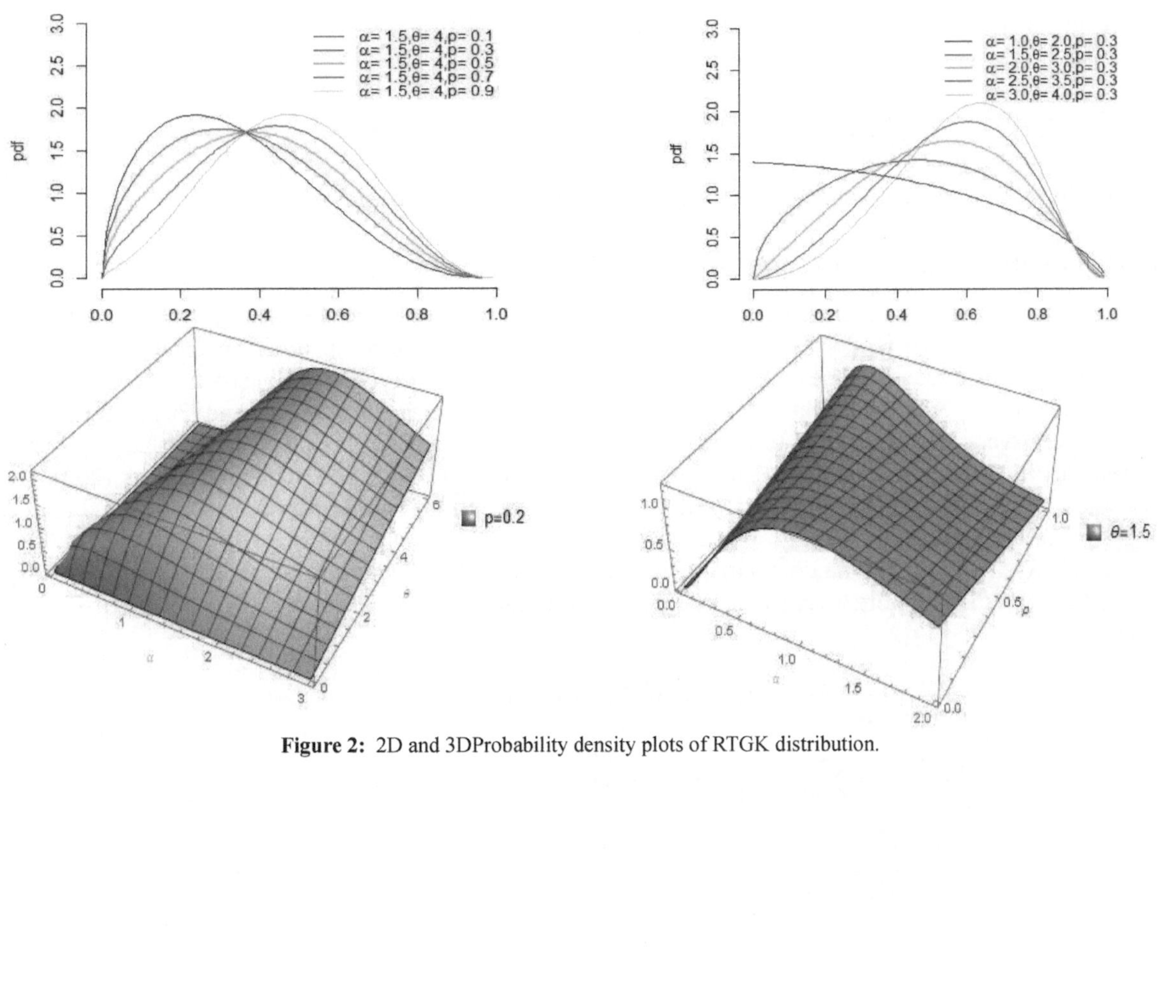

Figure 2: 2D and 3DProbability density plots of RTGK distribution.

3. Distributional properties

Theorem 3.1 *f(x) is the proper density function.*

Proof: The proof is straightforward and easy. So, we are skipping it.

Theorem 3.2 *The moment generating function (mgf) of a random variable X following RTGK distribution is defined as,*

$$M_X(t) = \sum_{i=0}^{\infty} \frac{t^i}{i!} \left\{ \theta Beta\left(\frac{i}{\alpha}+1, \theta\right) \left[1 - p\left(1 + \theta(\varphi(\theta) - \varphi\left(\frac{i}{\alpha}+1+\theta\right)\right)\right]\right\} \tag{8}$$

where, $\varphi(\cdot) = \frac{d}{dx}$.

Proof: From the formula of the moment generating function, we have,

$$M_X(t) = E[e^{tX}]$$

$$= \int_0^1 e^{tX}\{\alpha\theta x^{\alpha-1}[1-x^{\alpha}]^{\theta-1}\{1-p[1+\theta\log[1-x^{\alpha}]]\}\}dx$$

$$= \sum_{i=0}^{\infty} \frac{t^i}{i!} \left\{ \int_0^1 \alpha\theta x^{\alpha+i-1}[1-x^{\alpha}]^{\theta-1} \int_0^1 \alpha\theta x^{\alpha+i-1}[1-x^{\alpha}]^{\theta-1} [1+\theta\log[1-x^{\alpha}]]\,dx \right\}.$$

Now by taking $x^{\alpha} = t$ and $(1 - x^{\alpha}) = z$ and in the first and second integrals, respectively, we get the required result given in (8).

Theorem 3.3 *The rth moment of the random variable X about the origin following the RTGK distribution is defined as,*

$$\mu_r' = \theta Beta\left(\frac{r}{\alpha}+1, \theta\right)\left[1 - p\left(1+\theta\left(\varphi(\theta) - \varphi\left(\frac{r}{\alpha}+1+\theta\right)\right)\right)\right].$$

Proof: The proof of the theorem can be easily seen on similar lines to the proof of Theorem 3.2.

4. RTGK distribution in terms of order and record statistic

Order and record statistics are the building blocks of non-parametric inferences. This section provides the density functions of the random variable X having a RTGK distribution in order and record paradigm. Studying order and record values, and providing their analysis for different problems has been done by many researchers over the past years. Readers are advised to go through Khan and Arshad (2016), Devi et al. (2017), Chaturvedi et al. (2019a, 2019b), Sharma et al. (2019), Arshad and Baklizi (2019), Arshad and Jamal (2019a, 2019b, 2019c), Arshad et al. (2021a, 2021b), Azhad et al. (2021a, 2021b), Tripathi et al. (2021), to have more insight into the use of generalized distributions and further, the applications of order and record values in research problems.

Let $X_1, X_2, X_3, \ldots, X_n$ be a random variable following the RTGK distribution and let $X_{(1)}, X_{(2)}, X_{(3)}, \ldots, X_{(n)}$ denote the corresponding order statistics. The density function of rth order statistics $X_{(r)}$, $(r = 1,2,3,\cdots, n)$ is given as,

$$f_{X_{(r)}(x)} = \frac{1}{Beta(r,n-r+1)}[(F(x)^{r-1}(1-F(x))^{n-r} f(x)]$$

$$= \frac{f(x)}{Beta(r,n-r+1)} \sum_{i=0}^{r-1} \binom{r-1}{i} (-1)^i [1-F(x)]^{n+i-r}$$

$$= \frac{[\alpha\theta x^{\alpha-1}(1-x^{\alpha})^{\theta-1}\{1-p[1+\theta\log(1-x^{\alpha})]\}]}{Beta(r,n-r+1)} \sum_{i=0}^{r-1} \binom{r-1}{i} (-1)^i [(1-x^{\alpha})^{\theta}\{1-p\log(1-x^{\alpha})^{\theta}\}]^{n+i-r}.$$

Moreover, the densities of smallest and the largest order statistics are, given, respectively as,

$$f_{X_{(1)}}(x) = [n\alpha\theta x^{\alpha-1}(1-x^{\alpha})^{\theta-1}\{1-p[1+\theta\log(1-x^{\alpha})]\}][(1-x^{\alpha})^{\theta}$$
$$\{1-p\log(1-x^{\alpha})^{\theta}\}]^{n-1})$$

and,

$$f_{X_{(N)}}(x) = [n\alpha\theta x^{\alpha-1}(1-x^{\alpha})^{\theta-1}\{1-p[1+\theta\log(1-x^{\alpha})]\}]$$
$$\sum_{i=0}^{n-1}\binom{n-1}{i}(-1)^{i}[(1-x^{\alpha})^{\theta}\{1-p\log(1-x^{\alpha})^{\theta}\}]^{i}$$

For finding the density of record statistics, let $X_1, X_2, X_3,\ldots, X_n$ be a random variable following the RTGK distribution and let R_1, R_2, R_3,…, R_n denote the corresponding record statistics. Then, the density function of nth upper record statistics R_n is given as,

$$f_{R_n}(x) = [[-\ln[1-F(r_n)]]^{n-1}f(r_n)]/(n-1)!$$
$$= \frac{\{-\ln[(1-r_n^{\alpha})^{\theta}[1-p\theta\ln(1-r_n^{\alpha})]\}^{-1}}{(n-1)!}\alpha\theta r_n^{\alpha-1}(1-r_n^{\alpha})^{\theta-1}\{1-p[1+\theta\log(1-x_n^{\alpha})]\}, \quad r_n > 0.$$

5. Maximum likelihood estimation of parameters

Let $X_1, X_2, X_3,\ldots, X_n$ be a random sample of size taken from the RTGK distribution with parameters α, θ and p. The likelihood function is given as,

$$L(\alpha,\theta,p|\underline{x}) = \alpha^n\,\theta^n\prod_{i=1}^{n}\{x_i^{\alpha-1}(1-x_i^{\alpha})^{\theta-1}[1-p(1+\theta\ln(1-x_i^{\alpha}))]\}.$$

Taking logs on both the sides, we will have the likelihood as,

$$\log L(\alpha,\theta,p|\underline{x}) = n\log\alpha + n\log\theta + \sum_{i=1}^{n}\{(\alpha-1)\log x_i + (\theta-1)\log(1-x_i^{\alpha-1}) + \log[1-p(1+\theta\log(1-x_i^{\alpha-1}))]\} \quad (9)$$

Now, taking the derivatives of equation (9) with respect to α, θ and p, and equating them to 0, we get, respectively, the partial derivatives as,

$$\frac{\partial}{\partial\alpha}\log L = \frac{n}{\alpha} + \sum_{i=1}^{n}\left\{\log x_i + (\theta-1)\frac{(x_i^{\alpha}\log x_i)}{(x_i^{\alpha}-1)} + \frac{(p\theta x_i^{\alpha}\log x_i)}{(x_i^{\alpha}-1)(p(1+\theta\log(1-x_n^{\alpha})-1)}\right\}$$
$$\frac{\partial}{\partial\theta}\log L = \frac{n}{\theta} + \sum_{i=1}^{n}\left\{\log(1-x_i^{\alpha}) + \frac{p\log(1-x_i^{\alpha})}{p(1+\theta\log(1-x_i^{\alpha})-1}\right\}$$
$$\frac{\partial}{\partial p}\log L = \sum_{i=1}^{n}\frac{1+\theta\log(1-x_i^{\alpha})}{\left(p(1+\theta\log(1-x_i^{\alpha})-1\right)}.$$

Now, for maximum likelihood estimate (MLE), consider,

$$\frac{\partial}{\partial\alpha}\log L = 0, \quad \frac{\partial}{\partial\theta}\log L = 0, \quad \frac{\partial}{\partial p}\log L = 0.$$

As, the equations are nonlinear in nature, so manual solutions of the equations are very tedious and time consuming. Hence, we utilize the boon of simulations and adopt some computational technique like Newton Raphson for solving these equations for finding the roots.

6. Computational study

In this section, we consider the computational aspect of the problem under consideration. We have obtained the descriptive measures like mean, variance, skewness (γ_1) and kurtosis (γ_2) of the RTGK distribution for various configurations of parameters and sample sizes. We have also performed the Monte Carlo study to monitor the performances of MLEs of unknown quantities. All the computation parts have been carried out with the aid of R software (R Core Team (2021). For the numerical computation of MLE, we have used the BFGS (Broyden-Fletcher-Goldfarb-Shanno) algorithm developed by Broyden (1970), Fletcher (1970), Goldfarb (1970), and Shanno (1970). The BFGS method is an inbuilt algorithm in the optim function provided in R. Tables [1–2] represents descriptive measures of the RBTK distribution. From these tables,

Table 1: Descriptive Measures of RBTK distribution.

p	α	θ	$E(X)$	$E(X^2)$	$E(X^3)$	$E(X^4)$	*Variance*	γ_1	γ_2
0.4	0.5	0.5	0.6471	0.5343	0.4713	0.4290	0.1156	–0.6124	–1.0849
	1		0.7556	0.6471	0.5808	0.5343	0.0762	–1.1094	0.0722
	1.5		0.8114	0.7120	0.6471	0.6000	0.0536	–1.4190	1.1558
	2		0.8461	0.7556	0.6935	0.6471	0.0397	–1.6416	2.1252
	2.5		0.8699	0.7872	0.7283	0.6832	0.0306	–1.8125	2.9862
	3		0.8872	0.8114	0.7556	0.7120	0.0243	–1.9490	3.7514
	0.5	1	0.4444	0.3027	0.2339	0.1924	0.1051	0.1732	–1.3465
	1		0.6000	0.4444	0.3583	0.3027	0.0844	–0.3939	–1.0554
	1.5		0.6862	0.5355	0.4444	0.3826	0.0647	–0.7153	–0.5040
	2		0.7414	0.6000	0.5089	0.4444	0.0503	–0.9363	0.0542
	2.5		0.7800	0.6484	0.5593	0.4943	0.0399	–1.1016	0.5747
	3		0.8086	0.6862	0.6000	0.5355	0.0324	–1.2315	1.0492
	0.5	1.5	0.3226	0.1833	0.1245	0.0927	0.0792	0.6418	–0.7781
	1		0.4960	0.3226	0.2354	0.1833	0.0766	–0.0286	–1.1420
	1.5		0.5988	0.4220	0.3226	0.2592	0.0635	–0.3792	–0.8596
	2		0.6667	0.4960	0.3922	0.3226	0.0515	–0.6119	–0.4765
	2.5		0.7149	0.5532	0.4490	0.3762	0.0421	–0.7824	–0.0889
	3		0.7509	0.5988	0.4960	0.4220	0.0348	–0.9146	0.2774
	0.5	2	0.2444	0.1173	0.0705	0.0476	0.0576	0.9900	0.0275
	1		0.4222	0.2444	0.1627	0.1173	0.0662	0.2102	–1.0063
	1.5		0.5342	0.3443	0.2444	0.1843	0.0589	–0.1709	–0.9291
	2		0.6103	0.4222	0.3138	0.2444	0.0498	–0.4165	–0.6606
	2.5		0.6651	0.4841	0.3724	0.2975	0.0417	–0.5936	–0.3530
	3		0.7065	0.5342	0.4222	0.3443	0.0351	–0.7295	–0.0485
	0.5	2.5	0.1915	0.0786	0.0420	0.0258	0.0419	1.2719	0.9382
	1		0.3673	0.1915	0.1170	0.0786	0.0565	0.3837	–0.8061
	1.5		0.4845	0.2884	0.1915	0.1362	0.0537	–0.0258	–0.9056
	2		0.5660	0.3673	0.2583	0.1915	0.0470	–0.2835	–0.7249
	2.5		0.6255	0.4316	0.3166	0.2424	0.0403	–0.4669	–0.4744
	3		0.6708	0.4845	0.3673	0.2884	0.0345	–0.6065	–0.2120

Table 2: Descriptive Measures of RBTK distribution.

p	α	θ	$E(X)$	$E(X^2)$	$E(X^3)$	$E(X^4)$	*Variance*	γ_1	γ_2
0.6	0.5	0.5	0.7040	0.5983	0.5364	0.4938	0.1027	–0.8927	–0.5831
	1		0.8000	0.7040	0.6426	0.5983	0.0640	–1.4198	1.0287
	1.5		0.8475	0.7621	0.7040	0.6607	0.0438	–1.7537	2.4794
	2		0.8764	0.8000	0.7457	0.7040	0.0319	–1.9959	3.7624
	2.5		0.8960	0.8271	0.7763	0.7365	0.0243	–2.1829	4.8960
	3		0.9102	0.8475	0.8000	0.7621	0.0191	–2.3328	5.9006
	0.5	1	0.5000	0.3540	0.2794	0.2330	0.1040	–0.0481	–1.3574
	1		0.6500	0.5000	0.4125	0.3540	0.0775	–0.6141	-0.7555
	1.5		0.7292	0.5890	0.5000	0.4375	0.0572	–0.9429	0.0045
	2		0.7788	0.6500	0.5633	0.5000	0.0434	–1.1721	0.7313
	2.5		0.8129	0.6948	0.6117	0.5491	0.0340	–1.3449	1.3954
	3		0.8378	0.7292	0.6500	0.5890	0.0273	–1.4815	1.9947
	0.5	1.5	0.3696	0.2195	0.1527	0.1154	0.0829	0.4315	–1.0272
	1		0.5440	0.3696	0.2769	0.2195	0.0737	–0.2181	–1.0527
	1.5		0.6423	0.4710	0.3696	0.3026	0.0585	–0.5678	–0.5933
	2		0.7055	0.5440	0.4411	0.3696	0.0463	–0.8035	–0.0786
	2.5		0.7496	0.5991	0.4978	0.4248	0.0372	–0.9779	0.4170
	3		0.7822	0.6423	0.5440	0.4710	0.0304	–1.1140	0.8757
	0.5	2	0.2833	0.1427	0.0879	0.0603	0.0624	0.7785	–0.4095
	1		0.4667	0.2833	0.1940	0.1427	0.0656	0.0353	–1.0282
	1.5		0.5763	0.3879	0.2833	0.2180	0.0557	–0.3394	–0.7778
	2		0.6488	0.4667	0.3564	0.2833	0.0458	–0.5849	–0.3911
	2.5		0.7001	0.5277	0.4166	0.3395	0.0376	–0.7639	0.0098
	3		0.7383	0.5763	0.4667	0.3879	0.0312	–0.9022	0.3926
	0.5	2.5	0.2237	0.0966	0.0530	0.0331	0.0465	1.0549	0.3276
	1		0.4082	0.2237	0.1409	0.0966	0.0571	0.2171	–0.9009
	1.5		0.5247	0.3270	0.2237	0.1626	0.0517	–0.1821	–0.8216
	2		0.6035	0.4082	0.2953	0.2237	0.0440	–0.4377	–0.5282
	2.5		0.6601	0.4726	0.3563	0.2785	0.0369	–0.6216	–0.1917
	3		0.7026	0.5247	0.4082	0.3270	0.0311	–0.7626	0.1413

it can be seen clearly that as we increase the value of α for fixed values of θ and p, the variance of the distribution decreases. Also, it can be seen that for increasing values of θ, again the mean and the variance of the distribution decreases. We also observe that the distribution shows a negatively skewed nature for the given combinations of parameters mostly. Tables [3–5] represents the behavior of ML estimates of unknown quantities with the aid of bias and MSE. From these tables, we observe that mostly estimates are showing that for increasing sample sizes the MSEs are decreasing.

Table 3: Bias and MSE of ML estimates for $n = 50$.

p	α	θ	Bias			MSE		
			$\hat{\alpha}$	$\hat{\theta}$	$\hat{p}$	$\hat{\alpha}$	$\hat{\theta}$	$\hat{p}$
0.3	1.5	1.5	–0.2619	–0.5249	–0.1623	0.1444	0.4253	0.0348
	2	2	–1.3791	–1.4537	–0.1997	1.9187	2.1399	0.0400
	2	2.5	–1.3853	–1.9788	–0.2000	1.9331	3.9320	0.0400
	2.5	2	–1.8294	–1.3900	–0.1992	3.3699	1.9726	0.0399
	2.5	2.5	–1.8436	–1.9120	–0.1994	3.4164	3.6847	0.0399
	1.5	2	–0.8176	–1.7694	–0.0631	0.6837	3.1310	0.0156
	1.5	3	–0.7120	–2.6708	0.0612	0.5325	7.1333	0.0085
	2	2	–1.1473	–1.7640	–0.0976	1.3582	3.1120	0.0207
	2	3	–1.0405	–2.6684	0.0286	1.1330	7.1205	0.0094
0.6	1.5	1.5	0.0299	–0.1764	–0.1710	0.0405	0.1561	0.0924
	2	2	–1.2860	–1.3242	–0.4921	1.6798	1.8118	0.2458
	2	2.5	–1.3187	–1.8875	–0.4988	1.7571	3.5920	0.2495
	2.5	2	–1.7581	–1.2774	–0.4877	3.1182	1.7059	0.2440
	2.5	2.5	–1.7704	–1.8110	–0.4951	3.1586	3.3300	0.2476
	1.5	2	–0.7466	–1.7646	–0.3333	0.5782	3.1141	0.1292
	1.5	3	–0.6812	–2.6510	–0.1685	0.4881	7.0278	0.0510
	2	2	–1.4781	–1.7597	–0.2267	2.2738	3.0968	0.0905
	2	3	–1.3291	–2.6471	–0.0844	1.8893	7.0076	0.0472
	1.5	1.5	–0.4259	–1.3038	–0.7369	0.2542	1.7004	0.5516

Table 4: Bias and MSE of ML estimates for $n = 75$.

p	α	θ	Bias			MSE		
			$\hat{\alpha}$	$\hat{\theta}$	$\hat{p}$	$\hat{\alpha}$	$\hat{\theta}$	$\hat{p}$
0.3	1.5	1.5	–0.2723	–0.5734	–0.1834	0.1313	0.4218	0.0377
	2	2	–1.3887	–1.4699	–0.1997	1.9389	2.1753	0.0400
	2	2.5	–1.3959	–1.9850	–0.2000	1.9577	3.9500	0.0400
	2.5	2	–1.8446	–1.4124	–0.1997	3.4155	2.0131	0.0400
	2.5	2.5	–1.8540	–1.9353	–0.2000	3.4476	3.7582	0.0400
	1.5	2	–0.8223	–1.7704	–0.0589	0.6872	3.1343	0.0136
	1.5	3	–0.7202	–2.6700	0.0713	0.5334	7.1291	0.0082
	2	2	–1.1822	–1.7666	–0.0980	1.4308	3.1211	0.0190
	2	3	–1.0734	–2.6663	0.0354	1.1817	7.1092	0.0081
0.6	1.5	1.5	0.0358	–0.2099	–0.2001	0.0349	0.1523	0.1085
	2	2	–1.3075	–1.3560	–0.4979	1.7241	1.8657	0.2490
	2	2.5	–1.3272	–1.8971	–0.5000	1.7739	3.6182	0.2500
	2.5	2	–1.7638	–1.2990	–0.4944	3.1291	1.7271	0.2472
	2.5	2.5	–1.7891	–1.8434	–0.5000	3.2150	3.4195	0.2500
	1.5	2	–0.7493	–1.7659	–0.3415	0.5768	3.1185	0.1291
	1.5	3	–0.6766	–2.6501	–0.1826	0.4757	7.0232	0.0487
	2	2	–1.5184	–1.7599	–0.2220	2.3575	3.0975	0.0821
	2	3	–1.4034	–2.6450	–0.0460	2.0494	6.9963	0.0335
	1.5	1.5	–0.4450	–1.3069	–0.7435	0.2438	1.7081	0.5589

Table 5: Bias and MSE of ML estimates for $n = 100$.

p	α	θ	Bias			MSE		
			$\hat{\alpha}$	$\hat{\theta}$	$\hat{p}$	$\hat{\alpha}$	$\hat{\theta}$	$\hat{p}$
0.3	1.5	1.5	–0.2858	–0.6061	–0.1923	0.1263	0.4292	0.0389
	2	2	–1.3906	–1.4739	–0.2000	1.9418	2.1823	0.0400
	2	2.5	–1.4006	–1.9932	–0.2000	1.9676	3.9793	0.0400
	2.5	2	–1.8513	–1.4220	–0.2000	3.4374	2.0365	0.0400
	2.5	2.5	–1.8608	–1.9348	–0.2000	3.4704	3.7530	0.0400
	1.5	2	–0.8290	–1.7709	–0.0640	0.6948	3.1362	0.0126
	1.5	3	–0.7263	–2.6694	0.0779	0.5387	7.1259	0.0081
	2	2	–1.1905	–1.7676	–0.1026	1.4440	3.1244	0.0192
	2	3	–1.0892	–2.6663	0.0454	1.2106	7.1093	0.0071
0.6	1.5	1.5	0.0554	–0.2390	–0.2303	0.0323	0.1550	0.1215
	2	2	–1.3191	–1.3763	–0.5000	1.7501	1.9094	0.2500
	2	2.5	–1.3376	–1.9102	–0.5000	1.7971	3.6590	0.2500
	2.5	2	–1.7751	–1.3179	–0.4998	3.1645	1.7589	0.2498
	2.5	2.5	–1.7965	–1.8536	–0.5000	3.2381	3.4505	0.2500
	1.5	2	–0.7568	–1.7667	–0.3450	0.5835	3.1211	0.1284
	1.5	3	–0.6900	–2.6504	–0.1738	0.4886	7.0245	0.0430
	2	2	–1.5453	–1.7597	–0.2001	2.4224	3.0965	0.0686
	2	3	–1.4298	–2.6433	–0.0291	2.1084	6.9871	0.0275
	1.5	1.5	–0.4529	–1.3084	–0.7496	0.2375	1.7120	0.5666

7. Real data illustration

In this section, we provide a real dataset to show the application aspect of the RTGK distribution. Here we consider the SC16 dataset which is an algorithm for estimating the unit capacity factors. These datasets were given by Caramanis et al. (1983) and, Mazumdar and Gaver (1984) and further used by Khan and Arshad (2016). The datasets are presented in Table [6].

In order to show the fitting of the SC16 dataset with the RTGK distribution, we consider the Kolmogorov-Smirnov (KS) test. From the KS test, we find that the SC16 dataset supports the RTGK distribution for $\alpha = 0.5009893$, $\theta = 1.2976525$ and $p = 0.1$ KS distance , 0.18003 and *p-value*, 0.4452. This fitting is exhibited in Figure 3. The fitting is visualized based on the probability density and cumulative distribution functions.

The values of parameters for which fitting is visualized are also ML estimates of the unknown quantities of the RTGK distribution. For these sets of values, the first four raw moments of the distribution are 0.28969521, 0.16036179, 0.10818759, 0.08049182, respectively. The variance of the distribution is 0.07643848. The coefficients of skewness and kurtosis are 0.82541778 and –0.47634863, respectively.

Table 6: SC16 Dataset.

0.853	0.759	0.866	0.809	0.717	0.544	0.492	0.403	0.344
0.213	0.116	0.116	0.092	0.07	0.059	0.048	0.036	0.029
0.021	0.014	0.011	0.008	0.006				

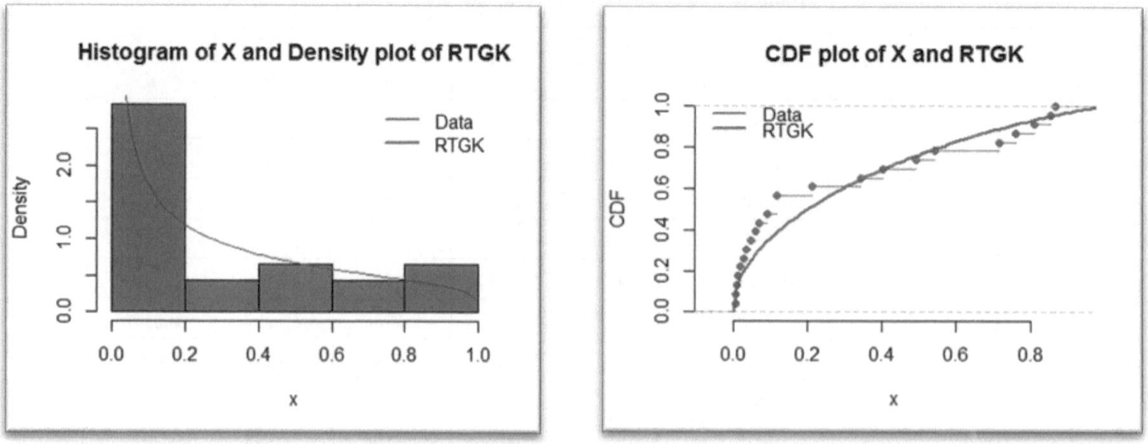

Figure 3: Plots depicting fitting of SC16 dataset.

8. Concluding remarks

In this chapter, we have proposed a new form of probability distribution called the RTGK distribution. We have derived different distributional properties of this distribution. The properties include descriptive statistics, shape of probability distribution, moment generating function and more. We have also considered point estimation for the unknown quantities of the RTGK distribution using the maximum likelihood technique. Further, we have reported these properties in tabular form with the aid of R software. The shape of the distribution is visualized for different configurations of unknown quantities. From the simulation study, we have monitored the performance of ML estimates using the criteria of Bias and MSE. Moreover, we have included a real dataset to show the aspect of application of the proposed distribution.

9. Funding information

Dr. M. Arshad would like to thank Science and Engineering Research Board, DST, Government of India [MATRICS Grant: MTR/2018/000084] for providing financial support.

References

Arshad, M. and Baklizi, A. (2019). Estimation of common location parameter of two exponential populations based on records. Communications in Statistics-Theory and Methods, 48(6): 1545–1552.

Arshad, M. and Jamal, Q. A. (2019a). Interval estimation for Topp-Leone generated family of distributions based on dual generalized order statistics. American Journal of Mathematical and Management Sciences, 38(3): 227–240.

Arshad, M. and Jamal, Q. A. (2019b). Statistical inference for Topp–Leone-generated family of distributions based on records. Journal of Statistical Theory and Applications, 18(1): 65–78.

Arshad, M. and Jamal, Q. A. (2019c). Estimation of common scale parameter of several heterogeneous Pareto populations based on records. Iranian Journal of Science and Technology, Transactions A: Science, 43(5): 2315–2323.

Arshad, M., Khetan, M., Kumar, V. and Pathak, A. K. (2021a). Record-based transmuted generalized linear exponential distribution with increasing, decreasing and bathtub shaped failure rates. Communications in Statistics—Simulation and Computation 2022: 1–25. https://doi.org/10.1080/03610918.2022.2106494.

Arshad, M., Azhad, Q. J., Gupta, N. and Pathak, A. K. (2021b). Bayesian inference of Unit Gompertz distribution based on dual generalized order statistics. Communications in Statistics-Simulation and Computation, 1–19.

Azhad, Q. J., Arshad, M. and Misra, A. K. (2021a). Estimation of common location parameter of several heterogeneous exponential populations based on generalized order statistics. Journal of Applied Statistics, 48(10): 1798–1815.

Azhad, Q. J., Arshad, M. and Khandelwal, N. (2021b). Statistical inference of reliability in multicomponent stress strength model for pareto distribution based on upper record values, International Journal of Modelling and Simulation, DOI: 10.1080/02286203.2021.1891496.

Balakrishnan, N. and Ristić, M. M. (2016). Multivariate families of gamma-generated distributions with finite or infinite support above or below the diagonal. Journal of Multivariate Analysis, 143: 194–207.

Balakrishnan, N. and He, M. (2021). A Record-Based Transmuted Family of Distributions. In Advances in Statistics-Theory and Applications (pp. 3–24). Springer, Cham.

Broyden, C. G. (1970). The convergence of a class of double-rank minimization algorithms 1. general considerations. IMA Journal of Applied Mathematics, 6(1): 76–90.

Caramanis, M., Stremel, J., Fleck, W. and Daniel, S. (1983). Probabilistic production costing: an investigation of alternative algorithms. International Journal of Electrical Power & Energy Systems, 5(2), 75–86.

Chaturvedi, A., Devi, B. and Gupta, R. (2019a). Robust Bayesian Analysis of Moore and Bilikam Family of Lifetime Distributions. International Journal of Agricultural and Statistical Sciences, 15(2): 497–522.

Chaturvedi, A., Devi, B. and Gupta, R. (2019b). Estimation and Testing Procedures of the Reliability Functions of Nakagami Distribution. Austrian Journal of Statistics, 48(3): 15–34. https://doi.org/https://doi.org/10.17713/ajs.v48i3.827.

Cordeiro Cordeiro, G. M. and de Castro, M. (2011). A new family of generalized distributions. Journal of statistical computation and simulation, 81(7): 883–898.

Courard-Hauri, D. (2007). Using Monte Carlo analysis to investigate the relationship between overconsumption and uncertain access to one's personal utility function. Ecological Economics, 64(1): 152–162.

Devi, B. Kumar, P. and Kour, K. (2017). Entropy of Lomax Probability Distribution and its Order Statistics. International Journal of Statistics and Systems. 12(2): 175–181.

Fletcher, R. (1970). A new approach to variable metric algorithms. The Computer Journal, 13(3): 317–322.

Ganji, A., Ponnambalam, K., Khalili, D. and Karamouz, M. (2006). Grain yield reliability analysis with crop water demand uncertainty. Stochastic Environmental Research and Risk Assessment, 20(4): 259–277.

Ghosh, S., Kataria, K. K. and Vellaisamy, P. (2021). On transmuted generalized linear exponential distribution. Communications in Statistics-Theory and Methods, 50(9): 1978–2000.

Gilchrist, W. G. (2001). Statistical Modelling with Quantile Functions, Chapman & Hall/CRC, Boca Raton, LA.

Goldfarb, D. (1970). A family of variable-metric methods derived by variational means. Mathematics of computation, 24(109): 23–26.

He, B., Cui, W. and Du, X. (2016). An additive modified Weibull distribution. Reliability Engineering & System Safety, 145: 28–37.

Khan, M. and Arshad, M. (2016). UMVU estimation of reliability function and stress–strength reliability from proportional reversed hazard family based on lower records. American Journal of Mathematical and Management Sciences, 35(2): 171–181.

Kumaraswamy, P. (1980). A generalized probability density function for double-bounded random processes. J. Hydrol. 46: 79–88.

Mazumdar, M. and Gaver, D. P. (1984). On the computation of power-generating system reliability indexes. Technometrics, 26(2): 173–185.

R: A language and environment for statistical computing. R Foundation for Statistical Computing, Vienna, Austria. URL https://www.R-project.org/.

Sanchez, S., Ancheyta, J. and McCaffrey, W. C. (2007). Comparison of probability distribution functions for fitting distillation curves of petroleum. Energy & Fuels, 21(5): 2955–2963.

Shanno, D. F. (1970). Conditioning of quasi-Newton methods for function minimization. Mathematics of Computation, 24(111): 647–656.

Sharma, A., Kumar, P. and Devi B. (2019). Entropy estimation of inverse rayleigh probability distribution and its order statistics. International Journal of Electronics Engineering, 11(2): 508–513.

Tahir, M. H., Hussain, M. A., Cordeiro, G. M., El-Morshedy, M., Eliwa, M. S. et al. (2020). A new Kumaraswamy generalized family of distributions with properties, applications, and bivariate extension. Mathematics, 8(11): 1989.

Tripathi, A., Singh, U. and Kumar Singh, S. (2021). Estimation of P (X < Y) for Gompertz distribution based on upper records. International Journal of Modelling and Simulation, 1–12.

Chapter 15

Finding an Efficient Distribution to Analyze Lifetime Data through Simulation Study

Anamul Haque Sajib,[1,*] *Trishna Saha*[2] and *M Sayedur Rahman*[3]

1. Introduction

Lifetime data or survival data arise vastly in our everyday life from different disciplines, most importantly, medical, engineering and actuarial science. The Gamma and the Weibull distributions are used over Exponential distributions for modelling time to event data as the former two distributions have flexible hazard functions while the latter one has a constant hazard function. Unfortunately, both the distributions have certain drawbacks. For example, computing the distribution function or survival (hazard) function of the Gamma distribution requires computer software or mathematical tables which are approximate. Although the distribution function or survival (hazard) function of the Weibull distribution can be computed directly, its hazard function increasing from zero to infinity when the shape parameter is greater than one. It may be unrealistic to use the Weibull distribution in some situations because of this property. For instance, in practical life the hazard rate should reach a stable value when population items are kept in regular follow up programs rather than increasing to infinity. On the contrary, the hazard of the Gamma density is increasing to finite numbers for shape parameter values greater than one. Therefore, the gamma distribution could be a more suitable alternative compared to the Weibull distribution. Furthermore, the MLE of Weibull parameters are not stable for all the values of parameters (Bain (1978)). However, the Weibull distribution is often considered for analyzing life time data instead of the Gamma distribution as the former one handles censored observations much more easily than the Gamma distribution (Gupta and Kundu, 1997).

Gupta and Kundu (1999) introduced the Generalized Exponential (GE) distribution with three-parameters as a substitute to the Gamma or Weibull distribution. They also showed that many properties of this family are very similar to those of the gamma family, but its distribution function has a closed form like Weibull. For example, like the gamma density likelihood ratio ordering property with respect to the shape parameter also holds for the GE. As a result, the UMP test can be constructed to test one-sided hypotheses related to the shape parameter. On the other hand, Weibull distribution does not have this likelihood ratio ordering property. From their data analysis, where one real data set was used, it was clear that the GE fits better than the Gamma and Weibull distributions with three-parameters.

Unfortunately, Gupta and Kundu (1999) did not consider any simulation study in their work to support their claim. Furthermore, it has not yet been explored how GE performs when the data contains censored observations compared to the Weibull or Gamma distribution. Motivated by this we aim to conduct an

[1] Associate Professor, Department of Statistics, University of Dhaka, Dhaka-1200.
[2] MSc Student, Department of Statistics, University of Dhaka, Dhaka-1200.
[3] Professor, Department of Statistics, Rajshahi University, Rajshahi-6205.
* Corresponding author: sajibstat@du.ac.bd

extensive simulation study to learn how GE performs compared to the Weibull or the Gamma distribution in cases with and without censoring.

2. Methodology

In this section, we will briefly introduce all the probability density functions which are used in this paper. The probability density function of the GE with three parameters is given by $f(x;\alpha,\lambda,\mu)=\frac{\alpha}{\lambda}\left(1-e^{-\left(\frac{x-\mu}{\lambda}\right)}\right)^{\alpha-1}e^{-\left(\frac{x-\mu}{\lambda}\right)}, x,\alpha,\lambda>0,$ where shape, scale and location parameters are α, λ and μ respectively. Similarly, the Gamma with three parameters and the Weibull with three-parameter densities are defined as $f(x;\alpha,\lambda,\mu)=\frac{1}{\Gamma(\alpha)\lambda^{\alpha}}(x-\mu)^{\alpha-1}e^{-\left(\frac{x-\mu}{\lambda}\right)}; x>\mu,\alpha,\lambda>0$ and $f(x;\alpha,\lambda,\mu)=\frac{\alpha}{\lambda}\left(\frac{x-\mu}{\lambda}\right)^{\alpha-1}e^{-\left(\frac{x-\mu}{\lambda}\right)^{\alpha}}; x>\mu,\alpha,\lambda>0,$ respectively. Here α, λ and μ used in both the Gamma and Weibull densities have the same meaning as α, λ and μ used in the GE density. Apart from these three density functions we have also used Lindley and Half-logistic distributions for simulation purposes. The probability density function of Lindley variate X (survival time), was introduced by Lindley (1958), is defined as $f(x;\theta)=\frac{\theta^2}{1+\theta}(1+x)e^{-\theta x}$, x, $\theta>0$, where θ is the scale parameter. Finally, the probability density function of the half-logistic distribution is given by $f(x)=\frac{2e^{-x}}{(1+e^{-x})^2}; x\geq 0.$

3. Simulation setting

We have used simulated data sets to investigate the performance of the Gamma, Weibull and Generalized Exponential distributions for analysing skewed or lifetime data. Two different situations are considered here to create simulated data sets: (i) data are simulated from one of the three distributions mentioned here, (ii) data are simulated from other lifetime distributions (Lindley and half-logistic) rather than the Gamma, Weibull and Generalized Exponential distributions. The purpose of creating these two settings is to investigate how the Gamma, Weibull and Generalized Exponential distributions perform when data came originally from either one of these three distributions or any other distribution rather than the Gamma, Weibull and Generalized Exponential distributions respectively. All the distributions considered here have one common property which is their hazard functions are increasing or decreasing.

For each setting, we have considered different sample sizes and different combinations of parameter values when data are simulated from a specific distribution. The purpose of considering different sample sizes and different combinations of parameter values is to see the effects of both sample sizes and different combinations of parameter values on the performance of each distribution. Finally, the performance of each method is evaluated based on the result obtained from the Kolmogorov-Smirnov Goodness-of-Fit (KS) test, Anderson- Darling (AD) test and Monte Carlo simulations. All the results presented in the results section will be reproducible as certain seed numbers are used when they are produced.

4. Results and discussions

In this section, we present all the results produced in our simulation study and a detailed discussion is made based on the findings in the simulation study. Tables' 1–3 shows the average KS distance and the total number of accepted H_0 against each method for different n and different parameters values. More specifically, column 3 of Table 1 shows the average KS distances for the Gamma, Weibull and GE densities when they are fitted on the sample data originally came from the *Weibull* ($\alpha = 2.5$, $\lambda = 1.2$, $\mu = 6$). The average KS distance is calculated based on 1000 simulated data sets for each sample size. We use the notation *KS* (*Gamma*), *KS* (*Weibull*) and *KS* (*GE*) to denote the KS distance for the Gamma, Weibull and GE densities respectively.

From the Table 1, it is observed that *KS* (*Gamma*) < *KS* (*Weibull*) < *KS* (*GE*) when $n \leq 250$ and as the sample size increases the relation becomes *KS* (*Weibull*) < *KS* (*Gamma*) < *KS* (*GE*).

In other words, we can conclude that the Gamma, Weibull and the GE perform roughly equally well (Gamma performs better than the other two densities marginally) for small sample sizes ($n \leq 250$) while for large sample sizes ($n > 250$) the Weibull performs better than the Gamma density marginally but far better than the GE.

The H_0 used in the columns 4–5 of Table 1 is defined as "H_0: the sample is drawn from the reference distribution" vs "H_1: the sample is drawn from any other distribution" under both *KS* and *AD* tests. For example, when Gamma density is considered for fitting the generated data then the null hypothesis is "H_0: the sample is drawn from the Gamma distribution". Similarly, when Weibull and GE are considered for fitting the sample data, the H_0 will be changed accordingly. From the KS and *AD* tests results, it is observed that total number of accepted H_0 for the Gamma and Weibull are roughly equal irrespective of all sample sizes but for the GE number of accepted H_0 is very low when the sample sizes increase ($n \geq 500$). Therefore, based on *KS* and *AD* tests results, it is concluded that when data came from the *Weibull* ($\alpha = 2.5, \lambda = 1.2, \mu = 6$), considering Gamma and Weibull to fit the sample data is roughly equally efficient irrespective of all sample sizes but considering GE to fit the sample data is a bad choice for large sample sizes, especially $n \geq 500$. These findings are consistent with the findings which are made based on the *KS* distance.

From Table 2, it is observed that the Gamma, Weibull and GE perform roughly equally well (Gamma performs better than the other two densities marginally) irrespective of all sample sizes when sample data originally came from the *Weibull* ($\alpha = 1.5, \lambda = 3.2, \mu = 2$) while Table 3 has findings similar to those of Table 1.

Figure 1 shows the overlaying of the fitted Weibull (red), Gamma (blue) and GE (green) densities on the sample histogram (data originally came from *Weibull* ($\alpha = 2.5, \lambda = 1.2, \mu = 6$)), and the empirical cdf versus fitted cdf plots for a sample size, $n = 1000$. From this plot, it is evident that the Weibull fits better than the other two densities which is consistent with the above findings.

Tables 4–6 and Figure 2 present exactly similar information like the information presented in Tables 1–3 and Figure 1. However, for Tables 4–6 and Figure 2, the sample data came originally from the Gamma density.

Table 1: Average KS distances and total number of accepted H_0 determined using 1000 synthetic data sets originally coming from the *Weibull* ($\alpha = 2.5, \lambda = 1.2, \mu = 6$), ($\alpha > \lambda$).

Sample size	Distribution	Average Distance	H_0 (*KS*)	H_0 (*AD*)
$n = 25$	Weibull	0.125	970	970
	Gamma	0.121	998	993
	GE	0.133	991	991
$n = 50$	Weibull	0.094	938	937
	Gamma	0.085	1000	999
	GE	0.100	997	999
$n = 100$	Weibull	0.066	950	950
	Gamma	0.061	1000	1000
	GE	0.078	996	998
$n = 250$	Weibull	0.043	970	970
	Gamma	0.041	999	999
	GE	0.061	966	962
$n = 500$	Weibull	0.028	987	987
	Gamma	0.031	999	999
	GE	0.053	778	665
$n = 1000$	Weibull	0.019	996	995
	Gamma	0.024	996	998
	GE	0.048	268	39

Table 2: Average KS distances and total number of accepted H_0 determined using 1000 synthetic data sets that originally came from the *Weibull* ($\alpha = 1.5, \lambda = 3.2, \mu = 2$), ($\alpha < \lambda$).

Sample size	Distribution	Average Distance	H_0 (*KS*)	H_0 (*AD*)
$n = 25$	Weibull	0.177	796	728
	Gamma	0.159	873	841
	GE	0.169	833	815
$n = 50$	Weibull	0.114	862	844
	Gamma	0.092	983	981
	GE	0.096	971	970
$n = 100$	Weibull	0.078	910	910
	Gamma	0.064	1000	1000
	GE	0.067	999	999
$n = 250$	Weibull	0.050	958	958
	Gamma	0.044	999	1000
	GE	0.047	997	999
$n = 500$	Weibull	0.038	962	962
	Gamma	0.034	998	1000
	GE	0.038	983	997
$n = 1000$	Weibull	0.024	984	984
	Gamma	0.028	988	999
	GE	0.033	933	963

Table 3: Average KS distances and total number of accepted H_0 determined using 1000 synthetic data sets that originally came from the *Weibull* ($\alpha = 3, \lambda = 3.2, \mu = 4$), ($\alpha \approx \lambda$).

Sample size	Distribution	Average Distance	H_0 (*KS*)	H_0 (*AD*)
$n = 25$	Weibull	0.117	996	994
	Gamma	0.122	992	992
	GE	0.138	992	987
$n = 50$	Weibull	0.085	997	995
	Gamma	0.087	1000	1000
	GE	0.108	998	999
$n = 100$	Weibull	0.061	991	987
	Gamma	0.063	1000	1000
	GE	0.087	990	991
$n = 250$	Weibull	0.040	984	984
	Gamma	0.043	999	1000
	GE	0.071	867	756
$n = 500$	Weibull	0.031	967	967
	Gamma	0.033	998	1000
	GE	0.064	410	106
$n = 1000$	Weibull	0.026	960	961
	Gamma	0.026	985	998
	GE	0.058	13	0

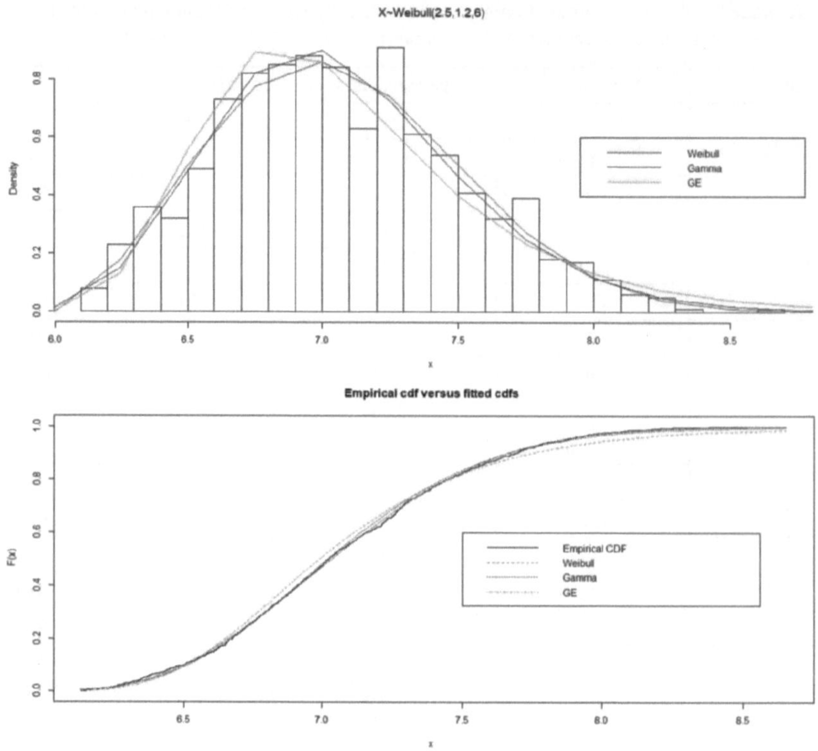

Figure 1: Overlaying the fitted Weibull (red), Gamma (blue) and GE (green) densities on the sample histogram (data originally came from the *Weibull* ($\alpha = 2.5$, $\lambda = 1.2$, $\mu = 6$)), and the empirical cdf versus fitted cdf plots.

Table 4: Average KS distances and total number of accepted H_0 determined using 1000 synthetic data sets that originally came from the *Gamma* ($\alpha = 2.5$, $\lambda = 1.2$, $\mu = 6$), ($\alpha > \lambda$).

Sample size	Distribution	Average Distance	H_0 (*KS*)	H_0 (*AD*)
$n = 25$	Weibull	0.162	879	870
	Gamma	0.137	981	956
	GE	0.139	982	956
$n = 50$	Weibull	0.108	908	901
	Gamma	0.088	993	997
	GE	0.089	991	992
$n = 100$	Weibull	0.073	938	935
	Gamma	0.061	997	997
	GE	0.062	997	999
$n = 250$	Weibull	0.051	934	930
	Gamma	0.038	1000	1000
	GE	0.040	999	1000
$n = 500$	Weibull	0.044	898	894
	Gamma	0.027	1000	999
	GE	0.030	999	1000
$n = 1000$	Weibull	0.037	863	831
	Gamma	0.020	999	999
	GE	0.023	993	987

Table 5: Average KS distances and total number of accepted H_0 determined using 1000 synthetic data sets that originally came from the *Gamma* ($\alpha = 1.5, \lambda = 3.2, \mu = 2$), ($\alpha < \lambda$).

Sample size	Distribution	Average Distance	H_0 (*KS*)	H_0 (*AD*)
$n = 25$	Weibull	0.248	542	481
	Gamma	0.181	822	769
	GE	0.182	825	759
$n = 50$	Weibull	0.187	614	605
	Gamma	0.106	918	900
	GE	0.109	909	901
$n = 100$	Weibull	0.145	670	669
	Gamma	0.065	993	992
	GE	0.067	987	986
$n = 250$	Weibull	0.124	703	703
	Gamma	0.040	1000	1000
	GE	0.041	1000	1000
$n = 500$	Weibull	0.080	826	827
	Gamma	0.028	998	997
	GE	0.029	1000	1000
$n = 1000$	Weibull	0.042	932	935
	Gamma	0.020	997	996
	GE	0.021	999	1000

Table 6: Average KS distances and number of accepted H_0 determined using 1000 synthetic data sets that originally came from the *Gamma* ($\alpha = 3, \lambda = 3.2, \mu = 4$), ($\alpha \approx \lambda$).

Sample size	Distribution	Average Distance	H_0 (*KS*)	H_0 (*AD*)
$n = 25$	Weibull	0.484	173	162
	Gamma	0.128	989	978
	GE	0.137	975	948
$n = 50$	Weibull	0.480	203	198
	Gamma	0.086	998	998
	GE	0.088	997	998
$n = 100$	Weibull	0.445	288	283
	Gamma	0.061	999	999
	GE	0.062	999	1000
$n = 250$	Weibull	0.454	270	266
	Gamma	0.040	996	986
	GE	0.040	1000	1000
$n = 500$	Weibull	0.471	262	259
	Gamma	0.028	985	981
	GE	0.029	1000	1000
$n = 1000$	Weibull	0.480	247	226
	Gamma	0.021	976	968
	GE	0.023	999	1000

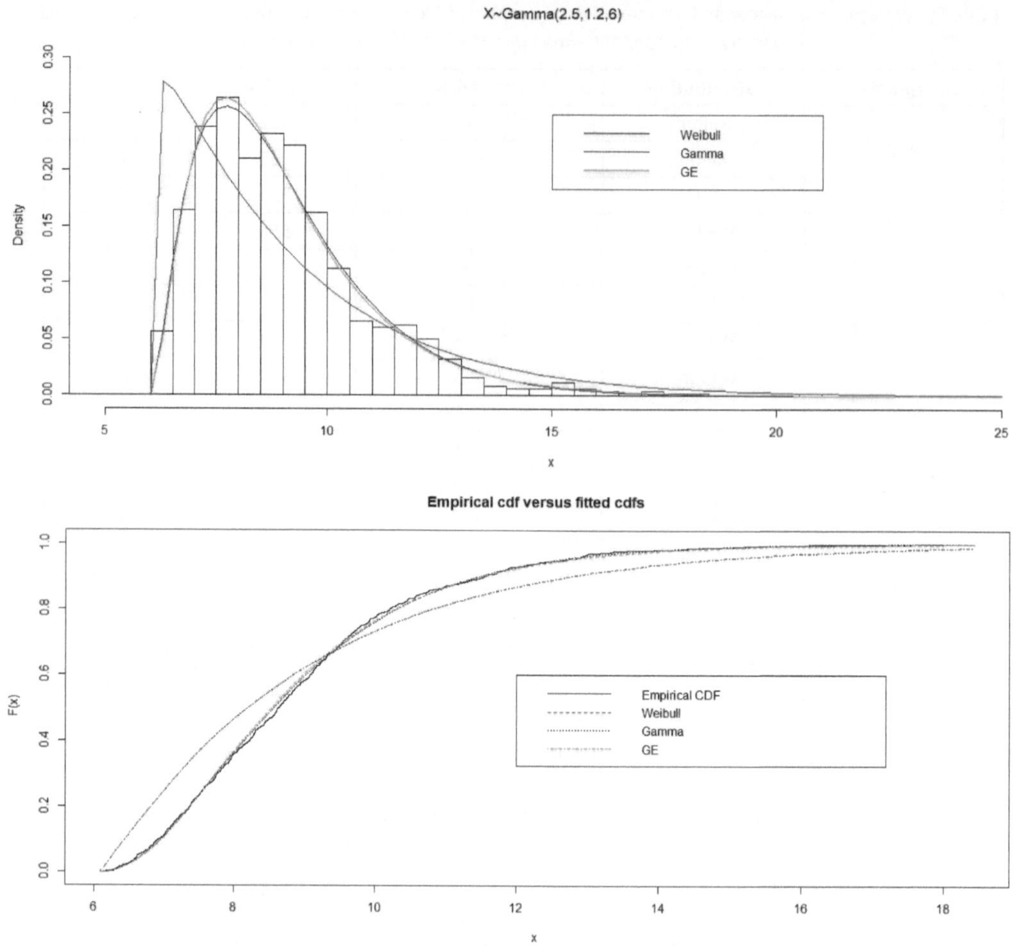

Figure 2: Overlaying the fitted Weibull (red), Gamma (blue) and GE (green) densities on the sample histogram (data that originally came from the *Gamma* ($\alpha = 2.5$, $\lambda = 1.2$, $\mu = 6$)), and the empirical cdf versus fitted cdf plots.

From these tables, it is clear that the performance of the Gamma and GE are better than the performance of the Weibull density for all sample sizes and the efficiency of the Gamma and GE are approximately equal for all situations. Especially, the Weibull density performs very badly when originally data came from the Gamma density with $\alpha \approx \lambda$. Figure 2 shows the performance of Gamma, Weibull and GE densities for a one data set with size $n = 1000$ originally that came from the *Gamma* ($\alpha = 2.5$, $\lambda = 1.2$, $\mu = 6$), and these findings are consistent with the findings from Tables 4–6.

When the data originally came from the *GE* density, irrespective of all sample sizes and combinations, the GE performs better than the Gamma and Weibull. In general, the Gamma marginally performs better than the Weibull irrespective of all sample sizes and combinations. However, the Weibull performs badly compared to Gamma when $\alpha \approx \lambda$ for GE density. These findings are shown in Tables 7–9 and in Figure 3.

Finally, when the data originally came from different distributions (other than these three distributions), the Weibull distribution performs better than both the Gamma and GE densities while the performance of the two latter ones are roughly equal. For example, when the data originally came from the Lindley and Half-logistic densities, the Weibull distribution fitted the sample data better than both the Gamma and GE for all scenarios which is shown in Tables 10–13 and in Figures 4–5.

To sum up, we can conclude that there is none among the Gamma, Weibull and GE densities which is the best to model the skewed or life time data for all situations. When the data came from different distributions

Table 7: Average KS distances and total number of accepted H_0 determined using 1000 synthetic data sets that originally came from the *GE* ($\alpha = 2.5, \lambda = 1.2, \mu = 6$), ($\alpha > \lambda$).

Sample size	Distribution	Average Distance	H_0 (*KS*)	H_0 (*AD*)
$n = 25$	Weibull	0.183	809	790
	Gamma	0.138	981	964
	GE	0.148	949	904
$n = 50$	Weibull	0.140	834	832
	Gamma	0.095	992	994
	GE	0.092	989	987
$n = 100$	Weibull	0.099	899	899
	Gamma	0.073	981	968
	GE	0.062	999	999
$n = 250$	Weibull	0.055	974	973
	Gamma	0.061	815	686
	GE	0.039	1000	1000
$n = 500$	Weibull	0.038	983	968
	Gamma	0.055	623	460
	GE	0.028	1000	1000
$n = 1000$	Weibull	0.033	900	764
	Gamma	0.054	387	285
	GE	0.020	1000	1000

Table 8: Average KS distances and total number of accepted H_0 determined using 1000 synthetic data sets that originally came from the *GE* ($\alpha = 1.5, \lambda = 3.2, \mu = 2$), ($\alpha < \lambda$) .

Sample size	Distribution	Average Distance	H_0 (*KS*)	H_0 (*AD*)
$n = 25$	Weibull	0.216	682	606
	Gamma	0.188	794	742
	GE	0.194	764	710
$n = 50$	Weibull	0.152	732	708
	Gamma	0.109	910	895
	GE	0.114	884	874
$n = 100$	Weibull	0.109	806	803
	Gamma	0.065	983	982
	GE	0.068	983	969
$n = 250$	Weibull	0.098	786	787
	Gamma	0.040	1000	1000
	GE	0.040	1000	999
$n = 500$	Weibull	0.103	746	745
	Gamma	0.029	1000	1000
	GE	0.029	999	997
$n = 1000$	Weibull	0.127	648	653
	Gamma	0.020	1000	1000
	GE	0.020	1000	1000

Table 9: Average KS distances and total number of accepted H_0 determined using 1000 synthetic data sets that originally came from the *GE* ($\alpha = 3, \lambda = 3.2, \mu = 4$), ($\alpha \approx \lambda$).

Sample size	Distribution	Average Distance	H_0 (*KS*)	H_0 (*AD*)
$n = 25$	Weibull	0.154	939	913
	Gamma	0.140	951	926
	GE	0.145	954	915
$n = 50$	Weibull	0.098	952	949
	Gamma	0.088	994	995
	GE	0.089	995	995
$n = 100$	Weibull	0.070	961	961
	Gamma	0.061	1000	1000
	GE	0.061	999	999
$n = 250$	Weibull	0.051	959	957
	Gamma	0.039	1000	1000
	GE	0.039	999	999
$n = 500$	Weibull	0.045	923	895
	Gamma	0.029	999	1000
	GE	0.028	1000	1000
$n = 1000$	Weibull	0.042	766	653
	Gamma	0.022	997	1000
	GE	0.020	997	995

Table 10: Average KS distances and total number of accepted H_0 determined using 1000 synthetic data sets that originally came from the *Lindley* ($\theta = .5$).

Sample size	Distribution	Average Distance	H_0 (*KS*)	H_0 (*AD*)
$n = 25$	Weibull	0.151	936	856
	Gamma	0.197	794	749
	GE	0.222	626	593
$n = 50$	Weibull	0.097	971	959
	Gamma	0.114	903	887
	GE	0.123	842	826
$n = 100$	Weibull	0.065	997	997
	Gamma	0.071	982	979
	GE	0.076	946	936
$n = 250$	Weibull	0.041	998	997
	Gamma	0.046	998	1000
	GE	0.048	995	999
$n = 500$	Weibull	0.030	1000	1000
	Gamma	0.036	996	998
	GE	0.038	983	995
$n = 1000$	Weibull	0.042	999	1000
	Gamma	0.022	974	992
	GE	0.020	945	997

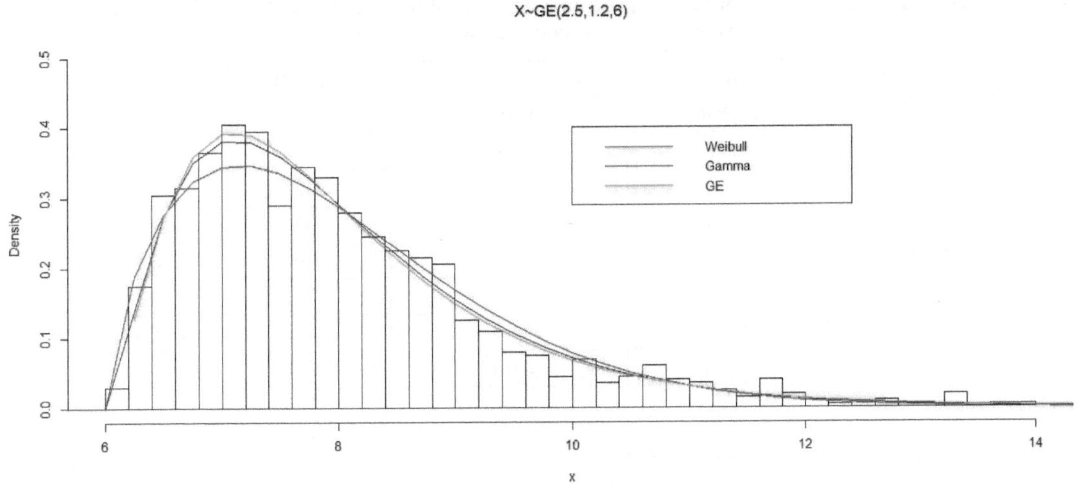

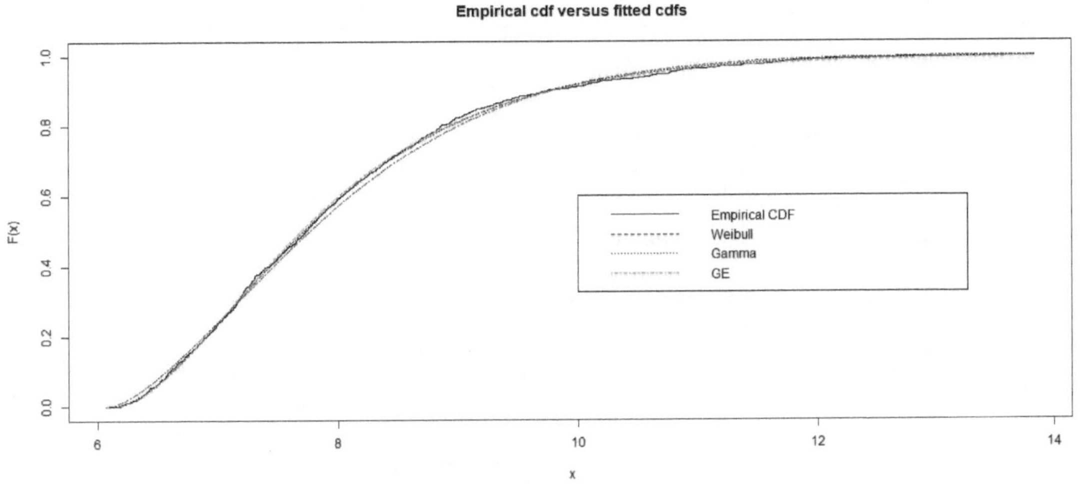

Figure 3: Overlaying the fitted Weibull (red), Gamma (blue) and GE (green) densities on sample histogram (data originally came from the $GE(\alpha = 2.5, \lambda = 1.2, \mu = 6)$), and the empirical cdf versus fitted cdf plots.

other than these three densities, the Weibull density performed marginally better than the Gamma and GE. Interestingly, the Gamma density performs marginally better than the Weibull even though originally the sample data came from the Weibull, especially for small sample sizes ($n < 250$). For large sample sizes both the Gamma and Weibull are equally efficient but the GE performs badly. On the other hand, the Weibull density performs very badly for some situations when originally the data came from the Gamma density. As expected, the performance of the Gamma and GE are the best, irrespective of all sample sizes and all combinations, when originally the data came from the Gamma and GE, respectively.

5. Conclusion

This chapter investigates the rationality of using the Generalized Exponential (GE) density as a substitute to the Gamma and Weibull densities to model the skewed or lifetime data. The simulation study suggests that in general the Weibull density performs marginally better than the Gamma and GE irrespective of all different parameters values and sample sizes when the sample data originally came from other skewed distributions rather than the Gamma, Weibull and GE. As expected, the performance of the Gamma and the GE are the best, irrespective of all sample sizes and all combinations of different parameter values, when originally the data

Table 11: Average KS distances and total number of accepted H_0 determined using 1000 synthetic data sets that originally came from the *Lindley* ($\theta = .8$).

Sample size	Distribution	Average Distance	H_0 (*KS*)	H_0 (*AD*)
$n = 25$	Weibull	0.153	931	853
	Gamma	0.179	843	806
	GE	0.203	739	678
$n = 50$	Weibull	0.103	965	941
	Gamma	0.121	890	873
	GE	0.137	766	749
$n = 100$	Weibull	0.068	981	978
	Gamma	0.076	960	953
	GE	0.082	924	896
$n = 250$	Weibull	0.043	983	982
	Gamma	0.047	991	995
	GE	0.049	983	989
$n = 500$	Weibull	0.030	1000	999
	Gamma	0.036	996	998
	GE	0.038	983	993
$n = 1000$	Weibull	0.022	997	999
	Gamma	0.029	967	992
	GE	0.031	943	979

Table 12: Average KS distances and total number of accepted H_0 determined using 1000 synthetic data sets that originally came from the *Lindley* ($\theta = .1$).

Sample size	Distribution	Average Distance	H_0 (*KS*)	H_0 (*AD*)
$n = 25$	Weibull	0.153	931	853
	Gamma	0.179	843	806
	GE	0.203	739	678
$n = 50$	Weibull	0.103	965	941
	Gamma	0.121	890	873
	GE	0.137	766	749
$n = 100$	Weibull	0.068	981	978
	Gamma	0.076	960	953
	GE	0.082	924	896
$n = 250$	Weibull	0.043	983	982
	Gamma	0.047	991	995
	GE	0.049	983	989
$n = 500$	Weibull	0.030	1000	999
	Gamma	0.036	996	998
	GE	0.038	983	993
$n = 1000$	Weibull	0.022	997	999
	Gamma	0.029	967	992
	GE	0.031	943	979

Table 13: Average KS distances and total number of accepted H_0 determined using 1000 synthetic data sets that originally came from the *Half-logistic.*

Sample size	Distribution	Average Distance	H_0 (*KS*)	H_0 (*AD*)
$n = 25$	Weibull	0.151	930	866
	Gamma	0.193	796	728
	GE	0.212	693	635
$n = 50$	Weibull	0.114	885	852
	Gamma	0.116	925	902
	GE	0.127	837	823
$n = 100$	Weibull	0.070	986	974
	Gamma	0.078	960	965
	GE	0.083	922	913
$n = 250$	Weibull	0.043	1000	999
	Gamma	0.050	996	999
	GE	0.052	985	989
$n = 500$	Weibull	0.032	998	1000
	Gamma	0.040	997	993
	GE	0.042	957	982
$n = 1000$	Weibull	0.026	998	993
	Gamma	0.034	889	937
	GE	0.036	824	874

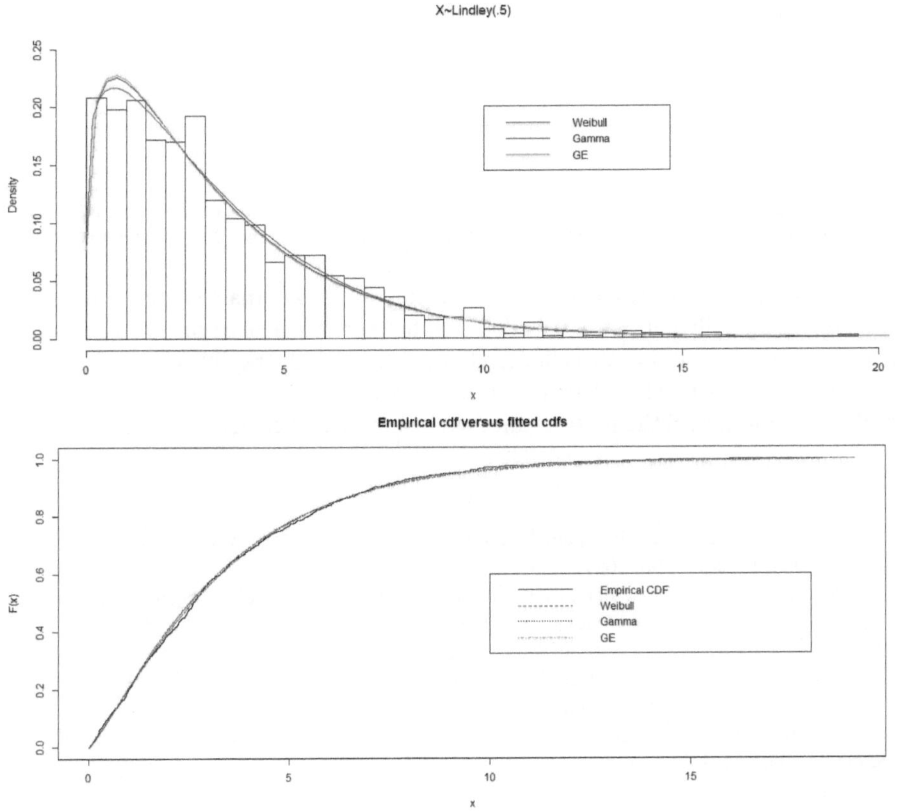

Figure 4: Overlaying the fitted Weibull (red), Gamma (blue) and GE (green) densities on the sample histogram (data that originally came from the *Lindley* ($\theta = .5$)), and the empirical cdf versus fitted cdf plots.

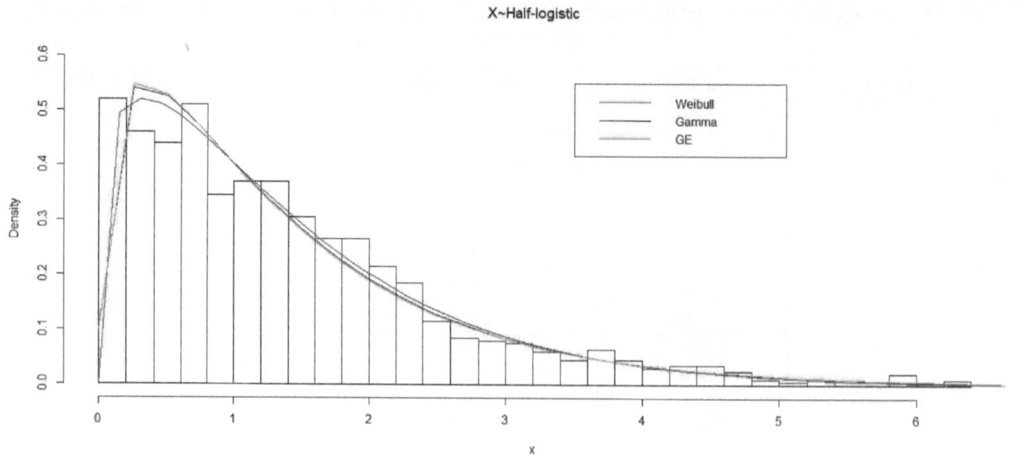

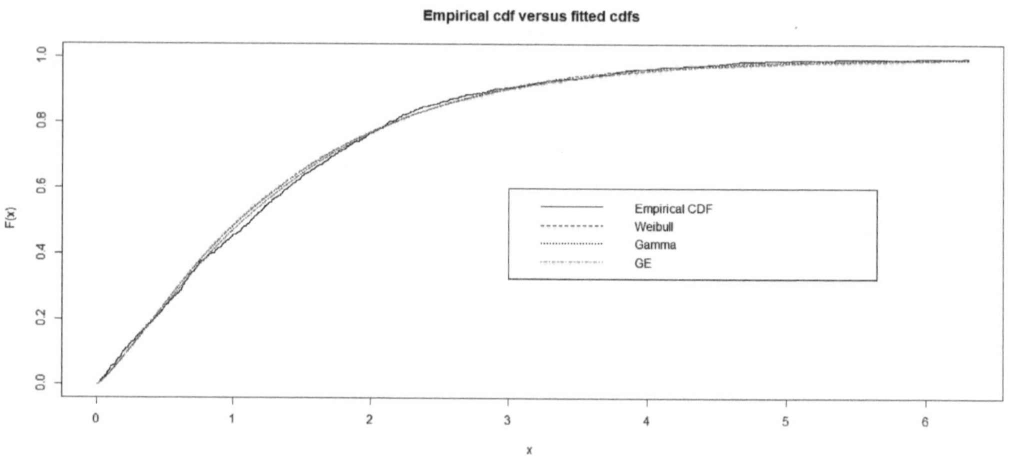

Figure 5: Overlaying the fitted Weibull (red), Gamma (blue) and GE (green) densities on the sample histogram (data that originally came from the *Half-logistic*), and the empirical cdf versus fitted cdf plots.

came from the Gamma and the GE, respectively. Interestingly, the Gamma also performed equally well like Weibull when the data originally came from Weibull but the performance of the GE is very poor, especially for large sample sizes. This research work is limited to investigating the performance of the Gamma, Weibull and GE to model the lifetime data when data do not have any censoring. Investigating the performance of the Gamma, Weibull and GE to model the lifetime data with censoring is currently being undertaken.

References

Bain, L. J. (1978). Statistical Analysis of Reliability and Life Testing Models. New York: Marcel Dekker, Inc. https://onlinelibrary.wiley.com/doi/abs/10.1002/zamm.19800601021.

Gupta, R. D. and Kundu, D. (1997). Exponentiated Exponential Family: An Alternative to gamma and Weibull Distribution. Technical Report. Department of Mathematics, Statistics and Computer Sciences., The University of New Brunswick, Saint John, NB, Canada. https://home.iitk.ac.in/~kundu/paper59.pdf.

Gupta, R. D. and Kundu, D. (1999). Generalized Exponential Distributions. The Australian & New Zealand Journal of Statistics. 41(2): 173–188. http://home.iitk.ac.in/~kundu/paper47.pdf.

Lindley, D. V. (1958). Fiducial distributions and Bayes' theorem. Journal of the Royal Statistical Society: Series B (Methodological), 20(1): 102–107. http://www.jstor.org/stable/2983909.

Chapter 16

Exponentiated Muth Distribution
Properties and Applications

R Maya,[1] *MR Irshad*[2,*] and *Anuresha Krishna*[2]

1. Introduction

In the backdrop of reliability theory, Muth (1977) restored a continuous probability distribution namely the Muth distribution (*MD*). However, until Jodrá et al. (2015), this distribution has been neglected in literature. A continuous random variable Z is said to have *MD* with the parameter α, if its probability density function (pdf) is given by,

$$f(z;\alpha) = (e^{\alpha z} - \alpha)e^{\left\{\alpha z - \frac{1}{\alpha}(e^{\alpha z} - 1)\right\}}, z > 0, \alpha \in (0,1]. \tag{1.1}$$

The cumulative distribution function (cdf) of the *MD* is obtained as,

$$F(z;\alpha) = 1 - e^{\left\{\alpha z - \frac{1}{\alpha}(e^{\alpha z} - 1)\right\}}, z > 0. \tag{1.2}$$

In order to demonstrate the significance of the model, the authors examined the scaled Muth distribution (*SMD*) and fitted it to the rainfall data set. They showed that the *SMD* is superior to well-known distributions such as Exponential, Gamma, Lognormal and Weibull. After the work of Jodrá et al. (2015), the study of the *SMD* has gained a commendable position in the literature both from theoretical and applied perspectives. Jodrá et al. (2017) studied the power version of the Muth random variable and highlighted its competence in the breaking stress of carbon fibres data and failure times of Kevlar 49/epoxy strands data. Inferential aspects of a geometric process with *SMD* are discussed by Biçer et al. (2021). Estimation of the scale parameter of *MD* and power *MD* are respectively discussed by Irshad et al. (2021) and Irshad et al. (2020). Meanwhile, a study on the scaled intermediate Muth distribution was carried out by Jodrá and Arshad (2021). As a result of the commendable performance of *MD* and its variants, we considered a more generalized version of *MD*. Hence in this paper, we consider a more generalized form of *MD* by incorporating an additional shape parameter β, and name the distribution as the Exponentiated Muth distribution (*EMD*) in which is its special case. Compared to the *MD*, one of the appealing characteristics of the *EMD* is its flexibility in modelling all forms of the hazard rate function, which are quite common in lifetime data analysis and reliability studies. Moreover, the distribution under study is identified to be suitable for rainfall data, glass fiber and carbon fiber data.

[1] Department of Statistics, Government College for Women, Trivandrum -695 014, India.
[2] Department of Statistics, Kochi-22, Kerala.
Emails: publicationsofmaya@gmail.com; anuresha.stat@gmail.com
* Corresponding author: irshadm24@gmail.com

The following is an order of organization for the remainder of the paper. The second section introduces the distribution and delineates some related functions like the survival function and the hazard rate function, as well as some of its important properties. Identifiability, moments, moment generating function (mgf) and various reliability measures including vitality function and mean residual life function (mrlf) are derived in Section 3. The uncertainty measures extropy and residual extropy is covered in Section 4. Among the topics discussed in Section 5, we discuss maximum likelihood estimation (MLE) of parameters, Fisher information matrix and asymptotic confidence interval. This section also describes the asymptotic behaviour of the *EMD* using several simulated data sets. MLE and Bayesian approaches are used to estimate unknown parameters of *EMD*. In Section 6, the proposed distribution is elucidated with three real data sets. Finally, the study is concluded in Section 7.

2. The exponentiated Muth distribution

The *EMD* is presented in this section, along with some of its properties.

Definition 2.1. A continuous random variable Z is said to follow an *EMD* with parameters $\alpha \in (0,1]$ and $\beta > 0$, if its cdf is of the following form, for $z \geq 0$,

$$F(z) = (1 - \phi(z;\alpha))^{\beta}, \tag{2.1}$$

where,

$$\phi(z;\alpha) = e^{\left\{\alpha z - \frac{1}{\alpha}(e^{\alpha z} - 1)\right\}}. \tag{2.2}$$

On differentiating (2.1), with respect to z, the pdf, $f(z)$ of the *EMD* is obtained in the following form (for $z > 0$),

$$f(z) = \beta(e^{\alpha z} - \alpha)\phi(z;\alpha)\,(1 - \phi(z;\alpha))^{\beta-1}, \tag{2.3}$$

where $\phi(z;\alpha)$ is given in (2.2).

Special cases

1. When $\beta = 1$ in (2.3), it will reduce to the pdf of the *MD*.
2. When $\beta = 1$ and $\alpha \to 0$ in (2.3), it will reduce to a unit exponential distribution.

Based on the pdf and cdf, the survival function $\bar{F}(z)$, hazard rate function $h_F(z)$ and reversed hazard rate function $\tau_F(z)$ of the *EMD* is obtained as follows (for $z > 0$);

$$\bar{F}(z) = 1 - (1 - \phi(z;\alpha))^{\beta}, \tag{2.4}$$

$$h_F(Z) = \frac{f(z)}{1 - F(z)} = \frac{\beta(e^{\alpha z} - \alpha)\phi(z;\alpha)(1 - \phi(z;\alpha))^{\beta-1}}{1 - (1 - \phi(z;\alpha))^{\beta}} \tag{2.5}$$

and

$$\tau_F(Z) = \frac{f(z)}{F(z)} = \frac{\beta(e^{\alpha z} - \alpha)\phi(z;\alpha)(1 - \phi(z;\alpha))^{\beta-1}}{(1 - \phi(z;\alpha))^{\beta}}. \tag{2.6}$$

On differentiating (2.3) and (2.5) with respect to z, we have

$$f'(z) = f(z)\left\{\frac{\Lambda_2(z;\alpha)}{\Lambda_1(z;\alpha)} - \frac{\Lambda_1(z;\alpha)\overline{\Psi_{\beta}}(z;\alpha)}{\overline{\Psi_1}(z;\alpha)}\right\} \tag{2.7}$$

and

$$h_F'(z) = h_F(z)\left\{\frac{\Lambda_2(z;\alpha)}{\Lambda_1(z;\alpha)} - \frac{\Lambda_1(z;\alpha)}{\overline{\Psi_1}(z;\alpha)}\right\}, \tag{2.8}$$

where, $\Lambda_1(z;\alpha) = (e^{\alpha z} - \alpha)$, $\Lambda_2(z;\alpha) = \alpha e^{\alpha z}$, $\overline{\Psi_\beta}(z;\alpha) = 1 - \beta e^{\left\{\alpha z - \frac{1}{\alpha}(e^{\alpha z}-1)\right\}}$ and $\overline{\Psi_1}(z;\alpha) = 1 - e^{\left\{\alpha z - \frac{1}{\alpha}(e^{\alpha z}-1)\right\}}$.
From (2.7) and (2.8), the *EMD* exhibits the following properties:

Remark 2.1 *From (2.7), one can infer that f(z) is an increasing (or decreasing) function of z if* $\Lambda_2(z;\alpha)\overline{\Psi_1}(z;\alpha)$ *is greater than (or less than)* $\Lambda_1^2(z;\alpha)\overline{\Psi_\beta}(z;\alpha)$.

Remark 2.2 *From (2.8), one can infer that* $h_F(z)$ *is an increasing (or decreasing) function of z if* $\Lambda_2(z;\alpha)\overline{\Psi_1}(z;\alpha)$ *is greater than (or less than)* $\Lambda_1^2(z;\alpha)$.

Remark 2.3 *By using the definition of log-concavity, we can infer that the EMD with cdf given in (2.1) is unimodal iff,*

$$\frac{d^2 \log f(z)}{dx^2} \le 0,$$

which implies that,

$$(\beta-1)\Lambda_1^2(z;\alpha)\phi(z;\alpha)\{\Lambda_2(z;\alpha)\overline{\Psi_1}(z;\alpha) - \Lambda_1^2(z;\alpha)\overline{\Psi_2}(z;\alpha)\}$$
$$\le \overline{\Psi_1}^2(z;\alpha)\Lambda_2(z;\alpha)\{\alpha^2 + \Lambda_1^2(z;\alpha)\},$$

where, $\Lambda_1(z;\alpha) = (e^{\alpha z} - \alpha)$, $\Lambda_2(z;\alpha) = \alpha e^{\alpha z}$, $\overline{\Psi_1}(z;\alpha) = 1 - e^{\left\{\alpha z - \frac{1}{\alpha}(e^{\alpha z}-1)\right\}}$, $\overline{\Psi_2}(z;\alpha) = 1 - 2e^{\left\{\alpha z - \frac{1}{\alpha}(e^{\alpha z}-1)\right\}}$ and $\phi(z;\alpha) = e^{\left\{\alpha z - \frac{1}{\alpha}(e^{\alpha z}-1)\right\}}$.
A series expansion of the cdf of the is given by the following theorem:

Theorem 2.1 *The cdf, F(z) of the EMD can be expressed as,*

$$F(z) = \sum_{k=0}^{\beta} \frac{(-1)^k(\beta+1-k)_k}{k!}(1-F(z;\alpha))^k, \tag{2.9}$$

where F(z;α) represents the cdf of MD and is given in (1.2).

Proof: We have,

$$(1+z)^a = \sum_{j=o}^{\infty}(a+1-j)j\frac{z^j}{j!} \tag{2.10}$$

for $a \in R$ and $(z)_k = z(z+1)\cdots(z+k-1)$ for $k \ge 1$ with $(z)_0 = 1$.
Applying (2.10) in (2.1), the proof is immediate.

Corollary 2.1 *By differentiating (2.9), a series expansion of the EMD pdf of the can be obtained as,*

$$f(z) = \sum_{k=0}^{\beta} \frac{k(-1)^{k+1}(\beta+1-k)_k}{k!}(1-F(z;\alpha))^{k-1}f(z;\alpha), \tag{2.11}$$

where f(z;α) is the pdf of the MD and is given in (1.1).

We may plot the pdf and hazard rate function of the EMD for different sets of parameters in Figures 1 and 2, respectively, to have a better idea of how they look. Observing the figures, the following facts can be inferred regarding the shapes of the distribution, thus making the *EMD* a more flexible model than the *MD* depending on how its parameters are chosen, particularly β.

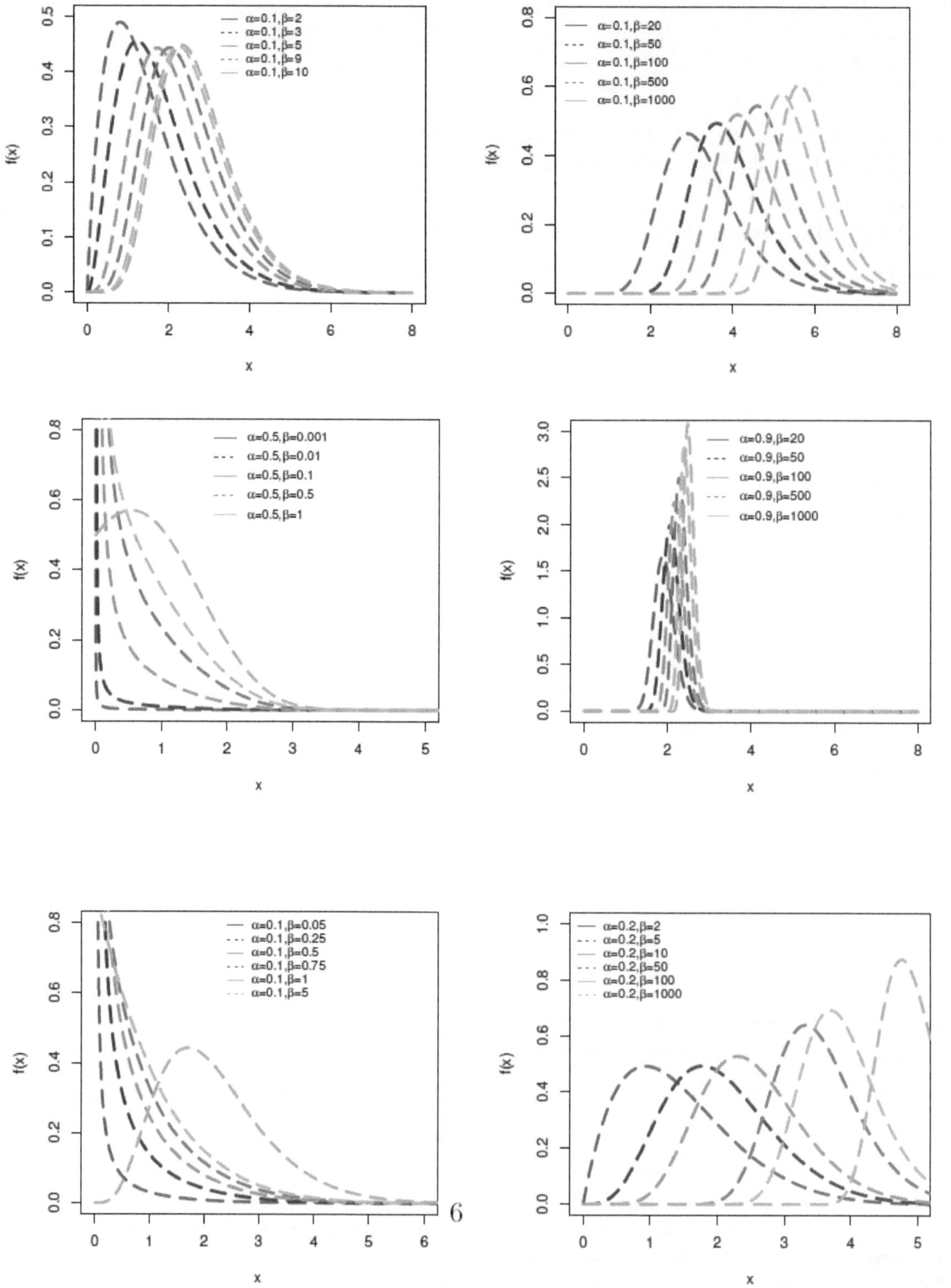

Figure 1: Different shapes of the pdf's of the *EMD*.

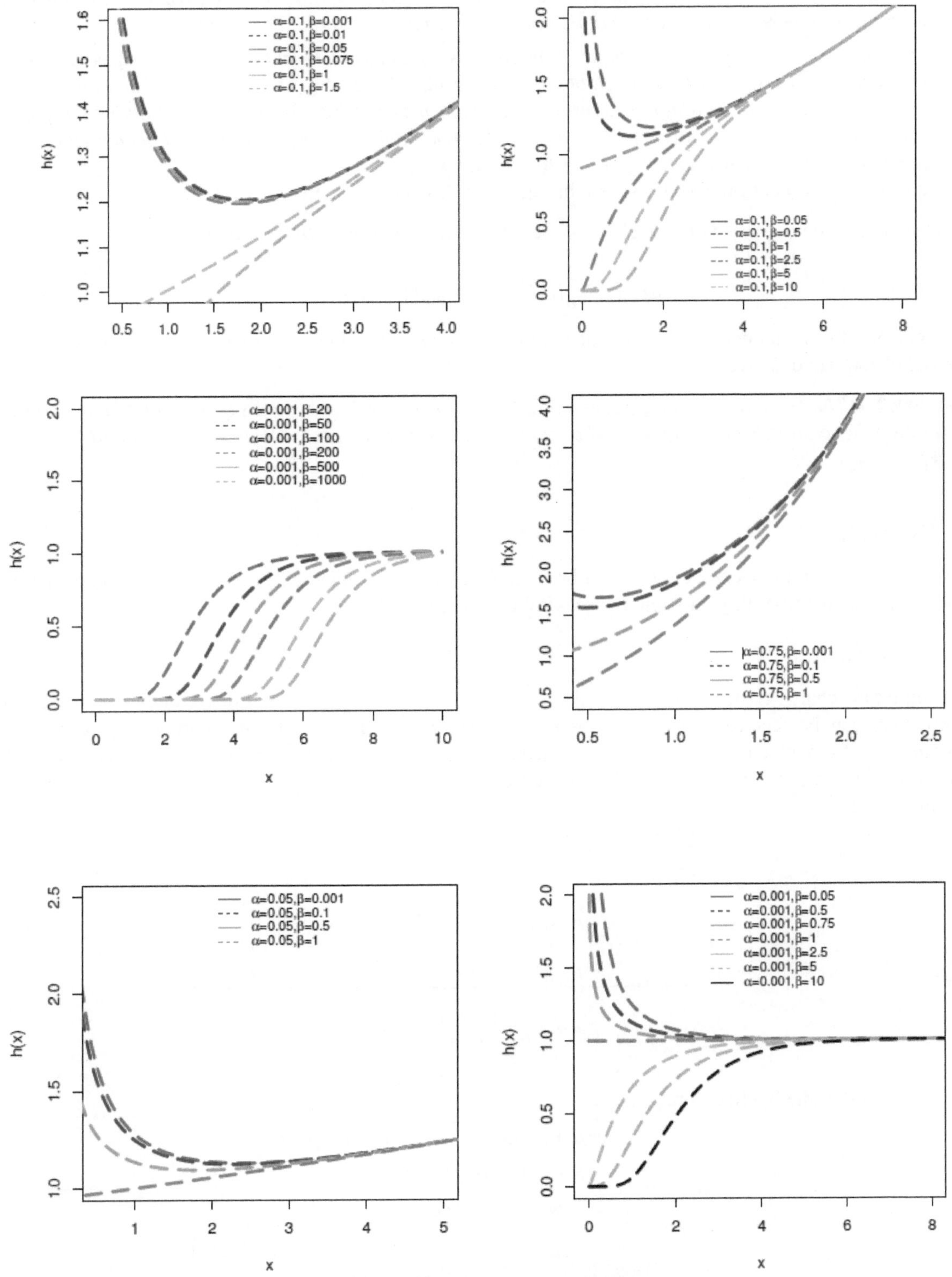

Figure 2: Different shapes of the hazard rate function of the *EMD*.

1. *For values of the parameters α and β that are relatively small, the EMD will be asymmetric or bimodal, whereas as β increases, the distribution approaches symmetry if α remains small.*
2. *The hazard rate graphs for various combinations of parameters show various shapes, including increasing, increasing and stable, stable and increasing, decreasing and stable, constant, and bathtub (decreasing-stable-increasing). As a result of the shape parameter β, the hazard rate function has a great deal of flexibility, which makes it very suitable for non-monotone empirical hazard behaviours, which are more likely to be encountered in real-life scenarios.*

The quantile function of the *EMD* is given through the following theorem.

Theorem 2.2 *The quantile function $\xi^*(\tau)$ of the EMD is the solution of the equation,*

$$\alpha\xi^*(\tau) - \frac{1}{\alpha}(e^{\alpha\xi^*(\tau)} - 1) - \log(1 - \tau^{\frac{1}{\beta}}) = 0. \tag{2.12}$$

The proof is the immediate consequence of the definition of the quantile function $\tau = F(z)$, where $F(z)$ is the cdf of *EMD* and is given by (2.1).

Remark 2.4 *The quantile function $\xi^*(\tau)$ of the EMD can also be written in terms of the qunatile function $\xi(\cdot)$ of through the relation $F(z) = (F(z;\alpha))^\beta$, where $F(z;\alpha)$ and $F(z)$ are respectively given by (1.2) and (2.1). That is $\xi^*(\tau) = \xi(\tau^{\frac{1}{\beta}})$.*

3. Properties

It is the purpose of this section to provide a brief summary of some statistical and reliability properties of the *EMD*, such as identifiability, moments, mgf, vitality function and mrlf.

3.1 *Identifiability*

An important characteristic of a statistical model is its identifiability, which determines whether model parameters can be reconstructed from observed data. Then, different sets of parameters give different distributions for a given x. That is, the parameters should uniquely determine the distribution. Let $f(\Theta_1)$ and $f(\Theta_2)$ be different members of *EMD* indexed by $\Theta_1 = (\alpha_1, \beta_1)$ and $\Theta_2 = (\alpha_2, \beta_2)$ respectively. Then the likelihood ratio,

$$L = \frac{f(\Theta_1)}{f(\Theta_2)}$$

$$= \frac{\beta_1(e^{\alpha_1 x} - \alpha_1)e^{(\alpha_1-\alpha_2)x - \frac{(e^{\alpha_1 x}-1)}{\alpha_1} + \frac{(e^{\alpha_2 x}-1)}{\alpha_2}}\left(1 - e^{\alpha_1 x - \frac{(e^{\alpha_1 x}-1)}{\alpha_1}}\right)^{\beta_1 - 1}}{\beta_2(e^{\alpha_2 x} - \alpha_2)\left(1 - e^{\alpha_2 x - \frac{(e^{\alpha_2 x}-1)}{\alpha_2}}\right)^{\beta_2 - 1}}.$$

The logarithm of the likelihood ratio is,

$$\begin{aligned}\log L &= \log(\beta_1) - \log(\beta_2) + \log(e^{\alpha_1 x} - \alpha_1) - \log(e^{\alpha_2 x} - \alpha_2) \\ &+ (\alpha_1 - \alpha_2)x - \frac{(e^{\alpha_1 x} - 1)}{\alpha_1} + \frac{(e^{\alpha_2 x} - 1)}{\alpha_2} \\ &+ (\beta_1 - 1)\log\left(1 - e^{\alpha_1 x - \frac{(e^{\alpha_1 x}-1)}{\alpha_1}}\right) - (\beta_2 - 1)\log\left(1 - e^{\alpha_2 x - \frac{(e^{\alpha_2 x}-1)}{\alpha_2}}\right).\end{aligned}$$

Taking the partial derivative of log L with respect to x and equating it to 0. That is,

$$\frac{\partial \log L}{\partial x} = 0$$

$$\frac{\alpha_1 e^{\alpha_1 x}}{(e^{\alpha_1 x} - \alpha_1)} + \alpha_1 - e^{\alpha_1 x} - \frac{(\beta_1 - 1)(\alpha_1 - e^{\alpha_1 x}) e^{\alpha_1 x - \left(\frac{e^{\alpha_1 x} - 1}{\alpha_1}\right)}}{\left(1 - e^{\alpha_1 x - \frac{(e^{\alpha_1 x} - 1)}{\alpha_1}}\right)}$$

$$= \frac{\alpha_2 e^{\alpha_2 x}}{(e^{\alpha_2 x} - \alpha_2)} + \alpha_2 - e^{\alpha_2 x} - \frac{(\beta_2 - 1)(\alpha_2 - e^{\alpha_2 x}) e^{\alpha_2 x - \left(\frac{e^{\alpha_2 x} - 1}{\alpha_2}\right)}}{\left(1 - e^{\alpha_2 x - \frac{(e^{\alpha_2 x} - 1)}{\alpha_2}}\right)}.$$

That is, RHS = LHS iff $\alpha_1 = \alpha_2$ and $\beta_1 = \beta_2$. Therefore we conclude that the *EMD* is identifiable in parameters. That is, $f(\Theta_1) = f(\Theta_2) \Leftarrow \Theta_1 = \Theta_2$.

3.2 Moments

Theorem 3.1 *If Z has the EMD with the pdf given in (2.3), then the* r^{th} *raw moment* μ_r' *about the origin is given by, for* $r = 1,2,...$

$$\mu_r' = \frac{\beta\Gamma(r+1)}{\alpha^{r+1}} \left\{ \sum_{k=0}^{\beta} \frac{(-1)^k (\beta - k)_k}{k!} e^{\frac{k+1}{\alpha}} \left(E_{-k-1}^r \left(\frac{k+1}{\alpha} \right) \right) - \alpha \sum_{k=0}^{\beta} \frac{(-1)^k (\beta - k)_k}{k!} e^{\frac{k+1}{\alpha}} \left(E_{-k}^r \left(\frac{k+1}{\alpha} \right) \right) \right\}, \tag{3.1}$$

where, $E_s^j(Z) = \frac{1}{\Gamma(j+1)} \int_1^{\infty} (\log t)^j t^{-s} e^{-zt} dt$ is the generalized integro-exponential function (see, Milgram (1985)), $\Gamma(a) = \int_0^{\infty} x^{a-1} e^{-x} dx$ is the gamma function and $(x)_k = x(x+1)...(x+k-1)$ for any $k \geq 1$ with $(x)_0 = 1$.

Proof. We have,

$$\mu_r' = \int_0^{\infty} z^r f(z) dz = M_1 - M_2, \tag{3.2}$$

where,

$$M_1 = \beta \int_0^{\infty} z^r e^{\alpha z} [1 - e^{\{\alpha z - \frac{1}{\alpha}(e^{\alpha z} - 1)\}}]^{\beta - 1} e^{\{\alpha z - \frac{1}{\alpha}(e^{\alpha z} - 1)\}} dz \tag{3.3}$$

and

$$M_2 = \alpha\beta \int_0^{\infty} z^r [1 - e^{\{\alpha z - \frac{1}{\alpha}(e^{\alpha z} - 1)\}}]^{\beta - 1} e^{\{\alpha z - \frac{1}{\alpha}(e^{\alpha z} - 1)\}} dz. \tag{3.4}$$

From (3.3),

$$M_1 = \beta \sum_{k=0}^{\beta} \frac{(-1)^k (\beta-k)_k}{k!} \int_0^{\infty} z^r e^{\{\alpha z - \frac{1}{\alpha}(e^{\alpha z}-1)\}(k+1)+\alpha z} dz$$

$$= \beta \sum_{k=0}^{\beta} \frac{(-1)^k (\beta-k)_k}{k!} \frac{e^{\frac{k+1}{\alpha}}}{\alpha^{r+1}} \int_1^{\infty} (\log y)^r y^{-(-k-1)} e^{-y\left(\frac{k+1}{\alpha}\right)} dy \quad (3.5)$$

$$= \frac{\beta \Gamma(r+1)}{\alpha^{r+1}} \sum_{k=0}^{\beta} \frac{(-1)^k (\beta-k)_k}{k!} e^{\frac{k+1}{\alpha}} \left(E_{-k-1}^{r}\left(\frac{k+1}{\alpha}\right)\right).$$

From (3.4),

$$M_2 = \alpha\beta \sum_{k=0}^{\beta} \frac{(-1)^k (\beta-k)_k}{k!} \int_0^{\infty} z^r e^{\{\alpha z - \frac{1}{\alpha}(e^{\alpha z}-1)\}(k+1)} dz$$

$$= \alpha\beta \sum_{k=0}^{\beta} \frac{(-1)^k (\beta-k)_k}{k!} \frac{e^{\frac{k+1}{\alpha}}}{\alpha^{r+1}} \int_1^{\infty} (\log y)^r y^{-(-k)} e^{-y\left(\frac{k+1}{\alpha}\right)} dy \quad (3.6)$$

$$= \frac{\alpha\beta \Gamma(r+1)}{\alpha^{r+1}} \sum_{k=0}^{\beta} \frac{(-1)^k (\beta-k)_k}{k!} e^{\frac{k+1}{\alpha}} \left(E_{-k}^{r}\left(\frac{k+1}{\alpha}\right)\right).$$

Substitute (3.5) and (3.6) in (3.2), we get (3.1). Hence the proof.

3.3 Moment generating function

Theorem 3.2 *If Z has the EMD with the pdf given in (2.3), then the moment generating function of Z is given by,*

$$M_z(t) = \beta \sum_{k=0}^{\beta} \frac{(-1)^k (\beta-k)_k}{k!} \frac{e^{\frac{k+1}{\alpha}} \alpha^{k+\frac{t}{\alpha}}}{(k+1)^{k+\frac{t}{\alpha}+1}} \Gamma\left(k+\frac{t}{\alpha}+1, \frac{k+1}{\alpha}\right)$$
$$- \alpha\beta \sum_{k=0}^{\beta} \frac{(-1)^k (\beta-k)_k}{k!} \frac{e^{\frac{k+1}{\alpha}} \alpha^{k+\frac{t}{\alpha}-1}}{(k+1)^{k+\frac{t}{\alpha}}} \Gamma\left(k+\frac{t}{\alpha}, \frac{k+1}{\alpha}\right), \quad (3.7)$$

where, $\Gamma(a,b) = \int_b^{\infty} t^{a-1} e^{-t} dt$ is the upper incomplete gamma function and $(x)_k = x(x+1)...(x+k-1)$ for any $k \geq 1$ with $(x)_0 = 1$.

Proof. We have,

$$M_Z(t) = E[e^{tZ}]$$
$$= B_1 - B_2, \quad (3.8)$$

where,

$$B_1 = \beta \int_0^{\infty} e^{tz} e^{\alpha z} [1 - e^{\{\alpha z - \frac{1}{\alpha}(e^{\alpha z}-1)\}}]^{\beta-1} e^{\{\alpha z - \frac{1}{\alpha}(e^{\alpha z}-1)\}} dz \quad (3.9)$$

and

$$B_2 = \alpha\beta \int_0^{\infty} e^{tz} [1 - e^{\{\alpha z - \frac{1}{\alpha}(e^{\alpha z}-1)\}}]^{\beta-1} e^{\{\alpha z - \frac{1}{\alpha}(e^{\alpha z}-1)\}} dz. \quad (3.10)$$

From (3.9),

$$B_1 = \beta \sum_{k=0}^{\beta} \frac{(-1)^k (\beta-k)_k}{k!} \int_0^{\infty} e^{\{\alpha z - \frac{1}{\alpha}(e^{\alpha z}-1)\}(k+1)+\alpha z + tz} dz$$

$$= \beta \sum_{k=0}^{\beta} \frac{(-1)^k (\beta-k)_k}{k!} \frac{e^{\frac{k+1}{\alpha}} \alpha^{k+\frac{t}{\alpha}}}{(k+1)^{k+\frac{t}{\alpha}+1}} \int_{\frac{k+1}{\alpha}}^{\infty} y^{k+\frac{t}{\alpha}} e^{-y} dy \tag{3.11}$$

$$= \beta \sum_{k=0}^{\beta} \frac{(-1)^k (\beta-k)_k}{k!} \frac{e^{\frac{k+1}{\alpha}} \alpha^{k+\frac{t}{\alpha}}}{(k+1)^{k+\frac{t}{\alpha}+1}} \Gamma\left(k+\frac{t}{\alpha}+1, \frac{k+1}{\alpha}\right).$$

From (3.10),

$$B_2 = \alpha\beta \sum_{k=0}^{\beta} \frac{(-1)^k (\beta-k)_k}{k!} \int_0^{\infty} e^{\{\alpha z - \frac{1}{\alpha}(e^{\alpha z}-1)\}(k+1)+tz} dz$$

$$= \alpha\beta \sum_{k=0}^{\beta} \frac{(-1)^k (\beta-k)_k}{k!} \frac{e^{\frac{k+1}{\alpha}} \alpha^{k+\frac{t}{\alpha}-1}}{(k+1)^{k+\frac{t}{\alpha}}} \int_{\frac{k+1}{\alpha}}^{\infty} y^{k+\frac{t}{\alpha}-1} e^{-y} dy \tag{3.12}$$

$$= \alpha\beta \sum_{k=0}^{\beta} \frac{(-1)^k (\beta-k)_k}{k!} \frac{e^{\frac{k+1}{\alpha}} \alpha^{k+\frac{t}{\alpha}-1}}{(k+1)^{k+\frac{t}{\alpha}}} \Gamma\left(k+\frac{t}{\alpha}, \frac{k+1}{\alpha}\right).$$

Substituting (3.11) and (3.12) in (3.8), we get (3.7). Thus the theorem is proved.

3.4 Reliability analysis

Quantitative evaluation of the mature product at each stage of its life cycle is an important aspect of reliability analysis. It also plays a significant role in reliability engineering to assure customer satisfaction. In the following subsections, we obtain certain reliability properties of the *EMD*.

3.4.1 Vitality function

In modeling lifetime data, the vitality function is crucial. In engineering and biomedical science, the vitality function, along with mrlf, play a major role. If Z is a non-negative random variable having cdf $F(z)$ with pdf $f(z)$, the vitality function associated with the random variable Z is defined as,

$$V(z) = E[Z|Z > z]. \tag{3.13}$$

In the reliability context, (3.13) can be interpreted as the average life span of components whose age exceeds z. Clearly the hazard rate reflects the risk of sudden death within a life span, where as the vitality function provides a more direct measure to describe the failure pattern in the sense that it is expressed in terms of an increased average life span.

3.4.2 Mean residual life function

If a unit survives an age z, the mrlf represents its remaining lifespan. It is the remaining lifetime of a unit from a certain point of time. The concise information by mrlf for establishing a warranty policy and for making maintenance decisions makes it an important measure in reliability modeling. If Z is a non-negative random variable having cdf $F(z)$ with pdf $f(z)$, then the mrlf is given by,

$$m(z) = E[Z - z|Z > z]. \tag{3.14}$$

In the following theorem, we provide the vitality function and mrlf of the *EMD*.

Theorem 3.3 *If Z has the EMD with the pdf given in (2.3), then,*

1. The vitality function of Z,

$$V(z)=\frac{1}{\{1-[1-\phi(z;\alpha)]^{\beta}\}}\left(\beta\sum_{k=0}^{\beta}\frac{(-1)^k(\beta-k)_k}{k!}\frac{\alpha^k e^{\frac{k+1}{\alpha}}}{(k+1)^{k+2}}\left\{\log\left(\frac{\alpha}{k+1}\right)\Gamma\left(k+2,\frac{e^{\alpha z}(k+1)}{\alpha}\right)+\Gamma'\left(k+2,\frac{e^{\alpha z}(k+1)}{\alpha}\right)\right\}\right. \tag{3.15}$$
$$\left.-\alpha\beta\sum_{k=0}^{\beta}\frac{(-1)^k(\beta-k)_k}{k!}\frac{\alpha^{k-1}e^{\frac{k+1}{\alpha}}}{(k+1)^{k+1}}\left\{\log\left(\frac{\alpha}{k+1}\right)\Gamma\left(k+1,\frac{e^{\alpha z}(k+1)}{\alpha}\right)+\Gamma'\left(k+1,\frac{e^{\alpha z}(k+1)}{\alpha}\right)\right\}\right),$$

where, $\Gamma(a,b)=\int_b^{\infty} t^{a-1}e^{-t}\,dt$ is the upper incomplete gamma function and $\Gamma'(a,b)=\int_b^{\infty} t^{a-1}\log(t)\,e^{-t}\,dt$ is the first derivative of the upper incomplete gamma function.

2. The mean residual life function of Z,

$$m(z)=V(z)-z, \tag{3.16}$$

where $V(z)$ is given in (3.15).

Proof. a) We have, the vitality function given by,

$$V(z)=E[Z|Z>z]=\frac{1}{\bar{F}(z)}\int_z^{\infty} tf(t)\,dt. \tag{3.17}$$

Now,

$$\int_z^{\infty} tf(t)\,dt=V_1-V_2 \tag{3.18}$$

where,

$$V_1=\beta\int_z^{\infty} te^{\alpha t}[1-e^{\{\alpha t-\frac{1}{\alpha}(e^{\alpha t}-1)\}}]^{\beta-1}e^{\{\alpha t-\frac{1}{\alpha}(e^{\alpha t}-1)\}}\,dt \tag{3.19}$$

and

$$V_2=\alpha\beta\int_z^{\infty} t[1-e^{\{\alpha t-\frac{1}{\alpha}(e^{\alpha t}-1)\}}]^{\beta-1}e^{\{\alpha t-\frac{1}{\alpha}(e^{\alpha t}-1)\}}\,dt. \tag{3.20}$$

From (3.19),

$$V_1 = \beta \sum_{k=0}^{\beta} \frac{(-1)^k (\beta-k)_k}{k!} \frac{\alpha^k e^{\frac{k+1}{\alpha}}}{(k+1)^{k+2}} \int_{\frac{e^{\alpha z}(k+1)}{\alpha}}^{\infty} \log\left(\frac{\alpha y}{k+1}\right) y^{k+1} e^{-y} dy$$

$$= \beta \sum_{k=0}^{\beta} \frac{(-1)^k (\beta-k)_k}{k!} \frac{\alpha^k e^{\frac{k+1}{\alpha}}}{(k+1)^{k+2}} \left\{ \log(\alpha) \int_{\frac{e^{\alpha z}(k+1)}{\alpha}}^{\infty} y^{k+1} e^{-y} dy \right.$$

$$\left. + \int_{\frac{e^{\alpha z}(k+1)}{\alpha}}^{\infty} \log(y) y^{k+1} e^{-y} dy - \log(k+1) \int_{\frac{e^{\alpha z}(k+1)}{\alpha}}^{\infty} y^{k+1} e^{-y} dy \right\} \tag{3.21}$$

$$= \beta \sum_{k=0}^{\beta} \frac{(-1)^k (\beta-k)_k}{k!} \frac{\alpha^k e^{\frac{k+1}{\alpha}}}{(k+1)^{k+2}} \left\{ \log\left(\frac{\alpha}{k+1}\right) \Gamma\left(k+2, \frac{e^{\alpha z}(k+1)}{\alpha}\right) \right.$$

$$\left. + \; \Gamma'\left(k+2, \frac{e^{\alpha z}(k+1)}{\alpha}\right) \right\}.$$

From (3.20),

$$V_2 = \alpha\beta \sum_{k=0}^{\beta} \frac{(-1)^k (\beta-k)_k}{k!} \frac{\alpha^{k-1} e^{\frac{k+1}{\alpha}}}{(k+1)^{k+1}} \int_{\frac{e^{\alpha z}(k+1)}{\alpha}}^{\infty} \log\left(\frac{\alpha y}{k+1}\right) y^{k} e^{-y} dy$$

$$= \alpha\beta \sum_{k=0}^{\beta} \frac{(-1)^k (\beta-k)_k}{k!} \frac{\alpha^{k-1} e^{\frac{k+1}{\alpha}}}{(k+1)^{k+1}} \left\{ \log(\alpha) \int_{\frac{e^{\alpha z}(k+1)}{\alpha}}^{\infty} y^{k} e^{-y} dy \right.$$

$$\left. + \int_{\frac{e^{\alpha z}(k+1)}{\alpha}}^{\infty} \log(y) y^{k} e^{-y} dy - \log(k+1) \int_{\frac{e^{\alpha z}(k+1)}{\alpha}}^{\infty} y^{k} e^{-y} dy \right\} \tag{3.22}$$

$$= \alpha\beta \sum_{k=0}^{\beta} \frac{(-1)^k (\beta-k)_k}{k!} \frac{\alpha^{k-1} e^{\frac{k+1}{\alpha}}}{(k+1)^{k+1}} \left\{ \log\left(\frac{\alpha}{k+1}\right) \Gamma\left(k+1, \frac{e^{\alpha z}(k+1)}{\alpha}\right) \right.$$

$$\left. + \; \Gamma'\left(k+1, \frac{e^{\alpha z}(k+1)}{\alpha}\right) \right\}.$$

Substitute (3.21) and (3.22) in (3.18) and using equation (3.17), we get (3.15). Hence the proof.
b) We have, the mrlf, given by,

$$\begin{aligned} m(z) &= E[Z - z | Z > z] \\ &= E[Z | Z > z] - z \\ &= V(z) - z, \end{aligned} \tag{3.23}$$

which is immediate from (3.15).

4. Uncertainty measures

In this section, we derive two recently developed uncertainty measures of *EMD* in two distinct contexts.

4.1 Extropy

The complementary dual of Shannon entropy, extropy, has a number of exciting applications, including the theory of proper scoring systems for alternate forecast distributions (see, Lad et al. (2015)). Entropy and extropy, two information metrics, are inextricably linked. For a random variable Z, its extropy is defined as

$$J(Z) = -\frac{1}{2} \int_0^\infty f^2(z)dz. \tag{4.1}$$

4.2 Residual extropy

The formulation of the concept of residual extropy was done by Qiu and Jia (2018) to measure the residual uncertainty of a random variable. For a random variable Z, its residual extropy is defined as (see, Qiu and Jia (2018)),

$$J(f;t) = \frac{-1}{2\overline{F}^2(t)} \int_t^\infty f^2(z)dz. \tag{4.2}$$

More recently, Maya and Irshad (2019) developed some non-parametric estimators of the residual extropy function under the α mixing dependence condition.

In the following theorems, respectively, given are the extropy and residual extropy functions of *EMD*.

Theorem 4.1 *The extropy function for the EMD has the following form:*

$$\begin{aligned} J(Z) = -\frac{\beta^2}{2}\Bigg\{ &\sum_{k=0}^{\beta} \frac{(-1)^k (2\beta-1-k)_k}{k!} \frac{\alpha^{k+3}}{(k+2)^{k+4}} e^{\frac{k+2}{\alpha}} \Gamma\left(k+4, \frac{k+2}{\alpha}\right) \\ &+ \alpha^2 \sum_{k=0}^{\beta} \frac{(-1)^k (2\beta-1-k)_k}{k!} \frac{\alpha^{k+1}}{(k+2)^{k+2}} e^{\frac{k+2}{\alpha}} \Gamma\left(k+2, \frac{k+2}{\alpha}\right) \\ &- 2\alpha \sum_{k=0}^{\beta} \frac{(-1)^k (2\beta-1-k)_k}{k!} \frac{\alpha^{k+2}}{(k+2)^{k+3}} e^{\frac{k+2}{\alpha}} \Gamma\left(k+3, \frac{k+2}{\alpha}\right) \Bigg\}. \end{aligned} \tag{4.3}$$

Proof. We have,

$$J(Z) = -\frac{1}{2} \int_0^\infty f^2(z)dz. \tag{4.4}$$

Now,

$$\int_0^\infty f^2(z)\, dz = \int_0^\infty \beta^2 [1 - e^{\{\alpha z - \frac{1}{\alpha}(e^{\alpha z}-1)\}}]^{2(\beta-1)} e^{2\{\alpha z - \frac{1}{\alpha}(e^{\alpha z}-1)\}} (e^{\alpha z} - \alpha)^2\, dz = A_1 + A_2 - A_3, \tag{4.5}$$

where,

$$A_1 = \beta^2 \int_0^\infty e^{2\alpha z} [1 - e^{\{\alpha z - \frac{1}{\alpha}(e^{\alpha z}-1)\}}]^{2(\beta-1)} e^{2\{\alpha z - \frac{1}{\alpha}(e^{\alpha z}-1)\}}\, dz, \tag{4.6}$$

$$A_2 = \alpha^2 \beta^2 \int_0^\infty [1 - e^{\{\alpha z - \frac{1}{\alpha}(e^{\alpha z}-1)\}}]^{2(\beta-1)} e^{2\{\alpha z - \frac{1}{\alpha}(e^{\alpha z}-1)\}}\, dz \tag{4.7}$$

and

$$A_3 = 2\alpha\beta^2 \int_0^\infty e^{\alpha z} [1 - e^{\{\alpha z - \frac{1}{\alpha}(e^{\alpha z}-1)\}}]^{2(\beta-1)} e^{2\{\alpha z - \frac{1}{\alpha}(e^{\alpha z}-1)\}}\, dz. \tag{4.8}$$

From (4.6),

$$\begin{aligned} A_1 &= \beta^2 \sum_{k=0}^{\beta} \frac{(-1)^k (2\beta-1-k)_k}{k!} \int_0^\infty e^{\alpha zk - \frac{ke^{\alpha z}}{\alpha} + \frac{k}{\alpha} + 2\alpha z - \frac{2e^{\alpha z}}{\alpha} + \frac{2}{\alpha} + 2\alpha z} dz \\ &= \beta^2 \sum_{k=0}^{\beta} \frac{(-1)^k (2\beta-1-k)_k}{k!} \frac{\alpha^{k+3}}{(k+2)^{k+4}} e^{\frac{k+2}{\alpha}} \Gamma\left(k+4, \frac{k+2}{\alpha}\right). \end{aligned} \tag{4.9}$$

From (4.7),

$$A_2 = \alpha^2\beta^2 \sum_{k=0}^{\beta} \frac{(-1)^k (2\beta-1-k)_k}{k!} \int_0^{\infty} e^{\alpha zk - \frac{ke^{\alpha z}}{\alpha} + \frac{k}{\alpha} + 2\alpha z - \frac{2e^{\alpha z}}{\alpha} + \frac{2}{\alpha}} dz$$
$$= \alpha^2\beta^2 \sum_{k=0}^{\beta} \frac{(-1)^k (2\beta-1-k)_k}{k!} \frac{\alpha^{k+1}}{(k+2)^{k+2}} e^{\frac{k+2}{\alpha}} \Gamma\left(k+2, \frac{k+2}{\alpha}\right). \tag{4.10}$$

From (4.8),

$$A_3 = 2\alpha\beta^2 \sum_{k=0}^{\beta} \frac{(-1)^k (2\beta-1-k)_k}{k!} \int_0^{\infty} e^{\alpha zk - \frac{ke^{\alpha z}}{\alpha} + \frac{k}{\alpha} + 2\alpha z - \frac{2e^{\alpha z}}{\alpha} + \frac{2}{\alpha} + \alpha z} dz$$
$$= 2\alpha\beta^2 \sum_{k=0}^{\beta} \frac{(-1)^k (2\beta-1-k)_k}{k!} \frac{\alpha^{k+2}}{(k+2)^{k+3}} e^{\frac{k+2}{\alpha}} \Gamma\left(k+3, \frac{k+2}{\alpha}\right). \tag{4.11}$$

Substitute (4.9), (4.10) and (4.11) in (4.4), we get (4.3).
Thus the theorem is proved.

Theorem 4.2 *The residual extropy function for EMD the has the following form:*

$$J(f;t) = \frac{-1}{2\{1-[1-\phi(z;\alpha)]^{\beta}\}^2} \left\{ \beta^2 \sum_{k=0}^{\beta} \frac{(-1)^k (2\beta-1-k)_k}{k!} \frac{\alpha^{k+3}}{(k+2)^{k+4}} \right.$$
$$\times e^{\frac{k+2}{\alpha}} \Gamma\left(k+4, \frac{e^{\alpha t}(k+2)}{\alpha}\right) + \alpha^2\beta^2 \sum_{k=0}^{\beta} \frac{(-1)^k (2\beta-1-k)_k}{k!} \frac{\alpha^{k+1}}{(k+2)^{k+2}}$$
$$\times e^{\frac{k+2}{\alpha}} \Gamma\left(k+2, \frac{e^{\alpha t}(k+2)}{\alpha}\right) + 2\alpha\beta^2 \sum_{k=0}^{\beta} \frac{(-1)^k (2\beta-1-k)_k}{k!} \frac{\alpha^{k+2}}{(k+2)^{k+3}}$$
$$\left. \times e^{\frac{k+2}{\alpha}} \Gamma\left(k+3, \frac{e^{\alpha t}(k+2)}{\alpha}\right) \right\}. \tag{4.12}$$

The proof is similar to that of theorem 4.1, hence omitted.

5. Estimation and inference

5.1 Method of maximum likelihood estimation

The most commonly used method of parameter estimation is the method of maximum likelihood. Its popularity is due to a number of desirable qualities, including consistency, asymptotic efficiency, invariance, and intuitive appeal. Let $Z_1, Z_2, \dots, Z_n$ be a random sample of size n from the *EMD* with unknown parameter vector $\theta = [\alpha, \beta]^T$. Then the likelihood function of θ is given by,

$$l(\theta) = \prod_{i=1}^{n} f_i(z;\alpha,\beta)$$
$$= \beta^n \prod_{i=1}^{n} \left(1 - e^{\left(\alpha z_i - \frac{1}{\alpha}(e^{\alpha z_i}-1)\right)}\right)^{\beta-1} e^{\sum_{i=1}^{n}\left(\alpha z_i - \frac{1}{\alpha}(e^{\alpha z_i}-1\right)} \prod_{i=1}^{n} (e^{\alpha z_i} - \alpha). \tag{5.1}$$

The partial derivatives of log $l(\theta)$ with respect to the parameters are

$$\frac{\partial \log l}{\partial \alpha} = \sum_{i=1}^{n} \frac{(\beta-1)e^{\alpha z_i - \frac{1}{\alpha}(e^{\alpha z_i}-1)}\left(\frac{1}{\alpha^2} - \frac{e^{\alpha z_i}}{\alpha^2} + \frac{z_i e^{\alpha z_i}}{\alpha} - z_i\right)}{(1-e^{\alpha z_i - \frac{1}{\alpha}(e^{\alpha z_i}-1)})} \quad (5.2)$$

$$+\sum_{i=1}^{n}\left(z_i - \frac{z_i e^{\alpha z_i}}{\alpha} + \frac{e^{\alpha z_i}}{\alpha^2} - \frac{1}{\alpha^2}\right) + \sum_{i=1}^{n} \frac{(z_i e^{\alpha z_i}-1)}{(e^{\alpha z_i}-\alpha)}$$

and

$$\frac{\partial \log l}{\partial \alpha} = \frac{n}{\beta} + \sum_{i=1}^{n} \log\left(1 - e^{\alpha z_i - \frac{1}{\alpha}(e^{\alpha z_i}-1)}\right). \quad (5.3)$$

The MLE of the parameters $\theta = (\alpha, \beta)$ say $\hat{\theta} = (\hat{\alpha}, \hat{\beta})$ say is obtained by solving the equations $\frac{\partial \log l}{\partial \alpha} = 0$ and $\frac{\partial \log l}{\partial \beta} = 0$. This can only be achieved by the numerical optimization technique such as the Newton-Raphson method and Fisher's scoring algorithm using mathematical packages like R and Mathematica. By using any optimization method to compute $\hat{\theta}$, we face the problem that (5.1) has more than one local maximum because the optimizer function with different initial values can lead to different maximums. To alleviate this problem first we plot the density function with some parameter values on the histogram of the data and find a good vector of parameters with a good fit to the histogram and choose this vector as the initial parameter vector to the optimization problem.

5.1.1 Fisher information matrix

In order to carry out statistical inference on the parameters of the *EMD*, we need to find the 2×2 expected Fisher information matrix $I(\theta)$. The expected Fisher information matrix of is given by,

$$I(\theta) = \begin{bmatrix} -E(\frac{\partial \log l}{\partial \alpha^2}) - E\frac{\partial^2 \log l}{\partial \alpha \partial \beta}) \\ -E(\frac{\partial^2 \log l}{\partial \beta \partial \alpha}) - E\frac{\partial^2 \log l}{\partial \beta^2}) \end{bmatrix}. \quad (5.4)$$

The expected Fisher information matrix can be approximated by the observed Fisher information matrix $J(\hat{\theta})$ given by,

$$I(\hat{\theta}) = \begin{bmatrix} -\frac{\partial^2 \log l}{\partial \alpha^2} - \frac{\partial^2 \log l}{\partial \alpha \partial \beta} \\ -\frac{\partial^2 \log l}{\partial \beta \partial \alpha} - \frac{\partial^2 \log l}{\partial \beta^2} \end{bmatrix}. \quad (5.5)$$

That is,

$$\lim_{n\to\infty} \frac{1}{n} J(\hat{\theta}) = I(\theta).$$

For large n, the following approximation can be used,

$$J(\hat{\theta}) = nI(\theta).$$

The elements of $J(\hat{\theta})$ are given in Appendix.

5.1.2 Asymptotic confidence interval

Here we present the asymptotic confidence interval for the parameters of the *EMD*. Let $\hat{\theta} = (\hat{\alpha}, \hat{\beta})$ be the MLE of $\theta = (\alpha, \beta)$. Under the usual regularity conditions and that the parameters are in the interior of the parameter space, but not on the boundary, we have $\sqrt{n}(\theta - \hat{\theta}) \xrightarrow{d} N_2(\underline{0}, I^{-1}(\theta))$, where $I(\theta)$ is the expected Fisher information matrix. The asymptotic behavior is still valid if $I(\theta)$ is replaced by the observed Fisher information matrix $J(\hat{\theta})$. The multivariate normal distribution, $N_2(0, I^{-1}(\theta))$ with mean vector $\underline{0} = (0,0)^{\tau}$ can be used to construct confidence intervals for the model parameters. The approximate $100(1 - \phi)\%$ two-sided confidence intervals for α and β are respectively given by $\hat{\alpha} \pm Z_{\frac{\phi}{2}}\sqrt{I_{\alpha\alpha}^{-1}(\hat{\theta})}$ and $\hat{\beta} \pm Z_{\frac{\phi}{2}}\sqrt{I_{\beta\beta}^{-1}(\hat{\theta})}$, where $I_{\alpha\alpha}^{-1}(\hat{\theta})$ and $I_{\beta\beta}^{-1}(\hat{\theta})$ are diagonal elements of $J^{-1}(\hat{\theta})$ and $Z_{\frac{\phi}{2}}$ is the upper $\frac{\phi^{th}}{2}$ percentile of a standard normal distribution.

5.2 Bayesian analysis

By the Bayesian approach, in this section, we are estimating the unknown parameters of the *EMD*. This approach can incorporate prior information about the problem at hand. Due to this fact, analysts consider this method as flexible to estimate parameters of a model. A statistical comparison between the maximum likelihood and Bayesian procedures for estimating parameters of the *EMD* is performed based on the real data sets in Section 6.

Various types of priors are available in the Bayesian approach to handle various situations. Here we focus on the weakly informative prior (*WIP*). For the numerical integration, prior distributions that are not completely flat provide enough information for the numerical approximation algorithm to continue to explore the target posterior density. The half-Cauchy distribution (*HCD*) have such shapes. For the *HCD*, the mean and variance don't exist, but its mode is equal to zero. When the parameter becomes 25 in the *HCD*, the density is partially flat. According to Gelman and Hill (2007), depending upon the information necessary, uniform or *HCD* is a better choice of prior. Assuming and $\alpha \sim N(\mu_0, \sigma_0^2)$ and $\beta \sim HCD(a)$, where N and *HCD* stands for normal and half- Cauchy distributions respectively. The parameters of the prior density can be fixed as $\mu_0 = 0$, $\sigma_0^2 = 1$ and $a = 25$. Suppose the parameters are apriori independent, posterior density of the parameters is given by,

$$\pi(\theta \mid z) \propto \beta^n \prod_{i=1}^{n}\left(1 - e^{\left(\alpha z_i \frac{1}{\alpha}(e^{\alpha z_i} - 1)\right)}\right)^{\beta-1} e^{\sum_{i=1}^{n}\left(\alpha z_i - \frac{1}{\alpha}(e^{\alpha z_i} - 1)\right)} \prod_{i=1}^{n}(e^{\alpha z_i} - \alpha)\frac{1}{\sqrt{2\pi}}e^{-\frac{\alpha^2}{2}} \frac{2 * 25}{\pi(\beta^2 + 25^2)}. \quad (5.6)$$

Since (5.6) is not in closed form, one may use numerical integration or MCMC methods. By 'Laplace Approximation' method, any Bayesian model can be fit for which likelihood and priors are specified (see, Khan et al. (2016)). By using R software, we can solve this problem by performing 100000 iterations and the Random walk Metropolis algorithm (RwM). For the three data sets, the posterior mode and posterior median estimates are computed as Bayesian estimates and compared with corresponding MLE's which is given in Table 1. It can be concluded that approximately equal estimates are obtained by both the MLE and Bayesian approaches.

Table 1: Comparison between MLE and Bayesian estimates.

Data set	Parameter	MLE	Posterior median	Posterior mode
1	α	1	1	0.98
	β	0.46	0.38	0.44
2	α	0.94	0.93	0.92
	β	4.07	3.95	4.14
3	α	0.53	0.52	0.52
	β	14.14	13.61	13.88

5.3 Simulation

Our simulation study is based on generating observations with R software using three combinations of parameters to study the asymptotic behaviour of the MLEs of the parameters of the *EMD*. To obtain a good estimate of an estimator's variance, Efron and Tibshirani (1991) recommended a maximum of 200 bootstrap samples. From *EMD*, we generated 200 bootstrap samples of sizes 25, 50, 75, 100, 150, 250, 500, and 1000 for different parameter combinations. Our results are shown in Table 2. From Table 2, one can infer those

Table 2: The results of the simulation study.

$\alpha = 0.25, \beta = 5$						
Parameter	n	**MLE**	**Bias**	**MSE**	**CP**	**AL**
α	25	0.2771	0.0271	0.0096	0.910	0.3669
	50	0.2694	0.0194	0.0045	0.935	0.2590
	75	0.2637	0.0137	0.0032	0.940	0.2104
	100	0.2609	0.0109	0.0022	0.945	0.1825
	150	0.2563	0.0063	0.0015	0.930	0.1490
	250	0.2535	0.0035	0.0008	0.950	0.1161
	500	0.2523	0.0023	0.0004	0.960	0.0821
	1000	0.2484	-0.0016	0.0002	0.980	0.0580
β	25	5.2977	0.2977	1.3807	0.955	4.1980
	50	5.1528	0.1528	0.5346	0.955	2.8715
	75	5.1734	0.1734	0.3618	0.975	2.3510
	100	5.1314	0.1314	0.2855	0.975	2.0178
	150	5.1067	0.1067	0.1567	0.970	1.6383
	250	5.0149	0.0149	0.0786	0.980	1.2451
	500	5.0144	0.0144	0.0461	0.955	0.8801
	1000	5.0072	0.0072	0.0272	0.960	0.6213
$\alpha = 0.5, \beta = 3$						
Parameter	n	**MLE**	**Bias**	**MSE**	**CP**	**AL**
α	25	0.5285	0.0285	0.0137	0.915	0.4360
	50	0.5219	0.0219	0.0063	0.935	0.3077
	75	0.5168	0.0168	0.0046	0.945	0.2502
	100	0.5125	0.0125	0.0031	0.945	0.2169
	150	0.5071	0.0071	0.0021	0.930	0.1770
	250	0.5042	0.0042	0.0011	0.955	0.1379
	500	0.5027	0.0027	0.0006	0.955	0.0975
	1000	0.4981	-0.0019	0.0002	0.985	0.0689
β	25	3.1617	0.1617	0.4830	0.950	2.4965
	50	3.0872	0.0872	0.1979	0.950	1.7166
	75	3.0940	0.0940	0.1264	0.970	1.4031
	100	3.0724	0.0724	0.1009	0.980	1.2062
	150	3.0602	0.0602	0.0549	0.975	0.9805
	250	3.0069	0.0069	0.0283	0.980	0.7461
	500	3.0073	0.0073	0.0165	0.955	0.5276
	1000	3.0050	0.0050	0.0098	0.950	0.3728

Table 2 contd. ...

...Table 2 contd.

$\alpha = 0.7, \beta = 4$						
Parameter	*n*	**MLE**	**Bias**	**MSE**	**CP**	**AL**
α	25	0.7733	0.0233	0.0108	0.920	0.4051
	50	0.7697	0.0197	0.0053	0.925	0.2853
	75	0.7639	0.0139	0.0036	0.955	0.2321
	100	0.7612	0.0112	0.0026	0.955	0.2012
	150	0.7563	0.0063	0.0018	0.925	0.1642
	250	0.7543	0.0043	0.0009	0.960	0.1277
	500	0.7530	0.0030	0.0005	0.950	0.0902
	1000	0.7483	-0.0017	0.0002	0.985	0.0638
β	25	4.2948	0.2948	1.0395	0.965	3.5646
	50	4.1771	0.1771	0.4342	0.965	2.4178
	75	4.1670	0.1670	0.2623	0.975	1.9615
	100	4.1283	0.1283	0.2045	0.975	1.6789
	150	4.0995	0.0995	0.1155	0.975	1.3575
	250	4.0208	0.0208	0.0546	0.980	1.0279
	500	4.0174	0.0174	0.0315	0.960	0.7256
	1000	4.0036	0.0036	0.0183	0.960	0.5107

estimates are quite stable and more precisely, are close to the true parameter values for these sample sizes while the MSEs of the estimators are in decreasing order. In addition, for each parameter, the coverage probabilities (CP) are fairly close to the 95% nominal level. As increases, the average lengths (ALs) for each parameter decrease to zero.

6. Application

In this section, three data sets are used to exemplify the excellence of the *EMD* (*EMD*(α,β)). The first data set considered here was taken from the website of the Bureue of Meteorology of the Australian Government. The data comprises total monthly rainfall (in mm) collected from January 2000 to February 2007 in the rain gauge station of Carrol, located in the state of New South Wales on the east coast of Australia (see, Jodrá et al. (2015)). The data set is listed below.

12.0, 22.7, 75.5, 28.6, 65.8, 39.4, 33.1, 84.0, 41.6, 62.3,52.5, 13.9, 15.4, 31.9, 32.5, 37.7, 9.5, 49.9, 31.8, 32.2, 50.2, 55.8, 20.4, 5.9, 10.1, 44.5, 19.7, 6.4, 29.2, 42.5, 19.4, 23.8, 55.2, 7.7, 0.8, 6.7, 4.8, 73.8, 5.1, 7.6, 25.7, 50.7, 59.7, 57.2, 29.7, 32.0, 24.5, 71.6, 15.0, 17.7, 8.2, 23.8, 46.3, 36.5, 55.2, 37.2, 33.9, 53.9, 51.6, 17.3, 85.7, 6.6, 4.7, 1.8, 98.7, 62.8, 59.0, 76.1, 67.9, 73.7, 27.2, 39.5, 6.9, 14.0, 3.0, 41.6, 49.5, 11.2, 17.9, 12.7, 0.8, 21.1, 24.5.

The second data set is the strength of glass fibers of length 1.5 cm from the National Physical Laboratory in England (see, Smith and Naylor (1987)). The data set is listed below.

0.55, 0.93, 1.25, 1.36, 1.49, 1.52, 1.58, 1.61, 1.64, 1.68, 1.73, 1.81, 2.00, 0.74, 1.04, 1.27, 1.39, 1.49, 1.53, 1.59, 1.61, 1.66, 1.68, 1.76, 1.82, 2.01, 0.77, 1.11, 1.28, 1.42, 1.50, 1.54, 1.60, 1.62, 1.66, 1.69, 1.76, 1.84, 2.24, 0.81, 1.13, 1.29, 1.48, 1.50, 1.55, 1.61, 1.62, 1.66, 1.70, 1.77, 1.84, 0.84, 1.24, 1.30, 1.48, 1.51, 1.55, 1.61, 1.63, 1.67, 1.70, 1.78, 1.89.

The third data presented here is originally proposed by Bader and Priest (1982) on failure stresses of single carbon fibers of length 50mm. The data set is listed below.

1.339, 1.434, 1.549, 1.574, 1.589, 1.613, 1.746, 1.753, 1.764, 1.807, 1.812, 1.84, 1.852, 1.852, 1.862,1.864, 1.931, 1.952, 1.974, 2.019, 2.051, 2.055, 2.058, 2.088, 2.125, 2.162, 2.171, 2.172, 2.18, 2.194, 2.211, 2.27, 2.272, 2.28, 2.299, 2.308, 2.335, 2.349, 2.356, 2.386, 2.39, 2.41, 2.43, 2.431, 2.458, 2.471, 2.497, 2.514, 2.558, 2.577, 2.593, 2.601, 2.604, 2.62, 2.633, 2.67, 2.682, 2.699, 2.705, 2.735, 2.785, 3.02, 3.042, 3.116, 3.174.

We shall compare the fits of Muth distribution ($MD(\alpha)$) with some competitive models such as Exponential distribution ($ED(\lambda)$), generalized exponential distribution ($GED(\alpha,\lambda)$, lognormal distribution ($LND(\mu,\sigma)$) scaled Muth distribution ($SMD(\alpha,\beta)$) and gamma distribution ($GD(a,b)$). For the above three data sets, we have computed MLEs of the parameters and calculated AIC (Akaike information criterion), AICc (Akaike information criterion corrected), CAIC (Consistent Akaike information criterion), BIC (Bayesian information criterion). All these criteria are calculated using R software. Each of these criteria takes into account the likelihood, the number of observations, and the number of parameters. The numerical values of AIC, AICc, CAIC and BIC of all fitted models for the above 3 data sets are respectively given in Tables 3–5.

Table 3: MLEs, log-likelihood, AIC, AICc, CAIC and BIC for the first data set.

Model	MLE	Loglikelihood	AIC	AICc	CAIC	BIC	KS
EMD(α,β)	$\tilde{\alpha}$=1 $\tilde{\beta}$=0.46	–43.3327	90.6654	90.8154	97.5031	95.5032	0.0445
MD(α)	$\tilde{\alpha}$=0.37	–55.7681	113.5362	113.5856	116.9550	115.9550	0.2019
SMD(α,β)	$\tilde{\alpha}$=0.46 $\tilde{\beta}$=0.68	–43.4821	90.9642	91.1142	97.8019	95.8019	0.0570
ED(λ)	$\tilde{\lambda}$=1.47	–50.8281	103.6562	103.7056	107.0750	106.0750	0.1186
GED(α,λ)	$\tilde{\alpha}$=1.51 $\tilde{\lambda}$= 1.89	–47.3560	98.7120	98.8620	105.5497	103.5497	0.0873
GD(a,b)	$\tilde{a}$=1.51 $\tilde{b}$= 2.23	–46.9565	97.9129	98.8620	104.7507	102.7506	0.0837
LND(μ,σ)	$\tilde{\mu}$=–0.75 $\tilde{\sigma}$=1.02	–56.8982	117.7964	117.9464	124.6340	122.6340	0.1217

Table 4: MLEs, log-likelihood, AIC, AICc, CAIC and BIC for the second data set.

Model	MLE	Loglikelihood	AIC	AICc	CAIC	BIC	KS
$EMD(\alpha,\beta)$	$\tilde{\alpha}$=0.94 $\tilde{\beta}$=4.07	–19.3903	42.7805	42.9805	49.0668	47.0669	0.2018
$MD(\alpha)$	$\tilde{\alpha}$=0.82	–58.7365	119.4730	119.5386	122.6161	121.6161	0.5357
$SMD(\alpha,\beta)$	$\tilde{\alpha}$=1 $\tilde{\beta}$=1.32	–39.3860	82.7720	82.9720	89.0583	87.0583	0.3265
$ED(\lambda)$	$\tilde{\lambda}$=0.66	–88.8303	179.6606	179.7262	182.7262	181.8037	0.4160
$GED(\alpha,\lambda)$	$\tilde{\alpha}$=31.35 $\tilde{\lambda}$=2.61	–31.3835	66.7669	66.9669	73.0532	71.0533	0.2282
$GD(a,b)$	$\tilde{a}$=17.44 $\tilde{b}$=11.57	–23.9515	51.9031	52.1031	58.1894	56.1893	0.2158
$LND(\mu,\sigma)$	$\tilde{\mu}$=0.38 $\tilde{\sigma}$=0.26	–28.0049	60.0099	60.2099	66.2961	64.2961	0.2328

Table 5: MLEs, log-likelihood, AIC, AICc, CAIC and BIC for the third data set.

Model	MLE	Loglikelihood	AIC	AICc	CAIC	BIC	KS
EMD(α,β)	$\tilde{\alpha}$=0.53 $\tilde{\beta}$=14.14	−34.9967	73.9934	74.1869	80.3421	78.3421	0.0723
MD(α)	$\tilde{\alpha}$=0.30	−134.7635	271.5270	271.5905	274.7014	273.7014	0.7477
SMD(α,β)	$\tilde{\alpha}$=1 $\tilde{\beta}$=1.96	−64.6493	133.2985	133.492	139.6473	137.6474	0.3294
ED(λ)	$\tilde{\lambda}$=0.45	−117.5382	237.0765	237.14	240.2509	239.2508	0.4678
GED(α,λ)	$\tilde{\alpha}$=176.35 $\tilde{\lambda}$= 2.54	−38.3621	80.7242	80.9178	85.0730	85.0730	0.0983
GD(a,b)	$\tilde{a}$=28.55 $\tilde{b}$= 12.72	−35.0701	74.1402	74.3337	80.4890	78.4889	0.0724
LND(μ,σ)	$\tilde{\mu}$=0.79 $\tilde{\sigma}$=0.19	−35.7676	75.5352	75.7287	81.8839	79.8840	0.0838

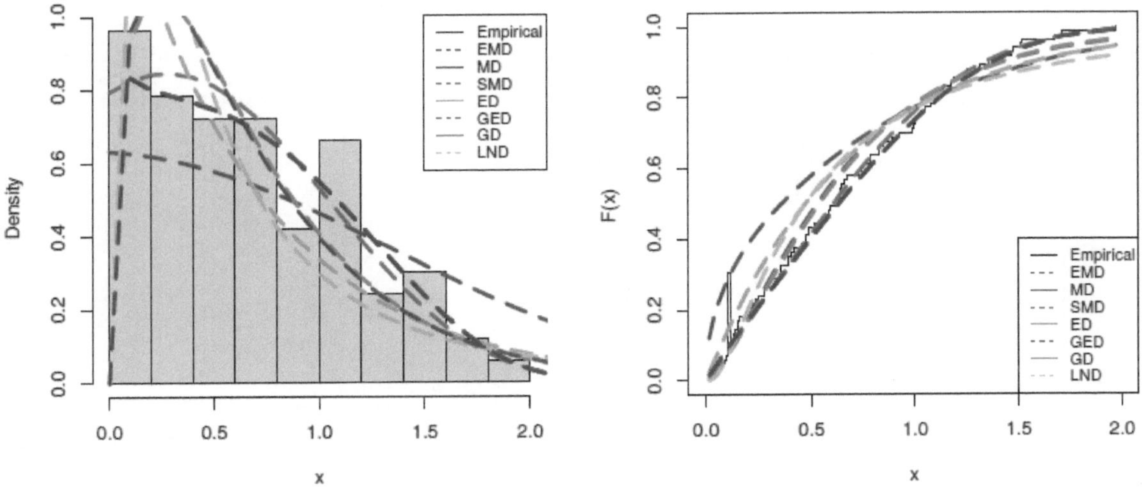

Figure 3: Estimated pdf and cdf plots of the fitted distribution of the rainfall data.

From each table it can be observed that *EMD* has the smallest value of AIC, AICc, CAIC and BIC, thus one can conclude that *EMD* has a better performance compared to the other competing models. Further, we have conducted the Kolmogorov Smirnov (KS) test to check the goodness of fit for all the data sets of the *EMD* as well as the other models. The value of the KS statistic is also included in the final columns of Tables 3–5.

The result of this study shows that *EMD* has high fitting ability compared to all other models. Once again the promising performance of the proposed distribution is visible from Figure 3, Figure 4 and Figure 5.

To test the null hypothesis H_0:*MD* versus H_1:*EMD* or equivalently $H_0:\beta = 1$ versus $H_1:\beta \neq 1$, we use the likelihood ratio (LR) test for each dataset. Table 6 includes the LR statistics and corresponding *p*-values for

Table 6: LR statistics and their -values on data sets 1, 2 and 3.

Data set	LR	*p*-value
1	24.8708	0.0000006
2	78.6920	$< 2.2\times10^{-16}$
3	199.5300	$< 2.2\times10^{-16}$

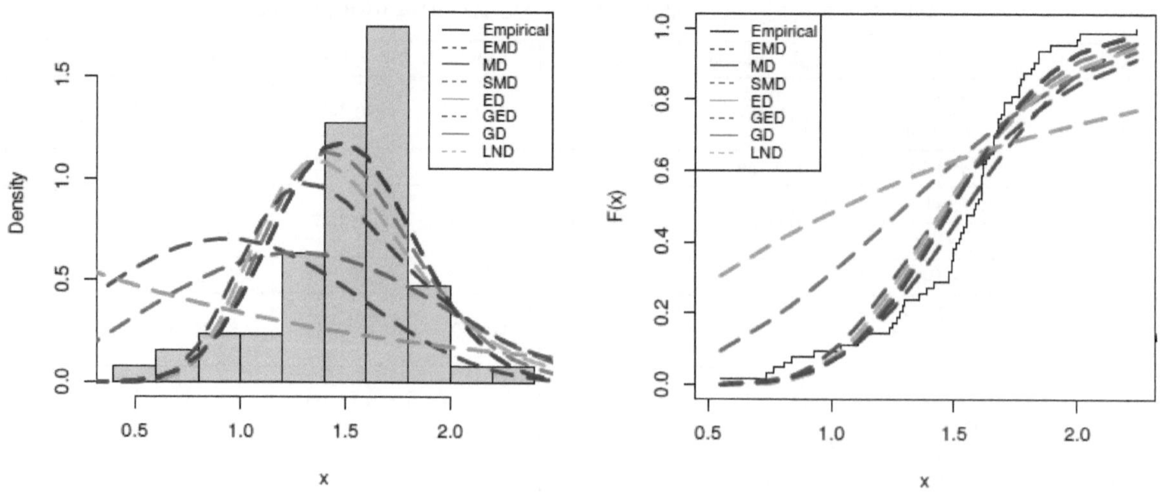

Figure 4: Estimated pdf and cdf plots of the fitted distribution of the glass fibres data.

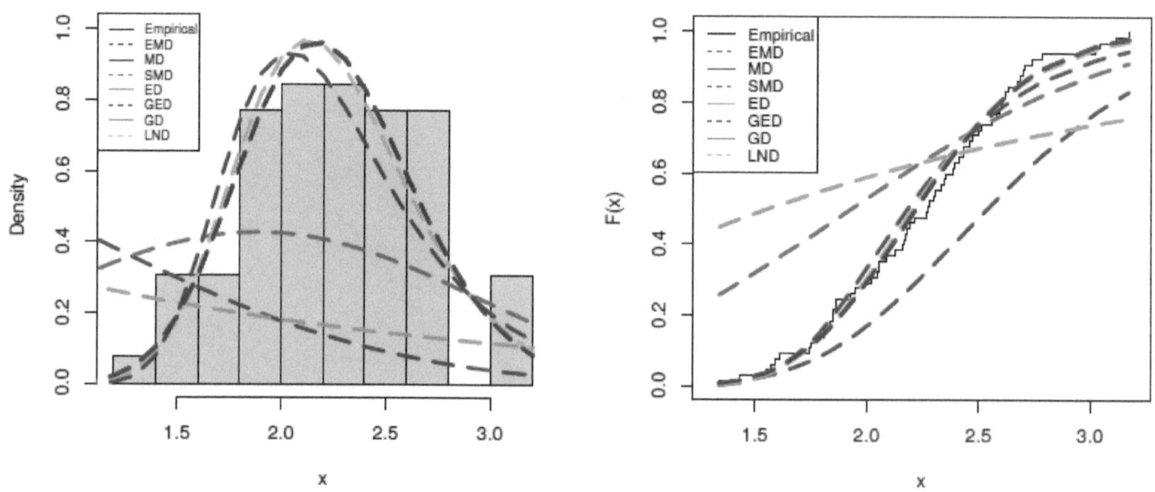

Figure 5: Estimated pdf and cdf plots of the fitted distribution of the carbon fibres data.

all the three data sets. Given the values of the test statistics and their associated p-values, we reject the null hypotheses for all data sets and conclude that the *EMD* model provides a significantly better representation of the distribution of these data sets than the *MD*.

7. Concluding remarks

The baseline model under-considered here was first coined by French biologist Teisser (1934). He introduced this model in order to study the mortality of animal species dying out of pure ageing, That is, not from accidents or disease. A modified version of the Teissier model was obtained and studied by Laurent (1975). Muth (1977) examined this model and found that it exhibited a heavier tail than the commonly used lifetime distributions like gamma, lognormal and Weibull. Using this model, Rinne (1981) estimated a German data set based on prices of used cars the lifetime distribution (with lifetime expressed in kilometres). Afterwards, Jodrá et al. (2015) termed it the Muth distribution and explored its significant properties through the Lambert W function. Through this study we introduced a more flexible form of the Muth distribution namely exponentiated Muth distribution by introducing an additional shape parameter β. Certain important distributional properties such as closed form expression for moments and moment generating function as well as analytical expression for some reliability measures such as vitality function and mrlf are obtained. Closed-form expressions for two recently developed uncertainty measures extropy and residual extropy are provided. Estimation of model parameters was established based on the maximum likelihood method and examined by simulation studies. For the three data sets, the posterior mode and posterior median estimates are computed as Bayesian estimates and compared with corresponding MLEs. Bayesian estimates are approximately equal to the MLEs. The dominance of the proposed model has been illustrated using certain real data sets where the LR test is also performed and it is concluded that it can be considered a good candidate for reliability analysis.

Aknowledgement

Thanks to the Editor and to the unidentified reviewers for their constructive comments, which greatly improved our article.

Conflicts of Interest

It is declared that the authors have no conflict of interest.

References

Bader, M. G. and Priest, A. M. (1982). Statistical aspects of fibre and bundle strength in hybrid composites. In Progress in Science and Engineering Composites, (Eds. Hayashi, T., Kawata, K. and Umekawa, S.), ICCM-IV: Tokyo, Japan, 1129–1136.

Biçer, C., Bakouch, H. S. and Biçer, H. D. (2021). Inference on Parameters of a Geometric Process with Scaled Muth Distribution. Fluctuation and Noise Letters, 20(01): 2150006.

Efron, B. and Tibshirani, R. (1991). Statistical data analysis in the computer age. Science, 253(5018): 390–395.

Gelman, A. and Hill, J. (2007). Data analysis using regression and multilevel/hierarchical models. Cambridge university press.

Irshad, M. R., Maya, R. and Arun, S. P. (2021). Muth distribution and estimation of a parameter using order statistics, Statistica (Accepted for publication).

Irshad, M. R., Maya, R., Sajeevkumar, N. K. and Arun, S. P. (2020). Estimation of the Scale Parameter of Power Muth Distribution by U-Statistics, Aligarh Jouranl of Statistics, 40: 76–96.

Jodrá, P., Jiménez-Gamero, M. D. and Alba-Fernández, M. V. (2015). On the Muth distribution. Mathematical Modelling and Analysis, 20(3): 291–310.

Jodrá, P., Gómez, H. W., Jiménez-Gamero, M. D. and Alba-Fernández, M. V. (2017). The power Muth distribution. Mathematical Modelling and Analysis, 22: 186–201.

Jodrá, P. and Arshad, M. (2021). An intermediate muth distribution with increasing failure rate. Communications in Statistics-Theory and Methods, 1–21.

Khan, N., Akhtar, M. T. and Khan, A. A. (2016). A Bayesian approach to survival analysis of inverse Gaussian model with Laplace approximation. International Journal of Statistics and Applications, 6(6): 391–398.

Lad, F., Sanfilippo, G. and Agrò, G. (2015). Extropy: Complementary dual of entropy. Statistical Science, 30: 40–58.

Laurent, A. G. (1975). Failure and mortality from wear and ageing- The Teissier model. A Modern Course on Statistical Distributions in Scientific Work, (Eds. Patil, G.P., Kotz, S. and Ord, J.K.), Springer, Heidelberg, 17: 301–320.

Maya, R. and Irshad, M. R. (2019). Kernel estimation of residual extropy function under -mixing dependence condition. South African Statistical Journal, 53(2): 65–72.

Milgram, M. S. (1985). The generalized integro-exponential function. Mathematics of computation, 44(170): 443–458.

Muth, E. J. (1977). Reliability models with positive memory derived from the mean residual life function. The Theory and Applications of Reliability, (Eds. Tsokos, C.P. and Shimi. I), New York: Academic Press, Inc., 401–435.

Qiu, G. and Jia, K. (2018). The residual extropy of order statistics. Statistics and Probability Letters, 133: 15–22.

Rinne, H. (1981). Estimating the lifetime distribution of private motor-cars using prices of used cars-the Teissier model. Statistiks Zwischen Theorie und Praxis, 172–184.

Smith, R. L. and Naylor, J. C. (1987). A comparison of maximum likelihood and Bayesian estimators for the three parameter Weibull distribution. Journal of the Royal statistical Society, Series C, 36(3): 358–369.

Teissier, G. (1934). Recherches sur le vieillissement et sur les lois de mortalite. Annales de Physiologie et de Physicochimie Biologique, 10: 237–284.

Appendix

$$\frac{\partial^2 \log l}{\partial \alpha^2} = \sum_{i=1}^{n} \{\frac{(1-\beta)[(1-\phi(z_i;\alpha)\}\phi(z_i;\alpha)\{\chi(z_i;\alpha)+\psi^2(z_i;\alpha)\}+\phi^2(z_i;\alpha)\psi^2(z_i;\alpha)]}{[1-\phi(z;\alpha)]^2} + \chi(z_i;\alpha) - \kappa(z_i;\alpha)\},$$

$$\frac{\partial^2 \log l}{\partial \beta^2} = -\frac{n}{\beta^2},$$

and

$$\frac{\partial^2 \log l}{\partial \alpha \partial \beta} = -\sum_{i=1}^{n} \frac{\phi(z_i;\alpha)\psi(z_i;\alpha)}{(1-\phi(z;\alpha))},$$

where,

$$\phi(z_i;\alpha) = e^{\left\{\alpha z_i \frac{1}{\alpha}(e^{\alpha z_i}-1)\right\}},$$

$$\psi(z_i;\alpha) = z_i - \frac{z_i e^{\alpha z_i}}{\alpha} + \frac{e^{\alpha z_i}}{\alpha^2} - \frac{1}{\alpha^2},$$

$$\chi(z_i;\alpha) = \frac{2z_i e^{\alpha z_i}}{\alpha^2} + \frac{z_i^2 e^{\alpha z_i}}{\alpha} - \frac{2e^{\alpha z_i}}{\alpha^3} + \frac{1}{\alpha^3}$$

and

$$\kappa(z_i;\alpha) = \frac{1+\alpha z_i^2 e^{\alpha z_i} - 2z_i e^{\alpha z_i}}{(e^{\alpha z_i}-\alpha)^2}.$$

Chapter 17

Exponentiated Discrete Modified Lindley Distribution and its Applications in the Healthcare Sector

Lishamol Tomy,[1] *Veena G*[2,*] and *Christophe Chesneau*[3]

1. Introduction

Statistical distributions that lie in the range [0, ∞] are widely used to explain and study real world data. Numerous traditional distributions and their generalizations have been widely used for studying data from a wide range of sectors in recent years, which include engineering, medical science, finance, human physiology, wind power, and reliability analysis.

Lifetimes must be quantified on a discrete basis instead of a continuous scale for certain scenarios. Such scenarios include the number of women who are working on shells for 5 weeks; the survival times in months of individuals infected with virus, are a few among many. Recent works in the area include, those of Eldeeb et al. (2021), Eliwa and El-Morshedy (2021), and Almetwally and Ibrahim (2020). Considering the numerous discrete distributions used in research, one may develop novel discrete distributions with different properties that appear to be suitable for a wide range of applications.

One such discrete distribution is the D-ML distribution introduced by Tomy et al. (2021). It is developed by discretizing the Modified Lindley distribution studied by Chesneau et al. (2021) via the survival discretization scheme. The distribution function (df) and the corresponding probability mass function (pmf) of the D-ML distribution can be expressed as follows:

$$W(y;\alpha) = 1-\alpha^{y+1} + \frac{\log\alpha}{1-\log\alpha}(y+1)\alpha^{2(y+1)};\, y \in N_0,$$

and

$$w(y;\alpha) = \frac{\alpha^y}{1-\log\alpha} \times [(1-\log\alpha)(1-\alpha) - \alpha^y \log\alpha(y-\alpha^2(y+1))];\, y \in N_0.$$

respectively, where N_0 is the set of whole numbers and $\alpha \in (0,1)$. One among the vital properties of this newly evolved model is that it can model not only positively skewed data sets, but it can also be utilized for modelling

[1] Department of Statistics, Deva Matha College, Kuravilangad, Kerala-686633, India.
[2] Department of Statistics, St.Thomas College, Palai, Kerala-686574, India.
[3] Université de Caen, LMNO, Campus II, Science 3, 14032, Caen, France.
Emails: lishatomy@gmail.com; christophe.chesneau@gmail.com
* Corresponding author: veenagpillai@hotmail.com

increasing, decreasing or unimodal failure rates. It can be noted that the distribution is efficient in modelling data compared to the discrete Lindley distribution, and other discrete compound Poisson distributions.

In this paper, we establish a novel count distribution with two parameters α and β, called the exponentiated discrete Modified Lindley (E-DML) distribution. The following are some of the features of this distribution: Both the reliability function (rf) and the hazard rate function (hrf) have closed forms. Furthermore, because its hrf may take on a variety of decreasing forms, the basic distribution's parameters can be changed to fit count data sets. Finally, the proposed E-DML distribution matches the count data the best, despite the fact that it only has two parameters. The E-DML distribution, we feel, is suitable for attracting a wide range of applications in disciplines such as medical, technology, and others.

The rest of the paper is laid out as follows: The origins of the E-DML distribution concept are discussed in Section 2. The moments associated with the E-DML distribution are studied in Section 3. The order statistics of the distribution, as well as the L-moment statistic, are investigated in Section 4. The maximum likelihood estimation technique is used to estimate the model parameters in Section 5. Section 6 discusses specific data set uses. Finally, the results are presented in Section 7.

2. The E-DML distribution

One of the most popular schemes used in generalizing distributions is the exponentiated-W with, df, $G(y;\alpha)$, when it comes to modelling lifetime data. Applying this approach, for $\beta > 0$, the df of exponentiated-G class of distributions is given by,

$$F(y;\alpha,\beta) = [G(y;\alpha)]^{\beta}.$$

For a detailed review on the exponentiated-G technique, we redirect the readers to Lehmann (1952).

A random variable (rv) Y is said to have the E-DML distribution with parameters α and β, if the df and pmf of the E-DML distribution are given by,

$$G_{E\text{-}DML}(y;\alpha,\beta) = \frac{\Delta(y+1;\alpha,\beta)}{(1-\log\alpha)^{\beta}},$$

and

$$g_{E\text{-}DML}(y;\alpha,\beta) = \frac{1}{(1-\log\alpha)^{\beta}}\,[\Delta(y+1;\alpha,\beta) - \Delta(y;\alpha,\beta)],$$

respectively, where $y \in N_0$ and $\Delta(y;\alpha,\beta) = [(1-\log\alpha)(1-\alpha^{y}) + \log\alpha(y\alpha^{2y})]^{\beta}$.
Figure 1 illustrates the plots of $g_{E\text{-}DML}(y;\alpha,\beta)$ for various values of parameters α and β.

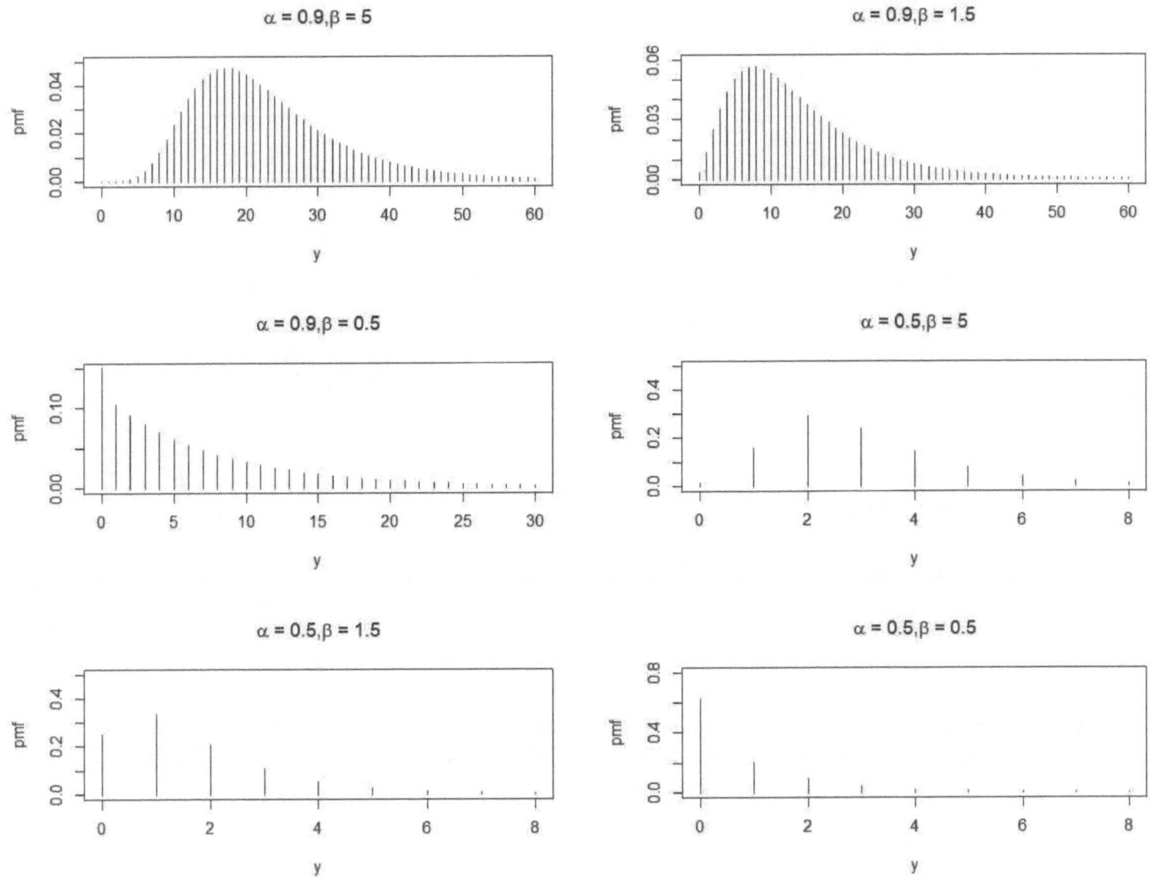

Figure 1: Shapes of $g_{E\text{-}DML}(y;\,\alpha,\,\beta)$ for selected values of α and β.

From the plots in Figure 1, as the pmf of the E-DML distribution is log-concave, thus, it can be deduced that the distribution is always unimodal. The distribution also has a long right tail for some values of α and β. The hrf can be defined as follows:

$$h_{E-DML}(y;\alpha,\beta)=\frac{g_{\text{E-DML}}(y;\alpha,\beta)}{R_{\text{E-DML}}(y;\alpha,\beta)}=\frac{\Delta(y+1;\alpha,\beta)-\Delta(y;\alpha,\beta)}{(1-\log\alpha)^{\beta}-\Delta(y+1;\alpha,\beta)};\mathrm{y}\in\mathrm{N}_0$$

where, $R_{E-DML}(y;\alpha,\beta)=\dfrac{(1-\log\alpha)^{\beta}-\Delta(y+1;\alpha,\beta)}{(1-\log\alpha)^{\beta}}$. .

In addition, the reversed hazard rate function (rhrf) is formulated as:

$$r_{E-DML}(y;\alpha,\beta)=1-\frac{\Delta(y;\alpha,\beta)}{\Delta(y+1;\alpha,\beta)};\ \mathrm{y}\in\mathrm{N}_0$$

Figure 2 shows the rhrf plots for varying parameter values of the E-DML distribution.

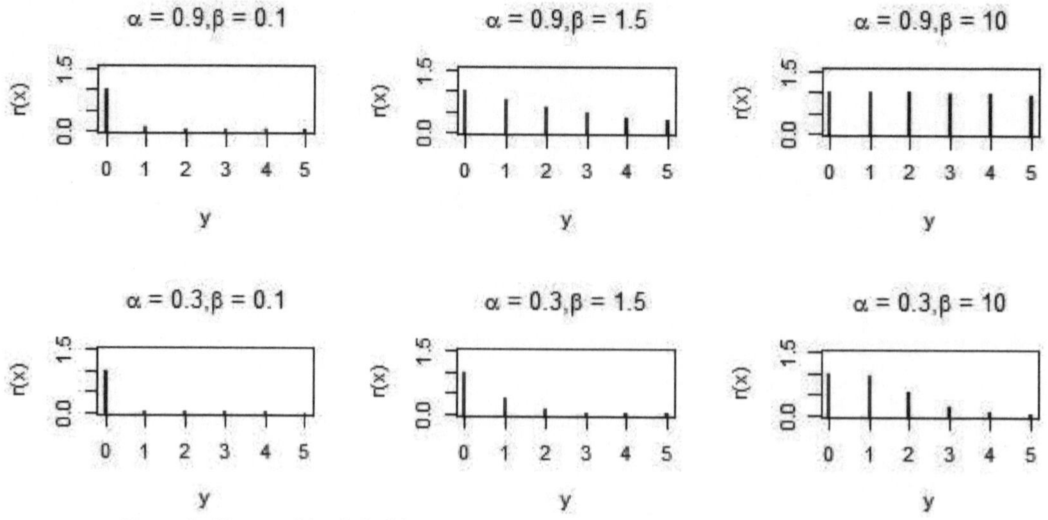

Figure 2: Shapes of the rhrf of the E-DML distribution for selected values of the parameters.

From Figure 2, we can observe that the rhrf is decreasing and constant in y for some values of the parameters.

3. Moments

In this section, the moments of the E-DML distribution are developed. Thanks to them, one can analyze the mean, distributional spread, symmetry, and peakedness of the distribution by determining its moments.
The rth moment of a rv Y following the E-DML distribution denoted by $\acute{\eta}_r$, is given as follows:

$$\acute{\eta}_r = E(Y^r) = \sum_{y=0}^{\infty} y^r g_{E\text{-}DML}(y;\alpha,\beta) = \frac{1}{(1-\log\alpha)^{\beta}} \sum_{y=0}^{\infty} y^r [\Delta(y+1;\alpha,\beta) - \Delta(y;\alpha,\beta)] \quad (1)$$

We recall that the reliability function of Y is defined by,

$$R_{E\text{-}DML}(y;\alpha,\beta) = \frac{(1-\log\alpha)^{\beta} - \Delta(y+1;\alpha,\beta)}{(1-\log\alpha)^{\beta}}.$$

The reliability function, $\acute{\eta}_r$ as given in Equation (1), can be defined as follows:

$$\acute{\eta}_r = \sum_{y=1}^{\infty} [y^r - (y-1)^r] R_{E\text{-}DML}(y;\alpha,\beta) = \frac{1}{(1-\log\alpha)^{\beta}} \sum_{y=1}^{\infty} [y^r - (y-1)^r][(1-\log\alpha)^{\beta} - \Delta(y+1;\alpha,\beta)]. \quad (2)$$

From Equation (2), the mean (μ) and variance (μ_2) of the E-DML distribution are as follows:

$$\mu = \frac{1}{(1-\log\alpha)^{\beta}} \sum_{y=1}^{\infty} [(1-\log\alpha)^{\beta} - \Delta(y+1;\alpha,\beta)]$$

and

$$\mu_2 = \frac{1}{(1-\log\alpha)^{\beta}} \sum_{y=1}^{\infty} [2y-1][(1-\log\alpha)^{\beta} - \Delta(y+1;\alpha,\beta)] - \mu^2.$$

The μ and μ_2 of the E-DML distribution can be calculated numerically since a closed form of the rth moment cannot be found. The metrics of central tendency and dispersion can be mathematically analysed using statistical tools.
Tables 1–2 list μ and μ_2 of the E-DML distribution as the parameters α and β are allowed to vary.

Table 1: Values of μ of the E-DML distribution for various values of α and β.

$\beta \downarrow$	$\alpha \rightarrow$	**0.1**	**0.2**	**0.3**	**0.4**	**0.5**	**0.6**
2		0.2249	0.5012	0.8432	1.282	1.8716	2.6770
3		0.3212	0.6848	1.1096	1.638	2.335	3.265
4		0.4083	0.8363	1.315	1.904	2.677	3.689
5		0.3212	0.9627	1.480	2.114	2.947	4.016

Table 2: Values of μ_2 of the E-DML distribution for various values of α and β.

$\beta \downarrow$	$\alpha \rightarrow$	**0.1**	**0.2**	**0.3**	**0.4**	**0.5**	**0.6**
2		0.2240	0.5051	0.8902	1.4771	2.413	3.790
3		0.2923	0.5929	0.9826	1.587	2.554	3.880
4		0.3402	0.6324	1.016	1.638	2.625	3.881
5		0.2923	0.6465	1.030	1.673	2.669	3.847

We formulate the following remarks:

- For a fixed β, μ and μ_2 increase as $\alpha \rightarrow 1$.
- The E-DML distribution is well-suited to model data that is either over or under-dispersed . Hence, the underlying distribution's parameters can be altered to fit most of the data sets.

The skewness (Sk) and kurtosis (Kurt) can be numerically attained as follows,

$$Sk = (\acute{\eta}_3 - 3\acute{\eta}_2\,\acute{\eta}_1 + \acute{\eta}_1^{3})/\mu_2^{3}$$

and

$$Kurt = (\acute{\eta}_4 - 4\acute{\eta}_2\,\acute{\eta}_1 - 3\acute{\eta}_2^{2} + 12\acute{\eta}_2\mu\acute{\eta}_1^{2} - 6\acute{\eta}_1^{4})/\mu_2^{2}.$$

Tables 3–4 report the Sk and Kurt of the E-DML distribution.

Table 3: Sk of the E-DML distribution for various values of α and β.

$\beta \downarrow$	$\alpha \rightarrow$	**0.1**	**0.2**	**0.3**	**0.4**	**0.5**	**0.6**
2		2.150	1.587	1.453	1.422	1.329	1.100
3		1.606	1.224	1.240	1.293	1.212	0.978
4		1.2711	1.047	1.1179	1.254	1.150	0.910
5		1.6065	0.9683	1.1751	1.232	1.102	0.867

Table 4: Kurt of the E-DML distribution for various values of α and β.

$\beta \downarrow$	$\alpha \rightarrow$	**0.1**	**0.2**	**0.3**	**0.4**	**0.5**	**0.6**
2		8.038	6.426	6.374	6.278	5.519	4.211
3		5.646	5.346	5.827	5.856	5.054	3.793
4		4.584	5.068	5.736	5.670	4.772	3.552
5		5.646	5.076	5.724	5.521	4.565	3.395

The following observations are made:

- The E-DML distribution has longer right tails for varying values of α and β.
- The Sk and Kurt of the distribution, decrease as $\beta \to \infty$ for a fixed value of α.

Following the E-DML distribution, the probability generating function of Y is defined by,

$$\omega_Y(q) = E(q^Y) = \sum_{y=0}^{\infty} q^y g_{E\text{-}DML}(y;\alpha,\beta)$$
$$= 1 + \frac{q-1}{(1-\log\alpha)^{\beta}} \sum_{y=1}^{\infty} q^{y-1}[(1-\log\alpha)^{\beta} - \Delta(y+1;\alpha,\beta)],$$

where, $q \in (-1,1)$. With the help of $\omega_Y(q)$, the mean can be obtained by differentiating $\omega_Y(q)$ once at $q = 1$, and the variance by using the following formula: $(d^2\omega_Y(q)/dq^2)|_{q=1} + (d\omega_Y(q)/dq)|_{q=1} - (d\omega_Y(q)/dq)|_{q=1}^2$.

4. Order statistics and L-moment statistics

Suppose $Y_1, Y_2..., Y_n$ be a random sample from the E-DML distribution, let $Y_{1:n}, Y_{2:n}, ..., Y_{n:n}$ be the relevant order statistics. Then, the df of $Y_{i:n,}$ can thus be represented as follows assuming an integer value of y:

$$G_{i:n}(y;\alpha,\beta) = \sum_{m=i}^{n} \binom{m}{k} [G_{E\text{-}DML}(y;\alpha,\beta)]^m [1 - G_{E\text{-}DML}(y;\alpha,\beta)]^{n-m}$$
$$= \sum_{m=i}^{n} \sum_{j=0}^{n-m} (-1)^j \binom{n}{m} \binom{n-m}{j} \frac{\Delta(y+1;\alpha,\beta(m+j))}{(1-\log\alpha)^{\beta(m+j)}}.$$

Furthermore, the pmf of $Y_{i:n}$ is given as follows:

$$g_{i:n}(y;\alpha,\beta) = \sum_{m=i}^{n} \sum_{j=0}^{n-m} (-1)^j \binom{n}{m} \binom{n-m}{j} \frac{\Delta(y+1;\alpha,\beta(m+j)) - \Delta(y;\alpha,\beta(m+j))}{(1-\log\alpha)^{\beta(m+j)}}.$$

and the lth moment of $Y_{i;n}$ is expressed as follows:

$$E(Y_{i:n}{}^l) = \sum_{y=0}^{\infty} \sum_{m=i}^{n} \sum_{j=0}^{n-m} (-1)^j \binom{n}{m} \binom{n-m}{j} y^l \frac{\Delta(y+1;\alpha,\beta(m+j)) - \Delta(y;\alpha,\beta(m+j))}{(1-\log\alpha)^{\beta(m+j)}}.$$

To summarise, theoretical distribution and actual samples, Hosking (1990) developed L-moments (LM). Additionally, it was also demonstrated that LM are a suitable indicator of the distribution's shape and can be used to fit distributions to data. These are the mean of certain mixtures of $Y_{i:n}$. The LM statistics of Y is given as,

$$\delta_t = \frac{1}{t} \sum_{l=0}^{t-1} (-1)^l \binom{t-1}{l} E(Y_{t-l:t}).$$

As LM statistics of Y were defined to be quantities, we may introduce some basic statistical metrics related to LM statistics for the E-DML distribution. These include the LM(μ) = δ_1, LM of coefficient of variation = δ_2/δ_1, LM coefficient of Sk = δ_3/δ_2 and LM coefficient of Kurt = δ_4/δ_2 can be formulated.

5. Estimation methods

We discuss the method of maximum likelihood estimation (MLE) in this section, to estimate the parameter vector, $\theta = (\alpha, \beta)$ of the E-DML model. Assuming $y_1, y_2, ..., y_n$ are random values from a sample of the E-DML distribution. The log-likelihood function (LL) can be expressed as,

$$LL(y;\alpha,\beta) = -n\beta\log(1-\log\alpha) + \sum_{i=1}^{n} \log[\Delta(y_i+1;\alpha,\beta) - \Delta(y_i;\alpha,\beta)]. \tag{4}$$

By means of the derivative of Equation (4) with respect to the parameters α and β, we obtain the normal nonlinear likelihood equations, which are as follows,

$$\frac{n\hat{\beta}}{\hat{\alpha}(1-\log\hat{\alpha})}+\hat{\beta}\sum_{i=1}^{n}\frac{[W_1(y_i+1;\hat{\alpha})]^{\hat{\beta}-1}W_2(y_i+1;\hat{\alpha})-[W_1(y_i;\alpha)]^{\hat{\beta}-1}W_2(y_i;\hat{\alpha})}{\Delta(y_i+1;\hat{\alpha},\hat{\beta})-\Delta(y_i;\hat{\alpha},\hat{\beta})}=0 \tag{5}$$

and

$$-n\log(1-\log\hat{\alpha})+\sum_{i=1}^{n}\frac{\Delta(y_i+1;\hat{\alpha},\hat{\beta})\log W_1(y_i+1;\hat{\alpha})-\Delta(y_i;\hat{\alpha},\hat{\beta})\log W_1(y_i;\hat{\alpha})}{\Delta(y_i+1;\hat{\alpha},\hat{\beta})-\Delta(y_i;\hat{\alpha},\hat{\beta})}=0 \tag{6}$$

respectively, where $W_1(y;\hat{\alpha}) = [\log(y\hat{\alpha}^{2y+1} + (1-\hat{\alpha}^y)(1-\log\hat{\alpha}))]$ and $W_2(y;\hat{\alpha}) = -y\hat{\alpha}^{y-1}(1-\log\hat{\alpha}-(1-\hat{\alpha}^{y-1}) + (2y+1)/\hat{\alpha})$.

The solutions of likelihood Equations (5) and (6) provide the MLEs of $\theta = (\alpha, \beta)^T$, say $\hat{\theta} = (\hat{\alpha}, \hat{\beta})^T$. This can be calculated using a mathematical approach like the Newton-Raphson method.

The E-DML distribution may be shown to satisfy the regularity criteria, which are satisfied by θ in their parameter space, but not on the boundary, as seen in Coy and Hinkley (1979). As a result, the MLE vector, $\hat{\theta}$ is stable and tends to a normal distribution as n tends to ∞.

6. Data analysis

The relevance of the E-DML distribution over other competitive distributions is highlighted in this section using a Healthcare data set. As an initial step towards data modelling,

- The MLEs and standard errors (se), are computed using the maxlogL function in R software.
- Model comparisons are made with the help of Akaike Information Criterion (AIC), Correct Akaike Information Criterion (AICC), chi-square (χ^2) and its p-value , computed in R.
- A visual representation of the data set with the estimated pmf is made.

The competitor models used to compare the E-DML distribution in this study are presented in Table 5.

Table 5: Models competing against the E-DML distribution.

Models	Abbreviations	References
Discrete Inverse Rayleigh	DLR	Hussain and Ahmad (2014)
Discrete Rayleigh	DR	Roy (2004)
Poisson	Poi	Poisson (1837)
Discrete Inverse Weibull	DIW	Jazi et al. (2010)
One parameter Discrete Lindley	DLi-I	Gomez-Déniz and Calderìn-Ojeda (2011)
Two parameter Discrete Lindley	DLi-II	Bakouch et al. (2014)
Three parameter Discrete Lindley	DLi-III	Eliwa et al. (2020)

The distribution having the highest p-value and lowest values of AIC and AICC is said to have the best fit compared to the other models.

The data comes from a study by Chan et al. (2010), who looked into the influence of a corticosteroid on cyst formation in mouse foetuses at University College London's Institute of Child Health. The kidneys of embryonic mice were cultivated, and a random sample was given steroids. Table 6 shows the number of cysts found in the kidneys after administering steroids.

Table 6: Number of cysts in kidneys.

Number	0	1	2	3	4	5	6	7	8	9	10	11
Observed Frequency	65	14	10	6	4	2	2	2	1	1	1	2

Table 7: MLE for number of cysts in kidneys

Model	MLE (se)
DIR	$\hat{\theta}$=0.554 (0.049)
DR	$\hat{\theta}$=0.901 (0.009)
Poi	$\hat{\alpha}$=1.390 (0.112)
DIW	$\hat{\alpha}$=1.049 (0.146); $\hat{\theta}$=0.581 (0.048)
DLi-I	$\hat{\alpha}$=0.436 (0.026)
DLi-II	$\hat{\alpha}$=0.581 (0.045); $\hat{\beta}$=0.001 (0.058)
DLi-III	$\hat{\alpha}$=0.582 (0.005); $\hat{\beta}$=358.728 (11863.37); $\hat{\theta}$=0.001 (20.698)
E-DML	$\hat{\alpha}$=0.775 (0.042); $\hat{\beta}$=0.232 (0.051)

Table 8: Goodness-of-fit metrics for the number of cysts in kidneys.

Model	-L	AIC	AICC	χ2	p-value
DIR	186.547	375.094	375.131	40.456	<0.001
DR	277.778	557.556	557.593	306.515	<0.001
Poi	246.210	494.420	494.457	89.277	<0.001
DIW	172.935	349.869	349.982	6.445	0.092
DLi-I	189.110	380.220	380.257	34.635	<0.001
DLi-II	178.767	361.534	361.646	19.091	0.0003
DLi-III	178.767	363.533	363.759	19.096	<0.0001
E-DML	167.192	338.385	338.550	6.398	0.7807

The MLE and se of the E-DML distribution and other competing distributions are reported in Table 7.

The goodness-of-fit metrics of the number of cysts in kidneys is reported in Table 8, using AIC, AICC, χ^2 and p-value.

From Table 8, we can see that the E-DML distribution has the highest p-value of 0.7807, with smaller values of AIC being 338.385, AICC being 338.550 and χ^2 value being 6.398. The histogram plot of the estimated pmf of the count data set of cysts in kidneys for the E-DML distribution is shown in Figure 3. Figure 3 illustrates the plot of the estimated pmf, which supports the result in Table 8.

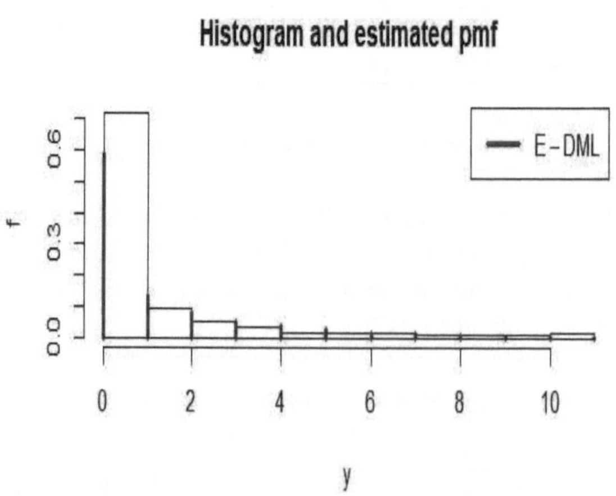

Figure 3: Histogram and estimated pmf of the number of cysts in kidneys using steroids.

7. Conclusion

A discrete modified Lindley-based two-parameter exponentiated model is proposed. In modelling data from many fields, it has been discovered that the new distribution is more versatile, has a longer right tail, and has a simpler shape than the parent distribution and the competitive distributions featured in the paper. There has been discussion of various distributional features. L-moments and order statistics are also investigated. The parameters are estimated using maximum likelihood estimation, and the distribution's relevance is compared to that of competing models using the goodness-of-fit approach. The importance is determined by the number of cysts found in the kidney. We discovered that the newly designed distribution is useful in data modelling and, unlike its counterpart models, could be used to model data from the medical sector.

References

Almetwally, E. M. and Ibrahim, G. M. (2020). Discrete Alpha Power Inverse Lomax Distribution with Application of COVID-19 Data. International Journal of Applied Mathematics, 9(6): 11–22.

Bakouch, H. S., Aghababaei, M. and Nadarajah, S. (2014). A new discrete distribution, Statistics, 48(1): 200–240. doi: 10.1080/02331888.2012.716677.

Chan, S. K., Riley, P. R., Price, K. L., McElduff, F., Winyard, P. J. et al. (2010). Corticosteroidinduced kidney dysmorphogenesis is associated with deregulated eypression of known cyst genic molecules, as well as Indian hedgehog, Am. J. Phys. Renal Phys., 298(2): 346–356. doi: 10.1152/ajprenal.00574.2009

Chesneau, C., Tomy, L. and Gillariose, J. (2021). A new modified Lindley distribution with properties and applications. Journal of Statistics and Management Systems. doi: 10.1080/09720570.2020.1824727.

Coy, D. R. and Hinkley, D. V. (1979). Theoretical statistics, Chapman and Hall/CRC press, London.

Eldeeb, A. S., Ahsan-Ul-Haq, M. and Babar, A. (2021). A Discrete Analog of Inverted Topp-Leone Distribution: Properties, Estimation and Applications. International Journal of Analysis and Applications, 19(5): 695–708. doi: 10.28924/2291-8639-19-2021-695.

Eliwa, M. S. and El-M orshedy, M. (2021). A one parameter discrete distributiion for over-dispersed data: statistical and reliability properties with applications. Journal of Applied Statistics, 1–21. doi: 10.1080/02664763.2021.1905787.

Eliwa, M. S., Altun, E., El-Dawoody, M. and El-Morshedy, M. (2020). A new three-parameter discrete distribution with associated INAR(1) process and applications, IEEE Access 8: 91150–91162. doi:10.1109/ACCESS.2020.2993593.

Gómez-Déniz, E. and Calderín-Ojeda, E. (2011). The discrete Lindley distribution: Properties and Applications. J. Stat. Comput. Simul., 81(11): 1405–1416. doi: 10.1080/00949655.2010.487825.

Hussain, T. and Ahmad, M. (2014). Discrete inverse Rayleigh distribution, Pakistan J. Statist., 30(2): 203–222.

Hosking, J. R. (1990). L-moments: Analysis and estimation of distributions using linear combinations of order statistics, J. R. Stat. Soc., 52: 105–124. doi: 10.1111/j.2517-6161.1990.tb01775.y.

Jazi, M. A., Lai, C. D. and Alamatsaz, M. H. (2010). A discrete inverse Weibull distribution and estimation of its parameters, Stat. Methodology, 7(2): 121–132. doi: 10.1016/j.stamet.2009.11.001.

Lehmann, E. L. (1952). The power of rank tests, Ann. Math. Stat., 24: 23–43.

Poisson, S. D. (1837). Probabilité des jugements en matiè re criminelle et en matière civile, précédées des règles générales du calcul des probabilitiés. Paris, France: Bachelier, 206–207.

Roy, D. (2004). Discrete Rayleigh distribution, IEEE Trans. Reliab., 53: 255–260. doi: 10.1109/TR.2004.829161.

Tomy, L., Veena, G. and Chesneau, C. (2021). The discrete Modified Lindley distribution. preprint.

Chapter 18

Length Biased Weighted New Quasi Lindley Distribution

Statistical Properties and Applications

Aafaq A Rather

1. Introduction

The study of weighted distributions is useful in distribution theory because it provides a new understanding of the existing standard probability distributions and methods for extending existing standard probability distributions for modeling lifetime data due to the introduction of additional parameters in the model which creates flexibility in them. Weighted distributions occur in modeling clustered sampling, heterogeneity, and extraneous variation in the dataset. The concept of weighted distributions was first introduced by Fisher (1934) to model ascertainment biases which were later formalized by Rao (1965) in a unifying theory for problems where the observations fell in a non-experimental, non-replicated and non-random manner. When observations are recorded by an investigator in nature according to certain stochastic models, the distribution of the recorded observations will not have the original distribution unless every observation is given an equal chance of being recorded. Weighted models were formulated in such situations to record the observations according to some weighted function. The weighted distribution reduces to a length biased distribution when the weight function considers only the length of the units. The concept of length biased sampling was first introduced by Cox (1969) and Zelen (1974). Warren (1975) was the first to apply the size biased distributions in connection with sampling wood cells. Patil and Rao (1978) studied weighted distributions and size biased sampling with applications to wildlife populations and human families. Van Deusen (1986) arrived at a size biased distribution theory independently and applied it in fitting assumed distributions to data arising from horizontal point sampling. More generally, when the sampling mechanism selects units with a probability proportional to some measure of unit size, the resulting distribution is called size-biased. There are various good sources which provide a detailed description of weighted distributions. Different authors have reviewed and studied the various weighted probability models and illustrated their applications in different fields. Weighted distributions are applied in various research areas related to reliability, biomedicine, ecology and branching processes. Afaq et al. (2016) have obtained the length biased weighted version of the Lomax distribution with properties and applications. Reyad et al. (2017) obtained the length biased weighted Frechet distribution with properties and estimation. Mudasir and Ahmad (2018) discussed the characterization and estimation of the length biased Nakagami distribution. Para and Jan (2018) introduced the Weighted Pareto

Department of Statistics, Faculty of Science and Technology, Vishwakarma University, Pune, India.
Email: aafaq7741@gmail.com

type II Distribution as a new model for handling medical science data and studied its statistical properties and applications. Rather and Subramanian (2019) obtained the length biased erlang truncated exponential distribution with lifetime data. Rather and Ozel (2020) discussed the weighted power Lindley distribution with application on real life time data. Hassan, Dar, Peer and Para (2019) obtained the weighted version of the Pranav distribution with real life data. Hassan, Wani and Para (2018) discussed the weighted three parameter quasi Lindley distribution with properties and applications. Ganaie, Rajagopalan and Rather (2019) discussed the length biased Aradhana distribution with applications. Recently, Ganaie, Rajagopalan and Rather (2020) discussed the weighted two parameter quasi Shanker distribution with its properties and applications, which shows more reliability and efficiency than the classical distribution.

Shanker and Ghebretsadik (2013) introduced a new quasi Lindley distribution, a newly proposed two parametric probability distribution and derived its various mathematical and statistical properties as moments, skewness, kurtosis, failure rate and mean residual life functions and stochastic ordering. It is observed that the expressions for the failure rate and mean residual life functions and stochastic ordering of the new quasi Lindley distribution show their flexibility over the Lindley and Exponential distributions and the quasi Lindley distribution. Also, the new quasi Lindley distribution is a particular case of a one parameter Lindley distribution. The parameter estimation is also discussed by using the methods of moments and the maximum likelihood estimation. The goodness of fit of the new quasi Lindley distribution has been fitted to a number of data sets related to survival times, grouped mortality data and waiting times to test its goodness of fit and it is observed that the new quasi Lindley distribution provides a closer fit than those of the Lindley and quasi Lindley distribution.

2. Length biased weighted new quasi lindley (LBWNQL) distribution

The probability density function of the new quasi Lindley distribution is given by,

$$f(x;\theta,\alpha)=\frac{\theta^2}{\theta^2+\alpha}(\theta+\alpha x)e^{-\theta x};x>0,\theta>0,\alpha<-\theta^2 \tag{1}$$

and the cumulative distribution function of the new quasi Lindley distribution is given by,

$$F(x;\theta,\alpha)=1-\frac{\theta^2+\beta+\theta\alpha x}{\theta^2+\alpha}e^{-\theta x};x>0,\theta>0,\alpha<-\theta^2 \tag{2}$$

Suppose X is a non-negative random variable with probability density function $f(x)$.Let $w(x)$ be the non-negative weight function, then, the probability density function of the weighted random variable X_w is given by, $f_W(x)=\dfrac{w(x)f(x)}{E(w(x)}$, $x>0$.
where $w(x)$ be a non negative weight function and $E(w(x))=\int w(x)f(x)dx<\infty$.

We should note that the different choices of the weight function $w(x)$ give different weighted distributions. When $w(x) = x^c$, the result is known as weighted distributions and when $w(x) = x$, the result is known as length biased distribution. In this paper, we have to obtain the length biased version of the new quasi Lindley distribution. The length biased weighted new quasi Lindley distribution is obtained by taking $c=1$ in the weights x^c to the distribution in order to obtain the length biased weighted distribution. Therefore, the probability density function of length biased weighted new quasi Lindley distribution is given by,

$$f_l(x;\theta,\alpha)=\frac{xf(x;\theta,\alpha)}{E(x)},x>0 \tag{3}$$

where,

$$E(x)=\int_0^\infty xf(x;\theta,\alpha)dx \tag{4}$$

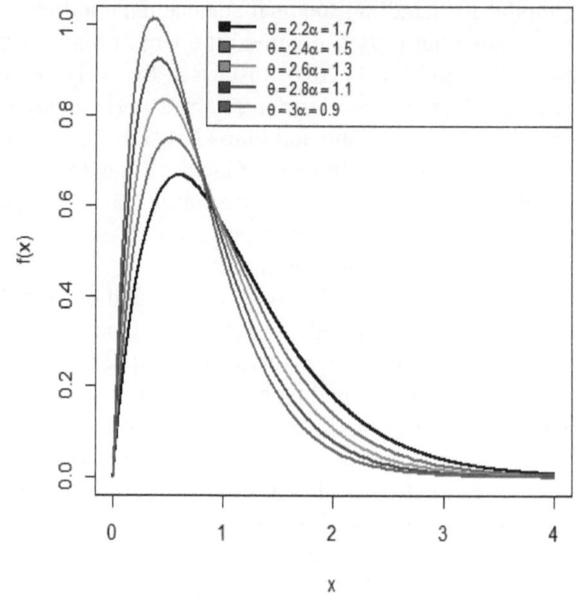

Figure 1: Pdf plot of length biased weighted new quasi Lindley distribution.

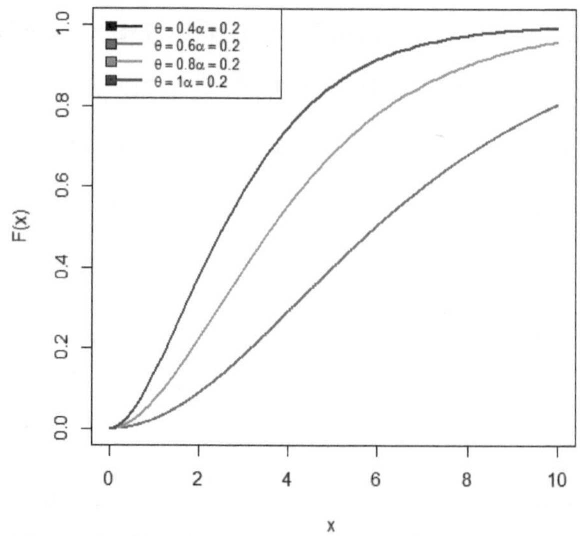

Figure 2: Cdf plot of length biased weighted new quasi Lindley distribution.

On substituting equations (1) and (4) in equation (3), we obtain the probability density function of the length biased weighted new quasi Lindley distribution,

$$f_l(x;\theta,\alpha)=\frac{x\theta^3}{(\theta^2+2\alpha)},(\theta+\alpha x)e^{-\theta x} \tag{5}$$

and the cumulative distribution function of the length biased weighted new quasi Lindley distribution is given by,

$$F_l\,(x;\theta,\alpha)=\int_0^x f_l\,(x;\,\theta,\,\alpha)dx$$

$$F_l(x;\theta,\alpha)=\int_0^x \frac{x\theta^3}{(\theta^2+2\alpha)},(\theta+\alpha x)e^{-\theta x}dx$$

$$F_l(x;\theta,\alpha)=\frac{\theta^3}{(\theta^2+2\alpha)},\int_0^x x(\theta+\alpha x)e^{-\theta x}dx \tag{6}$$

After the simplification of equation (6), we obtain the cumulative distribution function of the length biased weighted new quasi Lindley distribution,

$$F_l(x;\theta,\alpha)=\frac{1}{(\theta^2+2\alpha)},(\theta^2\gamma(2,\theta x)+\alpha\gamma(3,\theta x)) \tag{7}$$

3. Reliability analysis

In this sub section, we obtain the Reliability , hazard and Reverse hazard rate functions for the proposed length biased weighted new quasi Lindley distribution.

The reliability function or the survival function of the length biased weighted new quasi Lindley distribution is given by,

$$R(x)=1-F_l(x;\theta,\alpha)$$

$$R(x)=1-\frac{1}{(\theta^2+2\alpha)}(\theta^2\gamma(2,\theta x)+\alpha\gamma(3,\theta x))$$

The hazard function is also known as hazard rate or instantaneous failure rate or force of mortality and is given by,

$$h(x)=\frac{f_l(x;\theta,\alpha)}{R(x)}$$

$$h(x)=\frac{x\theta^3}{(\theta^2+2\alpha)-(\theta^2\gamma(2,\theta x)+\alpha\gamma(3,\theta x))}(\theta+\alpha x)e^{-\theta x}$$

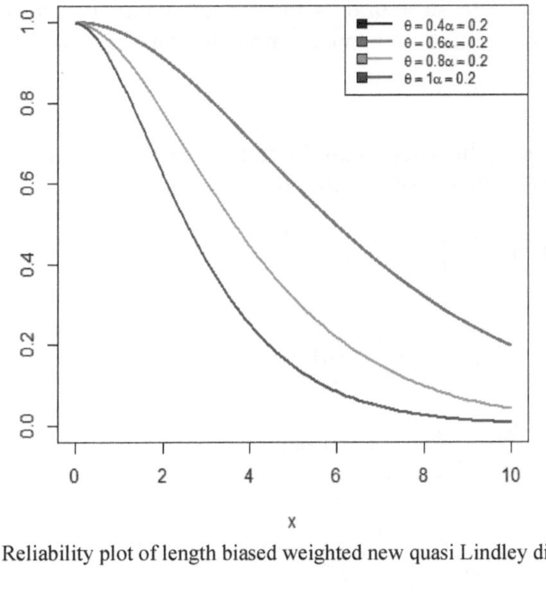

Figure 3: Reliability plot of length biased weighted new quasi Lindley distribution.

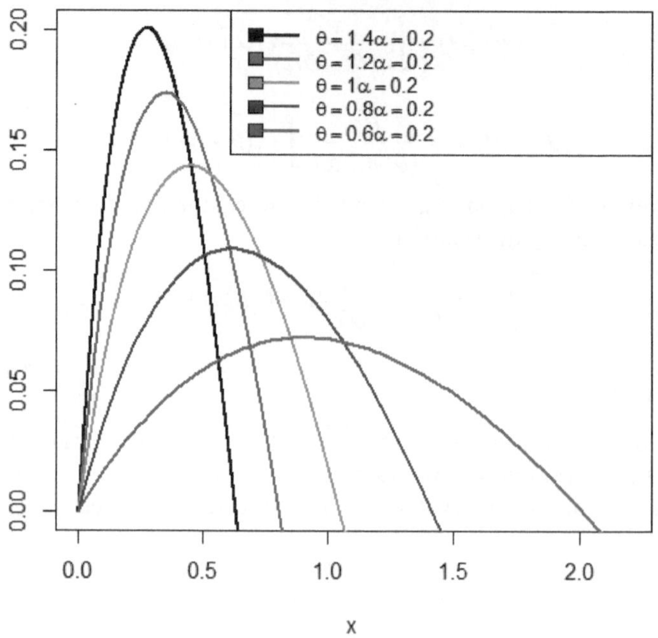

Figure 4: Hazard plot of length biased weighted new quasi Lindley distribution.

The reverse hazard function of the length biased weighted new quasi Lindley distribution is given by,

$$h_r(x) = \frac{f_l(x;\theta,\alpha)}{F_l(x;\theta,\alpha)}$$

$$h_r(x) = \frac{x\theta^3}{(\theta^2\gamma(2,\theta x) + \alpha\gamma(3,\theta x))}(\theta + \alpha x)e^{-\theta x}$$

4. Statistical properties

In this section we shall discuss the structural properties of length biased weighted new quasi Lindley distribution, especially its moments, harmonic mean, moment generating and characteristic functions.

4.1 Moments

Let X denote the random variable of the length biased weighted new quasi Lindley distribution with parameters θ and α, then the rth order moment $E(X^r)$ of the distribution is obtained as,

$$E(X^r) = \mu_r' \int_0^\infty x^r f_l(x;\theta,\alpha)dx$$

$$= \int_0^\infty x^r \frac{x\theta^3}{(\theta^2 + 2\alpha)}(\theta + \alpha x)e^{-\theta x}dx$$

$$= \frac{x\theta^3}{(\theta^2 + 2\alpha)}\int_0^\infty x^{r+1}(\theta + \alpha x)e^{-\theta x}dx$$

$$= \frac{x\theta^3}{(\theta^2 + 2\alpha)}\left(\theta\int_0^\infty x^{(r+2)-1}e^{-\theta x}dx + \alpha\int_0^\infty x^{(r+3)-1}e^{-\theta x}dx\right)$$

After simplification, we obtain,

$$E(X^r) = \mu_r' = \frac{\theta^2\Gamma(r+2) + \alpha\Gamma(r+3)}{\theta^r(\theta^2 + 2\alpha)} \tag{8}$$

Substituting, $r = 1, 2, 3, 4$ in equation (8), we obtain the first four moments of the distribution as,

$$E(X) = \mu_1' = \frac{2\theta^2 + 6\alpha}{\theta(\theta^2 + 2\alpha)}$$

$$E(X^2) = \mu_2' = \frac{6\theta^2 + 24\alpha}{\theta^2(\theta^2 + 2\alpha)}$$

$$E(X^3) = \mu_3' = \frac{24\theta^2 + 120\alpha}{\theta^3(\theta^2 + 2\alpha)}$$

$$E(X^4) = \mu_4' = \frac{120\theta^2 + 720\alpha}{\theta^4(\theta^2 + 2\alpha)}$$

$$\text{Variance}(\mu_2) = \frac{6\theta^2 + 24\alpha}{\theta^2(\theta^2 + 2\alpha)} - \left(\frac{2\theta^2 + 6\alpha}{\theta(\theta^2 + 2\alpha)}\right)^2$$

$$\text{Standard deviation } \sigma = \sqrt{\left(\frac{6\theta^2 + 24\alpha}{\theta^2(\theta^2 + 2\alpha)} - \frac{(2\theta^2 + 6\alpha)^2}{\theta^2(\theta^2 + 2\alpha)^2}\right)}$$

$$\text{Coefficient of variation} = \frac{\sigma}{\mu_1'} = \sqrt{\left(\frac{6\theta^2 + 24\alpha}{\theta^2(\theta^2 + 2\alpha)} - \frac{(2\theta^2 + 6\alpha)^2}{\theta^2(\theta^2 + 2\alpha)^2}\right)} \times \frac{\theta(\theta^2 + 2\alpha)}{(2\theta^2 + 6\alpha)}$$

4.2 Harmonic mean

The harmonic mean is the reciprocal of the arithmetic mean of the reciprocals. The harmonic mean for the proposed length biased weighted new quasi Lindley distribution is given by,

$$H.M = E\left(\frac{1}{x}\right) = \int_0^\infty \frac{1}{x} f_l(x;\theta,\alpha)dx$$

$$= \int_0^\infty \frac{x\theta^3}{(\theta^2 + 2\alpha)}(\theta + \alpha x)e^{-\theta x}$$

$$= \frac{x\theta^3}{(\theta^2 + 2\alpha)}\left(\theta\int_0^\infty e^{-\theta x}dx + \alpha\int_0^\infty xe^{-\theta x}dx\right)$$

$$= \frac{\theta^3}{(\theta^2 + 2\alpha)}\left(\theta\int_0^\infty x^{1-1}e^{-\theta x}dx + \alpha\int_0^\infty x^{2-1}e^{-\theta x}dx\right)$$

After simplification, we obtain,

$$H.M = \frac{x\theta^3}{(\theta^2 + 2\alpha)}(\theta\gamma(1,\theta x) + \alpha\gamma(2,\theta x))$$

4.3 Moment generating function

The moment generating function is the expected function of a continuous random variable and the moment generating function for the length biased weighted new quasi Lindley distribution is given by,

$$M_X(t) = E\left(e^{tx}\right) = \int_0^\infty e^{tx} f_l(x;\theta,\alpha)dx$$

$$= \int_0^\infty \left(1 + tx + \frac{(tx)^2}{2!} \cdots\right) f_l(x;\theta,\alpha)dx$$

$$= \int_0^\infty \sum_{j=0}^\infty \frac{t^j}{j!} x^j f_l(x;\theta,\alpha)dx$$

$$= \sum_{j=0}^\infty \frac{t^j}{j!} \mu_j'$$

$$= \sum_{j=0}^\infty \frac{t^j}{j!} \left(\frac{\theta^2 \Gamma(j+2) + \alpha\Gamma(j+3)}{\theta^j(\theta^2 + 2\alpha)}\right)$$

$$\Rightarrow M_X(t) = \frac{1}{(\theta^2 + 2\alpha)} \sum_{j=0}^\infty \frac{t^j}{j!} \left(\theta^2 \Gamma(j+2) + \alpha\Gamma(j+3)\right)$$

4.4 Characteristic function

The characteristic function is defined as the function of any real valued random variable and completely defines the probability distribution of a random variable and the characteristics function exists always even if the moment generating function does not exist. The characteristic function of the length biased weighted new quasi Lindley distribution is given by,

$$\varphi_X(t) = M_X(it)$$

$$\Rightarrow M_X(it) = \frac{1}{(\theta^2 + 2\alpha)} \sum_{j=0}^\infty \frac{t^j}{j!\theta^j} \left(\theta^2 \Gamma(j+2) + \alpha\Gamma(j+3)\right)$$

5. Order statistics

Let $X_{(1)}, X_{(2)}, \ldots, X_{(n)}$ denote the order statistics of a random sample $X_1, X_2, \ldots, X_n$ drawn from a continuous population with cumulative distribution function $F_x(x)$ and probability density function $fx(x)$, then the probability density function of the rth order statistics $X_{(r)}$ is given by,

$$fx(r)(x) = \frac{n!}{(r-1)!(n-r)!} f_X(x)(F_X(x))^{r-1}(1 - F_X(x))^r \tag{9}$$

for r = 1, 2… n.

Using equations (5) and (7) in equation (9), we obtain the probability density function of the rth order statistics of the length biased weighted new quasi Lindley distribution which is given by,

$$f_{x(r)}(x) = \frac{n!}{(r-1)!(n-r)!} \left(\frac{x\theta^3}{(\theta^2 + 2\alpha)}(\theta + \alpha x)e^{-\theta x}\right) \left(\frac{1}{(\theta^2 + 2\alpha)}(\theta^2 \gamma(2,\theta x) + \alpha\gamma(3,\theta x))\right)^{r-1}$$

$$\times \left(\frac{1}{(\theta^2 + 2\alpha)}(\theta^2 \gamma(2,\theta x) + \alpha\gamma(3,\theta x))\right)^{n-r}$$

Therefore, the probability density function of higher order statistics $X_{(n)}$ of the length biased weighted new quasi Lindley distribution is given by,

$$f_{x(n)}(x) = \frac{nx\theta^3}{(\theta^2+2\alpha)}(\theta+\alpha x)e^{-\theta x}\left(\frac{1}{(\theta^2+2\alpha)}(\theta^2\gamma(2,\theta x)+\alpha\gamma(3,\theta x))\right)^{n-1}$$

and the probability density function of the first order statistics $X_{(1)}$ of the length biased weighted new quasi Lindley distribution is given by,

$$f_{x(1)}(x) = \frac{nx\theta^3}{(\theta^2+2\alpha)}(\theta+\alpha x)e^{-\theta x}\left(1-\frac{1}{(\theta^2+2\alpha)}(\theta^2\gamma(2,\theta x)+\alpha\gamma(3,\theta x))\right)^{n-1}$$

6. Likelihood ratio test

Let $X_1, X_2, \ldots\ldots X_n$ be a random sample of size *n* from the new quasi Lindley distribution or length biased weighted new quasi Lindley distribution. We test the hypothesis,

$$H_o : f(x) = f(x;\theta,\alpha) \quad \text{against} \quad H_1 : f(x) = f_l(x;\theta,\alpha)$$

Thus for testing the hypothesis, whether the random sample of size *n* comes from new quasi Lindley distribution or length biased weighted new quasi Lindley distribution, the following test statistic is used.

$$\Delta = \frac{L_1}{L_o} = \prod_{i=1}^{n}\frac{f_l(x;\theta,\alpha)}{f(x;\theta,\alpha)}$$

$$\Delta = \frac{L_1}{L_o} = \prod_{i=1}^{n}\left(\frac{x_l\theta(\theta^2+\alpha)}{(\theta^2+2\alpha)}\right)$$

$$\Delta = \frac{L_1}{L_o} = \left(\frac{\theta(\theta^2+\alpha)}{(\theta^2+2\alpha)}\right)^n \prod_{i=1}^{n}x_i$$

We should reject the null hypothesis if,

$$\Delta = \left(\frac{\theta(\theta^2+\alpha)}{(\theta^2+2\alpha)}\right)^n \prod_{i=1}^{n}x_i > k$$

Equivalently, we reject the null hypothesis if,

$$\Delta^* = \prod_{i=1}^{n}x_i > k\left(\frac{(\theta^2+2\alpha)}{\theta(\theta^2+\alpha)}\right)^n$$

$$\Delta^* = \prod_{i=1}^{n}x_i > k^*, \text{Where } k^* = k\left(\frac{(\theta^2+2\alpha)}{\theta(\theta^2+\alpha)}\right)^n$$

Thus for a large sample of size *n*, *2log Δ* is distributed as a Chi-square distribution with one degree of freedom and the p- value is also obtained from it. Also, we reject the null hypothesis, when the probability value is given by,

$p(\Delta^* > \beta^*)$, Where $\beta^* = \prod_{i=1}^{n} x_i$ is less than a specified level of significance and $\prod_{i=1}^{n} x_i$ is the observed value of the statistic Δ^*.

7. Bonferroni and Lorenz curves

The Bonferroni and Lorenz curves are used not only in economics to study the distribution of income or wealth or income or poverty, but it is also being used in other fields like reliability, medicine, insurance and demography. The Bonferroni and Lorenz curves are given by,

$$B(p) = \frac{1}{p\mu_1'} \int_0^q x f_l(x;\theta,\alpha)dx$$

and $$L(p) = pB(p) = \frac{1}{\mu_1'} \int_0^q x f_l(x;\theta,\alpha)dx$$

Where $\mu_1' = E(X) = \dfrac{(2\theta^2 + 6\alpha)}{\theta(\theta^2 + 2\alpha)}$ and $q = F^{-1}(p)$

$$B(p) = \frac{\theta(\theta^2 + 2\alpha)}{p(2\theta^2 + 6\alpha)} \int_0^q \frac{\theta^3}{(\theta^2 + 2\alpha)} x^2 (\theta + \alpha x) e^{-\theta x} dx$$

$$B(p) = \frac{\theta(\theta^2 + 2\alpha)}{p(2\theta^2 + 6\alpha)} \frac{\theta^3}{(\theta^2 + 2\alpha)} \int_0^q x^2 (\theta + \alpha x) e^{-\theta x} dx$$

$$B(p) = \frac{\theta^4}{p(2\theta^2 + 6\alpha)} \left(\theta \int_0^q x^{3-1} e^{-\theta x} dx + \alpha \int_0^q x^{4-1} e^{-\theta x} dx \right)$$

After simplification, we obtain,

$$B(p) = \frac{\theta^4}{p(2\theta^2 + 6\alpha)} (\theta\gamma(3,\theta q) + \alpha\gamma(4,\theta q))$$

and $$B(p) = \frac{\theta^4}{p(2\theta^2 + 6\alpha)} (\theta\gamma(3,\theta q) + \alpha\gamma(4,\theta q))$$

8. Entropies

The concept of entropies is important in different areas such as probability and statistics, physics, communication theory and economics. Entropy is also called the degree of randomness or disorder in a system. Entropies quantify the diversity, uncertainty, or randomness of a system. Entropy of a random variable X is a measure of variation of the uncertainty.

8.1 Renyi entropy

The Renyi entropy is important in ecology and statistics as an index of diversity. The entropy is named after Alfred Renyi. The Renyi entropy is important in quantum information, where it can be used as a measure of entanglement. For a given probability distribution, Renyi entropy is given by,

$$e(\beta) = \frac{1}{1-\beta} \log\left(\int f_l^{\beta}(x;\theta,\alpha)dx \right)$$

where, $\beta > 0$ and $\beta \neq 1$

$$e(\beta) = \frac{1}{1-\beta} \log \int_0^\infty \left(\frac{x\theta^3}{(\theta^2 + 2\alpha)} (\theta + \alpha x) e^{-\theta x} \right)^\beta dx$$

$$e(\beta) = \frac{1}{1-\beta} \log \left(\left(\frac{\theta^3}{(\theta^2 + 2\alpha)} \right)^\beta \int_0^\infty x^\beta e^{-\theta\beta x} (\theta + \alpha x) e^\beta dx \right) \tag{10}$$

Using the Binomial expansion in (10), we obtain,

$$e(\beta) = \frac{1}{1-\beta} \log \left(\left(\frac{\theta^3}{(\theta^2 + 2\alpha)} \right)^\beta \sum_{j=0}^\infty \binom{\beta}{j} \theta^{\beta - j} (\alpha x)^j \int_0^\infty x^\beta e^{-\theta\beta x} dx \right)$$

$$e(\beta) = \frac{1}{1-\beta} \log \left(\left(\frac{\theta^3}{(\theta^2 + 2\alpha)} \right)^\beta \sum_{j=0}^\infty \binom{\beta}{j} \theta^{\beta - j} \alpha^j \int_0^\infty x^{(\beta + j + 1) - 1} e^{-\theta\beta x} dx \right)$$

$$e(\beta) = \frac{1}{1-\beta} \log \left(\left(\frac{\theta^3}{(\theta^2 + 2\alpha)} \right)^\beta \sum_{j=0}^\infty \binom{\beta}{j} \theta^{\beta - j} \alpha^j \frac{\Gamma(\beta + j + 1)}{(\theta\beta)^{\beta + j + 1}} \right)$$

8.2 Tsallis entropy

A generalization of Boltzmann-Gibbs (B.G) statistical properties initiated by Tsallis has focused a great deal on attention. This generalization of B-G statistics was proposed first by introducing the mathematical expression of Tsallis entropy (Tsallis, 1988) for a continuous random variable , defined as follows,

$$S_\lambda = \frac{1}{\lambda - 1} \left(- \int_0^\infty f_l^\lambda (x; \theta, \alpha) dx \right)$$

$$S_\lambda = \frac{1}{\lambda - 1} \left(1 - \int_0^\infty \left(\frac{x\theta^3}{(\theta^2 + 2\alpha)} (\theta + \alpha x) e^{-\theta x} \right)^\lambda dx \right) \tag{11}$$

$$S_\alpha = \frac{1}{\lambda - 1} \left(1 - \left(\frac{\theta^3}{(\theta^2 + 2\alpha)} \right)^\lambda \int_0^\infty x^\lambda e^{-\lambda\theta x} (\theta + \alpha x)^\lambda dx \right)$$

Using the Binomial expansion in equation (11), we get,

$$S_\lambda = \frac{1}{\lambda - 1} \left(1 - \left(\frac{\theta^3}{(\theta^2 + 2\alpha)} \right)^\lambda \sum_{j=0}^\infty \binom{\lambda}{j} \theta^{\lambda - j} (\alpha x)^j \int_0^\infty x^\lambda e^{-\lambda\theta x} dx \right)$$

$$S_\lambda = \frac{1}{\lambda - 1} \left(1 - \left(\frac{\theta^3}{(\theta^2 + 2\alpha)} \right)^\lambda \sum_{j=0}^\infty \binom{\lambda}{j} \theta^{\lambda - j} \alpha^j \int_0^\infty x^{(\lambda + j + 1) - 1} e^{-\lambda\theta x} dx \right)$$

$$S_\lambda = \frac{1}{\lambda - 1} \left(1 - \left(\frac{\theta^3}{(\theta^2 + 2\alpha)} \right)^\lambda \sum_{j=0}^\infty \binom{\lambda}{j} \theta^{\lambda - j} \alpha^j \frac{\Gamma(\lambda + j + 1)}{(\lambda\theta)^{\lambda + j + 1}} \right)$$

9. Maximum likelihood estimation and Fisher's information matrix

In this section, we will discuss the maximum likelihood estimation for estimating the parameters of length biased weighted new quasi Lindley distribution and also its Fisher's information matrix. Let $X_1, X_2, \ldots, X_n$ be a random sample of size n from the length biased weighted new quasi Lindley distribution, then the likelihood function of the length biased weighted new quasi Lindley distribution is given by,

$$L(x;\theta,\alpha) = \prod_{i=1}^{n} f_l(x;\theta,\alpha)$$

$$L(x;\theta,\alpha) = \prod_{i=1}^{n}\left(\frac{x_i\theta^3}{(\theta^2+2\alpha)}(\theta+\alpha x)e^{-\theta x_i}\right)$$

$$L(x;\theta,\alpha) = \frac{\theta^{3n}}{(\theta^2+2\alpha)^n}\prod_{i=1}^{n}\left(x_i(\theta+\alpha x_i)e^{-\theta x_i}\right)$$

The log likelihood function is given by,

$$\log L(x;\theta,\alpha) = 3n\log\theta - n\log(\theta^2+2\alpha) + \sum_{i=1}^{n}\log x_i + \sum_{i=1}^{n}\log(\theta+\alpha x_i) - \theta\sum_{i=1}^{n}x_i \tag{12}$$

Differentiating the log likelihood equation (12) with respect to θ and α and equating it to zero, we obtain the normal equations,

$$\frac{\partial \log L}{\partial\theta} = \frac{3n}{\theta} - n\left(\frac{2\theta}{(\theta^2+2\alpha)}\right) + \sum_{i=1}^{n}\left(\frac{1}{(\theta+\alpha x_i)}\right) - \sum_{i=1}^{n}x_i = 0$$

$$\frac{\partial \log L}{\partial\alpha} = -n\left(\frac{2}{(\theta^2+2\alpha)}\right) + \frac{\sum_{i=1}^{n}x_i}{(\theta+\alpha x_i)} = 0$$

Because of the complicated form of the likelihood equations, algebraically it is very difficult to solve the system of non-linear equations. Therefore we use R and wolfram mathematics for estimating the required parameters of the proposed distribution.

To obtain the confidence interval we use the asymptotic normality results. If $\hat{\lambda} = (\hat{\theta},\hat{\alpha})$ denotes the MLE of $\lambda = (\theta,\alpha)$ we can state the results as follows:

$$\sqrt{n}(\hat{\lambda}-\lambda) \to N_2(0, I^{-1}(\lambda)$$

where $I(\lambda)$ is the Fisher's Information matrix .i.e.,

$$I(\lambda) = -\frac{1}{n}\begin{pmatrix} E\left(\frac{\partial^2 \log L}{\partial\theta^2}\right) & E\left(\frac{\partial^2 \log L}{\partial\theta\,\partial\alpha}\right) \\ E\left(\frac{\partial^2 \log L}{\partial\alpha\,\partial\theta}\right) & E\left(\frac{\partial^2 \log L}{\partial\alpha^2}\right) \end{pmatrix}$$

where,

$$E\left(\frac{\partial^2 \log L}{\partial\theta^2}\right) = -\frac{3n}{\theta^2} - n\left(\frac{2(\theta^2+2\alpha)-4\theta^2}{(\theta^2+2\alpha)^2}\right) - \sum_{i=1}^{n}\left(\frac{1}{(\theta+\alpha x_i)^2}\right)$$

$$E\left(\frac{\partial^2 \log L}{\partial \alpha^2}\right) = n\left(\frac{4}{(\theta^2 + 2\alpha)^2}\right) - \sum_{i=1}^{n}\left(\frac{E(x_i^{\,2})}{(\theta + \alpha x_i)^2}\right)$$

$$E\left(\frac{\partial^2 \log L}{\partial \theta\, \partial \alpha}\right) = E\left(\frac{\partial^2 \log L}{\partial \alpha\, \partial \theta}\right) = n\left(\frac{4\theta}{(\theta^2 + 2\alpha)^2}\right) - \sum_{i=1}^{n}\left(\frac{E(x_i)}{(\theta + \alpha x_i)^2}\right)$$

Since λ being unknown, we estimate $I^{-1}(\lambda)$ by $I^{-1}(\hat{\lambda})$ and this can be used to obtain asymptotic confidence intervals for θ and α.

10. Applications

In this section, here we analyse and evaluate two real life data sets for fitting length biased weighted new quasi Lindley distribution and the model has been compared with the new quasi Lindley, quasi Lindley, Lindley and exponential distributions. In order to show that the length biased weighted new quasi Lindley distribution is better than the new quasi Lindley, quasi Lindley, Lindley and exponential distributions, the results obtained from the two real life data sets are used. The two real life data sets are given below as:

Data set 1: The first data set denotes the time to failure of turbocharger (*103h*) of one type of engine studied by Xu et al. (2003). The first data set is given as follows:

1.6, 8.4, 8.1, 7.9, 3.5, 2, 8.4, 8.3, 4.8, 3.9, 2.6, 8.5, 5.4, 5, 4.5, 3, 6.0, 5.6, 5.1, 4.6, 6.5, 6.1, 5.8, 5.3, 7, 6.5, 6.3, 6, 7.3, 7.1, 6.7, 8.7, 7.7, 7.3, 7.3, 8.8, 8, 7.8, 7.7, 9

Data set 2: The second data set represents 40 patients suffering from blood cancer (leukemia) from one of the ministry of Health Hospitals in Saudi Arabia (see Abouammah et al. 2000).The ordered lifetimes (in years) are given as follows:

0.315, 0.496, 0.616, 1.145, 1.208, 1.263, 1.414, 2.025, 2.036, 2.162, 2.211, 2.37, 2.532, 2.693, 2.805, 2.91, 2.912, 2.192, 3.263, 3.348, 3.348, 3.427, 3.499, 3.534, 3.767, 3.751, 3.858, 3.986, 4.049, 4.244, 4.323, 4.381, 4.392, 4.397, 4.647, 4.753, 4.929, 4.973, 5.074, 5.381

Table 1: Shows maximum likelihood estimates, corresponding Standard errors, criterion values AIC, BIC, AICC and –2logL and comparison of fitted distribution .

Data Set 1						
Distribution AICC	**MLE**	**S.E**		**–2logL**	**AIC**	**BIC**
LBWNQL	$\hat{\alpha} = 3.6391$ $\hat{\theta} = 4.7974$	$\hat{\alpha} = 4.8440$ $\hat{\theta} = 4.3797$	189.011	193.011	196.3888	193.3353
NQL	$\hat{\alpha} = 3.3626$ $\hat{\theta} = 3.1980$	$\hat{\alpha} = 3.3560$ $\hat{\theta} = 3.5756$	201.0361	205.0361	208.4138	205.3604
QL	$\hat{\alpha} = 0.0010$ $\hat{\theta} = 0.3197$	$\hat{\alpha} = 0.2123$ $\hat{\theta} = 0.0311$	201.0589	205.0589	208.4367	205.3832
Lindley	$\hat{\theta} = 0.2844$	$\hat{\theta} = 0.0321$	208.5708	210.5708	212.2597	210.6760
Exponential	$\hat{\theta} = 0.1599$	$\hat{\theta} = 0.0252$	226.6385	228.6385	230.3274	228.7437

Table 2: Shows maximum likelihood estimates, corresponding Standard errors, criterion values AIC, BIC, AICC and -2logL and comparison of fitted distribution.

Data Set 2						
Distribution AICC	**MLE**	**S.E**	**–2logL**		**AIC**	**BIC**
LBWNQL	$\hat{\alpha} = 8.9514$ $\hat{\theta} = 0.9402$	$\hat{\alpha} = 38.4471$ $\hat{\theta} = 0.1044$	147.461	151.461	154.8388	151.7853
NQL	$\hat{\alpha} = 1.6173$ $\hat{\theta} = 6.3670$	$\hat{\alpha} = 1.6778$ $\hat{\theta} = 7.1183$	152.7528	156.7528	160.1306	157.0771
QL	$\hat{\alpha} = 0.0010000$ $\hat{\theta} = 0.6365481$	$\hat{\alpha} = 0.5536723$ $\hat{\theta} = 0.1545506$	152.766	156.766	160.1437	157.0903
Lindley	$\hat{\theta} = 0.5269$	$\hat{\theta} = 0.0607$	160.5012	162.5012	164.19	162.6064
Exponential	$\hat{\theta} = 0.3183$	$\hat{\theta} = 0.0503$	171.5563	173.5563	175.2452	173.6615

R software is used to carry out the numerical analysis of two data sets and is also used for estimating the unknown parameters and model comparison criterion values. In order to compare the length biased weighted new quasi Lindley distribution with new quasi Lindley, quasi Lindley, Lindley and exponential distributions, we consider the criterion values like AIC (Akaike information criterion), AICC (corrected Akaike information criterion) and BIC (Bayesian information criterion). The better distribution corresponds to lesser values of AIC, BIC, AICC and –2logL. The formulas for calculation of AIC, AICC and BIC values are,

$$AIC = 2k - 2\log L \qquad AICC = AIC + \frac{2k(k+1)}{n-k-1} \quad \text{and} \quad BIC = k\log n - 2\log L$$

where k is the number of parameters in the statistical model, n is the sample size and -2log$L$ is the maximized value of the log-likelihood function under the considered model.

From Tables 1 and 2 given above, it has been observed that the length biased weighted new quasi Lindley distribution have the lower AIC, AICC, BIC and –2logL values as compared to the new quasi Lindley, quasi Lindley, Lindley and exponential distributions. Hence, we conclude that the length biased weighted new quasi Lindley distribution leads to a better fit than the new quasi Lindley, quasi Lindley, Lindley and exponential distributions.

11. Conclusion

In the present study, we have introduced the length biased weighted new quasi Lindley distribution as a new generalization of the new quasi Lindley distribution. The subject distribution is generated by using the length biased technique and taking the two parameter new quasi Lindley distribution as the base distribution. The different mathematical and statistical properties of the newly executed distribution along with the reliability measures are discussed. The parameters of the proposed distribution are obtained by using the methods of maximum likelihood estimator and the Fisher's information matrix which have been discussed. Finally, the application of the new distribution has also been illustrated by demonstrating with two real life data sets. The results of the two data sets are used by comparing the length biased weighted new quasi Lindley distribution to the new quasi Lindley, quasi Lindley, Lindley and exponential distributions and the results indicate that the length biased weighted new quasi Lindley distribution provides a better fit than the new quasi Lindley, quasi Lindley, Lindley and exponential distributions.

References

Abouammoh, A. M., Ahmed, R. and Khalique, A. (2000). On new renewal better than used classes of life distribution. Statistics and Probability Letters, 48: 189–194.

Afaq, A., Ahmad, S. P. and Ahmed, A. (2016). Length-Biased weighted Lomax distribution: Statistical properties and applications. Pak.j.Stat.Oper.res, 12(2): 245–255.

Cox, D. R. (1969). Some sampling problems in technology in New Development in Survey Sampling. Johnson, N. L. and Smith, H., Jr. (eds.) New York Wiley- Interscience, 506–527.

Fisher, R. A. (1934). The effects of methods of ascertainment upon the estimation of frequencies. Ann. Eugenics, 6: 13–25.

Ganaie, R. A., Rajagopalan, V. and Rather, A. A. (2019). A New Length Biased distribution With Applications. Science, Technology and Development, 8(8): 161–174.

Ganaie, R. A., Rajagopalan, V. and Rather, A. A. (2020). On weighted two parameter quasi Shanker distribution with properties and its applications. International Journal of Statistics and Reliability Engineering, 7(1): 1–12.

Hassan, A., Wani, S. A. and Para, B. A. (2018). On three Parameter Weighted Quasi Lindley Distribution: Properties and Applications. International journal of scientific research in mathematical and statistical sciences, 5(5): 210–224.

Hassan, A., Dar, M. A., Peer, B. A. and Para, B. A. (2019). A new generalization of Pranav distribution using weighted technique. International Journal of Scientific Research in Mathematical and Statistical Sciences, 6(1): 25–32.

Mudasir, S. and Ahmad, S. P. (2018). Characterization and estimation of length biased Nakagami distribution. Pak.j.stat.oper. res., 14(3): 697–715.

Para, B. A. and Jan, T. R. (2018). On three Parameter Weighted Pareto Type II Distribution: Properties and Applications in Medical Sciences. Applied Mathematics and Information Sciences Letters, 6 (1): 13–26.

Patil, G. P. and Rao, C. R. (1978). Weighted Distributions and Size-Biased Sampling with Applications to Wildlife Populations and Human Families. Biometrics, 34(2): 179.doi:10.2307/2530008.

Reyad, M. H., Hashish, M. A., Othman, A. S. and Allam, A. S. (2017). The length-biased weighted frechet distribution: properties and estimation. International Journal of Statistics and Applied Mathematics, 3(1): 189–200.

Rao, C. R. (1965). On discrete distributions arising out of methods of ascertainment *In*: Patil, G. P. (eds.). Classical and Contagious Discrete Distributions. Statistical Publishing Society, Calcutta, 320–332.

Rather, A. A. and Subramanian, C. (2019). Length biased erlang truncated exponential distribution with lifetime data. Journal of Information and Computational Science, 9(8): 340–355.

Rather, A. A. and Ozel, G. (2020). The weighted power Lindley distribution with applications on the life time data. Pak.j.stat. oper.res., 16(2): 225–237.

Shanker, R. and Ghebretsadik, A. H. (2013). A New Quasi Lindley distribution. International Journal of Statistics and Systems, 8(2): 143–156.

Tsallis, C. (1988). Possible generalization of Boltzmann-Gibbs statistics. J. Stat. Phys., 52: 479–487. https://doi.org/10.1007/BF01016429.

Van Deusen, P. C. (1986). Fitting assumed distributions to horizontal point sample diameters. Forest Science, 32: 146–148.

Warren, W. (1975). Statistical distributions in forestry and forest products research. *In*: Patil, G. P., Kotz, S. and Ord, J. K. (eds.). Statistical Distributions in Scientific Work. D. Reidel, Dordrecht, The Netherlands, 2: 369–384.

Xu, K., Xie, M., Tang, L. C. and Ho, S. L. (2003). Application of neural networks in forecasting engine systems reliability. Applied Soft Computing, 2(4): 255–268.

Zelen, M. (1974). Problems in cell kinetic and the early detection of disease, in Reliability and Biometry, F. Proschan & R. J. Sering, eds, SIAM, Philadelphia, 701–706.

Chapter 19

A New Alpha Power Transformed Weibull Distribution

Properties and Applications

Ashok Kumar Pathak,[1] *Mohd Arshad,*[2,*] *Sanjeev Bakshi,*[3]
Mukti Khetan[4] and *Sherry Mangla*[5]

1. Introduction

Exponential and Rayleigh distributions are two important commonly used lifetime distributions in modeling lifetime data with broad applications in reliability and survival analysis. Due to constant and increasing failure rate functions, these distributions have limited applications in a wide class of real data where complexity arises in it. Applicability of statistical distributions in modelling real-world phenomena attracts statisticians to construct new flexible families of distributions with applications in the diverse areas of the applied sciences like reliability, engineering, medicine, energy, finance and insurance. The Linear exponential (LE) distribution includes exponential and Rayleigh distributions as two important sub-models for modelling lifetime data that encounter linearly increasing failure rates. However, the LE distribution does not provide a good fit to the data that arises in reliability analysis and biological studies, where hazard rates decrease, non-linearity increases, and bathtub shape behavior is observed. Several generalizations of the exponential, Rayleigh, and LE distributions have been studied in the recent past. For some useful references, one may refer to Gupta and Kundu (1999), Mahmoud and Alam (2010), Sarhan et al. (2013), Tian et al. (2014), Khan et al. (2017), Pathak et al. (2021).

The Weibull distribution is one of the most popular lifetime distributions with diverse applications in different disciplines (see Mudholkar and Srivastava (1993), Lai (2014)). We say that a random variable X has a Weibull distribution with parameters β and γ if its cumulative distribution function (CDF) and probability density function (PDF) are given by,

$$G_X(x) = 1 - e^{-\beta x^{\lambda}}, x > 0$$

and

$$g_X(x) = \beta\lambda x^{\lambda-1}\, e^{-\beta x^{\lambda}}, x > 0,$$

where $\beta > 0$ and $\lambda > 0$.

[1] Department of Mathematics and Statistics, Central University of Punjab, Bathinda, India.
[2] Department of Mathematics, Indian Institute of Technology Indore, Simrol, Indore, India.
[3] Department of Statistics, Indira Gandhi National Tribal University, Amarkantak, India.
[4] Department of Mathematics, Amity University Mumbai, Maharashtra, India.
[5] International Institute for Population Sciences, Mumbai, Maharashtra, India.
* Corresponding author: arshad.iitk@gmail.com

The Weibull distribution is a natural extension of the exponential and Rayleigh distributions. The hazard rate function of the Weibull distribution can be increasing, decreasing, or constant depending on its choice of parameters and may be used as an alternative to the exponential and Rayleigh models in lifetime data modelling. But this model may not be useful in real data with bathtub hazard rate functions. Several generalizations have been proposed by adopting some new state-of-the-art techniques to overcome the arising issues with the Weibull distribution. Some important methodologies involve linear, log and inverse transformations, enhancing the number of shape and location parameters, and power transformation of the Weibull distribution. Apart from these approaches, some new families of the generalized Weibull distribution have also been proposed by mixing two or more Weibull distributions to enhance the applicability of the model in dealing with lifetime data. Mudholkar and Srivastava (1993) proposed a three-parameter exponentiated Weibull distribution which includes generalized exponential and Weibull distributions as sub-models. This model can accommodate bathtub-shaped, unimodal, monotonically increasing, and monotonically decreasing hazard rates. A three-parameter extended Weibull distribution has also been studied by Marshall and Olkin (1997). Xie et al. (2002) considered a new extension of the Weibull distribution and estimated the model parameters using graphical techniques. Several families in these generalizations of the Weibull distributions include two or three parameters in the model and provide flexibility in the models. For a good source of literature review of these models, one may refer to Pham and Lai (2007) and Lai (2014).

Apart from these extensions, Weibull models with four and five parameters are also studied in the literature. Xie and Lai (1996) introduced a four-parameter additive Weibull distribution by mixing two Weibull survival functions, with one having an increasing and the other a decreasing failure rate function and discussed its parameter estimation. This distribution is flexible in modelling lifetime data with bathtub hazard rate. A four-parameter beta Weibull distribution was studied by Famoye et al. (2005). Another new four-parameter generalization of the Weibull distribution using a power transformation is proposed by Carrasco et al. (2008). Sarhan and Zaindin (2009) proposed a new four-parameter modified Weibull distribution and demonstrated its statistical properties. Recently, some other modified Weibull distributions with five parameters have been proposed and studied by Silva et al. (2010), Nadarajah et al. (2011), Sarhan et al. (2013), Almalki and Yuan (2013), He et al. (2016), and Abd EL-Baset and Ghazal (2020).

The main aim of this chapter is to introduce a three-parameter new alpha power transformed Weibull distribution and study its various important statistical properties. We denote the new alpha power transformed Weibull distribution by the NAPTW distribution. The NAPTW family includes a large class of well-known distributions and their generalizations, including exponential, Rayleigh, new alpha power transformed exponential (NAPTE) distributions, and more. The hazard rate function of the NAPTW distribution takes different shapes. It can be used in the analysis of a wide class of lifetime data.

The organization of the chapter is as follows: In Section 2, we present a new alpha power transformed Weibull (NAPTW) distribution and deduce some known families of the distributions and their extensions from it. We present the expressions for survival function, hazard rate function, and moments for the NAPTW distribution. Then, we numerically tabulate several measures of descriptive statistics for the NAPTW family. We also calculate the distribution of order statistics for the proposed distribution. In Section 3, we discuss the estimation of the model parameters using maximum likelihood (ML), least squares (LS), weighted least squares (WLS), Anderson Darling (AD) and Cramer von Mises (CvM) methods. We also perform the simulation study to demonstrate the performance of the estimators. Finally, two real data sets are examined by the NAPTW distribution to demonstrate the applicability of the proposed model in real-life applications.

2. A new alpha power transformed weibull distribution

A random variable X is said to follow the new alpha power transformed Weibull (NAPTW) distribution if its density function is given by,

$$f(x;\alpha,\beta,\lambda)=\frac{\log(\alpha)\beta\lambda x^{\lambda-1}\alpha^{\log(1-e^{-\beta x^{\lambda}})}}{(e^{\beta x^{\lambda}}-1)},x>0,\alpha>1,\beta>0,\lambda>0. \tag{1}$$

The family proposed in (1) includes a large class of well-known distributions. Some important well-known sub-models are listed below:

(i) For $\alpha = e$, model (1) reduces to a Weibull distribution with parameters β and λ.
(ii) When $\alpha = e$ and $\lambda = 2$, (1) leads to a Rayleigh distribution with parameters β.
(iii) For $\alpha = e$ and $\lambda = 1$, (1) reduces to an exponential distribution with parameters β. In particular, for $\beta = 1$, it reduces to the standard exponential distribution.
(iv) For $\lambda = 1$, (1) leads to a new alpha power transformed exponential distribution proposed by Ijaz et al. (2021).

The cumulative distribution function and survival function of the NAPTW random variable X is given by,

$$F(x;\alpha,\beta,\lambda) = \alpha^{\log(1-e^{-\beta x^{\lambda}})},\ x > 0,\ \alpha > 1,\ \beta > 0,\ and\ \lambda > 0$$

and

$$S(x;\alpha,\beta,\lambda) = P(X > x) = 1 - \alpha^{\log(1-e^{-\beta x^{\lambda}})}),\ x > 0,\ \alpha > 1,\ \beta > 0,\ and\ \lambda > 0,$$

respectively.

The hazard rate function of the NAPTW distribution is,

$$h(x;\alpha,\beta,\lambda) = \frac{\log(\alpha)\beta\lambda x^{\lambda-1}\alpha^{\log(1-e^{-\beta x^{\lambda}})}}{(e^{\beta x^{\lambda}}-1)(1-\alpha^{\log(1-e^{-\beta x^{\lambda}})})}. \quad (2)$$

For different values of the parameters, various density and hazard rates plots are presented in Figure 1 and Figure 2.

From Figure 2, we see that the hazard rate function takes different shapes for several sets of values of the model parameters. This provides the usefulness of the model in practical situations.

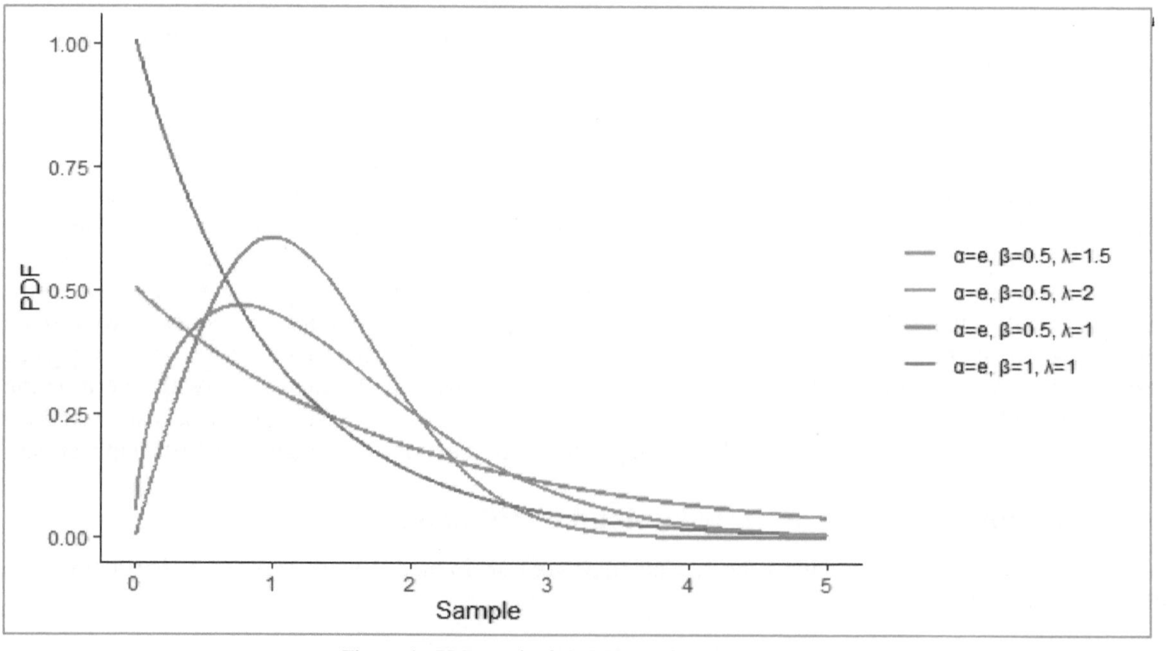

Figure 1: PDF graph of NAPTW (α,β,λ) distribution.

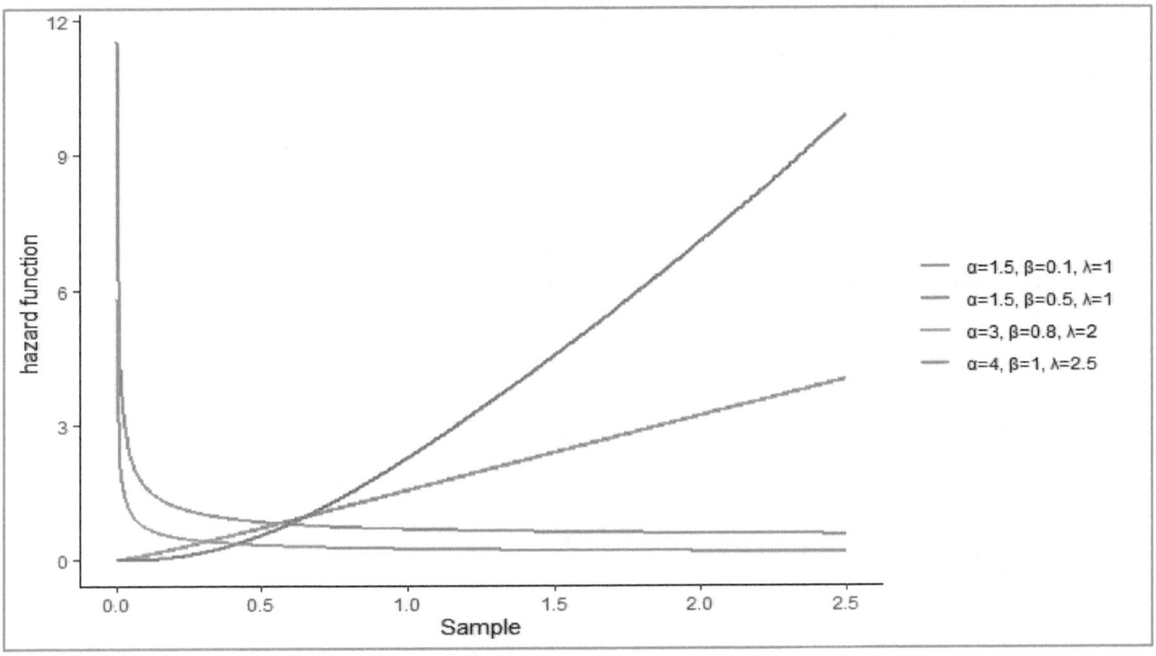

Figure 2: Hazard function graph of NAPTW (α,β,λ) distribution.

2.1 Moments

The evaluation of the moments is essential in order to study the various important statistical properties of the distribution. Several measures of tendency, dispersion, skewness, and kurtosis are expressed in terms of the moments. Let X be a random variable with density function $f(x)$. Then k^{th} order raw moments are defined by,

$$E(X^k) = \int_0^{\infty} x^k f(x)dx.$$

For the NAPTW distribution, the kth order raw moments are given by,

$$E(X^k) = log(\alpha)\beta\lambda \int_0^{\infty} \frac{x^{k+\lambda-1}\alpha^{log(1-e^{-\beta x^{\lambda}})}}{(e^{\beta x^{\lambda}}-1)}dx \tag{3}$$

The integrand in (3) is not an explicit function of x and is complex in nature. Therefore, an algebraic calculation of (3) is quite difficult. We evaluate equation (3) numerically for different values of k and parameters. With the help of these values, we present various measures of descriptive statistics like mean, variance, skewness (γ_1) and kurtosis (β_2) of the NAPTW distribution in Table 1 and study its nature .

2.2 Quantile function, skewness and kurtosis

The quantile function represents the inverse of the cumulative distribution function and determines the number of values in a distribution that are below and above certain limits. It is also a basic unit for random data generation from non-uniform random variables. For a random variable X with distribution function $F(x)$, it is defined by,

$$Q(q) = F^{-1}(q) = \inf\{x \in R: F(x) \geq q\}, for\ 0 < q < 1.$$

Table 1: Mean, variance, skewness and kurtosis of NAPTW distribution.

α	B	λ	Mean	Variance	γ_1	β_2
2.72	1.5	0.5	0.89	3.95	6.62	87.72
2.72	1.5	1	0.67	0.44	2.00	9.00
2.72	1.5	1.5	0.69	0.22	1.07	4.39
2.72	1.5	2	0.72	0.14	0.63	3.25
2.72	1	2	0.89	0.21	0.63	3.25
2.72	2	2	0.63	0.11	0.63	3.25
2.72	3	2	0.51	0.07	0.63	3.25
2.72	4	2	0.44	0.05	0.63	3.25
1.5	1.5	2	1.50	0.45	0.14	1.06
2.5	1.5	2	2.50	0.70	0.14	0.66
3.5	1.5	2	3.50	0.79	0.14	0.57
4.5	1.5	2	4.50	0.82	0.14	0.55

The qth quantile of the distribution is obtained by solving,

$$F(x) = q.$$

For the NAPTW distribution, it is given by,

$$Q(q) = x = x\left\{-\frac{1}{\beta}\ln\left(1-q^{\frac{1}{\ln(\alpha)}}\right)\right\}^{\frac{1}{\lambda}}, \tag{4}$$

where $q \sim U(0,1)$ uniform distribution.

Three quantiles $Q_1 = Q(0.25)$, $Q_2 = Q(0.5)$, and $Q_3 = Q(0.75)$, are very useful in describing the summary of data. $Q_2 = Q(0.5)$ corresponds to the median of the distribution. Interquartile range is given by $Q_3 - Q_1$.

With the help of (4), we can calculate Bowley's skewness Ψ and Moors kurtosis (see Arshad et al. (2021)) by,

$$\psi = \frac{Q(0.750) + Q(0.250) - 2Q(0.50)}{Q(0.750) - Q(0.250)}$$

and

$$\Phi = \frac{Q(0.875) + Q(0.375) - Q(0.625) - Q(0.125)}{Q(0.750) - Q(0.250)}$$

2.3 Order statistics

In this subsection, we calculate the distributions and density of various order statistics from the NAPTW distribution. Let $X_1, X_2, \ldots, X_n$ be a random sample from a population with CDF ($F_x(x)$) and PDF ($f_x(x)$). If $X_{(1)} \le X_{(2)} \le \ldots \le X_{(n)}$ denotes the order statistics of $X_1, X_2, \ldots, X_n$, then distribution and density function of the sth order statistics $X_{(s)}$, ($s = 1,2,\ldots, n$) are given by,

$$F_{X_{(s)}}(x) = \sum_{j=s}^{n}\binom{n}{j}[F_X(x)]^j[1-F_X(x)]^{(n-j)}$$

and

$$f_{X_{(s)}}(x) = \frac{n!}{(j-1)!(n-j)!}[F_X(x)]^{j-1}[1-F_X(x)]^{(n-j)} f_X(x).$$

For the NAPTW variates, the distributions of order statistics are given by,

$$F_{X_{(s)}}(x) = \sum_{j=s}^{n} \binom{n}{j} \alpha^{j \log(1-e^{-\beta x^{\lambda}})}[1-\alpha^{\log(1-e^{-\beta x^{\lambda}})}]^{n-j}$$

and

$$f_{X_{(s)}}(x) = \frac{n!\log(\alpha)\beta\lambda x^{\lambda-1}}{(j-1)!(n-j)!} \frac{\alpha^{j \log(1-e^{-\beta x^{\lambda}})}[1-\alpha^{\log(1-e^{-\beta x^{\lambda}})}]^{n-j}}{(e^{\beta x^{\lambda}}-1)}.$$

3. Parameter estimation and numerical experiments

In this section, we discuss the parameter estimation of the model using various methods and perform two sets of numerical experiments. The inverse probability integral transform has been utilized to simulate sets of 5000 (N) samples of predetermined sizes 200, 500, 800, 1000 and 1200. These samples were simulated for the selected combination of parameters $\alpha = 2.4$, $\beta = 4.0$, $\lambda = 2.0$ (Table 2) and $\alpha = 3.2$, $\beta = 1.5$, $\lambda = 2.6$ (Table 3). The estimates for parameters of NAPTW, namely, maximum likelihood (ML), least squares (LS), weighted least squares (WLS), Anderson Darling (AD) and Cramer von Mises (CvM) for each of these samples are obtained. Determining ML estimates requires the maximization of the likelihood function of the NAPTW distribution. Let $X_1, X_2, \ldots, X_n$ be a random sample from the NAPTW distribution. The likelihood function of the NAPTW distribution is given by,

$$L(\boldsymbol{\theta};\boldsymbol{x}) = \prod_{i=1}^{n} f(x_i; \alpha, \beta, \lambda) = \prod_{i=1}^{n} \log(\alpha)\, \beta\, \lambda\, x_i^{\lambda-1} \frac{\alpha^{\log(1-e^{-\beta x_i^{\lambda}})}}{(e^{-\beta x_i^{\lambda}}-1)}.$$

After taking logs on both the sides, the log-likelihood function of the NAPTW distribution is given by,

$$l(\boldsymbol{\theta}) = n \log\{\lambda\beta \log(\alpha)\} + (\lambda-1)\sum_{i=1}^{n} \log x_i + \sum_{i=1}^{n} \log(1-e^{-\beta x_i^{\lambda}}) \log(\alpha) - \sum_{i=1}^{n} \log(e^{-\beta x_i^{\lambda}}-1)$$

Here $\boldsymbol{\theta} = (\alpha, \beta, \lambda)$.

Table 2: Estimated average bias (AB) and root mean square error (RMSE) for maximum likelihood (ML) estimator, least squares (LS) estimator, weighted least squares (WLS) estimator, Anderson Darling (AD) estimator and Cramer-von Mises (CvM) estimator for estimation of parameters of the NAPTW distribution with parameters $\alpha = 2.4$, $\beta = 4.0$, $\lambda = 2.0$.

Parameter	Sample size	ML Estimates		LS Estimates		WLS Estimates		AD Estimates		CvM Estimates	
		AB	RMSE	AB	RMSE	AB	RMSE	AB	RMSE	AB	RMSE
α	200	0.28	1.35	3.70	59.31	0.56	2.83	0.66	8.76	3.61	53.42
	500	0.07	0.50	0.27	1.48	0.12	0.66	0.13	0.65	0.28	1.51
	800	0.04	0.36	0.13	0.69	0.07	0.45	0.07	0.45	0.14	0.70
	1000	0.03	0.32	0.10	0.55	0.05	0.38	0.06	0.38	0.10	0.55
	1200	0.03	0.28	0.08	0.47	0.04	0.34	0.04	0.34	0.08	0.47
β	200	0.11	0.46	0.16	0.67	0.10	0.50	0.10	0.47	0.22	0.70
	500	0.04	0.26	0.06	0.32	0.03	0.27	0.04	0.27	0.08	0.33
	800	0.03	0.20	0.04	0.25	0.02	0.21	0.02	0.21	0.05	0.25
	1000	0.02	0.18	0.03	0.22	0.02	0.19	0.02	0.19	0.04	0.22
	1200	0.02	0.16	0.02	0.19	0.01	0.17	0.02	0.17	0.03	0.20
λ	200	0.08	0.45	0.11	0.68	0.07	0.52	0.06	0.48	0.11	0.69
	500	0.03	0.26	0.05	0.40	0.03	0.30	0.03	0.30	0.05	0.40
	800	0.02	0.20	0.03	0.30	0.02	0.23	0.02	0.23	0.03	0.31
	1000	0.02	0.18	0.02	0.27	0.02	0.21	0.01	0.20	0.02	0.27
	1200	0.01	0.16	0.02	0.24	0.01	0.18	0.01	0.18	0.02	0.24

Table 3: Estimated average bias (AB) and root mean square error (RMSE) for maximum likelihood (ML) estimator, least squares (LS) estimator, weighted least squares (WLS) estimator, Anderson Darling (AD) estimator and Cramer - von Mises (CvM) estimator for estimation of parameters of NAPTW distribution with parameters $\alpha = 3.2, \beta = 1.5, \lambda = 2.6$.

Parameter	Sample size	ML Estimates		LS Estimates		WLS Estimates		AD Estimates		CvM Estimates	
		AB	RMSE	AB	RMSE	AB	RMSE	AB	RMSE	AB	RMSE
α	200	0.93	5.08	3.38	596.26	2.68	31.15	7.22	334.31	34.08	644.26
	500	0.20	1.06	0.98	12.26	0.33	1.59	0.33	1.51	1.00	11.95
	800	0.11	0.72	0.36	1.80	0.17	0.95	0.18	0.93	0.37	1.81
	1000	0.09	0.63	0.26	1.26	0.13	0.78	0.14	0.78	0.27	1.27
	1200	0.07	0.55	0.20	1.00	0.10	0.67	0.11	0.67	0.21	1.00
β	200	0.02	0.36	0.05	0.52	0.03	0.41	0.04	0.40	0.07	0.52
	500	0.00	0.22	0.01	0.32	0.01	0.25	0.01	0.25	0.02	0.32
	800	0.00	0.17	0.01	0.24	0.00	0.19	0.00	0.19	0.01	0.25
	1000	0.00	0.15	0.01	0.22	0.00	0.17	0.00	0.17	0.01	0.22
	1200	0.00	0.14	0.01	0.19	0.00	0.16	0.00	0.15	0.01	0.20
λ	200	0.09	0.57	0.12	0.85	0.08	0.65	0.07	0.62	0.13	0.85
	500	0.04	0.33	0.04	0.51	0.04	0.39	0.03	0.38	0.06	0.51
	800	0.03	0.25	0.04	0.39	0.03	0.30	0.02	0.29	0.04	0.39
	1000	0.02	0.23	0.03	0.34	0.02	0.26	0.02	0.26	0.03	0.34
	1200	0.02	0.20	0.02	0.31	0.02	0.24	0.01	0.23	0.02	0.31

The LS, WLS, AD and CvM estimates are obtained by minimizing, with respect to $\boldsymbol{\theta}$, the functions $S(\boldsymbol{\theta})$, $W(\boldsymbol{\theta})$, $A(\boldsymbol{\theta})$ an $C(\boldsymbol{\theta})$ respectively. These functions are defined as follows:

$$S(\theta) = \sum_{i=1}^{n} \left\{ F(x_{(i)}) - \frac{1}{n+1} \right\}^2$$

$$W(\theta) = \sum_{i=1}^{n} \tau_i \left\{ F(x_{(i)}) - \frac{1}{n+1} \right\}^2, \text{where } \tau_i = \frac{1}{Var(F(x_{(i)}))} = \frac{(n+1)^2(n+2)}{i(n-i+1)}$$

$$A(\theta) = -n - \frac{1}{n} \sum_{i=1}^{n} \left[(2i-1)\left\{ \ln\left(F(x_{(i)})\right) + \ln\left(1 - F\left(x_{(n+1-i)}\right)\right) \right\} \right]$$

$$C(\theta) = \frac{1}{12n} + \sum_{i=1}^{n} \left\{ F\left(x_{(i)}\right) - \left(\frac{2i-1}{2n}\right) \right\}^2,$$

where $F(\cdot)$ is the CDF of NAPTW and $x_{(1)} \leq x_{(2)} \leq \cdots \leq x_{(n)}$ denote ordered observations for a given sample. The functions ML, LS, WLS, AD and CvM are optimized using the optim function in R (R Core Team 2017). For data simulation, one may refer to the paper by Arshad et al. (2021).

Average bias (AB) and root mean square error (RMSE) are estimated for all selected sample sizes for each of the estimators discussed above. Estimates of AB and RMSE of the MLE $\hat{\theta}$ of a parameter θ are given as:

$$AB = \frac{1}{N} \sum_{i=1}^{N} |\hat{\theta}_t - \theta|$$

$$RMSE = \sqrt{\frac{1}{N} \sum_{i=1}^{N} (\hat{\theta}_t - \theta)^2}.$$

The change in AB and RMSE for an increasing sample size is investigated. For a given sample size, the AB and RMSE of the estimators are compared. Tables 2 and 3 provide the estimated AB and RMSE for estimators of α, β and λ. From the results, it can be verified that the AB and the RMSE decrease with an increase in the sample size. Further, the AB and RMSE for MLE are found to be lower than that of LS, WLS, AD and CvM estimators. Hence, MLE is a better choice for the estimation of the parameters of the NAPTW distribution.

3.1 Real data applications

The NAPTW distribution admits the exponential, the Weibull and the novel alpha power transformed exponential (NAPTE) as some of its sub-models. The fit of the NAPTW is compared with the fit of these selected sub-models for two data sets. These data sets are the bladder cancer patients' (BCP) data and the bank customers' (BC) data. The model fit for each selected data set is assessed using different goodness of fit statistics pertaining to popular goodness of fit tests. The goodness of fit statistics utilized for this purpose are -2ln (L_M), the Akaike's Information Criterion (AIC), the Bayesian Information Criterion (BIC), AIC corrected for small samples (AIC$_c$), the Consistent Akaike's Information Criterion (CAIC) and the Hannan-Quinn Information Criterion (HQIC). Here, L_M denotes the maximized value of the likelihood function. These measures utilize the maximized value of the likelihood function. Other goodness of fit statistics are W^*, A^* (Chen and Balakrishnan (1995)) and the sum of squares (SS). The smaller the values of these statistics, the better is the fit of the model. In addition, the statistics, namely, Kolmogorov-Smirnov (KS), Cramer von Mises (CvM) and Anderson-Darling (AD), along with their p-values, have also been utilized for the purpose. Let L_M, p and n denote the maximized value of the likelihood function, the number of estimated parameters and the size of the sample, respectively. The statistics discussed above are defined as follows:

1. $AIC = -2\ ln(L_M) + 2p$
2. $AIC_c = -2\ ln(L_M) + 2p + \dfrac{2p(p+1)}{m-p-1}$
3. $BIC = --2\ ln(L_M) + p\ \ln(n)$
4. $CAIC = -2\ ln(L_M) + p\ \{\ln(n) + 1\}$
5. $HQIC = -2\ ln(L_M) + 2p\{\ln(\ln(n))\}$
6. $W^* = W^2\left(1+\dfrac{0.5}{n}\right)$
7. $A^* = A^2\left(1+\dfrac{0.75}{n}+\dfrac{2.25}{n^2}\right)$.

The steps involved in the computation of W^2 and A^2 are explained in what follows. Consider the sample $(x_1, x_2,\ldots, x_n)$ where $x_1, x_2,\ldots, x_n$ are arranged in ascending order. Let $F(x; \theta)$ be the CDF of the population from which the sample is drawn. Further, let $\hat{\theta}$ be the MLE of θ based on the given sample. We define the following statistics:

$$u_i = F(x_i; \hat{\theta})$$

$$x_i = \Phi^{-1}(u_i)$$

$$\bar{x} = \frac{1}{n}\sum_{i=1}^{n} x_i$$

$$s_x^2 = \frac{1}{n-1}\sum_{i=1}^{n}(x_i - \bar{x})^2$$

$$v_i = \Phi\left(\frac{x_i - \bar{x}}{s_x}\right),$$

where Φ denotes the CDF of a standard normal variable and Φ^{-1} denotes its inverse. Utilizing the above statistics, W^2 and A^2 are defined as follows:

$$W^2 = \frac{1}{12n} + \sum_{i=1}^{n} \left\{ v_i - \left(\frac{2i-1}{2n} \right) \right\}^2,$$

$$A^2 = -n - \frac{1}{n} \sum_{i=1}^{n} \{(2i-1)\ln(v_i) + (2n+1-2i)\ln(1-v_i)\}.$$

Besides these measures, LR tests are conducted where the sub models are tested against the full model, i.e., NAPTW for each selected data set.

3.1.1 Bladder cancer patients (BCP) data

BCP data consists of a sample of 128 remission times (in months) of bladder cancer patients (Aldeni et al. (2017)). The sample remission times are below:

0.08	0.20	0.40	0.50	0.51	0.81	0.90	1.05	1.19	1.26
1.35	1.40	2.02	2.07	2.09	2.23	2.26	2.46	2.54	2.62
2.64	2.69	2.75	2.83	3.31	3.36	3.36	3.48	3.52	3.57
3.64	3.70	3.82	3.88	4.18	4.23	4.40	4.50	4.51	4.87
4.98	5.06	5.09	5.17	5.32	5.32	5.34	5.41	5.71	5.85
6.25	6.54	6.76	6.93	6.94	6.97	7.09	7.26	7.28	7.32
7.63	7.66	7.87	7.93	8.26	8.37	8.53	8.65	8.66	9.02
9.22	9.47	10.66	10.75	11.25	11.64	11.79	11.98	12.02	12.03
12.07	12.63	13.11	13.29	14.77	14.83	15.96	16.62	17.12	17.14
17.36	18.10	19.13	20.28	21.73	22.69	26.31	32.15	34.26	36.66
43.01	46.12	79.05	1.46	1.76	2.02	5.41	5.49	2.87	3.02
3.25	7.39	7.59	5.62	13.80	14.24	14.76	10.34	25.74	25.82
4.26	4.33	4.34	9.74	10.06	7.62	23.63	2.69		

The mean remission time is 9.37 months (Table 4). It has also been used by Ijaz et al. (2021) in their study on the NAPTE distribution. The observations range from 0.80 months to 79.05 months, with a median remission time of 6.40 months. The first and the third quartiles are found to be 3.35 months and 11.84 months, respectively (Table 4). Based on the BCP data, the maximum likelihood estimates of NAPTW and its sub-models are given in Table 5. The values of the selected goodness of fit statistics, pertaining to the fitting of the NAPTW model, namely, $-2\ln(L_M)$, AIC, BIC, AIC_c, CAIC, HQIC, W^*, A^* and SS are found to be 821.36, 825.36, 831.06, 825.46, 833.06, 827.68, 0.044, 0.288 and 0.038 respectively. Each of these values is lower than that of corresponding values for the selected NAPTW sub-models. Hence, NAPTW is a better model for BCP data when compared to the NAPTE, the Weibull or the exponential models. The LR tests for hypotheses H_0: exponential against H_a: NAPTW, H_0: Weibull against H_a: NAPTW and H_0: NAPTE against H_a: NAPTW resulted in statistics with p-values 0.03, 0.01 and 0.03, respectively, indicating NAPTW to be a preferred choice for modeling of the BCP data. The KS, AD and CvM statistics (Table 6) also substantiate the finding that NAPTW provides a better fit to the BCP data when compared to its sub-models.

Table 4: Descriptive statistics for bladder cancer patients' data and bank customers' data.

Data	Sample size	minimum	maximum	1st quartile	3rd quartile	median	mean
BCP	128	0.80	79.05	3.35	11.84	6.40	9.37
BC	100	0.80	38.50	4.68	13.03	8.10	9.89

Table 5: Maximum likelihood estimates for NAPTW and its selected sub-models and the model fit under various goodness of fit criteria for bladder cancer patients' data and bank customers' data.

Data	Model	$\hat{\alpha}$	$\hat{\beta}$	$\hat{\lambda}$	$-2log$ (L_{max})	AIC	BIC	AIC_c	CAIC	HQIC	W^*	A^*	SS
BCP Data	Exponential	-	0.1068	-	828.68	830.68	833.54	830.72	834.54	831.84	0.119	0.716	0.174
	Weibull	-	0.0939	1.0478	828.17	832.17	837.88	832.27	839.88	834.49	0.131	0.786	0.150
	NAPTE	3.3804	0.1212	-	826.16	830.16	835.86	830.25	837.86	832.47	0.112	0.674	0.125
	NAPTW	**16.3820**	**0.4537**	**0.6544**	**821.36**	**825.36**	**831.06**	**825.46**	**833.06**	**827.68**	**0.044**	**0.288**	**0.038**
BC Data	Exponential	-	0.1012	-	658.04	660.04	662.65	660.08	663.65	661.10	0.027	0.179	0.076
	Weibull	-	0.0306	1.4573	637.46	641.46	646.67	641.59	648.67	643.57	0.063	0.396	0.058
	NAPTE	8.8766	0.1592	-	634.19	638.19	643.40	638.31	645.40	640.30	0.021	0.143	0.021
	NAPTW	**14.5255**	**0.2229**	**0.9054**	**634.07**	**638.07**	**643.28**	**638.19**	**645.28**	**640.18**	**0.017**	**0.127**	**0.017**

3.1.2 Bank customers (BC) data

BC data is a sample of waiting times (in minutes) of 100 bank customers (Ghitany et al. 2008). The sample waiting times are below:

0.8	0.8	1.3	1.5	1.8	1.9	1.9	2.1	2.6	2.7
2.9	3.1	3.2	3.3	3.5	3.6	4.0	4.1	4.2	4.2
4.3	4.3	4.4	4.4	4.6	4.7	4.7	4.8	4.9	4.9
5.0	5.3	5.5	5.7	5.7	6.1	6.2	6.2	6.2	6.3
6.7	6.9	7.1	7.1	7.1	7.1	7.4	7.6	7.7	8.0
8.2	8.6	8.6	8.6	8.8	8.8	8.9	8.9	9.5	9.6
9.7	9.8	10.7	10.9	11.0	11.0	11.1	11.2	11.2	11.5
11.9	12.4	12.5	12.9	13.0	13.1	13.3	13.6	13.7	13.9
14.1	15.4	15.4	17.3	17.3	18.1	18.2	18.4	18.9	19.0
19.9	20.6	21.3	21.4	21.9	23.0	27.0	31.6	33.1	38.5

The sample mean waiting time is found to be 9.89 minutes. Waiting times are found to range between 0.80 minutes to 38.50 minutes, with a median waiting time of 8.10 minutes. The first and the third quartiles are found to be 4.68 minutes and 13.03 minutes (Table 4), respectively. For the BC data the values for -2ln (L_M), AIC, BIC, AIC_c, CAIC, HQIC, W^*, A^* and SS are found to be 634.07, 638.07, 643.28, 638.19, 645.28, 640.18, 0.017, 0.127 and 0.017 respectively (Table 5). These are lower than the corresponding values for

Table 6: Kolmogorov-Smirnov (KS), Anderson Darling (AD) and Cramer von Mises goodness of fit test statistics and corresponding p-values for NAPTW and its selected sub-models for fitting the bladder cancer patients' data and bank customers' data.

Data	Model	KS	AD	CvM
BCP	Exponential	0.0846 (0.3183)	1.1736 (0.2777)	0.1788 (0.3129)
	Weibull	0.0700 (0.5570)	0.9579 (0.3799)	0.1537 (0.3788)
	NAPTE	0.0725 (0.5115)	0.7138 (0.5472)	0.1279 (0.4652)
	NAPTW	**0.0450 (0.9576)**	**0.2704 (0.9586)**	**0.0403 (0.9321)**
BC	Exponential	0.1730 (0.0050)	4.2293 (0.0068)	0.7154 (0.0115)
	Weibull	0.0576 (0.8947)	0.4049 (0.8436)	0.0607 (0.8108)
	NAPTE	0.0403 (0.9969)	0.1457 (0.9990)	0.0214 (0.9957)
	NAPTW	**0.0365 (0.9994)**	**0.1279 (0.9996)**	**0.0176 (0.9988)**

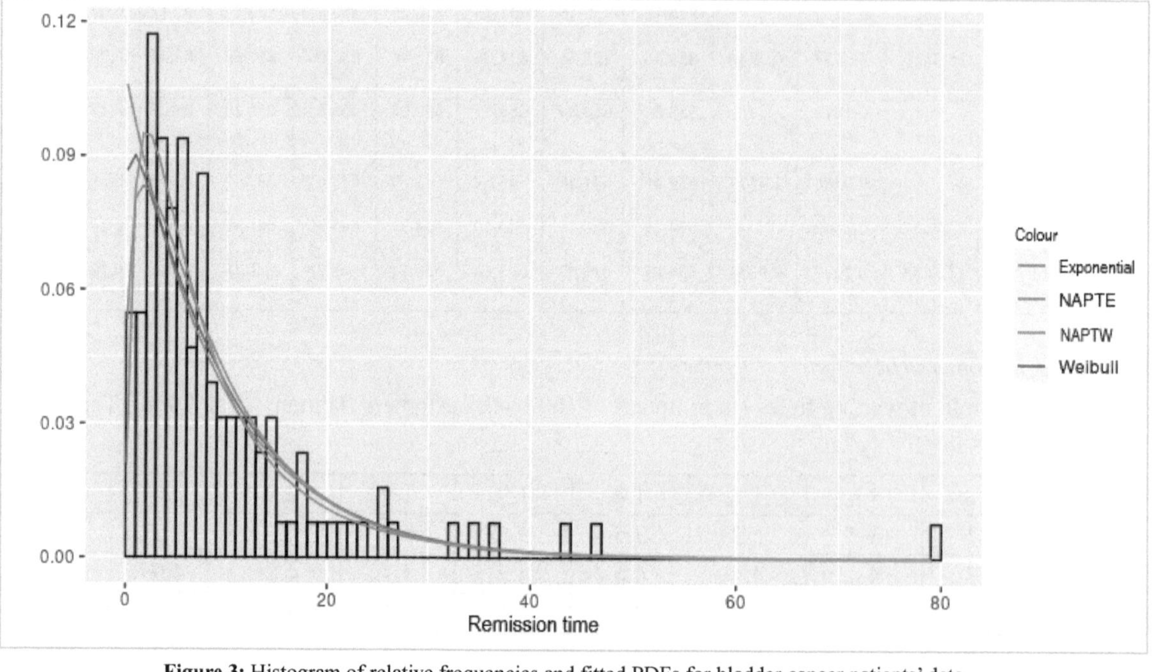

Figure 3: Histogram of relative frequencies and fitted PDFs for bladder cancer patients' data.

selected NAPTW sub-models . Hence NAPTW is found to provide a better fit for BC data. In their study on the NAPTE model, Ijaz et al. (2021) have shown that the NAPTE model fits better to the BC data when compared to the exponential, the Rayleigh, the Weibull and the Weibull Exponential models. The present study found the NAPTW to be a better choice to model BC data when compared to NAPTE or the other selected sub-models of NAPTW. The LR tests for hypotheses H_0: exponential against H_a: NAPTW, H_0: Weibull against H_a: NAPTW and H_0: NAPTE against H_a: NAPTW resulted in statistics with p-values 0.01, 0.07 and 0.72, respectively. It implies that NAPTW may be preferred over the exponential or Weibull models for modeling the BC data. However, NAPTE may be considered a good model for the BC data, as is the NAPTW model. Based on the values obtained for the KS, Ad and CvM statistics, it can be inferred that NAPTW provides a good fit to the BC data compared to its sub-models or is a better choice for modeling BC data.

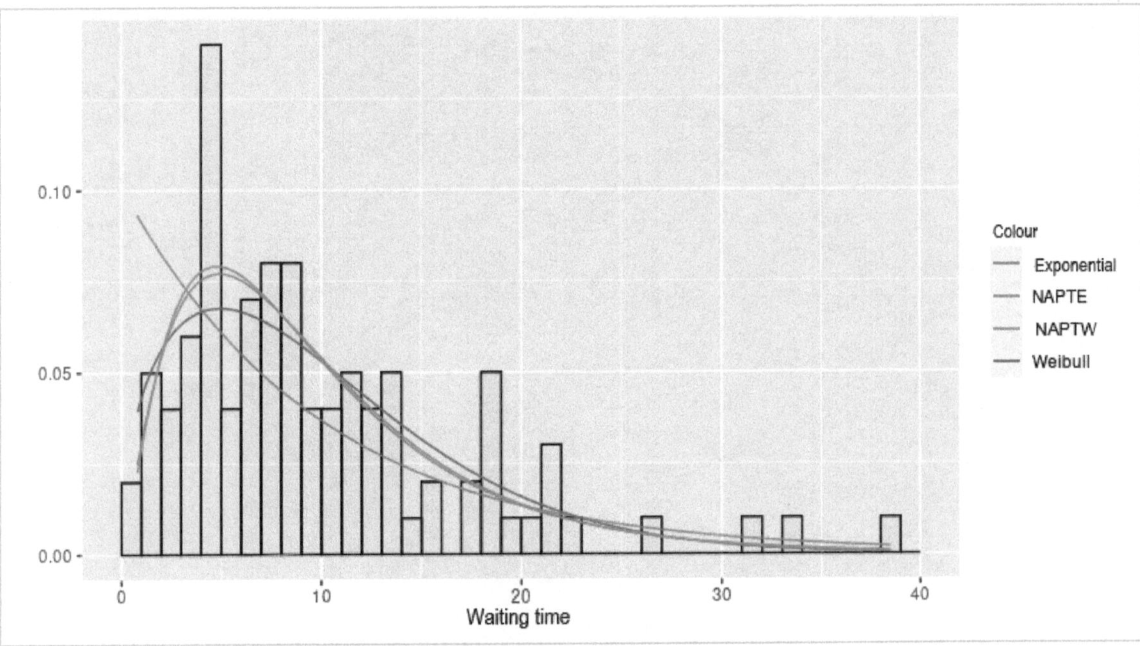

Figure 4: Histogram of relative frequencies and fitted PDFs for bank customers' data.

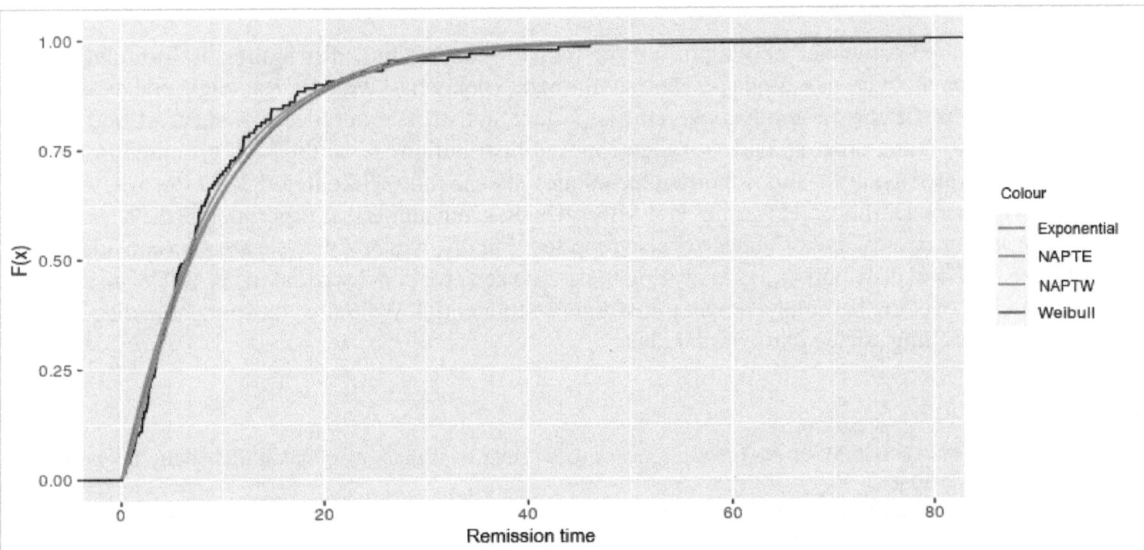

Figure 5: Empirical and fitted CDFs for bladder cancer patients' data.

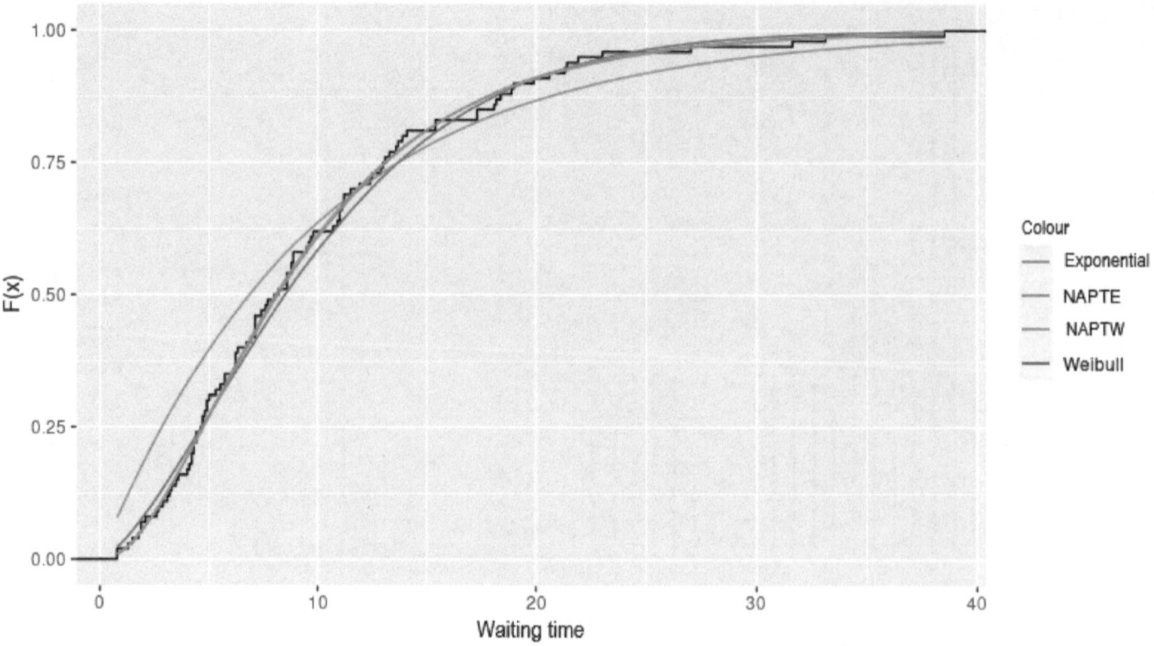

Figure 6: Empirical and fitted CDFs for bank customers' data.

4. Discussion and conclusion

In this chapter, a three-parameter new alpha power transformed Weibull distribution is formulated. The proposed distribution includes new alpha power transformed exponential, Weibull, Rayleigh, and exponential distributions as its important sub-models. Several important statistical properties like survival rate, hazard rate, quantile function, and order statistics of the proposed distribution are studied. Th estimation of model parameters is preformed using various important techniques like maximum likelihood, least squares, weighted least squares, Anderson Darling, and Cramer von Mises. Using some numerical experiments, the average bias and root mean square error of the estimates are also reported. Finally, two real data set are also considered and fitted using the NAPTW distribution. By analyzing these data sets, we conclude that the NAPTW distribution provides a better fit over the new alpha power transformed exponential, Weibull and exponential distributions and is useful in modelling a wide class of real data.

References

Abd EL-Baset, A. A. and Ghazal, M. G. M. (2020). Exponentiated additive Weibull distribution. Reliability Engineering & System Safety, 193: 106663.

Aldeni, M., Lee, C. and Famoye, F. (2017). Families of distributions arising from the quantile of generalized lambda distribution. Journal of Statistical Distributions and Applications, 4(1): 1–18.

Almalki, S. J. and Yuan, J. (2013). A new modified Weibull distribution. Reliability Engineering & System Safety, 111: 164–170.

Arshad, M., Azhad, Q. J., Gupta, N. and Pathak, A. K. (2021). Bayesian inference of unit gompertz distribution based on dual generalized order statistics. Communications in Statistics-Simulation and Computation, 1–19.

Arshad, M., Khetan, M., Kumar, V. and Pathak, A. K. (2021). Record-Based transmuted generalized linear exponential distribution with increasing, decreasing and bathtub shaped failure rates. arXiv preprint arXiv:2107.09316.

Carrasco, J. M., Ortega, E. M. and Cordeiro, G. M. (2008). A generalized modified Weibull distribution for lifetime modeling. Computational Statistics & Data Analysis, 53: 450–462.

Chen, G. and Balakrishnan, N. (1995). A general purpose approximate goodness-of-fit test. Journal of Quality Technology, 27: 154–161.

Famoye, F., Lee, C. and Olumolade, O. (2005). The beta-Weibull distribution. Journal of Statistical Theory and Applications, 4: 121–136.

Ghitany, M. E., Atieh, B. and Nadarajah, S. (2008). Lindley distribution and its application. Mathematics and Computers in Simulation, 78(4): 493–506.

Gupta, R. D. and Kundu, D. (1999). Generalized exponential distributions. Australian and New Zealand Journal of Statistics, 41: 173–188.

He, B., Cui, W. and Du, X. (2016). An additive modified Weibull distribution. Reliability Engineering & System Safety, 145: 28–37.

Ijaz, M. Mashwani, W. K., Göktaş, A. and Unvan, Y. A. (2021). RETRACTED ARTICLE: A novel alpha power transformed exponential distribution with real-life applications, Journal of Applied Statistics, 48: 11, I–XVI.

Khan, M. S., King, R. and Hudson, I. L. (2017). Transmuted generalized exponential distribution: A generalization of the exponential distribution with applications to survival data. Communications in Statistics-Simulation and Computation, 46(6): 4377–4398.

Lai, C. D. (2014). Generalized Weibull distributions. Generalized Weibull Distributions (pp. 23–75). Springer, Berlin, Heidelberg.

Mahmoud, M. A. and Alam, F. M. A. (2010). The generalized linear exponential distribution. Statistics & Probability Letters, 80: 1005–1014.

Marshall, A. W. and Olkin, I. (1997). A new method for adding a parameter to a family of distributions with application to the exponential and Weibull families. Biometrika, 84: 641–652.

Merovci, F. and Elsatal, I. (2015). Weibull Rayleigh Distribution. Theory and Applications Appl. Math. Inf. Sci., 9: 2127–2137.

Mudholkar, G. S. and Srivastava, D. K. (1993). Exponentiated Weibull family for analyzing bathtub failure-rate data. IEEE Transactions on Reliability, 42: 299–302.

Nadarajah, S., Cordeiro, G. M. and Ortega, E. M. (2011). General results for the beta-modified Weibull distribution. Journal of Statistical Computation and Simulation, 81(10): 1211–1232.

Pathak, A. K., Arshad, M., Azhad, Q. J., Khetan, M., Pandey, A. et al. (2021). A novel bivariate generalized weibull distribution with properties and applications. arXiv preprint arXiv:2107.11998.

Pham, H. and Lai, C. D. (2007). On recent generalizations of the weibull distribution. IEEE transactions on reliability, 56(3): 454–458.

R Core Team (2017). R: A language and environment for statistical computing. R Foundation for Statistical Computing, Vienna, Austria. URL https://www.R-project.org/.

Sarhan, A. M., Abd EL-Baset, A. A. and Alasbahi, I. A. (2013). Exponentiated generalized linear exponential distribution. Applied Mathematical Modelling, 37(5): 2838–2849.

Sarhan, A. M. and Zaindin, M. (2009). Modified weibull distribution. Applied Sciences, 11: 123–136.

Silva, G. O., Ortega, E. M. and Cordeiro, G. M. (2010). The beta modified Weibull distribution. Lifetime Data Analysis, 16: 409–430.

Tian, Y., Tian, M. and Zhu, Q. (2014). Transmuted linear exponential distribution: A new generalization of the linear exponential distribution. Communications in Statistics-Simulation and Computation, 43(10): 2661–2677.

Xie, M. and Lai, C. D. (1996). Reliability analysis using an additive weibull model with bathtub-shaped failure rate function. Reliability Engineering & System Safety, 52: 87–93.

Xie, M., Tang, Y. and Goh, T. N. (2002). A modified weibull extension with bathtub-shaped failure rate function. Reliability Engineering & System Safety, 76: 279–285.

Chapter 20

An Extension of Topp-Leone Distribution with Increasing, Decreasing and Bathtub Hazard Functions

Unnati Nigam and *Arun Kaushik**

1. Introduction

In the recent times, various lifetime probability distributions have been proposed that are defined on the positive real line. They have high applicability in various fields like biomedicine and social sciences among others. However, such models are not useful for modeling random variables with a bounded range. The most used application of modeling random variables with abounded range is to model proportion and percentage data measured on the unit interval . The Beta distribution is one of the most commonly used distributions to model unit interval data. Two important unit distributions, Johnson (see Johnson, 1949) and Kumaraswamy distribution (see Kumaraswamy, 1980) are also recommended to model unit interval data. However, these classical models may be inadequate and may pose problems for accurate data analysis. For this, various lifetime distributions have been transformed to unit-intervals. Some of these are, unit-Gamma or log-Gamma (Consul and Jain, 1971), unit-Weibull (see Mazucheli et al., 2018b), log-Lindley (see Gómez-Déniz et al., 2014), unit Gompertz (see Mazucheli et al., 2019), unit Birnbaum-Saunders (see Mazucheli et al., 2018a), unit Burr-XII (see Korkmaz and Chesneau, 2021) and more.

One important distribution which is widely used for fitting unit range data is the one parameter Topp-Leone distribution given by, Topp and Leone, 1995. Nadarajah and Kotz, 2003, presented the closed form expressions of the moments of Topp-Leone (TL) distribution. The Topp-Leone distribution has a J-shaped frequency curve and has been used by various researchers for their studies. The TL distribution is a mixture of generalized triangular distribution and uniform distribution.

The Topp-Leone distribution has also proved to be effective in generating new flexible families of distributions. Sharma, 2018, introduced the Topp-Leone normal distribution and showed its application to three real data sets. Shekhawat and Sharma, 2021, present a two-parameter extension of Topp-Leone distribution by adding a skewness parameter and showing it's application on tissue damage proportions data. Sangsanit and Bodhisuwan, 2016, proposed the Topp-Leone generalized exponential distribution. Alizadeh et al., 2018, proposed a Topp-Leone odd log-logistic family of distributions with its application on a regression model.

Department of Statistics, Banaras Hindu University, Varanasi, India.
* Corresponding author: arunkaushik@bhu.ac.in

In this article, we introduce a two-parameter extension of the Topp-Leone distribution using the transformation given below. If F(x) and f(x) be the cdf and pdf of the proposed distribution respectively, then F(x) is defined as,

$$F(x)=\frac{a^{G(x)}-1}{a-1};a>1. \tag{1}$$

and the corresponding pdf is,

$$f(x)=\frac{ln(a)g(x)a^{G(x)}}{a-1};\ a>1. \tag{2}$$

This transformation gives the DUS transformation if we take $a = e$, given by, Kumar et al., 2015.

We consider the baseline distribution One-Parameter Topp-Leone distribution with cdf $G(x) = x^{\alpha}(2-x)^{\alpha}$; $0 < x < 1$, $\alpha > 0$ and corresponding pdf as $g(x) = 2\alpha x^{\alpha-1}(1-x)(2-x)^{\alpha-1}$; $0 < x < 1$; $\alpha > 0$. Using the transformation proposed in equation (1), the cdf and pdf of the resulting distribution, hereafter referred to as the Power Exponentiated Topp-Leone (PETL) distribution can easily be obtained as:

$$F(x)=\frac{a^{(x^{\alpha}(2-x)^{\alpha})}-1}{a-1};\ 0<x<1;\ \alpha>0;\ a>1. \tag{3}$$

$$F(x)=\frac{2\alpha ln(a)x^{\alpha-1}(1-x)\,(2-x)^{\alpha-1}\,a^{(x^{\alpha}(2-x)^{\alpha})}}{(a-1)};\ 0<x<1;\ \alpha>0;\ a>1. \tag{4}$$

The Power Exponentiated Topp-Leone (PETL) distribution is very flexible as it can accommodate a variety of shapes of hazard rate and density functions. We use PETL (α, a) to denote the distribution given in equations (3) and (4).

In this article, we present the attractive statistical properties of the proposed PETL (α,γ) distribution and present its effective use for modeling failure time (in days) data of air conditioning system of an airplane (as presented by Linhart and Zucchini, 1986) over the existing distributions defined on the unit interval.

The shapes of density, distribution, reliability and hazard functions are explored in Section 2. The Statistical Properties of the distribution are obtained in Section 3. These include ordinary moments, conditional moments, quantile function, order statistics, mean deviation about mean and median, entropy, stress-strength reliability, identifiability, stochastic ordering and differential equations. We derive maximum likelihood estimators (MLEs) and their asymptotic confidence intervals in Section 4. A simulation study is carried out to study the behavior of the mean squared error and mean absolute bias of the MLEs. In Section 5, the proposed distribution is used to model a real dataset of maximum flood levels of Susquehenna River at Harrisburg, Pennsylvania over the existing unit-interval distributions like Beta, Kumaraswamy, unit-Gamma, Topp-Leone and Generalized Topp-Leone (given by Shekhawat and V.K. Sharma,2021) distributions. The findings of the paper are highlighted in Section 6.

2. Shapes of the distribution

Figure 1 shows the probability density and cumulative density plots for different values of parameters as per equations (3) and (4). The associated reliability function is,

$$R(x)=\frac{a-a^{(x^{\alpha}(2-x)^{\alpha})}}{a-1};\ 0<x<1;\ \alpha>0;\ a>1. \tag{5}$$

The associated hazard rate is,

$$h(x)=\frac{2\alpha ln(a)x^{\alpha-1}(1-x)\,(2-x)^{\alpha-1}\,a^{(x^{\alpha}(2-x)^{\alpha})}}{a-a^{(x^{\alpha}(2-x)^{\alpha})}};\ 0<x<1;\ \alpha>0;\ a>1. \tag{6}$$

Figure 2 shows the reliability and hazard rate functions for different values of the parameter.

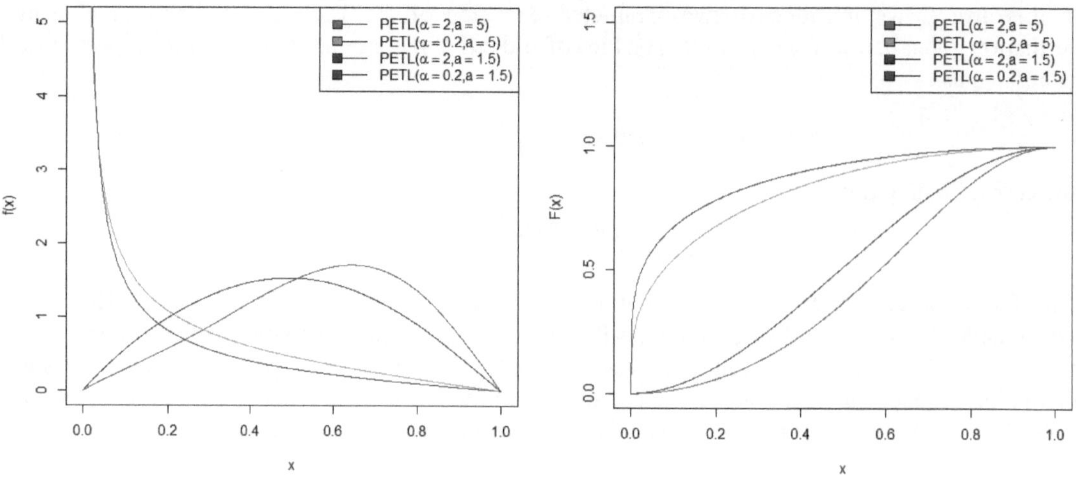

Figure 1: PDFs and CDFs of PETL(α,a).

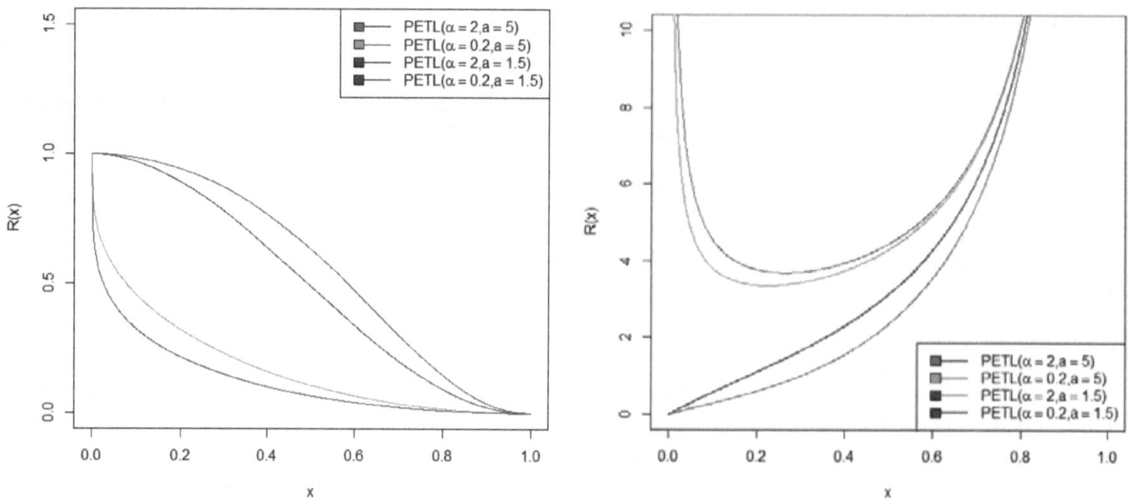

Figure 2: Reliability and Hazard Rate Functions of PETL(α,a).

3. Statistical properties

3.1 Moments

Theorem 3.1

$$E(X^r) = \frac{2\alpha \ln(a)}{a-1} \sum_{i=0}^{\infty} \sum_{j=0}^{\infty} \frac{(ln(a))^i}{i!} (-1)^j 2^{\alpha+\alpha i-j-1} \binom{\alpha+\alpha i-1}{j} \times \left(\frac{1}{r+\alpha+\alpha i+j} - \frac{1}{r+\alpha+\alpha i+j+1} \right) \tag{7}$$

Proof.

$$E(X^r) = \int_0^1 x^r f(x)dx,$$

$$= \int_0^1 x^r \frac{2\alpha ln(a) x^{\alpha-1}(1-x)\,(2-x)^{\alpha-1}\, a^{x^{\alpha}(2-x)^{\alpha}}}{(a-1)}; \; 0 < x < 1; \; \alpha > 0; \; a > 1,$$

Expanding the term $a^{(x^\alpha(2-x)^\alpha)} = \sum_{i=0}^{\infty}(x^\alpha(2-x)^\alpha \ln(a))^i/i!$ as a convergent sum of infinite terms, we get,

$$E(X^r) = \frac{2\ln(a)\alpha}{a-1}\sum_{i=0}^{\infty}\frac{(\ln(a))^i}{i!}\int_0^1 (1-x)\,(2-x)^{\alpha+\alpha i-1}\,x^{r+\alpha+\alpha i-1}\,dx,$$

$$= \frac{2n(a)\alpha}{a-1}\sum_{i=0}^{\infty}\frac{(\ln(a))^i}{i!}\int_0^1 (x^{r+\alpha+\alpha i-1} - x^{r+\alpha+\alpha i})\,(2-x)^{\alpha+\alpha i-1}\,dx$$

using the expansion of series,

$$(2-x)^A = 2^A\left(\frac{2-x}{2}\right)^A$$

$$= 2^A\left(1-\frac{x}{2}\right)^A$$

$$= 2^A\sum_{i=0}^{\infty}\left(\frac{-1}{2}\right)^j\binom{A}{j}(x)^j$$

where, $A = \alpha + \alpha i - 1$ and after simplifying, we get,

$$E(X^r) = \frac{2\alpha\ln(a)}{a-1}\sum_{i=0}^{\infty}\sum_{j=0}^{\infty}\frac{(\ln(a))^i}{i!}(-1)^j 2^{\alpha+\alpha i-j-1}\binom{\alpha+\alpha i-1}{j}$$
$$\times\left(\frac{1}{r+\alpha+\alpha i+j} - \frac{1}{r+\alpha+\alpha i+j+1}\right)$$

The mean, variance, skewness and kurtosis of the distribution of X can now be easily obtained using their respective formulae.

3.2 Conditional moments

The expressions of conditional moments, can be derived by using the following theorem.

Theorem 3.2

$$E(X^r \mid X > t) = \frac{2\alpha\ln(a)}{a-1}\sum_{i=0}^{\infty}\sum_{j=0}^{\infty}\frac{(\ln(a))^i}{i!}(-1)^j 2^{\alpha+\alpha i-j-1}\binom{\alpha+\alpha i-1}{j}(1-t)$$
$$\times\left(\frac{1}{r+\alpha+\alpha i+j} - \frac{1}{r+\alpha+\alpha i+j+1}\right) \quad (8)$$

Proof. By proceeding in the same way as mentioned in Theorem 2.1,

$$E(X^r|X > t) = \int_t^1 x^r f(x)dx,$$

$$= \int_t^1 x^r \frac{2\alpha\ln(a)x^{\alpha-1}(1-x)\,(2-x)^{\alpha-1}\,a^{x^\alpha(2-x)^\alpha}}{(a-1)}dx,$$

$$= \frac{2\alpha\ln(a)}{a-1}\sum_{i=0}^{\infty}\frac{(\log(a)^i)^i}{i!}\int_t^1(1-x)\,(2-x)^{\alpha+\alpha i-1}x^{r+\alpha+\alpha i-1}\,dx,$$

$$= \frac{2\alpha\ln(\alpha)}{a-1}\sum_{i=0}^{\infty}\frac{(\ln(a))^i}{i!}\int_t^1(x^{r+\alpha+\alpha i-1} - x^{r+\alpha+\alpha i})\,(2-x)^{\alpha+\alpha i-1}\,dx$$

$$E(X^r \mid X > t) = \frac{2\alpha ln(a)}{a-1}\sum_{i=0}^{\infty}\sum_{j=0}^{\infty}\frac{(ln(a))^i}{i!}(-1)^j 2^{\alpha+\alpha i-j-1}\binom{\alpha+\alpha i-1}{j}(1-t)$$
$$\times\left(\frac{1}{r+\alpha+\alpha i+j}-\frac{1}{r+\alpha+\alpha i+j+1}\right)$$

3.3 Quantile function

The pth quantile function $Q(p)$ is obtained by solving $F(Q(p)) = p$. Hence from equation (3), we get,

$$(Q(p))^2 - 2Q(p) + (log_a(1 + p(a-1)))^{1/\alpha} \quad (9)$$

This equation can be further solved using the quadratic formula or by numerical methods. We have also explored the changing behavior of the median with varying values of parameters α and a. For this, we put $p = 0.5$ in equation 9 and it becomes,

$$(Q(p))^2 - 2Q(p) + (ln(0.5(a-1))^{1/\alpha}$$

The behaviour of the median with respect to the changing parameters is visualized in Figure 3.

3.4 Order statistics

Let $X_1, X_2,\ldots, X_n$ be a random sample of size n, from the proposed distribution and $X_{1:n} < X_{2:n} < \cdots < X_{n:n}$ denote the corresponding order statistics. It is well known that the pdf $f_r(x)$ of rth (for $r = 1,2,\ldots, n$) order statistics $X_{r:n}$, when the population cdf and pdf are F(x) and f(x) respectively, is given as,

$$f_r(x) = \frac{n!}{(r-1)!(n-r)!}F^{r-1}(x)\,[1-F(x)]^{n-r} f(x)$$
$$= \frac{n!}{(r-1)!(n-r)!}\sum_{l=0}^{n-r}(-1)^l\binom{n-r}{l}F^{r+l-1}(x)\,f(x)$$

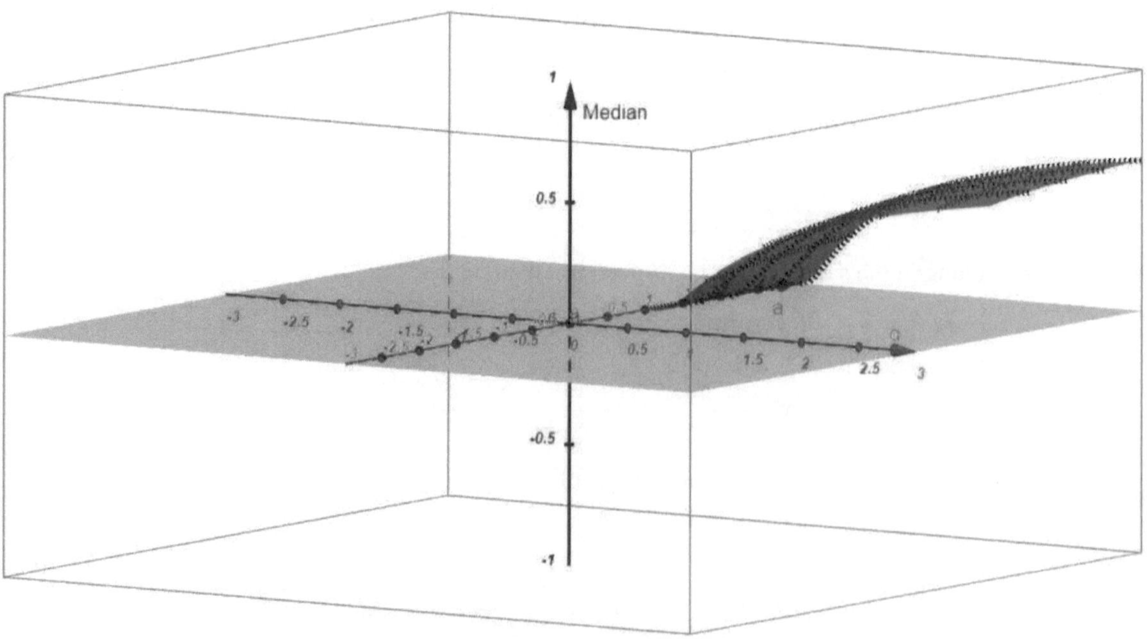

Figure 3: Variation in median with respect to changing parameters (α,γ).

and r^{th} cdf $F_r(x)$ as,

$$F_r(x) = \sum_{j=r}^{n} \binom{n}{j} F^j(x)\,[1 - F(x)]^{n-j}$$

$$= \sum_{j=r}^{n} \sum_{l=0}^{n-j} \binom{n}{j} \binom{n-r}{l} (-1)^l F^{j+l}(x)$$

Thus using equation (3) and (4) the pdf $f_r(x)$ and cdf $F_r(x)$ of the rth order statistics based on a random sample of size n from the proposed distribution can be easily obtained and is given below as,

$$f_r(x) = \frac{n!2\alpha x^{\alpha-1}(1-x)(2-x)}{(r-1)!(n-r)!} a^{x^{\alpha}(2-x)^{\alpha}} \sum_{l=0}^{n-r} \sum_{k=0}^{\infty} \frac{(-1)^{2r+3l-k-2}}{(1-a)^{r+l}} \times \binom{n-r}{l} \binom{r+l-1}{k} a^{(x^{\alpha}(2-x))^{\alpha}(r+l-k-1)} \tag{10}$$

and

$$F_r(x) = \sum_{j=r}^{n} \sum_{l=0}^{n-j} \sum_{k=0}^{\infty} \sum_{k=0}^{\infty} \binom{\mathrm{n}}{\mathrm{j}} \binom{\mathrm{n-r}}{\mathrm{l}} \binom{\mathrm{j+1}}{\mathrm{k}} \frac{(-1)^{2j+3l-k}}{(1-e)^{j+l}} a^{(x(2-x))^{\alpha(j-k+l)}} \tag{11}$$

3.5 Mean deviation

The mean deviation about the mean is defined by,

$$\delta_1(X) = \int_0^1 |x - \mu| f(x)dx,$$

where μ is the mean which can be rewritten as follows $\delta_1(X) = \int_0^{\mu}(\mu - x)\, f(x)dx + \int_{\mu}^{1}(|x - \mu)\, f(x)dx$. Using integration by parts and putting $E(X) = \int_0^1 xf(x)dx = \mu$, it simplifies to

$$\delta_1(X) = 2\mu F(\mu) - 2\mu + 2\int_{\mu}^{1} xf(x)dx,$$

where $F(\cdot)$ denotes the proposed cdf. Hence, from Theorem 3.2,

$$\int_{\mu}^{1} f(x)dx = \frac{4\alpha ln(a)}{a-1} \sum_{i=0}^{\infty} \sum_{j=0}^{\infty} \frac{(ln(a))^i}{i!} (-1)^j 2^{\alpha+\alpha i-j-1} \binom{\alpha+\alpha \mathrm{i}-1}{\mathrm{j}} (1-\mu) \times \left(\frac{1}{r+\alpha+\alpha i+j} - \frac{1}{r+\alpha+\alpha i+j+1} \right)$$

and thus,

$$\delta_1(X) = 2\mu F(\mu) - 2\mu + \frac{4\alpha ln(a)}{a-1} \sum_{i=0}^{\infty} \sum_{j=0}^{\infty} \frac{(ln(a))^i}{i!} (-1)^j 2^{\alpha+\alpha i-j-1} \binom{\alpha+\alpha \mathrm{i}-1}{\mathrm{j}} (1-\mu) \times \left(\frac{1}{r+\alpha+\alpha i+j} - \frac{1}{r+\alpha+\alpha i+j+1} \right) \tag{12}$$

The mean deviation about the median is defined as,

$$\delta_2(X) = \int_0^1 |x - M|\, f(x)dx$$
$$= \int_0^M (M - x)\, f(x)dx + \int_M^1 (x - M) f(x)dx,$$

where M stands for median, then after simplification, by putting $F(M) = \frac{1}{2}$, we get,

$$\delta_2(X) = -\mu + 2\int_M^1 x f(x)dx.$$

By Theorem 3.2,

$$\int_M^1 x f(x)dx = \frac{2\alpha ln(a)}{a-1}\sum_{i=0}^{\infty}\sum_{j=0}^{\infty}\frac{(ln(a))^i}{i!}(-1)^j 2^{\alpha+\alpha i-j-1}\binom{\alpha+\alpha i-1}{j}(1-M)$$
$$\times\left(\frac{1}{r+\alpha+\alpha i+j}-\frac{1}{r+\alpha+\alpha i+j+1}\right)$$

Thus, the expression for the mean deviation about the median can easily be written as,

$$\delta_2(X) = -\mu + \frac{2\alpha\gamma}{e-1}\sum_{i=0}^{\infty}\sum_{j=0}^{\infty}\frac{(ln(a))^i}{i!}(-1)^j 2^{\alpha+\alpha i-j-1}\binom{\alpha+\alpha i-1}{j}(1-M)$$
$$\times\left(\frac{1}{r+\alpha+\alpha i+j}-\frac{1}{r+\alpha+\alpha i+j+1}\right) \tag{13}$$

3.6 Entropy

An entropy is a measure of randomness or uncertainty of any system. We derive the expression of Reńyi entropy (see Renyi, 1961) which generalizes the Hartley and Shannon entropies. Let X have the pdf f(x) then Reńyi entropy is defined as,

$$\mathcal{J}_R(\beta) = \frac{1}{1-\beta}\log[\int f^\beta(x)dx] \quad \text{where } \beta > 0 \text{ and } \beta \neq 1.$$

From equation (4) we get,

$$\int_0^1 f^\beta(x)dx = \int_0^1\left[\frac{2\alpha\ln(a)}{a-1}(1-x)(x(2-x)^{\alpha-1}a^{x(2-x))^\alpha}\right]^\beta dx$$

after simplification,

$$\int_0^1 f^\beta(x)dx = \left[\frac{2\alpha\ln(a)}{a-1}\right]^\beta\sum_{i=0}^{\infty}\sum_{j=0}^{\infty}\sum_{k=0}^{\infty}(-1)^{j+k}(2)^{\alpha\beta+\alpha\beta i-\beta}$$

hence we get,

$$\mathcal{J}_R(\beta) = \frac{\beta}{1-\beta}\log\left(\frac{2\alpha\ln(a)}{a-1}\right)+\frac{1}{1-\beta}\log\left\{\sum_{i=0}^{\infty}\sum_{j=0}^{\infty}\sum_{k=0}^{\infty}(-1)^{j+k}(2)^{\alpha\beta+\alpha\beta i-\beta}\right.$$
$$\left.\times\binom{\alpha\beta+\alpha\beta i-\beta}{j}\binom{\beta}{k}\times\left(\frac{1}{\alpha\beta+\alpha\beta i-\beta+j+k+1}\right)\right\} \tag{14}$$

3.7 Stress-strength reliability

The stress-strength reliability has been widely used in reliability analysis as a measure of the system performance under stress. In terms of probability, the stress-strength reliability can be obtained as:

$$R = P[X > Y],$$

where, X denotes strength of the system and Y denotes the stress applied on the system. The probability R can be used to compare two random variables encountered in various applied disciplines.

The stress-strength reliability, R for PETL random variables $X \sim PETL\,(\alpha_1, \gamma_1)$ and $Y \sim PETL\,(\alpha_2, \gamma_2)$ is given by,

$$R = \int_0^1 \left(\int_0^x f_Y(y, \alpha_2, \gamma_2)dy\right) f_X(x, \alpha_1, \gamma_1)dx$$

$$R = \int_0^1 f_X(x, \alpha_1, \gamma_1) f_Y(x, \alpha_2, \gamma_2)$$

Here, we assume that $a_1 = a_2$ and on simplifying, we have,

$$R = \frac{2\alpha_1 \ln(a)}{(a-1)^2}\left[\frac{\Gamma(\frac{\alpha_1}{\alpha_1+\alpha_2}) - \Gamma(\frac{\alpha_1}{\alpha_1+\alpha_2}, 2(\alpha_1+\alpha_2)ln(a))}{2(\alpha_1+\alpha_2)(-1)^{\frac{\alpha_1}{\alpha_1+\alpha_2}}} - \frac{a-1}{2\alpha_1 \ln(a)}\right] \tag{15}$$

where, Γ represents the gamma integral and $\Gamma(.,.)$ represents the incomplete gamma integral.

3.8 Identifiability

A family of distributions is said to be identifiable in parameters if the distributions of two members of the family are equal, i.e., $f_1(x, \Theta_1) = f_2\,(x, \Theta_2)$, then $\Theta_1 = \Theta_2$ for all x. Theorem 1 of Basu and Ghosh, 1980, states that the density ratio $\frac{f_1(x, \Theta_1)}{f_1(x, \Theta_1)}$, of two distinct members of the family defined on the interval (a,b), either converges to 0 or diverges to ∞, as $x \to a$. For the PETL distribution, we have,

$$\lim_{x\to 0}\frac{f_1(x,\alpha_1,\gamma_1)}{f_2(x,\alpha_2,\gamma_2)} = \frac{\alpha_1 \ln(a_1)}{\alpha_2 \ln(a_2)}\frac{a_1-1}{a_1-1}(x(2-x))^{\alpha_1-\alpha_2}\, a_1^{(x(2-x)\alpha_1}\, a_2^{(x(2-x)\alpha_1}$$

$$\lim_{x\to 0}\frac{f_1(x,\alpha_1,a_1)}{f_2(x,\alpha_2,a_2)} = \begin{cases} 0 & \text{if } \alpha_1 > \alpha_2 \\ \infty & \text{if } \alpha_1 < \alpha_2 \\ 1 & \text{if } \alpha_1 = \alpha_2 \end{cases} \tag{16}$$

Thus, the parameters of the PETL distribution are indentified since two members of the family have different densities for different values of α.

3.9 Stochastic ordering

A random variable X is said to be stochastically greater $(X \geq_{st} Y)$ than Y if $F_X\,(x) \geq F_Y\,(x)$ for all x. In a similar manner, X is said to be greater than Y in the

- hazard rate order $(X \geq_{hr} Y)$ if $h_Y\,(x) \geq h_X\,(x)$ for all x.
- mean residual life order $(X \geq_{mlr} Y)$ if $m_X\,(x) \geq m_Y\,(x)$ for all x.
- likelihood ratio order $(X \geq_{lr} Y)$ if $\frac{f_X(x)}{f_Y(x)}$ is an increasing function of x.

Theorem 3.3 *Let $X \sim PETL\,(\alpha_1, \alpha_1)$ and $X \sim PETL\,(\alpha_2, \alpha_2)$. Then, we have the following conditions:*

1. For $a_1 = a_2 = a$ and $\alpha_1 \geq \alpha_2$, $(X \geq_{lr} Y)$, $(X \geq_{mlr} Y)$, $(X \geq_{hr} Y)$ and $(X \geq_{st} Y)$ for all x.
2. For $a_1 = a_2 = a$ and $\alpha_1 \geq \alpha_2$, $(X \geq_{lr} Y)$, $(X \geq_{mlr} Y)$, $(X \geq_{hr} Y)$ and $(X \geq_{st} Y)$ for all x.

Proof. The likelihood ratio of RV X and Y is given by,

$$\frac{f_X(x;\alpha_1,a_1)}{f_Y(x;\alpha_2,a_2)} = \frac{\alpha_1 \ln(a_1)}{\alpha_2 \ln(a_2)}(x(2-x))^{\alpha_1-\alpha_2}\, a_1^{(x(2-x))\alpha_1}\, a_2^{(x(2-x))\alpha_1}$$

We first take $a_1 = a_2 = a$ and differentiate the likelihood ratio with respect to x. It gives,

$$\frac{d}{dx}\left(\frac{f_X(x;\alpha_1,a_1)}{f_Y(x;\alpha_2,a_2)}\right)=\frac{2\alpha_1}{\alpha_2}\frac{(x(2-x))^{\alpha_1-\alpha_2}\,a_1^{(x(2-x))\alpha_1}\,a_2^{(x(2-x))\alpha_1}}{x(2-x)} \tag{17}$$
$$\times[\ln(a)\alpha_1(x(2-x))^{\alpha_1}]-\ln(a)\alpha_2(x(2-x))^{\alpha_2+\alpha_1=-\alpha_2}$$

For $\alpha_1 > \alpha_2$, $\frac{d}{dx}(\frac{f_X(x;\,\alpha_1,\,\gamma)}{f_Y(x;\,\alpha_2,\,\gamma)}) > 0$, for all x. Clearly, the likelihood ratio is an increasing function of x. Hence, for $a_1 = a_2 = a$, $(X \geq_{lr} Y)$.

By Shaked and Shanthikumar, 1994, $(X \geq_{lr} Y) \Rightarrow (X \geq_{hr} Y)$ $(X \geq_{mlr} Y)$ and $(X \geq_{st} Y)$.

Similarly, for $\alpha_1 > \alpha_2 = \alpha$ we have,

$$\frac{d}{dx}\left(\frac{f_X(x;\alpha,a_1)}{f_Y(x;\alpha,a_2)}\right)=\frac{2\alpha\ln(a_1)(a_2-1)}{\ln(a_2)(a_1-1)}\frac{(\ln(a_1)-\ln(a_2))(1-x)(x(2-x))^{\alpha}}{x(2-x)}\frac{a_1^{(x(2-x))\alpha}}{a_2^{(x(2-x))\alpha}} \tag{18}$$

For $\alpha_1 > \alpha_2$, $\frac{d}{dx}(\frac{f_X(x;\,\alpha,\,\gamma_1)}{f_Y(x;\,\alpha,\,\gamma_2)}) > 0$, for all x. Clearly, the likelihood ratio is an increasing function of x. Hence, for $\alpha_1 = \alpha_2 = \alpha$, $(X \geq_{lr} Y)$.

By, Shaked and Shanthikumar, 1994, $(X \geq_{lr} Y) \Rightarrow (X \geq_{hr} Y)$ $(X \geq_{mlr} Y)$ and $(X \geq_{st} Y)$.

3.10 Ordinary differential equations for density and survival functions

We provide the first order differential equations of density and survival functions of the PETL distribution. The first order derivative of the pdf is,

$$f'(x)=\left(\frac{2\alpha\ln(a)}{a}\right)(x(2-x))^{\alpha-1}(1-x)a^{(x(2-x))^{\alpha}}\left[\frac{2\alpha\ln(a)(x(2-x))^{\alpha}(1-x)^2+(2\alpha-1)(1-x)^2-1}{x(2-x)(1-x)}\right]$$

Thus, the first order ODE for the density function is,

$$y'-\left[\frac{2\alpha\ln(a)(x(2-x))^{\alpha}(1-x)^2+(2\alpha-1)(1-x)^2-1}{x(2-x)(1-x)}\right]y=0 \tag{19}$$

where, $y = f(x)$ and $Y' = \frac{df(x)}{dx}$. For some parameter values, first ODEs of the pdf are given in Table 1.

The survival function of the PETL distribution given by $z = S_X(x) = 1 - \frac{a^{(x(2-x))^{\alpha}-1}}{a-1}$. On differentiating it w.r.t. x, we have,

$$z'=\frac{-2\alpha\ln(a)(1-x)(x(2-x))^{\alpha}\,a^{(x(2-x))^{\alpha}}}{(a-1)x(2-x)}$$

Thus, on simplification we have,

$$z' + 2\alpha\ln(\alpha)(1-x)\log_{\alpha}[(1-z)(a-1)+1][(1-z)(a-1)+1]=0 \tag{20}$$

where, $z = S_X(x)$ and $z' = \frac{dS_X(x)}{dx}$. For some parameter values, first order ODEs of the survival function are given in Table 2.

Table 1: Ordinary Differential Equations of PETL density.

α	a	**First Order ODE**
1	2	$y' - \left\{ \dfrac{2\ln(2)(x(2-x))(1-x)^2 + (1-x)^2 - 1}{x(2-x)(1-x)} \right\} y = 0$
2	2	$y' - \left\{ \dfrac{4\ln(2)(x(2-x))^2(1-x)^2 + 3(1-x)^2 - 1}{x(2-x)(1-x)} \right\} y = 0$
3	2	$y' - \left\{ \dfrac{6\ln(2)(x(2-x))^3(1-x)^2 + 5(1-x)^2 - 1}{x(2-x)(1-x)} \right\} y = 0$

Table 2: Ordinary Differential Equations of the PETL survival function.

α	a	**First Order ODE**
1	2	$z' + 2\ln(2)(1-x)(2-z)\log_2(2-z) = 0$
2	2	$z' + 4\ln(2)(1-x)(2-z)\log_2(2-z) = 0$
3	2	$z' + 6\ln(2)(1-x)(2-z)\log_2(2-z) = 0$

4. Maximum likelihood estimation and simulation

4.1 Point estimation

Maximum likelihood estimates of parameters and of the proposed distribution are obtained by maximizing the logarithm of the likelihood function. The logarithm likelihood function is,

$$\begin{aligned}\log L = n\log\frac{2\ln(a)}{(a-1)} + n\log(\alpha) + \sum_{i=1}^{n} log(1-x_i) + (\alpha-1) \\ \times \sum_{i=1}^{n} log(x_i(2-x_i)) + \log(a)\sum_{i=1}^{n}(x_i(2-x_i))^{\alpha}\ln(x_i(2-x_i)).\end{aligned} \tag{21}$$

Differentiating it with respect to the parameters we get,

$$\frac{d\log L}{d\alpha} = \frac{n}{\alpha} + \sum_{i=1}^{n} log(x_i(2-x_i)) + \ln(a)\sum_{i=1}^{n}(x_i(2-x_i))^{\alpha} log(x_i(2-x_i)) \text{ and}$$

$$\frac{d\log L}{d\alpha} = \left\{\frac{a - a\log(a) - 1}{(a-1)a\log(a)}\right\} + \frac{1}{a}\sum_{i=1}^{n}(x_i(2-x_i)^{\alpha}).$$

After equating these equations to zero, we get two non-linear equations. Solving these simultaneously, we get MLEs $\hat{\alpha}$ and $\hat{a}$ of parameters α and a respectively. It may be noted here that these equations cannot be solved analytically. However one can use some numerical technique for their solution therefore, we propose the use of the Newton Raphson method. For the choice of an initial guess the contour plot technique is used.

4.2 Asymptotic confidence intervals

For large samples, we can obtain the confidence intervals based on the diagonal elements of the Fisher information matrix $I^{-1}(\hat{\alpha}, \hat{a})$ which provides the estimated asymptotic variance for the parameters and respectively. Thus the two sided $100(1-\beta)$ confidence interval of and can be defined as $\hat{\alpha} \pm Z_{\beta/2}\sqrt{var(\hat{\alpha})}$ and $\hat{a} \pm Z_{\beta/2}\sqrt{var(\hat{a})}$

respectively where denotes the upper point of standard normal distribution. The Fisher Information matrix can be estimated by,

$$I(\hat{\alpha},\hat{a})=\begin{bmatrix}\dfrac{-d^2\log L}{d\alpha^2} & \dfrac{-d^2\log L}{d\alpha da}\\ \dfrac{-d^2\log L}{d\alpha da} & \dfrac{-d^2\log L}{da^2}\end{bmatrix}_{(\hat{\alpha},\hat{a})} \tag{22}$$

where,

$$\frac{d^2\log L}{d\alpha^2}=\frac{-n}{\alpha^2}+\log(a)\sum_{i=1}^{n}(x_i(2-x_i))^{\alpha}[\log x_i(2-x_i)]^2$$

$$\frac{d^2\log L}{da^2}=\frac{a^2(\log(a))^2+(-a^2+2a-1)\log(a)-a^2+2a-1}{(a-1)^2a^2(\log(a))^2}-\frac{1}{a^2}\sum_{i=1}^{n}(x_i(2-x_i))^{\alpha}$$

$$\frac{d^2\log L}{d\alpha a}=\frac{1}{a}\sum_{i=1}^{n}(x_i(2-x_i))^{\alpha}\log(x_i(2-x_i))$$

4.3 Random number generation

The steps to generate random numbers from the proposed $PETL(\alpha,a)$ distribution are,

1. Select n,α and a.
2. Generate a standard uniform random number, $u\sim U(0,1)$.
3. Using the quantile function, compute $x=1-\sqrt{2-\{\log_a(1+u(a-1))\}^{1/\alpha}}$.
4. Repeat steps 2 and 3, n times to get a sample of size n, $\{x_1,x_2,\ldots,x_n\}$ from $PETL(\alpha,a)$.

Illustration:

1. We fix $n=10$, $\alpha=2$ and $a=3$.

The random sample generated is: {0.5589, 0.4685, 0.8239, 0.8712, 0.4002, 0.5428, 0.6220, 0.8125, 0.2229, 0.5197}.

2. We fix $n=20$, $\alpha=0.5$ and $a=1.5$.

The random sample generated is: {0.7579, 0.2430, 0.7586, 0.0027, 0.5100, 0.5114, 0.1640, 0.0124, 0.5053, 0.0049, 0.0014, 0.6928, 0.0062, 0.0211, 0.6638, 0.6572, 0.0193, 0.0399, 0.3028, 0.4389}.

4.4 Simulation study

We compute MSE and Mean Absolute Bias of the MLEs on the basis of 10000 simulated samples for the given values of parameters and sample size n. The MSE and Mean Absolute Bias (AB) are computed using the following formulae,

$$MSE(\hat{\alpha})=\frac{1}{10000}\sum_{j=1}^{10000}(\hat{\alpha}_j-\alpha)^2,\ AB(\hat{\alpha})=\frac{1}{10000}\sum_{j=1}^{10000}|\hat{\alpha}_j-\alpha|,$$

$$MSE(\hat{\alpha})=\frac{1}{10000}\sum_{j=1}^{10000}(\hat{\alpha}_j-\alpha)^2,\ AB(\hat{\alpha})=\frac{1}{10000}\sum_{j=1}^{10000}|\hat{\alpha}_j-\alpha|.$$

We present the results of the simulation study for parameters, $(\alpha,a)=(3,1.5)$ in Table 3 and the results are visualized in Figure 4.

It can be seen that the MLEs are consistent since the MSEs of $\hat{\alpha}$ and $\hat{a}$ decrease with increasing sample size. The mean absolute bias also decreases as the sample size increases.

Table 3: MSE and Mean Absolute Bias (AB) for (α, a) for simulated samples.

$(\alpha,a) = (3,1.5)$				
Sample Size	MSE ($\hat{\alpha}$)	AB($\hat{\alpha}$)	MSE($\hat{a}$)	AB($\hat{a}$)
50	1.0764	0.8013	28406.9643	19.0250
100	0.5616	0.5797	17420.9930	12.2759
150	0.3631	0.4640	14402.0671	8.0635
200	0.2595	0.3969	5507.2345	3.8488
250	0.2075	0.3565	3556.9818	2.7741
300	0.1694	0.3222	4811.6824	2.6653
350	0.1392	0.2952	5019.0465	1.5187
400	0.1254	0.2797	592.9395	0.8561
450	0.1063	0.2575	55.3961	0.5489
500	0.1001	0.2501	1737.2620	0.9461

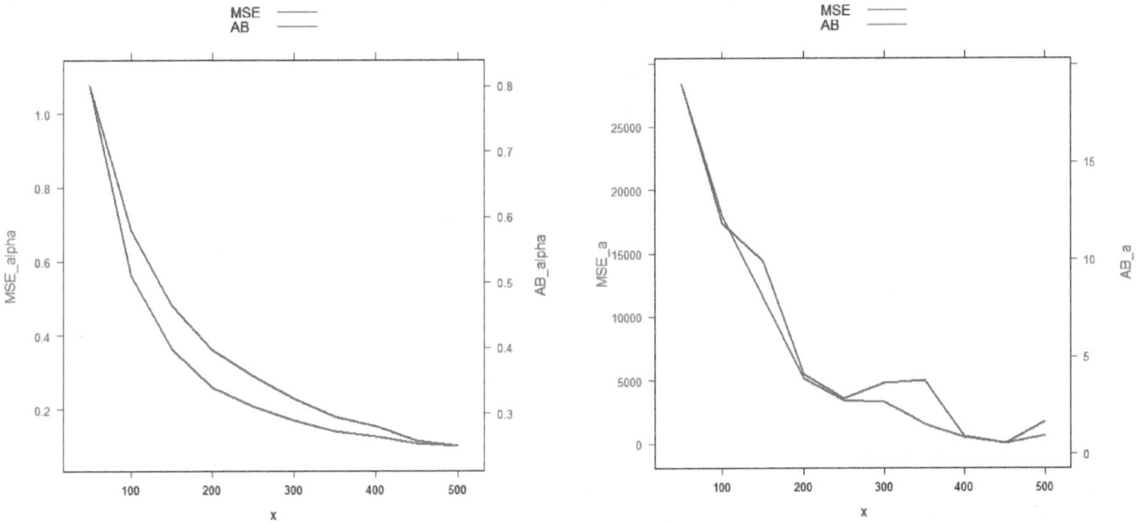

Figure 4: MSE and Mean Absolute Bias of $\hat{\alpha}$ and $\hat{a}$ for simulated samples with $\alpha = 3$ and $a = 1.5$

5. Real data fitting

In this section, we analyze a real dataset in order to illustrate the good performance of the proposed PETL(α, a) distribution. The data set that is considered here is given by Linhart and Zucchini, 1986, and it represents the failure time (in days) of the air conditioning system of an airplane as: 0.46, 0.46, 0.46, 0.5, 0.58, 0.58, 0.58, 0.67, 0.67, 0.83, 0.87. The initial values of the iterative algorithm are: $\alpha = 5$ and $a = 1.65$. The MLEs of the parameters of the PETL distribution are: $\hat{\alpha} = 5.3064$ and $\hat{a}$ =0.0017.

Figure 5 gives the graphical representation of the fitting of different models that are considered over the given data set (refer Sharma, 2020 and R Core Team, 2013). The K-S statistic for the fitting is 0.2727, with p-value 0.8071. These figures suggest that the PETL distribution is very suitable for this data. Further, we compare the goodness-of-fit statistics with Beta, Kumaraswamy, unit-Gamma, Topp-Leone and Generalized Topp-Leone (given by Shekhawat and Sharma, 2021) distributions which are also defined on the unit interval.

We use Akaike Information Criterion (AIC) and Bayesian Information Criterion (BIC) to identify the best possible model for the given data set. The model that has the smallest values of AIC and BIC statistics

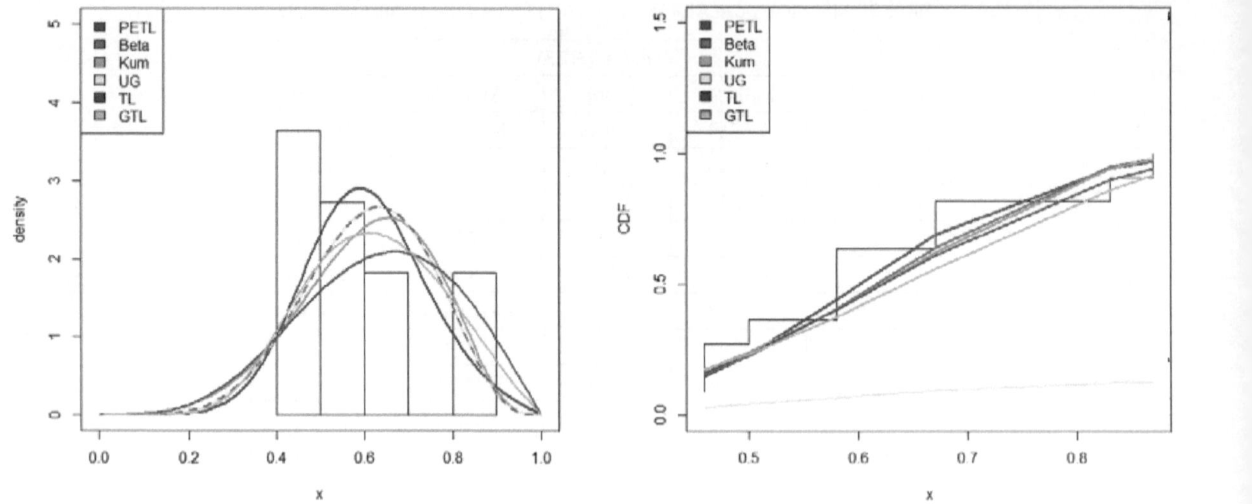

Figure 5: Fitting of different models on the given dataset.

is considered to be the best possible model among the distributions under comparison. The statistics are computed by,

$$AIC = 2k - 2\log(L)$$

$$BIC = k\log(n) - 2\log(L)$$

where, k is the number of estimated parameters, L is the maximum value of the likelihood function and n is the number of observations.

In Table 4, we show the MLEs and goodness-of-fit statistics for the PETL, Beta, Kumaraswamy, unit-Gamma, Topp-Leone and GTL (given by Shekhawat and Sharma, 2021) distributions for the considered dataset. We also present the K-S statistic, corresponding p-values and the values of the maximum likelihood functions. From Table 4, we observe that the proposed Power Exponentiated Topp-Leone distribution is statistically fitted for the considered dataset and has the smallest AIC & BIC along with highest log(L) value among all the distributions. Therefore, we recommend the use of the PETL distribution for modeling the considered dataset over some other existing distributions.

Table 4: MLEs, AIC, BIC, K-S statistic and p-values for fitted models.

Comparison of Distributions						
Model	$log(L)$	MLEs	AIC	BIC	K-S statistic	p-value
PETL (α,a)	6.6380	(8.4832,0.0614)	−9.2761	−8.4803	0.2727	0.8071
Beta (α,β)	6.0499	(6.6369,4.2526)	−8.0998	−7.3040	0.2727	0.8079
Kumaraswamy (α,β)	5.6788	(4.1107,4.5691)	−7.3577	−6.5619	0.2727	0.8079
Unit-Gamma (α,β)	6.0774	(4.2104,8.0081)	−8.1549	−7.3591	0.9090	0.0002
Topp-Leone (α)	6.0774	(5.0433)	−8.8614	−8.4635	0.2727	0.8079
GTL (α,β)	6.3733	(0.0220,6350.2850)	−8.7467	−7.9509	0.2727	0.8079

6. Conclusion

A two parameter extension of the J-shaped Topp-Leone distribution called as Power Exponentiated Topp-Leone (PETL) distribution is introduced for a possible application to model failure times (in days) of the air conditioning system of an airplane. The proposed distribution has increasing, decreasing and bathtub shaped hazard functions. The expressions of ordinary moments, conditional moments, quantile function, mean deviation, order statistics and entropy are discussed. Other important properties of the distribution such as identifiability, ordinary differential equations, stochastic orderings and stress-strength reliability are also discussed.

The estimation techniques for the parameters are also discussed. The simulation study that was conducted proved the consistency of the ML estimators of the parameters. Further, an algorithm for the generation of random samples from the proposed distribution is also given to facilitate future studies.

According to the various model selection criteria, AIC & BIC and KS goodness-of-fit test the proposed PETL distribution is a better model for fitting the maximum flood levels of Susquehenna River data over the Beta, Kumaraswamy, unit-Gamma, Topp-Leone and Generalized Topp-Leone (given by Shekhawat and Sharma, 2021) distributions. Summing up, it can be concluded that the proposed PETL distribution can be effectively used for modeling real life data defined on the unit interval.

References

Alizadeh, M., Lak, F., Rasekhi, M., Ramires, T. G., Yousof, H. M. et al. (2018). The odd log-logistic topp-leone g family of distributions: heteroscedastic regression models and applications. Comput. Stat., 33, 3: 1217–1244.

Basu, A. P. and Ghosh, J. K. (1980). Identifiability of distributions under competing risks and complementary risks model. Communications in Statistics - Theory and Methods, 9: 1515–1525.

Consul, P. C. and G. C. Jain. (1971). On the log-gamma distribution and its properties. Stat Pap., 12: 100–106.

Gómez-Déniz, E., Sordo, M. A. and Calderín-Ojeda, E. (2014). The log-lindley distribution as an alternative to the beta regression model with applications in insurance. Insur Math Econ, 54: 49–57.

Johnson, N. L. (1949). Systems of frequency curves generated by methods of translation. Biometrika, 36(1/2): 149–176.

Korkmaz, M. Ç. and Chesneau, C. (2021). On the unit burr-xii distribution with the quantile regression modeling and applications. Comp. Appl. Math., 40(1).

Kumaraswamy, P. (1980). A generalized probability density function for double-bounded random processes. J. Hydrol., 46(1-2): 79–88.

Kumar, D., Singh, U. and Singh, S. K. (2015). A method of proposing new distribution and its application to bladder cancer patient data. JSAPL, 2: 235–245.

Linhart, H. and Zucchini, W. (1986). Model selection. Wiley, New York.

Mazucheli, J., Dey, S. and Menezes, A. F. (2018a). The unit-birnbaum-saunders distribution with applications. Chile J. Stat., 9(1): 47–57.

Mazucheli, J., Menezes, A. F. B. and Ghitany, M. E. (2018b). The unit-weibull distribution and associated inference. J. Appl. Probab. Stat., 13: 1–22.

Mazucheli, J., Menezes, A. F. and Dey, S. (2019). Unit-gompertz distribution with applications. Statistica, 79(1): 25–43.

Nadarajah, S. and Kotz, S. (2003). Moments of some j-shaped distributions. J. Appl. Stat., 35, 10: 1115–1129.

R Core Team. R: A Language and Environment for Statistical Computing. R Foundation for Statistical Computing, Vienna, Austria, 2013.

Renyi, A. (1961). On measures of entropy and information. In Proceedings of the 4th Berkeley symposium on mathematical statistics and probability. Berkeley :University of California Press., 1: 547–561.

Sangsanit, Y. and Bodhisuwan, W. (2016). The topp-leone generator of distributions: properties and inferences. Songklanakarin. J. Sci. Technol., 38: 537–548.

Shaked, M. and Shanthikumar, J. (1994). Stochastic orders and their applications. Academic Press, Boston.

Sharma, V. K. (2018). Topp-leone normal distribution with application to increasing failure rate data. Journal of Statistical Computation and Simulation, 83, 2: 326–339.

Sharma, V. K. (2020). R for lifetime data modeling via probability distributions. Handbook of Probabilistic Models, Elsevier Inc.

Shekhawat, K. and Sharma, V. K. (2021). An extension of j-shaped distribution with application to tissue damage proportions in blood. Sankhya B, 83: 543–574.

Topp, C. W. and Leone, F. C. (1995). A family of j-shaped frequency functions. J. Am. Stat. Assoc., 50: 209–219.

Chapter 21

Testing the Goodness of Fit in Instrumental Variables Models

*Shalabh** and *Subhra Sankar Dhar*

1. Introduction

An important application of any statistical modeling is that the fitted models are used in different applications. For example, the fitted statistical model is used to make various types of forecasts. The success of such applications depends upon how good the model is, i.e., how the model is well fitted the given set of data. Only a good fitted model can provide valid results in further applications. The goodness of fit of a model is usually judged by a statistic which itself is a random variable and its value is computed on the basis of a given data set. This estimated value reflects the value of the population parameter responsible for measuring the goodness of fit. So different samples will generate different estimated values of the parameter. Moreover, in order to further validate the statistical inferences, the testing of hypothesis concerning the parameter being estimated is also required. Only estimating the statistical parameter related to the goodness of fit may not suffice.

Considering the set up of a multiple linear regression, the coefficient of determination, popularly known as R^2, is used to judge the goodness of fit of the model based on the set of observations on a study variable and a set of explanatory or independent variables. The goodness of fit through R^2 is measured by estimating the squared population multiple correlation coefficient between the study and explanatory variables. The R^2 is based on the ordinary least squares (OLS) estimator of regression coefficients and is a consistent estimator of the squared multiple correlation coefficient. Hence the test of hypothesis concerning the squared population multiple correlation coefficient between the study and explanatory variables is conducted based on the distribution of R^2. The suitable test statistics and test procedures are available in the literature, see, Anderson (2003, Chap. 4).

Note that the OLS estimator (OLSE) is the best linear unbiased estimator of the regression coefficients. Hence the coefficient of determination is expected to give good results under the assumptions of the multiple linear regression model. In real data analysis, one or more assumptions of the multiple linear regression model are often violated. Suppose the assumption that the explanatory variables and the random errors are statistically independent is violated. Such a violation is possible when the explanatory variables are stochastic which happens under different econometric models, e.g., errors-in-variables or error measurement models, simultaneous equation models, time series models and more. When the explanatory variables and the random errors are correlated, the properties of the OLSE to remain the best linear unbiased estimator of regression coefficients are lost. In fact, the OLSE becomes not only biased but also an inconsistent estimator of the regression coefficient under such a violation of assumptions. Consequently, the coefficient of determination also becomes an inconsistent estimator of the squared population multiple correlation coefficient, see Cheng

Department of Mathematics and Statistics, Indian Institute of Technology Kanpur, Kanpur 208016 (India).
Email: subhra@iitk.ac.in
* Corresponding author: shalab@iitk.ac.in

et al. (2014, 2016). Under this situation, the hypothesis testing related to the squared multiple correlation coefficient will be based on an inconsistent estimator of the multiple correlation coefficient which may provide invalid and erroneous statistical inferences. A natural question arises on how to check the goodness of fit in such a situation when such assumptions are violated and to conduct the hypothesis test.

The instrumental variable (IV) estimation provides a consistent estimator of the regression parameters under the multiple linear regression model when the explanatory variables and the random errors are not statistically independent, and consequently, we expect that it may provide consistent tests for the relevant hypothesis testing problems. The IVs are a set of variables which are highly correlated with the explanatory variables, at least in limit and uncorrelated with the random errors, at least in limit, see Bowden and Turkington (1984), Wansbeek and Meijer (2000, Chap. 6) and others, for an interesting exposition on IV estimation. It is important to note about the goodness of fit that it is a value obtained by computing a statistic based on a data sample and the it is obtained through the estimation of a corresponding relevant population parameter. The population parameter in this case is the squared multiple correlation coefficient between the study and explanatory variables in the multiple linear regression model. Nevertheless the validity of statistical analysis and inferences remains incomplete without the hypothesis test. Measuring the degree of goodness of fit is the fundamental requirement in any model fitting but models fitted using the IV estimation posses more challenges. So we address the important questions, how to check the goodness of fit in the IV model and then how to conduct the test of hypothesis for the squared multiple correlation coefficient based on IV estimates.

It is evident from the developments in the area of IV estimation that such issues are very pertinent to a user in real life applications. For example, the choice of IV is not unique and different choices of IVs provide different models for the same data. The goodness of fit statistic and its related hypothesis test help in choosing the appropriate IVs to provide a better fitted model. For example, consider the three popular techniques to choose the IVs, see Rao et al. (2008, Chap. 4, pp. 208–209). The Wald instrument technique divides the observations on the explanatory variable into two groups and chooses the IVs as +1 and -1 for the two groups. Similarly, the Bartlett instrument technique divides the observations on the explanatory variable into three groups and choose the IV as +1, 0 and -1 for the upper, middle and lower groups respectively, and the Durbin instrument technique uses the ranks of the observations on the explanatory variable as instruments. The question is now how to decide which choice will give a better fitted model? A goodness of fit statistic is needed to answer such questions and to decide which choice of IV provides a better fitted model. More complicated situations arise when more than one choice of IVs are used for the same explanatory variables in multiple linear regression model, then what choice will yield a better model is a question in which an experimenter will always be interested. The goodness of fit statistic and relevant hypothesis test based on IV estimation can answer such queries.

The issue about how to measure the goodness of fit in the IV model is addressed in Dhar and Shalabh (2021) and a goodness of fit statistic is obtained which is based on the use of IV estimators but there is no knowledge available on how to conduct the hypothesis test. In this context, the proposed goodness of fit statistic consistently estimates the squared multiple correlation coefficient, which measures the goodness of fit of the model based on IV estimation using a data set on the study and explanatory variables. Now, one may be interested in testing the significance of the squared multiple correlation coefficient. For example, testing the significance that the squared multiple correlation coefficient equals any given value or more than any given value will shed more light on the status of goodness of fit. Such tests can be carried out using the goodness of fit statistics based on IV estimation. How to address this issue is thoroughly studied in this chapter in Section 5.

The plan of the paper is as follows. We consider a general setup of the multiple linear regression model in which the covariance matrix of the random errors is assumed to be unknown. The set ups of multiple linear regression models and IV estimation are described in Section 2. A motivation to develop the goodness of fit statistic is presented in Section 3. The development of goodness of fit statistics under such a set up is briefly presented in Section 4 from Dhar and Shalabh (2021) for the sake of completeness and better understanding. A consistent hypothesis test for testing the significance of squared multiple correlation coefficient is developed in Section 5 followed by some conclusions in Section 6. The proof of the results are presented in the Appendix in Section 7.

2. Instrumental variable (IV) estimation

First we consider the multiple linear regression model with an intercept term,

$$y = X\beta + \epsilon, \tag{2.1}$$

where y is the $(n \times 1)$ vector of observations on the study variable, X is the $(n \times (p+1))$ matrix of n observations on each of the p explanatory variables and an intercept term, β is the $((p+1) \times 1)$ vector of regression coefficients associated with the $(p+1)$ explanatory variables, and ϵ is the $(n \times 1)$ vector of random errors. The study variable y is linearly related to the p explanatory variables $X_1, X_2, \ldots, X_p$ and the first column in X has all the unity elements representing the intercept term. We consider a very general framework for the random errors ϵ in terms of their covariance matrix. We assume that $E(\epsilon) = 0$ and the random errors are non-spherically distributed, and $E(\epsilon\epsilon') = \sigma^2\Omega^{-1}$, where Ω is an unknown positive definite matrix. Suppose Ω is consistently estimated by an estimator $\hat{\Omega}$. Such a specification will also allow $\Omega = I$ as the special case of identically and independently distributed random errors as in the case of standard multiple linear regression model.

We assume that the assumptions of the multiple linear regression model (see Rao et al. (2008, p. 34)) are satisfied but the assumption that X and ϵ are uncorrelated, at least in limit, is violated. It is assumed in the usual multiple linear regression model that $\frac{X'X}{n} \xrightarrow{p} \Sigma_X$ and $\frac{X'\epsilon}{n} \xrightarrow{p} 0$ as $n \to \infty$, where $\xrightarrow{p}$ denotes the convergence in probability. So here we assume that X and ϵ are correlated in the sense that,

$$\frac{X'\epsilon}{n} \xrightarrow{p} X^* \tag{2.2}$$

as $n \to \infty$ where X^* is an arbitrary non-zero random variable. Such an assumption holds, e.g., when X is stochastic in nature which arises in several econometric models. As the presence of the intercept term in the usual multiple linear regression model is needed for the validity of the coefficient of determination, without the loss of generality, we also assume the presence of an intercept term in the model for the validity of the proposed goodness of fit statistics based on IV estimation which is studied in Section 4.

It is well known that the OLSE $b = (X'X)^{-1}X'y$ is the best linear unbiased estimator of β under the standard assumptions of a multiple linear regression model. It remains a consistent estimator of β as long as X and ϵ are uncorrelated, at least in the limit. When X and ϵ are correlated, at least in the limit, the same OLSE b becomes an inconsistent estimator of β. Consequently, all the model diagnostic tools and statistics based on OLSE then may not provide the correct statistical inferences. For example, the coefficient of determination is based on OLSE and measures the degree of goodness of fit. It is a consistent estimator of the squared population multiple correlation coefficient as long as X and ϵ are uncorrelated, at least in the limit. The coefficient of determination becomes an inconsistent estimator of the squared population multiple correlation coefficient when X and ϵ become correlated, at least in the limit. There can be different approaches to solve such issues. One approach is to use a consistent estimator of β in place of b. The instrumental variable estimation provides a consistent estimator of β when X and ϵ become correlated, in the limit.

The instruments in the IV estimation are a set of variables which are highly correlated with the explanatory variables and least correlated with the random errors, in the limit. Suppose $Z_1, Z_2, \ldots, Z_p$ is a set of p instrumental variables such that they are correlated with X, in the limit and uncorrelated with ϵ, in the limit. Similar to X, the observations on these instrumental variables are arranged in a $(n \times (p+1))$ matrix, with an intercept term Z_0 as $Z = (Z_0, Z_1, \ldots, Z_p)$. We assume that,

$$\frac{Z'\epsilon}{n} \xrightarrow{p} 0 \tag{2.3}$$

$$\frac{Z'X}{n} \xrightarrow{p} \Sigma_{ZX} > 0 \tag{2.4}$$

$$\frac{X'X}{n} \xrightarrow{p} \Sigma_{XX} > 0 \tag{2.5}$$

$$\frac{Z'Z}{n} \xrightarrow{p} \Sigma_{ZZ} > 0, \tag{2.6}$$

where, Σ_{ZX}, Σ_{XX} and Σ_{ZZ} are non-singular positive definite matrices of constants.

The instrumental variable estimator is obtained in two stages as follows. Consider and express X in the set up of multiple linear regression model as,

$$X = Z\alpha^* + \phi, \tag{2.7}$$

where, α^* is a coefficient vector associated with Z, ϕ is the associated random error term with $E(\phi) = 0$ and $E(\phi\phi') = \sigma^2\Omega^{-1}$ where Ω is an unknown positive definite matrix. Suppose Ω is consistently estimated by an estimator $\hat{\Omega}$. The generalized least squares estimate of α^* is obtained in the first stage as,

$$\hat{\alpha}^*_{IV} = (Z'\hat{\Omega}Z)^{-1}Z'\hat{\Omega}X$$

from (2.7) and is used in the second stage. We obtain the predicted value of X as $\hat{X} = Z\hat{\alpha}^*_{IV} = P_{Z\hat{\Omega}}X$ where $P_{Z\hat{\Omega}} = Z(Z'\hat{\Omega}Z)^{-1}Z'$ in,

$$y^* = P_{Z\hat{\Omega}}X\beta + \epsilon. \tag{2.8}$$

Applying generalized least squares on (2.8) yields the two stage feasible generalized least squares estimator of β as,

$$\hat{\beta}_{IV} = (X'P_{Z\hat{\Omega}}X)^{-1}X'P_{Z\hat{\Omega}}y^*. \tag{2.9}$$

The IV estimators have been an attractive choice for the theoretical and applied researchers from various perspectives in parametric, nonparametric, semiparametric, Bayesian and frequentist frameworks. An important application of IV estimation is in handling the measurement error models which goes back to Sargan (1958, 1971), Mallios (1969) and Leamer (1978); see also Iwata (1992) and Abarin and Wang (2012). The group mean ordinary least squares estimator with a IV estimator is considered in Batistatou and McNamee (2008), a generalized IV estimator containing several common methods used in measurement errors is discussed in Söderström (2011). The method of moments estimation using IVs in generalized linear measurement error models under not necessarily normally distributed measurement errors in parametric and nonparametric setups is considered in Abrin and Wang (2012). The IV estimation in an error-components model is considered in Amemiya and MaCurdy (1986) and is illustrated under nonlinear measurement error models in Amemiya (1990). The IV estimation in nonparametric and semiparametric models is considered in, e.g., Park (2003), Newey and Powell (2003), Conley et al. (2008), Chib and Greenberg (2007), Horowitz (2011), Carroll (2004) and others. The IV estimation with some applications from an econometrician's view and for causal inferences are discussed in Imbens (2014) and Baiocchi et al. (2014), respectively. The relationships between the Bayesian and classical approaches to IV regression in simultaneous equation models is established in Kleibergen and Zivot (2003). The bayesian IV estimation is discussed in Wiesenfarth et al. (2014), Zellner et al. (2014), Lopes and Polson (2014), Gustafson (2007) etc. The IV estimation in a random coefficient model is studied in Clarke and Windmeijer (2012) and Chesher and Rosen (2014). The IV estimation in various other models, e.g., in varying-coefficient models is discussed in Zhao and Xue (2013); quantile regression is investigated in Chernozhukov et al. (2007) and Horowitz and Lee (2007); time series models are discussed in Kuersteiner (2001). The application of two stage IV estimation has been extensively used in real data, see, e.g., Angrist (1990), Card (1995), Acemoglu et al. (2001), Kim et al. (2011), Fish et al. (2010), Davies et al. (2013), Pokropek (2016) and many others.

Next we briefly describe how the goodness of fit statistics using the IV estimates can be found but before that we present some details about the coefficient of determination to motivate the development of goodness of fit statistic in IV model.

3. Goodness of fit in multiple linear regression model

Consider the multiple linear regression model under the standard assumptions (see Rao et al. (2008, p. 34)) in which the random errors are identically and independently distributed having mean zero and identity covariance matrix as, $\sigma^2 I$. The coefficient of determination, popularly denoted as R^2, is based on the OLSE $b = (X'X)^{-1}X'y$ of β. The R^2 measures the goodness of fit in the classical multiple linear regression model under one of the assumptions that explanatory variables and random errors are uncorrelated. The R^2 is defined as the ratio of sum of squares due to regression to the total sum of squares. It measures the proportion of variation in the data explained by the fitted model based on OLSE with respect to the total variation. The total sum of squares is orthogonally partitioned into sum of two orthogonal components, viz., the sum of squares due to regression and the sum of squares due to errors in the context of analysis of variance in the multiple linear regression model, see Rao et al. (2008, p. 57). Such a partitioning into orthogonal components is possible only when the explanatory variables and the random errors are uncorrelated.

Assuming that the explanatory variables and random errors are uncorrelated and $\frac{X'P_IX}{n} \xrightarrow{p} \Sigma_X$, the squared population multiple correlation coefficient between the study variable y and explanatory variables in X is given by,

$$\theta = \frac{\beta'\Sigma_X\beta}{\beta'\Sigma_X\beta + \sigma^2}, \quad 0 \leq \theta \leq 1, \tag{3.1}$$

where Σ_X is a positive definite finite matrix, $P_I = I_n - \frac{1}{n}e_ne_n'$, and $e_n = (1, 1, \ldots, 1)'$ is a $(n \times 1)$ vector of all unity elements.

When $\sigma^2 = 0$, then $\theta = 1$ and it indicates that the model is best fitted. On the other hand, if all the β's are zero, then $\theta = 0$ which indicates that the model is worst fitted. Similarly, any other value of θ will measure the goodness of fit of the model in terms of the squared population multiple correlation coefficient.

Obviously, the population multiple correlation coefficient is based on unknown parameters and is not usable in real data applications. We need a suitable estimator to estimate θ. When explanatory variables and random errors are uncorrelated, an estimator of squared population multiple correlation coefficient is defined as,

$$\begin{aligned} R_n^2 &= \frac{b'X'P_IXb}{y'P_Iy}; \quad 0 \leq R_n^2 \leq 1 \\ &= \frac{y'P_IX(X'P_IX)^{-1}X'P_Iy}{y'P_Iy} \end{aligned} \tag{3.2}$$

which is known as the coefficient of determination and measures the goodness of fit in terms of the ratio of sum of squares due to regression to the total sum of squares. The R_n^2 measures the proportion of variability explained by the fitted model using OLSE with respect to the total variability in the data. On the lines of interpretation of θ, the values of $R_n^2 = 0$ indicates that the model is worst fitted and $R_n^2 = 1$ indicates that the model is best fitted. Any other value of R_n^2 between 0 and 1 will suitably reflect the degree of goodness of fit. For example, if $R_n^2 = 0.7$, then the model is considered to be nearly 70% good fitted.

It is important to note that under the assumptions in multiple linear regression model that $\frac{X'P_I\epsilon}{n} \xrightarrow{p} 0$ and $\frac{X'P_IX}{n} \xrightarrow{p} \Sigma_X > 0$ is a finite matrix, R_n^2 is a biased but consistent estimator of θ, i.e., $E(R_n^2) \neq \theta$ and $R_n^2 \xrightarrow{p} \theta$, see, Anderson (2003).

The concept of coefficient of determination has been extensively studied and extended for various models in the literature. For example, the coefficient of determination in entropy form for generalized linear models is proposed by Eshima and Tabata (2010, 2011); the logistic regression model is discussed in Tjur (2009), Hong, Ham and Kim (2005), Liao and McGee (2003); a local polynomial model is proposed in Huang and Chen (2008), a mixed regression model is discussed in Hössjer (2008); multivariate normal repeated measure data is presented in Lipsitz et al. (2001); simultaneous equation models are discussed in Knight (1980); see also Renaud and Victoria-Feser (2010), van der Linde and Tutz (2008), Marchand (2001), Srivastava and Shobhit (2002), Marchand (1997, 2001), Nagelkerke (1991) and more for other developments including generalization of coefficient of determination in various directions.

4. Goodness of fit in instrumental variable model

Considering the basic philosophy of goodness of fit behind the definition of R_n^2, we formulate a statistic to measure the goodness of fit in IV model. It may be noted that R_n^2 is based on consistent OLSE b of β and the assumption that explanatory variables and random errors are uncorrelated. When this assumption does not hold true in case of IV model, b becomes an inconsistent estimator of β and the total sum of squares can not be partitioned into orthogonal components- the sum of squares due to regression and the sum of squares due to error. Also, the IV estimators are not the best linear unbiased estimators of β like the OLSE. A statistic for quantitatively measuring the goodness of fit in the IV models is developed in Dhar and Shalabh (2021) and is discussed briefly as follows for the sake of completeness and better understanding. We will develop the test of hypothesis based on this statistic later in Section 5

We consider model (2.8) and express the total sum of squares as,

$$\begin{aligned} y^{*'} P_I y^* &= (P_{Z\hat{\Omega}} X\beta + \epsilon)' P_I (P_{Z\hat{\Omega}} X\beta + \epsilon) \\ &= \beta' X' P_{Z\hat{\Omega}} P_I P_{Z\hat{\Omega}} X\beta + 2\beta' X' P_{Z\hat{\Omega}} P_I \epsilon + \epsilon' P_I \epsilon. \end{aligned} \tag{4.1}$$

The total sum of squares is partitioned into sum of squares due to regression and due to errors just like the case of the multiple linear regression model. So comparing (4.1) with the total sum of squares in multiple linear regression model, it can be considered that the first two terms in (4.1), viz., $\beta' X' P_{Z\hat{\Omega}} P_I P_{Z\hat{\Omega}} X\beta$ and $2\beta' X' P_{Z\hat{\Omega}} P_I \epsilon$ jointly constitute the sum of squares due to regression and $\epsilon' P_I \epsilon$ is the sum of squares due to errors. Since β and ϵ are unknown in $\beta' X' P_{Z\hat{\Omega}} P_I P_{Z\hat{\Omega}} X\beta$ and $2\beta' X' P_{Z\hat{\Omega}} P_I$, so we replace them by IV estimates $\hat{\beta}_{IV}(\hat{\Omega}) \equiv \hat{\beta}_{IV} = (X' P_{Z\hat{\Omega}} X)^{-1} X' P_{Z\hat{\Omega}} y^*$ and the corresponding residuals $\hat{\epsilon} = y^* - P_{Z\hat{\Omega}} X \hat{\beta}_{IV}(\hat{\Omega})$, respectively. A statistics measuring the goodness of fit in IV models can then be constructed as the proportion of sum of squares due to regression and total sum of squares as,

$$G_{IV}^2(\hat{\Omega}) = \frac{\hat{\beta}_{IV}' X' P_{Z\hat{\Omega}} P_I P_{Z\hat{\Omega}} X \hat{\beta}_{IV} + 2\hat{\beta}_{IV}' X' P_{Z\hat{\Omega}} P_I \hat{\epsilon}}{y^{*'} P_I y^*}, \quad 0 \leq G_{IV}^2(\hat{\Omega}) \leq 1 \tag{4.2}$$

The statistic (4.2) can be used to measure the goodness of fit in the linear regression model obtained through IV estimation. It can be used to measure the goodness of fit in the IV model with nonspherical random errors and unknown covariance matrix. It is termed as Goodness of Instrumental Variable Estimates (GIVE) statistic in IV models. Note that the presence of the term $2\beta' X' P_{Z\hat{\Omega}} P_I$ in (4.1) and $2\hat{\beta}_{IV}' X' P_{Z\hat{\Omega}} P_I \hat{\epsilon} y^{*\prime} P_I y^*$ in (4.2) makes it different compared to using R_n^2 to judge the goodness of fit in IV models.

Let $\theta_{IV}(\Omega)$ be the squared population multiple correlation coefficient between the study and explanatory variables in the IV model and is the counterpart of the squared population multiple correlation coefficient (3.1),

$$\theta_{IV}(\Omega) = \frac{\beta' \Sigma_{X\Omega X}^Z \beta}{\beta' \Sigma_{X\Omega X}^Z \beta + \sigma^2}, \quad 0 \leq \theta_{IV}(\Omega) \leq 1 \tag{4.3}$$

where, $\Sigma_{X\Omega X}^{Z} = X'P_{Z\Omega}P_I P_{Z\Omega}X$ and $P_{Z\Omega} = Z(Z'\Omega Z)^{-1}Z'$. When the model is best fitted, then $\sigma^2 = 0$, and consequently $\theta_{IV}(\Omega) = 1$. Similarly when the model is worst fitted, then all the β's will be ideally zero (or close to zero) indicating that none of the explanatory variables are important for contribution in the model, and consequently $\theta_{IV}(\Omega) = 0$. Any other value of $\theta_{IV}(\Omega)$ lying between 0 and 1 will indicate the goodness of the fitted model as measured by the squared multiple correlation coefficient.

Consider,

$$\frac{X'P_{Z\Omega}u}{n} = \left(\frac{X'Z}{n}\right).\left(\frac{Z'\Omega Z}{n}\right)^{-1}.\left(\frac{Z'u}{n}\right) \xrightarrow{p} 0 \tag{4.4}$$

$$\begin{aligned}\frac{X'P_{Z\Omega}X}{n} &= \left(\frac{X'Z}{n}\right).\left(\frac{Z'\Omega Z}{n}\right)^{-1}.\left(\frac{Z'X}{n}\right)\\ &\xrightarrow{p} \Sigma_{XZ}\Sigma_{Z\Omega Z}^{-1}\Sigma_{ZX} = \Sigma_{XX}^{\Omega} > 0.\end{aligned} \tag{4.5}$$

Since $\hat{\Omega}$ is a consistent estimator of Ω, under (4.4) and (4.5), we observe that,

$$\begin{aligned}plim\frac{Z'P_{Z\hat{\Omega}}X}{n} &= plim\left(\frac{Z'Z}{n}\right).plim\left(\frac{Z'\hat{\Omega}Z}{n}\right)^{-1}.plim\left(\frac{Z'X}{n}\right)\\ &= \Sigma_{ZZ}\Sigma_{Z\hat{\Omega}Z}\Sigma_{ZX} = \Sigma_{ZX}^{\hat{\Omega}} > 0\end{aligned} \tag{4.6}$$

$$\begin{aligned}plim\frac{Z'P_{Z\hat{\Omega}}Z}{n} &= plim\left(\frac{Z'Z}{n}\right).plim\left(\frac{Z'\hat{\Omega}Z}{n}\right)^{-1}.plim\left(\frac{Z'Z}{n}\right)\\ &= = \Sigma_{ZZ}\Sigma_{Z\hat{\Omega}Z}^{-1}\Sigma_{ZZ} = \Sigma_{ZZ}^{\hat{\Omega}} > 0.\end{aligned} \tag{4.7}$$

Hence, $G_{IV}^2(\hat{\Omega}) \xrightarrow{p} \theta_{IV}(\Omega)$ as $n \to \infty$, i.e., $G_{IV}^2(\hat{\Omega})$ is a consistent estimator of $\theta_{IV}(\Omega)$. Here $\theta_{IV}(\Omega)$ measures the goodness of fit of the model, therefore, an estimate of $\theta_{IV}(\Omega)$ is also expected to have the same interpretation in the fitted model obtained through IV estimation. When $\sigma^2 = 0$, then $G_{IV}^2(\hat{\Omega}) = 1$ indicating that the model is best fitted. On the other hand, if all estimated regression coefficients are close to zero or say, exactly zero then this indicates that the corresponding explanatory variables are not significant meaning thereby that the model is worst fitted, then $G_{IV}^2(\hat{\Omega}) = 0$. Any other value of $G_{IV}^2(\hat{\Omega})$ lying between 0 and 1 can be considered as measuring the degree of goodness of fit of model using instrumental variables for the given explanatory variables and sample size. For example, if $G_{IV}^2(\hat{\Omega}) = 0.95$, then it would mean that 95% of the variation in the values of the study variable are explained by the fitted IV model. Alternatively, in simple language, the fitted IV model is approximately 95% good.

5. Test based on GIVE statistic

In order to carry out any hypothesis testing of a problem based on the GIVE statistic, one needs to know the distribution of the GIVE statistic. In this context, it is indeed true that deriving the exact distributions of those $G_{IV}^2(.)$ is intractable in most of the cases. Even if it is possible to derive them in a few cases, the expressions may be so complicated that they may not shed any light on their behaviour. Therefore, we use the asymptotic distributions of $G_{IV}^2(.)$ in this section. The results for the following three cases are considered here: (i) when the random errors are identically and independently distributed and their covariance matrix is of the form $\sigma^2 I_n$, (ii) when the random errors are not identically and/or independently distributed and their covariance matrix is an off-diagonal known matrix of the form Ω, and (iii) when Ω is unknown. The results for case (iii) are more general and to concise the presentation, we state the result for (iii), and derive the results of (i) and (ii) directly from the results of (iii). The asymptotic distribution of $G_{IV}^2(\hat{\Omega})$ after appropriate normalization is stated in Theorem 1. The other two cases, i.e., (i) and (ii) are discussed in the remark followed by Theorem 1.

To derive the the asymptotic distribution of the GIVE statistic, we need to assume that the following conditions:
(A1) The parameter space of β is compact.
(A2) X is a bounded random variable almost surely.
(A3) Z is a bounded random variable almost surely.
(A4) The random variables Z and u are independent.
(A5) The random variables Z and ϵ are independent.
(A6) The correlation between ϵ and u does not equal zero.
(A7) Let $\hat{\Omega} = ((\hat{\sigma}_{i,j}))$, where $1 \leq i, j \leq p$, and $\hat{\sigma}_{i,j} > 0$ with probability one for all i, j, and $\hat{\sigma}_{i,j} - \sigma_{i,j} = O_p\left(\frac{1}{\sqrt{n}}\right)$ for all i and j.

Let $a = (a_1, a_2, \ldots, a_d) \in \mathbb{R}^d$ be an arbitrary d-dimensional vector,

$$\begin{aligned} V &= (X^T P_{Z\Omega} X)^{-1}(X^T P_{Z\Omega}\Omega P_{Z\Omega} X)((X^T P_{Z\Omega} X)^T)^{-1}, \\ g(a) &= \frac{a^T X^T P_{Z\Omega}(2(Xa+\epsilon) - P_{Z\Omega}Xa)}{(P_{Z\Omega}Xa+u)^T P_I (P_{Z\Omega}Xa+u)} \quad \text{and} \\ \nabla g(a) &= \left(\frac{\partial g(a)}{\partial a_1}, \frac{\partial g(a)}{\partial a_2}, \ldots, \frac{\partial g(a)}{\partial a_d}\right). \end{aligned}$$

Theorem 1 *Under conditions (A1)–(A7), $\sqrt{n}\{G^2_{IV}(\hat{\Omega}) - \theta_{IV}(\Omega)\}$ converges weakly to a normal distribution with mean 0 and variance $\{\nabla g(a)\}^T V \{\nabla g(a)\}|_{a=\beta_{IV}(\Omega)}$.*

Remark: The proof of Theorem 1 is provided in the Appendix for the sake of completeness and better reading, although it is available in Dhar and Shalabh (2021). Note that for a known Ω, i.e., case (ii), the result remains the same as long as $\hat{\Omega}$ is a consistent estimator of Ω. Case (i) is a special case of case (ii), and hence, the asymptotic normality of $\sqrt{n}(G^2_{IV}(I) - \theta_{IV}(I))$ will directly form the assertion in Theorem 1 by replacing Ω by I. The readers may refer to Lemma 3 and Corollary 1 in Section 7 for the precise statements of the results for cases (i) and (ii).

After having the asymptotic distribution of the GIVE statistic, we now want to formulate a hypothesis testing problem related to squared multiple correlation coefficient $\theta_{IV}(\Omega)$ based on $G^2_{IV}(\hat{\Omega})$. Suppose that we want to test,

$$H_0 : \theta_{IV}(\Omega) = \theta_0$$

against

$$H_1 : \theta_{IV}(\Omega) > \theta_0,$$

where, $\theta_0 \in [0, 1]$ is known and specified. In order to test H_0 against H_1, let us consider the test statistic $T_n = \sqrt{n}(G^2_{IV}(\hat{\Omega}) - \theta_0)$. The following theorem describes the large sample property of the test based on T_n.

Theorem 2 *Let c_α be the $(1-\alpha)$-th quantile of a normal distribution with mean 0 and variance $\{\nabla g(a)\}^T V \{\nabla g(a)\}|_{a=\beta_{IV}(\Omega)}$, where $\alpha \in (0, 1)$. Then under conditions (A1)–(A7), $P_{H_0}[T_n > c_\alpha] \to \alpha$ as $n \to \infty$, and $P_{H_1}[T_n > c_\alpha] \to 1$ as $n \to \infty$.*

Theorem 2 asserts that the test based on T_n can achieve the test level, α as $n \to \infty$, and moreover, the power of the test will converge to one as $n \to \infty$, i.e., the the test based on T_n is a consistent test.

Proof of Theorem 2: First note that $P_{H_0}[T_n > c_\alpha] \to \alpha$ as $n \to \infty$, which directly follows from the assertion of Theorem 1 as $\theta_{IV}(\Omega) = \theta_0$ under H_0.

Next, consider under $H_1 : \theta_{IV}(\Omega) = \theta_1$, where $\theta_1 > \theta_0$. Hence,

$$\begin{aligned}
\lim_{n\to\infty} P_{H_1}[T_n > c_\alpha] &= \lim_{n\to\infty} P_{H_1}[\sqrt{n}(G^2_{IV}(\hat{\Omega}) - \theta_0) > c_\alpha] \\
&= \lim_{n\to\infty} P_{H_1}[\sqrt{n}(G^2_{IV}(\hat{\Omega}) - \theta_1 + \theta_1 - \theta_0) > c_\alpha] \\
&= \lim_{n\to\infty} P_{H_1}[\sqrt{n}(G^2_{IV}(\hat{\Omega}) - \theta_1) + \sqrt{n}(\theta_1 - \theta_0) > c_\alpha] \\
&= \lim_{n\to\infty} P_{H_1}[\sqrt{n}(G^2_{IV}(\hat{\Omega}) - \theta_1) > c_\alpha - \sqrt{n}(\theta_1 - \theta_0)] \\
&= P[Z > -\infty] = 1,
\end{aligned}$$

where Z is a random variable associated with a normal distribution with mean 0 and variance $\{\nabla g(a)\}^T V\{\nabla g(a)\}|_{a=\beta_{IV}(\Omega)}$, and the last fact follows in view of $\sqrt{n}(\theta_1 - \theta_0) \to \infty$ as $n \to \infty$. This completes the proof. □

Remark: Note that the similar testing of hypotheses problems like $H_0^* : \theta_{IV}(\Omega) = \theta_0$ against $H_1^* : \theta_{IV}(\Omega) < \theta_0$ and $H_0^{**} : \theta_{IV}(\Omega) = \theta_0$ against $H_1^{**} : \theta_{IV}(\Omega) \neq \theta_0$ can also be resolved based on the GIVE statistic $G^2_{IV}(\Omega)$. In order to test H_0^* against H_1^*, one can use the test statistic $T_n^* = \sqrt{n}(\theta_0 - G^2_{IV}(\hat{\Omega}))$, and for testing H_0^{**} against H_1^{**}, one may use the test statistic as $T_n^{**} = \sqrt{n}|G^2_{IV}(\hat{\Omega}) - \theta_0|$. Here it should be mentioned that the tests based on T_n^* and T_n^{**} will also be consistent.

6. Conclusions

We have addressed the issue of hypothesis testing of the squared multiple correlation coefficient based on the IV estimation. A statistic, viz., GIVE statistic is formulated to measure the goodness of fit in the IV models. It is very difficult to obtain the exact distribution of test statistics. So using the asymptotic distribution of the GIVE statistic, a test statistic for a consistent hypothesis test is derived. The test statistic is quite general and it can be used when the covariance matrix of the random errors is a diagonal matrix or any known or unknown positive definite matrix.

7. Appendix: Some technicalities

We state some results here which are used in the results in the earlier sections.

First we show that the IV estimator is a consistent estimator of regression coefficients.

Lemma 1 *$\hat{\beta}_{IV} \xrightarrow{p} \beta$ as $n \to \infty$, where $\hat{\beta}_{IV}$ is the same as defined in* (2.9).

Proof of Lemma 1: Note that,

$$\begin{aligned}
(\hat{\beta}_{IV} - \beta) &= (X'P_{Z\hat{\Omega}}X)^{-1}X'P_{Z\hat{\Omega}}(P_{Z\hat{\Omega}}X\beta + \epsilon) - \beta \\
&= (X'P_{Z\hat{\Omega}}X)^{-1}X'P_{Z\hat{\Omega}}\epsilon. \qquad (7.1)
\end{aligned}$$

Using (7.1), we have,

$$\begin{aligned}
(\hat{\beta}_{IV} - \beta) &= \left(\frac{X'P_{Z\hat{\Omega}}X}{n}\right)^{-1}\left(\frac{X'P_{Z\hat{\Omega}}\epsilon}{n}\right) \\
&\xrightarrow{p} (\Sigma^{\Omega}_{XX})^{-1}.0 \qquad (7.2) \\
&= 0.
\end{aligned}$$

It completes the proof. □

Now we show that the GIVE statistics is a consistent estimator of the squared population multiple correlation coefficient between the study and explanatory variables.

Lemma 2 $G^2_{IV}(\hat{\Omega}) \xrightarrow{p} \theta_{IV}(\Omega)$ *as* $n \to \infty$, *where* $G^2_{IV}(\hat{\Omega})$ *and* $\theta_{IV}(\Omega)$ *are the same as defined in* (4.2) *and* (4.3), *respectively.*

Proof of Lemma 2: We first note that,

$$\begin{aligned} \beta' X' P_{Z\hat{\Omega}} P_I P_{Z\hat{\Omega}} X\beta &= \beta' \Sigma_{XZ\hat{\Omega}} \Sigma^{-1}_{ZZ\hat{\Omega}} \Sigma_{ZX\hat{\Omega}} \beta \\ &\xrightarrow{p} \beta' \Sigma^{Z}_{X\Omega X} \beta. \end{aligned}$$

Next, consider,

$$\begin{aligned} \left[\hat{\beta}'_{IV} X' P_{Z\hat{\Omega}} P_I \hat{\epsilon}_{\hat{\Omega}}\right] &= \left[\hat{\beta}'_{IV} X' P_{Z\hat{\Omega}} P_I (y^* - P_{Z\hat{\Omega}} X \hat{\beta}_{IV})\right] \\ &= \left[\hat{\beta}'_{IV} X' P_{Z\hat{\Omega}} P_I (2P_{Z\hat{\Omega}} X\beta + \epsilon - P_{Z\hat{\Omega}} X \hat{\beta}_{IV})\right] \\ &= \left[\hat{\beta}'_{IV} X' P_{Z\hat{\Omega}} P_I P_{Z\hat{\Omega}} X\beta\right] + \left[\hat{\beta}'_{IV} X' P_{Z\hat{\Omega}} P_I \epsilon\right] \\ &\quad - \left[\hat{\beta}'_{IV} X' P_{Z\hat{\Omega}} P_I P_{Z\hat{\Omega}} X \hat{\beta}_{IV}\right] \\ &\xrightarrow{p} \beta' \Sigma_{X\Omega Z} \Sigma^{-1}_{Z\Omega Z} \Sigma_{Z\Omega Z} \Sigma^{-1}_{Z\Omega Z} \Sigma_{Z\Omega X} \beta + 0 - \beta' \Sigma_{X\Omega Z} \Sigma^{-1}_{Z\Omega Z} \Sigma_{Z\Omega X} \beta \\ &= 0, \end{aligned}$$

and

$$\begin{aligned} (y^{*\prime} P_I y^*) &= (P_{Z\hat{\Omega}} X\beta + \epsilon)' P_I (P_{Z\hat{\Omega}} X\beta + \epsilon) \\ &= (\beta' X' P_{Z\hat{\Omega}} P_I P_{Z\hat{\Omega}} X\beta + 2\beta' X' P_{Z\hat{\Omega}} P_I \epsilon + \epsilon' P_I \epsilon) \\ &\xrightarrow{p} \beta' \Sigma_{XZ\Omega} \Sigma^{-1}_{ZZ\Omega} \Sigma_{ZX\Omega} \beta + 0 + \sigma^2 \\ &= \beta' \Sigma^{Z}_{X\Omega X} \beta + \sigma^2. \end{aligned}$$

Thus, we have,

$$\begin{aligned} G^2_{IV}(\hat{\Omega}) &= \frac{\left(\hat{\beta}'_{IV} X' P_{Z\hat{\Omega}} P_I P_{Z\hat{\Omega}} X \hat{\beta}_{IV} + 2\hat{\beta}'_{IV} X' P_{Z\hat{\Omega}} P_I \hat{\epsilon}_{\hat{\Omega}}\right)}{(y^{*\prime} P_I y^*)} \\ &\xrightarrow{p} \frac{\beta' \Sigma^{Z}_{X\Omega X} \beta}{\beta' \Sigma^{Z}_{X\Omega X} \beta + \sigma^2} \\ &= \theta_{IV}(\Omega), \;\; 0 \le \theta_{IV}(\Omega) \le 1. \end{aligned}$$

It completes the proof in view of the fact that the first three results of the proof, and $\hat{\Omega}$ is a consistent estimator of Ω. □

In order to prove Theorem 1, one needs to prove the following lemma.

Lemma 3 *Under conditions (A1)-(A6),* $\sqrt{n}\{G^2_{IV}(\Omega) - \theta_{IV}(\Omega)\}$ *converges weakly to a normal distribution with mean* 0 *and variance* $\{\nabla g(a)\}^T V \{\nabla g(a)\}|_{a=\beta_{IV}(\Omega)}$.

Proof of Lemma 3: To prove this lemma, we first note that,

$$\begin{aligned} G^2_{IV}(\Omega) &= \frac{\hat{\beta}_{IV}(\Omega)' X' P_{Z\Omega} (2y^* - P_{Z\Omega} X \hat{\beta}_{IV}(\Omega))}{y^{*\prime} P_I y^*} \\ &= \frac{\hat{\beta}_{IV}(\Omega)' X' P_{Z\Omega} (2y^* - P_{Z\Omega} X \hat{\beta}_{IV}(\Omega))}{(P_{Z\Omega} X \hat{\beta}_{IV}(\Omega) + \epsilon)' P_I (P_{Z\Omega} X \hat{\beta}_{IV}(\Omega) + \epsilon)}, \end{aligned}$$

and we have, $\theta_{IV}(\Omega) = \frac{\beta' \Sigma^{Z}_{X\Omega X} \beta}{\beta' \Sigma^{Z}_{X\Omega X} \beta + \sigma^2}$.

Hence, $G^2_{IV}(\Omega) = g(\hat{\beta}_{IV}(\Omega))$ and $\theta_{IV}(\Omega) = g(\beta_{IV}(\Omega))$, where for any $a = (a_1, a_2, \ldots, a_d) \in \mathbb{R}^d$ $(d \geq 1)$, $g(a) = \frac{a'X'P_{Z\Omega}(2(Xa+\epsilon)-P_{Z\Omega}Xa)}{(P_{Z\Omega}Xa+\epsilon)'P_I(P_{Z\Omega}Xa+\epsilon)}$, i.e., $g : \mathbb{R}^d \to \mathbb{R}$.

Now, Proposition 2.27 of van der Vaart (1998) implies that $\sqrt{n}(\hat{\beta}_{IV}(\Omega) - \beta)$ converges weakly to a normal distribution with mean 0 and variance V, where,

$$V = (X'P_{Z\Omega}X)^{-1}(X'P_{Z\Omega}\Omega P_{Z\Omega}X)((X'P_{Z\Omega}X)')^{-1}.$$

Therefore, by the delta method (see, e.g., van der Vaart (1998)), one can conclude that $\sqrt{n}(G^2_{IV}(\Omega) - \theta_{IV}(\Omega))$ converges weakly to a Gaussian distribution with mean 0 and variance $\{\nabla g(a)\}'V\{\nabla g(a)\}|_{a=\beta_{IV}(\Omega)}$, where, $\nabla g(a) = \left(\frac{\partial g(a)}{\partial a_1}, \frac{\partial g(a)}{\partial a_2}, \ldots, \frac{\partial g(a)}{\partial a_d}\right)$, which completes the proof. □

Proof of Theorem 1: Note that $G^2_{IV}(\hat{\Omega}) = \frac{\hat{\beta}_{IV}(\hat{\Omega})'X'P_{Z\hat{\Omega}}(2y^* - P_{Z\hat{\Omega}}X\hat{\beta}(\hat{\Omega}))}{y^{*\prime}P_I y^*}$ and $G^2_{IV}(\Omega) = \frac{\hat{\beta}_{IV}(\Omega)'X'P_{Z\Omega}(2y^* - P_{Z\Omega}X\hat{\beta}(\Omega))}{y^{*\prime}P_I y^*}$, hence,

$$\begin{aligned} & \sqrt{n}\{G^2_{IV}(\hat{\Omega}) - \theta_{IV}(\Omega)\} \\ = \; & \sqrt{n}\{G^2_{IV}(\hat{\Omega}) - G^2_{IV}(\Omega) + G^2_{IV}(\Omega) - \theta_{IV}(\Omega)\} \\ = \; & \sqrt{n}\{G^2_{IV}(\hat{\Omega}) - G^2_{IV}(\Omega)\} + \sqrt{n}\{G^2_{IV}(\Omega) - \theta_{IV}(\Omega)\} \end{aligned}$$

The assertion in Lemma 3 implies that $\sqrt{n}\{G^2_{IV}(\Omega) - \theta_{IV}(\Omega)\}$ converges weakly to a normal distribution with mean 0 and variance $\{\nabla g(a)\}^T V\{\nabla g(a)\}|_{a=\beta_{IV}(\Omega)}$. So, we now check if $\sqrt{n}\{G^2_{IV}(\hat{\Omega}) - G^2_{IV}(\Omega)\} \xrightarrow{p} 0$ as $n \to \infty$ or not. If this holds true, then the proof is completed.

Further, note that $\hat{\beta}_{IV}(\Omega) = (X'P_{Z\Omega}X)^{-1}X'P_{Z\Omega}y^*$, $\hat{\beta}_{IV}(\hat{\Omega}) = (X'P_{Z\hat{\Omega}}X)^{-1}X'P_{Z\hat{\Omega}}y^*$, and $\hat{\Omega}$ is a consistent estimator of Ω.

Using these expressions and aforementioned fact, we have,

$$\begin{aligned} & \sqrt{n}\{G^2_{IV}(\hat{\Omega}) - G^2_{IV}(\Omega)\} \\ = \; & \frac{\sqrt{n}}{y^{*\prime}P_I y^* + o_p(1)}[\hat{\beta}_{IV}(\Omega)'X'(P_{Z\hat{\omega}} - P_{Z\Omega})2y^* + \hat{\beta}_{IV}(\Omega)'X'P_{Z\Omega}2y^* \\ + \; & \{\hat{\beta}_{IV}(\hat{\Omega})\} - \hat{\beta}_{IV}(\Omega)X'\{P_{Z\hat{\Omega}} - P_{Z\Omega}\}2y^*]. \end{aligned}$$

For a detailed derivation, the readers may look at Dhar and Shalabh (2021).

Now, note that since the parameter space of β is compact (see (A1)), and $\hat{\sigma}_{i,j} - \sigma_{i,j} = O_p\left(\frac{1}{\sqrt{n}}\right)$ for all i and j (see (A7)), we have $\sqrt{n}(P_{Z\hat{\Omega}} - P_{Z\Omega}) = O_p\left(\frac{1}{\sqrt{n}}\right)$. Moreover, since $\{\hat{\beta}_{IV}(\hat{\Omega})\} - \hat{\beta}_{IV}(\Omega)\} = o_p(1)$, and X and Z are bounded random variables, we have,

$$\frac{\sqrt{n}[\hat{\beta}_{IV}(\Omega)'X'(P_{Z\hat{\omega}} - P_{Z\Omega})2y^* + \hat{\beta}_{IV}(\Omega)'X'P_{Z\Omega}2y^* + \{\hat{\beta}_{IV}(\hat{\Omega})\} - \hat{\beta}_{IV}(\Omega)\}X'\{P_{Z\hat{\Omega}} - P_{Z\Omega}\}2y^*]}{y^{*\prime}P_I y^* + o_p(1)} \xrightarrow{p} 0$$

as $n \to \infty$.

Thus the condition holds and along with the fact of Lemma 3, it completes the proof. □

Corollary 1: Under conditions (A1)-(A6), $\sqrt{n}\{G^2_{IV}(I) - \theta_{IV}(I)\}$ converges weakly to a normal distribution with mean 0 and variance $\{\nabla g(a)\}^T V\{\nabla g(a)\}|_{a=\beta_{IV}(\sigma^2 I_n)}$.

Proof of Corollary 1: The proof follows using the same arguments as the proof of Lemma 3. □

Acknowledgement

The authors are partially supported by their MATRICS grant of SERB (Files no: MTR/2019/000039 and MTR/2019/000033, respectively), Government of India.

References

Abarin Taraneh and Wang Liqun. (2012). Instrumental variable approach to covariate measurement error in generalized linear models. Annals Institute of Statistical Mathematics, 64(3): 475–493.

Acemoglu Daron, Simon Johnson and James A. Robinson. (2001). The colonial origins of comparative development: An empirical investigation. American Economic Review, 91(5): 1369–1401.

Amemiya Takeshi and MaCurdy Thomas E. (1986). Instrumental-variable estimation of an error-components model. Econometrica, 54(4): 869–880.

Amemiya Yasuo. (1990). Instrumental variable estimation of the nonlinear measurement error model in Statistical analysis of measurement error models and applications (Arcata, CA, 1989), 147–156, Contemporary Mathematics, 112, American Mathematical Society, Providence.

Anderson, T. W. (2003). Multivariate Analysis, John Wiley.

Angrist Joshua D. (1990). Lifetime earnings and the Vietnam era draft lottery: Evidence from social security administrative records. The American Economic Review, 80(3): 313–336.

Baiocchi Michael, Cheng Jing, Small Dylan, S. et al. (2014). Instrumental variable methods for causal inference. Statistics in Medicine, 33(13): 2297–2340.

Batistatou Evridiki and McNamee Roseanne. (2008). Instrumental variables vs. grouping approach for reducing bias due to measurement error. International Journal of Biostatistics, 4, Art., 8(24): 1–22.

Bowden Roger, L. and Turkington Darrell, A. (1984). Instrumental Variables, Cambridge University Press, Melbourne.

Card David. (1995). Using geographic variation in college proximity to estimate the return to schooling. *In*: Christodes, L. N., Grant, E. K. and Swidinsky, R. (eds.). Aspects of Labor Market Behaviour: Essays in Honour of John Vanderkamp. Toronto: University of Toronto Press.

Carroll Raymond, J., Ruppert David, Crainiceanu Ciprian, M., Tosteson Tor, D., Karagas Margaret, R. et al. (2004). Nonlinear and nonparametric regression and instrumental variables. Journal of American Statistical Association, 99(467): 736–750.

Cheng Shalabh, C. L. and Garg, G. (2014). Coefficient of determination for multiple measurement error models. Journal of Multivariate Analysis, 123: 137–152.

Cheng Shalabh, C. L. and Garg, G. (2016). Goodness of fit in restricted measurement error models. Journal of Multivariate Analysis, 145: 101–116.

Chernozhukov Victor, Hansen Christian and Jansson Michael. (2007). Inference approaches for instrumental variable quantile regression. Economic Letters, 95(2): 272–277.

Chesher Andrew and Rosen Adam, M. (2014). An instrumental variable random-coefficients model for binary outcomes. Econometric Journal, 17(2): S1–S19.

Chib Siddhartha and Greenberg Edward. (2007). Semiparametric modeling and estimation of instrumental variable models. Journal of Computational and Graphical Statistics, 16(1): 86–114.

Clarke Paul, S. and Windmeijer Frank. (2012). Instrumental variable estimators for binary outcomes. Journal of American Statistical Association, 107(500): 1638–1652.

Conley Timothy, G., Hansen Christian, B., McCulloch Robert, E. and Rossi Peter, E. (2008). A semi-parametric Bayesian approach to the instrumental variable problem. Journal of Econometrics, 144(1): 276–305.

Davies Neil, George Davey Smith, Frank Windmeijer and Richard M. Martina. (2013). COX-2 selective nonstereoidal anti-inflammatory drugs and risk of gastrointestinal tract complications and myocardial infarction: An instrumental variables analysis. Epidemiology, 24(3): 352–362.

Dhar, S. S. and Shalabh. (2021). Give Statistic for Goodness of Fit in Instrumental Variables Models with Application to COVID Data. Available at https://www.biorxiv.org/content/10.1101/2021.04.18.440376v1. (doi: https://doi.org/10.1101/2021.04.18.440376).

Eshima Nobuoki and Tabata Minoru. (2011). Three predictive power measures for generalized linear models: The entropy coefficient of determination, the entropy correlation coefficient and the regression correlation coefficient. Comput. Statist. Data Anal., 55(11): 3049–3058.

Eshima Nobuoki and Tabata Minoru. (2010). Entropy coefficient of determination for generalized linear models. Comput. Statist. Data Anal., 54(5): 1381–1389.

Fish Jason, S., Susan Ettner, Alfonso Ang and Arleen F. Brown. (2010). Association of perceived neighborhood safety on body mass index. American Journal of Public Health, 100(11): 2296–2303.

Gustafson Paul. (2007). Measurement error modelling with an approximate instrumental variable. Journal of Royal Statistical Society, 69(5): 797–815.

Horowitz Joel, L. (2011). Applied nonparametric instrumental variables estimation. Econometrica, 79(2): 347–394.

Horowitz Joel L. and Lee Sokbae. (2007). Nonparametric instrumental variables estimation of a quantile regression model. Econometrica, 75(4): 1191–1208.

Joel, L. and Lee Sokbae. (2007). Nonparametric instrumental variables estimation of a quantile regression model. Econometrica, 75(4): 1191–1208.

Hong, C. S., Ham, J. H. and Kim, H. I. (2005). Variable selection for logistic regression model using adjusted coefficients of determination. (Korean) Korean J. Appl. Statist., 18(2): 435–443.

Hössjer, Ola. (2008). On the coefficient of determination for mixed regression models. J. Statist. Plann. Inference., 138(10): 3022–3038.

Huang Li-Shan and Chen Jianwei. (2008). Analysis of variance, coefficient of determination and F-test for local polynomial regression. Ann. Statist., 36(5): 2085–2109.

Imbens Guido, W. (2014). Instrumental variables: An econometrician's perspective. Statistical Science, 29(3): 323–358.

Iwata Shigeru. (1992). Instrumental variables estimation in errors-in-variables models when instruments are correlated with errors. Journal of Econometrics, 53(1-3): 297–322.

Kim Daniel, Christopher F. Baum, Michael L. Ganz, Subramanian, S. V. and Ichiro Kawachi (2011). The contextual effects of social capital on health: A cross-national instrumental variable analysis. Social Science and Medicine, 73: 1689–1697.

Kleibergen Frank and Zivot Eric. (2003). Bayesian and classical approaches to instrumental variable regression. Journal of Econometrics, 114(1): 29–72.

Knight John, L. (1980). The coefficient of determination and simultaneous equation systems. Journal of Econometrics, 142: 265–270.

Kuersteiner Guido, M. (2001). Optimal instrumental variables estimation for ARMA models. Journal of Econometrics, 104(2): 359–405.

Leamer Edward, E. (1978). Least-squares versus instrumental variables estimation in a simple errors in variables model. Econometrica, 46(4): 961–968.

Liao, J. G. and McGee Dan. (2003). Adjusted coefficients of determination for logistic regression. Amer. Statist., 57(3): 161–165.

Lipsitz Stuart, R., Leong Traci, Ibrahim Joseph and Lipshultz Steven. (2001). A partial correlation coefficient and coefficient of determination for multivariate normal repeated measures data. The Statistician, 50(1): 87–95.

Lopes Hedibert, F. and Polson Nicholas, G. (2014). Bayesian instrumental variables: Priors and likelihoods. Econometric Reviews, 33(1-4): 100–121.

Mallios William, S. (1969). A generalized application of instrumental variable estimation to straight-line relations when both variables are subject to error. Technometrics, 11: 255–263.

Marchand Eric. (1997). On moments of beta mixtures, the noncentral beta distribution, and the coefficient of determination. J. Statist. Comput. Simulation, 59(2): 161–178.

Marchand Eric. (2001). Point estimation of the coefficient of determination. Statist. Decisions, 19(2): 137–154.

Nagelkerke, N. J. D. (1991). A note on a general definition of the coefficient of determination. Biometrika, 78(3): 691–692.

Newey Whitney, K. and Powell James, L. (2003). Instrumental variable estimation of nonparametric models. Econometrica, 71(5): 1565–1578.

Park Sangin. (2003). Semiparametric instrumental variables estimation. Journal of Econometrics, 112(2): 381–399.

Pokropek Artur. (2016). Introduction to instrumental variables and their application to large-scale assessment data. Large-scale Assessments in Education, volume 4, Article number: 4.

Rao C. Radhakrishna, Toutenburg Helge, Shalabh and Heumann Christian. (2008). Linear Models and Generalizations, Least Squares and Alternatives, 3rd edition, Springer, Berlin, Heidelberg.

Renaud Olivier, Victoria-Feser and Maria-Pia. 2010. A robust coefficient of determination for regression. J. Statist. Plann. Inference, 140(7): 1852–1862.

Sargan, J. D. (1958). The estimation of economic relationships using instrumental variables. Econometrica, 26: 393–415.

Sargan, J. D. and Mikhail, W. M. (1971). A general approximation to the distribution of instrumental variable estimates. Econometrica, 39: 131–169.

Srivastava Anil K. and Shobhit. (2002). A family of estimators for the coefficient of determination in linear regression models. J. Appl. Statist. Sci., 11(2): 133–144.

Söderström Torsten. (2011). A generalized instrumental variable estimation method for errors-in-variables identification problems. Automatica—A Journal of the International Federation of Automatic Control, 47(8): 1656–1666.

Tjur Tue. (2009). Coefficients of determination in logistic regression models-a new proposal: The coefficient of discrimination. Amer. Statist., 63(4): 366–372.

van der Linde Angelika and Tutz Gerhard. (2008). On association in regression: The coefficient of determination revisited. Statistics, 42(1): 1–24.

van der Vaart, A. W. (1998). Asymptotic Statistics, Cambridge Series in Statistical and Probabilistic Mathematics.

Wansbeek Tom and Meijer Erik. (2000). Measurement Error and Latent Variables in Econometrics, Elsevier Science, Amsterdam.

Wiesenfarth Manuel, Hisgen Carlos Matas, Kneib Thomas and Cadarso-Suarez Carmen. (2014). Bayesian nonparametric instrumental variables regression based on penalized splines and Dirichlet process mixtures. Journal of Business and Economic Statistics, 32(3): 468–482.

Zellner Arnold, Ando Tomohiro, Bastürk Nalan, Hoogerheide Lennart, van Dijk, Herman K. et al. (2014). Bayesian analysis of instrumental variable models: Acceptance rejection within direct Monte Carlo. Econometric Reviews, 33(1-4): 3–35.

Zhao Peixin and Xue Liugen. (2013). Instrumental variable-based empirical likelihood inferences for varying-coefficient models with error-prone covariates. Journal of Applied Statistics, 40(2): 380–396.

Chapter 22

Probability Distribution Analysis for Rainfall Scenarios

A Case Study

Md Abdul Khalek,[1,*] *Md Mostafizur Rahaman,*[1] *Md Kamruzzaman,*[2]
Md Mesbahul Alam[1] and *M Sayedur Rahman*[1]

1. Introduction

Bangladesh, like many other countries, is expected to observe variations in climatic variables over time. Estimating the amount of water available to suit the diverse needs of agriculture, industry, storm water design, and other human activities requires a thorough understanding of rainfall. However, knowing the nature or characteristics of these factors is critical for making it easier to find concealed data that could have substantial procedure ramifications in a country's short and long term. A country's climate is directly influenced by its geographical location and physical circumstances. Bangladesh is surrounded on the north by the Himalayas and on the south by the Bay of Bengal. Its latitude ranges from 20°34′N to 26°38′N and its longitude is from 88°01′E to 92°41′E. Our country receives an average of 2200 mm of rain per year, with yearly rainfall ranging from 339 mm to 5000 mm. In NW Bangladesh, the months of June to September often receive 80 to 85 percent of the total normal annual rainfall. An appropriate distribution for rainfall data is a task that involves selecting a probability distribution for model variables and estimating selected distribution parameters under the conditions of effective capability and effective decision, which usually necessitates defining quality-of-fit valuation.

According to statistical theory, the frequency of extreme rainfall occurrences is more dependent on changes in variability (generally, the scale parameter) than on the climate mean (Al Mamun et al., 2021; Alamgir et al., 2020; Katz and Brown, 1992). Rainfall probability distributions have been investigated by a number of scholars. The Gamma distribution was used to document the probability distribution of rainfall over monthly and annual periods (McCuen, 2016; Maity, 2018). The LN2 distribution is the best-fit probability distribution for India's annual maximum rainfall, according to Kumar (2000), Tabish et al. (2015), and Singh (2001). Amin et al. (2016) discovered that the log-Pearson type III (LP3) distribution was found to be most suitable in the northern Pakistan annual maximum rainfall data on a daily sample.

[1] Department of Statistics, Rajshahi University.
Emails: mostafiz_bd21@yahoo.com; mesbahru@gmail.com; rmsayedur@gmail.com
[2] Institute of Bangladesh Studies, Rajshahi University.
Email: mkzaman@ru.ac.bd
[*] Corresponding author: mak.stat09@gmail.com

According to Kwaku et al. (2007), the LN2 distribution was the best fit for one to five consecutive days of maximum rainfall in Accra, Ghana. According to Olofintoye et al. (2009), Al Mamoon and Rahman (2017), Alam et al. (2018), and Ogarekpe et al. (2020), 50 percent of stations in Nigeria follow LP3 distributions and 40 percent follow Pearson type III (P3) distributions for peak daily rainfall. The annual maximum rainfall distribution was used to study the annual maxima of daily rainfall in five locations in South Korea from 1961 to 2001, and it was discovered that the Gumbel distribution offered the most reasonable distribution in four of the five locations analyzed (Kamal et al., 2021; Mohsenipour et al., 2020; Nadarajah and Choi, 2007). Deka et al. (2009) used only five extreme worth distributions to determine the most suited probability to define the annual legacy of maximum rainfall for 42 years at nine remote locations in north-east India. The normal and gamma distributions were found to be the best-fitted probability distributions for annual rainfall data for fourteen Sudanese rainfall stations from 1971 to 2010. (Mahmud et al., 2020; Mahgoub et al., 2017).

Sharma and Singh (2010) used the least squares method to study the daily maximum rainfall data of Pantnagar, India, which was well-known, for a period of 37 years and the best-fitted probability distribution among the sixteen compared distributions. The log-Pearson type III distribution fitted for 50% of the total stations was used for the rainfall distribution characteristic on the Chinese plain, according to Lee (2005). The log-person type III distribution, according to Ogunlela (2001), best pronounces the stochastic investigation of peak daily rainfall. Choi et al. (2021) and Baskar et al. (2006) found that the gamma distribution was more appropriate than the other distributions in the frequency analysis of successive days peaked rainfall at Banswara, Rajasthan, India.

It is critical to assist planners in forming vital policy decisions by providing information on proper rainfall distribution in various locations of Bangladesh. Islam et al. (2021) and Rahman and Lateh (2016) investigated the assessment of Bangladesh's climatic attributes based on standard rainfall series and geographic data for the period 1971–2010. Hossian et al. (2016) fitted many forms of probability distributions for monthly maximum temperature in Dhaka, Bangladesh, using several types of distributions for climatic factors, with the skew logistic distribution proving to be the most acceptable. Ghosh et al. (2016) and Khudri and Sadia (2013) developed the generalized extreme value distribution, which provides the best-fitted distribution for monthly rainfall data in Chittagong, Rajshahi, Sylhet, and Dhaka.

There are primarily three objectives in this study: (1) describing the nature of rainfall at three different locations in northwest (NW) Bangladesh, namely Bogura, Ishurdi and Rajshahi, (2) fitting probability distributions of maximum monthly rainfall and (3) determining the best-fit distribution of seasonal and annual rainfall for each loçation. In this analysis, we will implement the following steps—data collection, exploratory analysis, detection of outliers, trimming in the particular fitting feasibility measurement . In addition to taking the statistics into account, and summarizing them, a number of descriptive methods are employed to set forth the nature of the statistics, or to fit probability distributions for maximum rainfall for altered sites, to examine the truthfulness of the best-fit distribution, or to assess the best-fit distribution based on the Akaike Information Criterion (AIC), or to use the Bayesian Information Criterion (BIC).

The study includes several sections, including this one. In the second section, we review the theory and methodology of different techniques related to fitting probability distributions. We have discussed the basic concepts of probability distributions in this section. For the fitted implemented potential probability distribution, various goodness-of-fit tests have also been estimated in this section. Finally, a brief summary is provided and some areas for further study are suggested.

2. Methods and materials

A common probability distribution which is also the best fit is applied in this section to analyze the rainfall data of a location.

2.1 Data source

This research is based on maximum annual, seasonal and monthly rainfall. The study area for this study is NW Bangladesh, which is located adjacent to Bangladesh with coordinates 20°34′N to 26°38′N latitude and

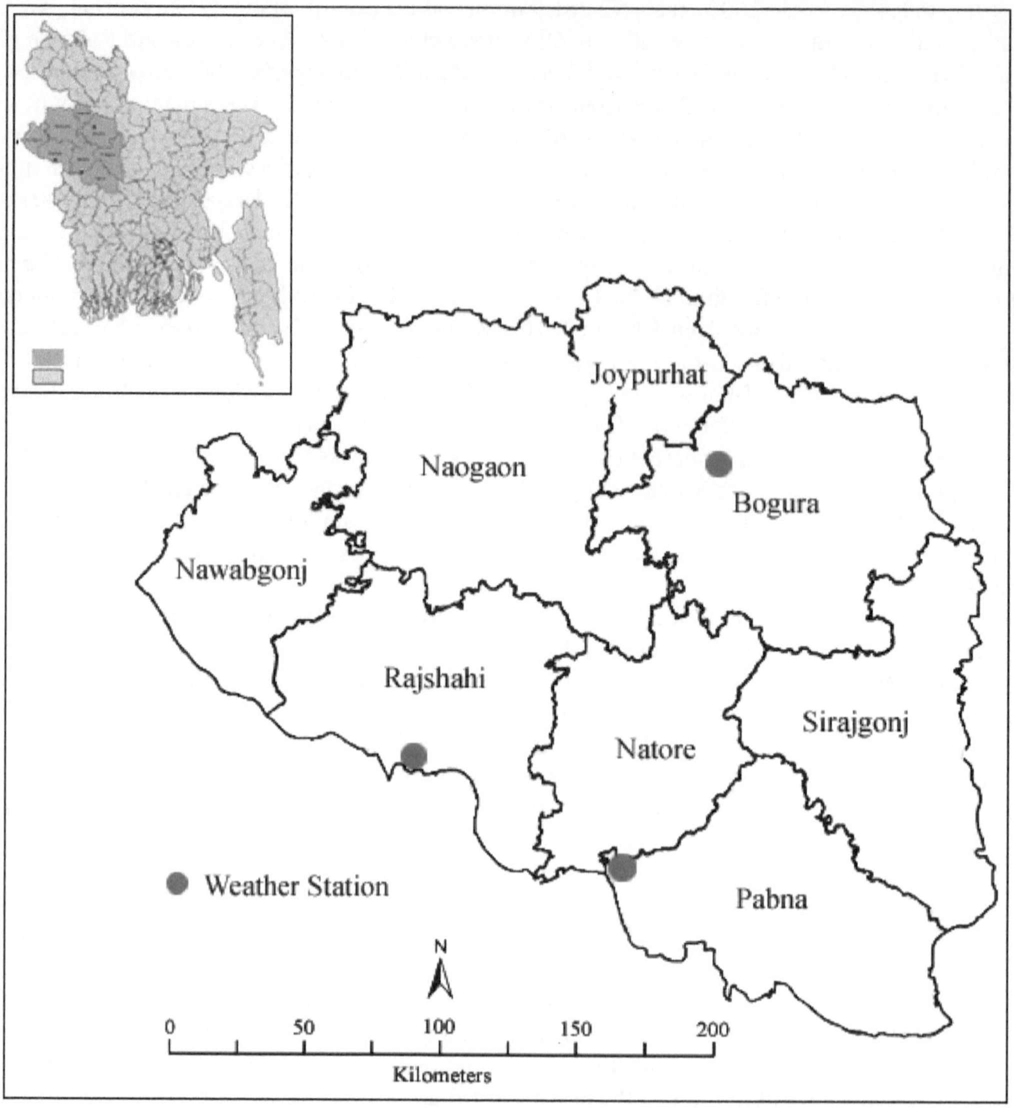

Figure 1: Location of rainfall stations used in this study; Source: Author.

88°01′E to 92°41′E longitude. The area consists of three meteorological stations (namely, Bogura, Ishurdi and Rajshahi) which were critically analyzed. These three stations were selected based on their geographical locations, which were expected to represent the characteristics of NW Bangladesh as a whole. The names and locations of rainfall stations are shown in Figure 1. For achieving the purpose of these investigations numerical analysis has been performed, based on secondary data. For reaching the objective, 55 years (1964–2018) of climatic indicator (monthly maximum rainfall) data for three selected stations was collected from the Bangladesh Meteorological Department.

2.2 Fitting probability distributions

In the article, we fit and compare the performance of maximum rainfall using six different distributions include 2-parameter log-normal (LN2), Exponential (Exp), Weibull (W2), Pearson type III (P3), log-Pearson type III (LP3) and Generalized Extreme Value (GEV) distributions. The advantages of these probability distributions make it easier and more popular to analyze the frequency of extreme events (Li et al., 2015). The parameters of the candidate distributions are estimated by the maximum likelihood method. In this study, the average weighted distance (AWD) technique is used, Kroll and Vogel (2002), for measuring the differences between sample and theoretical L-moment ratios. The AWD technique is defined by:

$$\text{AWD} = \frac{\sum_{i=1}^{N} n_i d_i}{\sum_{i=1}^{N} n_i} \quad \text{here, } d_i = \text{difference of two theoretical L-moments}$$

A distribution with the smallest AWD value will provides the best fit for the sample data.

Table 1: Probability distributions considered in the survey.

Distribution	Probability Distribution Function	Parameter
2-parameter Log-Normal (LN2)	$f(x) = \frac{1}{x\sigma_y\sqrt{2\pi}}\exp\left[-\frac{1}{2\sigma_y^2}(\ln(x)-\mu_y)^2\right]$ $y = \ln(x)$ $\mu_y = \ln(\mu_x) - \frac{\sigma_y^2}{2}, \sigma y = \left[\ln\left(1+\frac{\sigma_x^2}{\mu_x^2}\right)\right]^{\frac{1}{2}}$	μ: Location parameter σ: Scale parameter
Exponential (2-parameter)	$f(x) = \frac{1}{\lambda}\exp\left[-\left(\frac{x-\xi}{\lambda}\right)\right]$	ξ: Location parameter λ: Scale parameter
Weibull (W2)	$f(x) = \frac{\alpha}{\beta}\left(\frac{x}{\beta}\right)^{\alpha-1}\exp\left[-\left(\frac{x}{\beta}\right)^{\alpha}\right], x > 0$	β: Scale parameter α: Shape parameter
Pearson Type III (P3)	$f(x) = \frac{1}{\lvert\beta\rvert\Gamma(\alpha)}\left(\frac{x-\xi}{\beta}\right)^{\alpha-1}\exp\left[-\left(\frac{x-\xi}{\beta}\right)\right]$	ξ: Location parameter β: Scale parameter α: Shape parameter
Log-Pearson Type III (LP3)	$f(x) = \frac{1}{\lvert\beta\rvert x\Gamma(\alpha)}\left(\frac{\ln(x)-\xi}{\beta}\right)^{\alpha-1}\exp\left[-\left(\frac{\ln(x)-\xi}{\beta}\right)\right]$	ξ: Location parameter β: Scale parameter α: Shape parameter
Generalized Extreme Value (GEV)	$f(x) = \frac{1}{\alpha}\exp[-(1-\kappa)y - \exp(-y)]$ $y = -\frac{1}{\kappa}\log\left\{1-\frac{\kappa(x-\xi)}{\alpha}\right\}, \kappa \neq 0$	ξ: Location parameter α: Scale parameter κ: Shape parameter

2.3 *Goodness-of-fit test*

The goodness-of-fit test examines the validity of an estimated probability distribution model for the rainfall data. Graphical methods, numerical methods and formal normality tests are three techniques of checking the normality. The empirical distribution function (EDF) and the normality tests measure the discrepancy between the empirical and theoretical distributions (Dufour et al., 1998).

Kolmogorov-Smirnov (K-S) test, Anderson-Darling (A-D) and Cramer Von Mises (CvM) tests are the most popular EDF tests that were applied in this study (Arshad et al., 2003; Seier, 2002). Q-Q plot, a graphical test, and the root mean square error (RMSE) examined the best fitted model. AIC and BIC, and, the log-likelihood tests, were applied to compare the observed and estimated values (Table 2).

Table 2: Goodness-of-fit statistic.

Statistic	General Formula	Computational Formula
Anderson-Darling (AD)	$n\int_{-\infty}^{\infty}\frac{[F_n(x)-F(x)]^2}{F(x)[1-F(x)]}dx$	$-n-\frac{1}{n}\sum_{i=1}^{n}(2i-1)\log[F_i(1-F_{n+1-i})]$ where $F_i \cong F(x_i)$
Cramer-von Mise (CvM)	$n\int_{-\infty}^{\infty}[F_n(x)-F(x)]^2dx$	$\frac{1}{12n}+\sum_{i=1}^{n}\left(F_i-\frac{2i-1}{n}\right)^2$
Kolmogorov-Smirnov (KS)	$\sup\lvert F_n(x)-F(x)\rvert$	max (D^+, D^-) with $D^+=\max_{i=1,\ldots,n}\left(\frac{i}{n}-F_i\right)$ $D^-=\max_{i=1,\ldots,n}\left(F_i-\frac{i-1}{n}\right)$

3. Results and discussions

The present study determines the maximum monthly rainfall of each station by using the best fit probability distribution. These assumptions provide valuable direction in policy making and making proper judgments.

3.1 Data description

The simple statistical features of the monthly maximum rainfall for each station are considered. The coefficient of skewness is used to ensure an asymmetrical level of distribution around the average. The summary statistics mean, standard deviation (SD), skewness, kurtosis, coefficient of variation (CV), maximum and minimum values of monthly maximum rainfall are presented in Table 3 where, the annual mean of maximum rainfall is 395.312 mm. The seasonal means of maximum pre-monsoon , monsoon, post-monsoon and winter rainfall are 267.352 mm, 862.364 mm, 211.182 mm and 29.429 mm respectively. The maximum monthly rainfall in a year is 2013 mm and monthly maximum monthly rainfall in the monsoon season varies from 763 mm to 2013 mm. From the result, rainfall data for all seasons and stations showed positively skewed distributions.

The values of kurtosis in Bogura and Rajshahi stations are 1.299 and 1.281 respectively, which is less than 3 but in Ishurdi station it is 4.589. In pre-monsoon and winter seasons opposite results are observed as given in Table 3. Monthly rainfall data was skewed positively. It showed strong positive skewedness for Rajshahi station. It was leptokurtic in Bogura and Rajshahi stations and platykurtic in Ishurdi station. There is almost no rainfall, especially in the dry season, and in winter, and so the study area's data showed a platykurtic distribution. The maximum rainfall occurred in Ishurdi (1167 mm) and the minimum occurred in Bogura, Ishurdi and Rajshahi (0 mm) for the all seasons, but in the monsoon season, maximum rainfall occurred in Ishurdi (1167 mm) and minimum rainfall occurred in Bogura (84 mm) and Rajshahi (46 mm). The maximum value of CV (178.93%) indicated a large fluctuation in the rainfall data set in the winter season.

Three meteorological stations, Bogura, Ishurdi and Rajshahi were selected with monthly and annual maximum rainfall data through the period 1964 to 2018. These stations were selected judiciously based on long records of monthly rainfall data in locations as shown in (Figure 2 and Figure 3). As in Figure 2, an overall downward trends were observed for all the selected stations as well as NW Bangladesh.

From Figure 3, the pre-monsoon rainfall showed an upward trend. On the contrary, the monsoon and winter rainfall had downward trends, but a constant one in the post-monsoon season. This result revealed that climatic change is observed from station to station as well as according to seasonal variation. To adopt this situation, that is the change in climate, a change in crop calendar would be beneficial for the agriculture practitioners and policy makers of the region.

Table 3: Summary Statistics for rainfall of the three different locations in Bangladesh.

Stations	Min	Max	Mean	SD	CV	Skewness	Kurtosis
Overall							
Bogura	0	835	143.821	165.331	114.956	1.291	1.299
Ishurdi	0	2013	128.765	149.338	115.977	1.703	4.589
Rajshahi	0	763	122.726	140.675	114.625	1.300	1.281
Bangladesh	**0**	**2013**	**395.312**	**426.761**	**107.956**	**1.096**	**0.545**
Pre-monsoon							
Bogura	0	416	99.212	100.923	101.725	1.172	0.784
Ishurdi	0	470	93.758	91.465	97.554	1.401	2.089
Rajshahi	0	301	74.382	70.316	94.534	1.068	0.501
Bangladesh	**0**	**915**	**267.352**	**234.094**	**87.560**	**0.851**	**0.266**
Monsoon							
Bogura	84	835	313.945	150.667	47.991	0.969	0.617
Ishurdi	0	2013	275.586	145.765	52.893	1.737	6.279
Rajshahi	46	763	272.832	129.942	47.627	0.752	0.437
Bangladesh	**320**	**2013**	**862.364**	**351.371**	**40.745**	**0.792**	**0.091**
Post-monsoon							
Bogura	0	523	75.873	103.074	135.852	2.039	4.722
Ishurdi	0	664	68.555	93.756	136.761	3.189	15.673
Rajshahi	0	358	66.755	81.418	121.966	1.369	1.328
Bangladesh	**0**	**1462**	**211.182**	**258.054**	**122.195**	**1.884**	**4.714**
Winter							
Bogura	0	112	9.147	16.367	178.929	3.223	13.280
Ishurdi	0	81	10.067	17.069	169.548	2.213	4.678
Rajshahi	0	92	10.215	15.962	156.269	2.503	8.209
Bangladesh	**0**	**282**	**29.429**	**43.852**	**149.008**	**2.449**	**8.261**

According to Figure 4, the distribution of monthly maximum rainfall of all the stations for the period 1964–2018, showed that the maximum values are evident from June to September, while a dry period occurred from October to March, except that the highest maximum rainfall occurred in Ishurdi in June 1977, Rajshahi in July 1997 and September 2000, and in Bogura both in June 1973 and 1988.

From Table 4, it seems that the LN2, P3 and LP3 distributions all are possible candidates for representing pre-monsoon, monsoon, post-monsoon and winter rainfall; and LP3 are possible candidate distributions for annual maximum rainfall. It is challenging to identify the best distribution of the observations from the moment ratio alone. Table 4 presents AWD values and ranks of the fitted distributions.

The LP3 distribution is the best one for monsoon and winter rainfall with the P3 as a possible alternative because there is not much difference in the AWD value of the LP3 and P3 distributions. The LN2 distribution is the best for post-monsoon rainfall with the W2 as a potential alternative. The P3 distribution is the best for pre-monsoon rainfall with the GEV as a potential alternative. These results are largely consistent with the distribution that represents annual rainfall, which is expected because annual rainfall is the summation of rainfall over four seasons, and the sum type distribution of several variables following the same probability distribution will not change.

Table 4 presents the AWD values of 12 months and the ranks of the distributions. It indicates the P3 distribution as the best one to describe seasonal rainfall statistics from June to September. The LP3 distribution is best for October and December with P3 and GEV is the potential alternative. LN2 distribution is the best

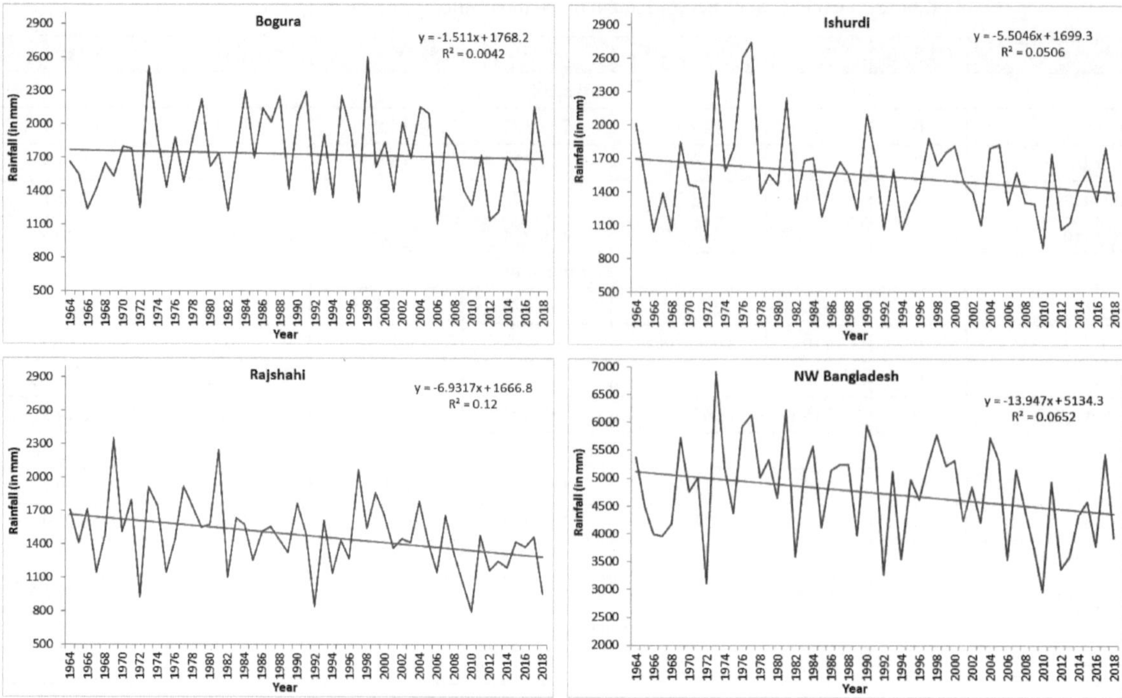

Figure 2: TS plot of yearly maximum rainfall (in mm) for location variation; Source: Author.

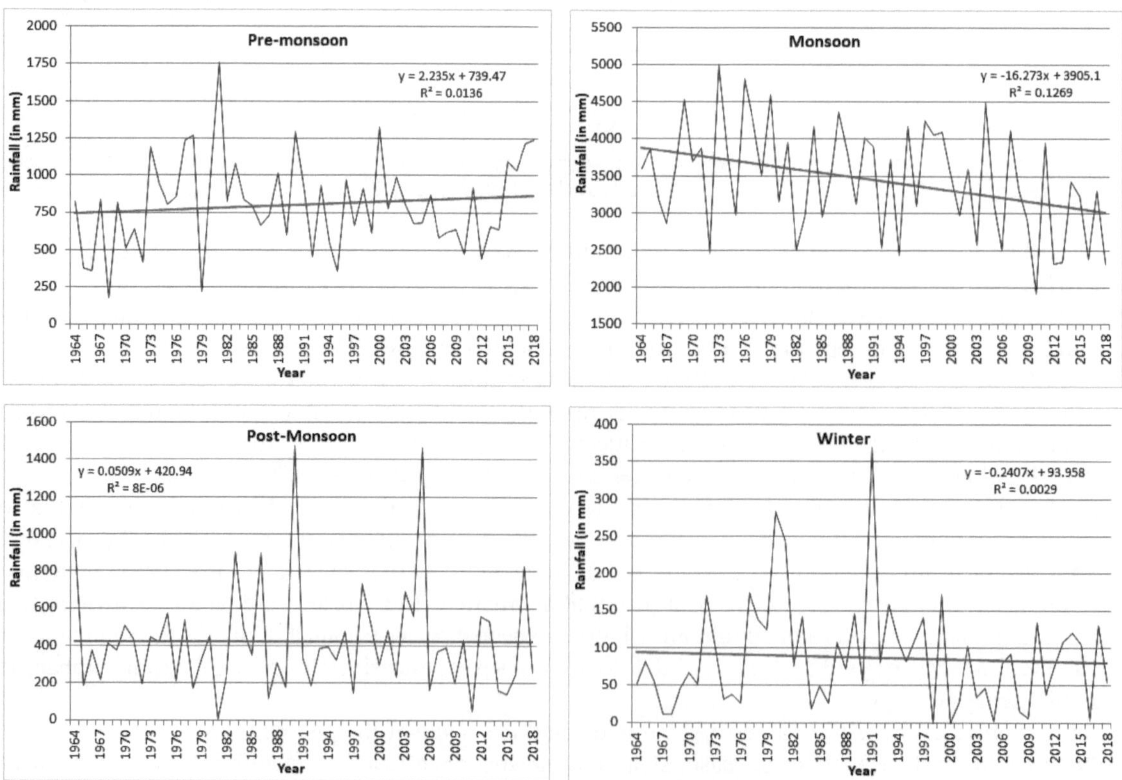

Figure 3: TS plot of yearly maximum rainfall (in mm) for seasonal variation; Source: Author.

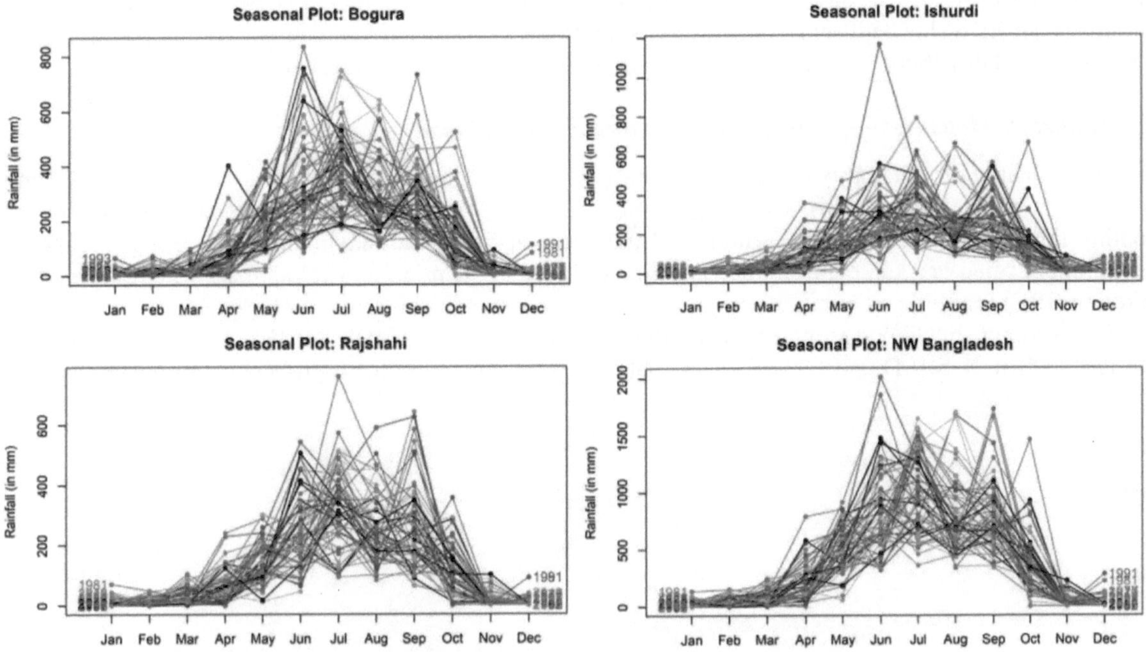

Figure 4: Distribution of monthly maximum rainfall for all stations; Source: Author.

Table 4: The AWD values of different probability distributions and their ranks.

Period	AWD values						Ranks					
	LN2	Exp	W2	P3	LP3	GEV	LN2	Exp	W2	P3	LP3	GEV
Year	**0.035**	0.086	0.175	0.035	**0.034**	0.048	**2**	5	6	3	**1**	4
Pre-monsoon	0.079	0.083	0.166	**0.026**	0.031	**0.028**	4	5	6	**1**	3	**2**
Monsoon	0.105	0.088	0.116	**0.029**	**0.029**	0.048	5	4	6	**2**	**1**	3
Post-monsoon	**0.028**	0.071	**0.029**	0.031	0.085	0.144	**1**	4	**2**	3	5	6
Winter	0.109	0.065	0.115	**0.043**	**0.041**	0.048	5	4	6	**2**	**1**	3
Jan	0.153	0.091	0.096	0.046	**0.044**	**0.038**	6	4	5	3	**2**	**1**
Feb	0.099	**0.035**	0.115	0.073	**0.038**	0.042	5	**1**	6	4	**2**	3
Mar	**0.029**	0.108	0.070	**0.031**	0.047	0.045	**1**	6	5	**2**	4	3
Apr	**0.036**	0.099	0.095	0.052	0.132	**0.040**	**1**	5	4	3	6	**2**
May	**0.031**	0.034	0.124	**0.033**	0.081	0.093	**1**	3	6	**2**	4	5
Jun	**0.030**	0.083	0.103	**0.025**	0.110	0.035	**2**	4	5	**1**	6	3
Jul	0.045	0.083	0.046	**0.033**	0.053	**0.037**	3	6	4	**1**	5	**2**
Aug	**0.045**	0.081	0.046	**0.041**	0.059	0.056	**2**	6	3	**1**	5	4
Sep	**0.035**	0.080	0.080	**0.034**	0.042	0.132	**2**	5	4	**1**	3	6
Oct	0.133	0.057	0.072	**0.038**	**0.027**	0.039	6	4	5	**2**	**1**	3
Nov	0.152	**0.024**	**0.023**	0.098	0.028	0.026	6	**2**	**1**	5	4	3
Dec	0.178	0.098	0.075	0.050	**0.041**	**0.044**	6	5	4	3	**1**	**2**

Shaded bold: Best-fit distribution; Bold regular: Alternative distribution for the months of March to May. The P3 and GEV distribution is a potential alternative for March to May rainfall. The AWD value delivers an objective method to select a distribution type in a region. Probability of monthly rainfall distribution types is mainly consistent with the type of annual and seasonal rainfall.

3.2 Accuracy Measures—goodness-of-fit statistic and criteria

The fitting quality of probability distributions is tested using the goodness-of-fit statistic. Anderson-Darling (AD), Cramer-von Mise (CvM), Kolmogorov-Smirnov (KS) and best-fitted probability distribution of monthly rainfall data, based on the criterion of goodness-of-fit for the maximum value of log likelihood, the minimum value of Akaike information criterion (AIC), and the minimum value of Bayesian Information Criterion (BIC) (Table 5) are used.

The maximum likelihood estimation method estimated the parameters of the fitted probability distributions. The test statistic K-S, AD and CvM for each data set were computed for 6 probability distributions. The

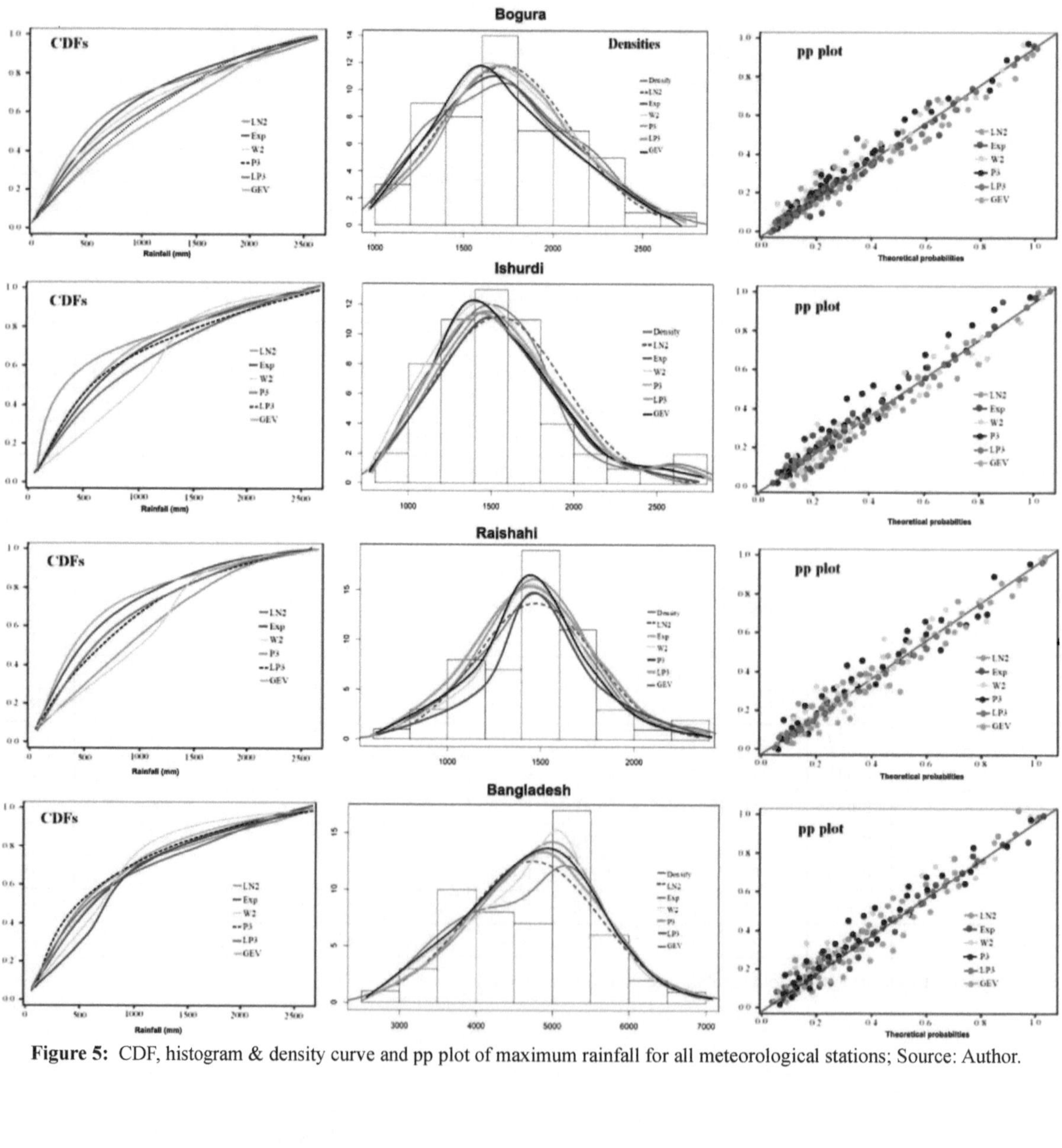

Figure 5: CDF, histogram & density curve and pp plot of maximum rainfall for all meteorological stations; Source: Author.

Table 5: Results of goodness-of-fit test statistic for different distributions of monthly maximum rainfall of three locations in NW Bangladesh.

SL	Station	Mean	SD	Skewness	Goodness-of-fit test results					
					K-S test		AD test		CvM test	
1	Bogura	143.82	165.33	1.291	LP3	(0.068)	LP3	(0.217)	LP3	(2.052)
2	Ishurdi	128.77	149.34	1.703	LN2	(0.072)	LP3	(0.212)	LP3	(1.107)
3	Rajshahi	122.73	140.67	1.300	GEV	(0.081)	GEV	(0.283)	LP3	(3.281)
4	Bangladesh	395.31	426.76	1.096	LP3	(0.172)	LP3	(0.147)	LP3	(3.002)

probability distribution having the first rank along with their test statistic is presented in Table 4. LP3 using K-S test, GEV using AD test and P3 and LP3 using CvM test obtained the first rank for maximum monthly rainfall. Thus based on these three tests, three probable distributions became the best independently.

Commonly, the three goodness-of-fit tests produced diverse categories of distribution. Finally, a ranking system including rank 1 as the best, rank 2 as the second best and so on selects the best-fit distribution for each station. The distribution type that contains the lowest sum of the three rankings is the best fit distribution for each station.

According to all goodness-of-fit tests, monthly maximum rainfall of the three locations in NW Bangladesh revealed that the LP3 distribution is the best fitted distribution for Bogura and the K-S test proposed the LN2, AD distributions and the CvM test proposed the LP3 distribution for Ishurdi. An overall finding of monthly maximum rainfall in NW Bangladesh is LP3, which is the best-fitted distribution according to all goodness-of-fit tests.

The rainfall of Bogura, Ishurdi and Rajshahi stations in NW Bangladesh provided the LP3 distribution and the rainfall of these stations best fitted the P3 and LN2 distributions. The findings will help in future planning as well as ensure welfare of country's population .

4. Summary and conclusion

We have studied a method to identify the best fit probability distribution for monthly maximum rainfall data in the selected stations of NW Bangladesh. The data shows that the maximum monthly rainfall at any given time ranged 0 mm (minimum) to 1167 mm (maximum), indicating a very large range of fluctuations during the monsoon season. The location wise minimum monsoon rainfall is less than ninety (mm) and maximum rainfall is greater than 300 (mm) for the period of 1964–2018. The maximum monsoon rainfall was 1167 (mm) for Ishurdi in 1977 during the 1964–2018 period. The six distributions are fitted for monthly rainfall of three locations and the parameters are estimated using the maximum likelihood method. The results of the rainfall study for identifying the best probability distribution have revealed that there is a variety of the best probability distributions for monthly maximum rainfall data sets in different places.

The LP3 and LN2 distributions were found to be the best fitted probability distribution models for the annual rainfall study. On the basis of goodness-of-fit criterion the P3 distribution is the best one for describing rainfall statistics of June to September. The LP3 distribution is best for October and December with P3 and GEV being the potential alternatives. The LN2 distribution is best for March to May. The P3 and GEV distribution is a potential alternative for March to May rainfall. The probability distribution of annual rainfall distribution types are mainly compatible with the distribution identified for annual and seasonal rainfall. The methodological results are clearly well-known and the analytical strategies formulated and established in this study may be accurately effective in documenting the best fitted probability distribution of climate parameters. These results show that climate change is observed from station to station as well as according to the change of seasons. To address this, climate change and crop calendar changes will be beneficial to agricultural practitioners and policy makers in the region, and it has been discovered that regional approaches will be more beneficial in allocating maximum rainfall in other parts of the country. We hope that this distribution may play an important role for sustainable development of agriculture practitioners in Bangladesh.

References

Al Mamun, A. and Rahman, A. (2017). Selection of the best fit probability distribution in rainfall frequency analysis for Qatar. *Natural Hazards*, 86(1): 281–296.

Al Mamun, M., Rahman, S. M., Uddin, M., Sultan-Ul-Islam, M. and Khairul, A. M. (2021). Rainfall and drought tendencies in Rajshahi division, Bangladesh. Geography, Environment, Sustainability, 14(1): 209–218.

Alam, M. A., Emura, K., Farnham, C. and Yuan, J. (2018). Best-fit probability distributions and return periods for maximum monthly rainfall in Bangladesh. Climate, 6(1): 9.

Alamgir, M., Khan, N., Shahid, S., Yaseen, Z. M., Dewan, A. et al. (2020). Evaluating severity–area–frequency (SAF) of seasonal droughts in Bangladesh under climate change scenarios. Stochastic environmental research and risk assessment, 34(2): 447–464.

Amin, M. T., Rizwan, M. and Alazba, A. A. (2016). A best-fit probability distribution for the estimation of rainfall in northern regions of Pakistan. Open Life Sciences, 11(1): 432–440.

Arshad, M., Rasool, M. T. and Ahmad, M. I. (2003). Anderson darling and modified anderson darling tests for. Pakistan Journal of Applied Sciences, 3(2): 85–88.

Choi, Y. W., Campbell, D. J., Aldridge, J. C. and Eltahir, E. A. (2021). Near-term regional climate change over Bangladesh. Climate Dynamics, 57(11): 3055–3073.

Deka, S., Borah, M. and Kakaty, S. C. (2009). Distributions of annual maximum rainfall series of north-east India. European Water, 27(28): 3–14.

Dufour, J. M., Farhat, A., Gardiol, L. and Khalaf, L. (1998). Simulation-based finite sample normality tests in linear regressions. The Econometrics Journal, 1(1): 154–173.

Ghosh, S., Roy, M. K. and Biswas, S. C. (2016). Determination of the best fit probability distribution for monthly rainfall data in Bangladesh. Am. J. Math. Stat, 6: 170–174.

Hossian, M. M., Abdulla, F. and Rahman, M. H. (2016). Selecting the probability distribution of monthly maximum temperature of Dhaka (capital city) in Bangladesh. Jahangirnagar University Journal of Statistical Studies, 33: 33–45.

Islam, A. R. M., Salam, R., Yeasmin, N., Kamruzzaman, M., Shahid, S. et al. (2021). Spatiotemporal distribution of drought and its possible associations with ENSO indices in Bangladesh. Arabian Journal of Geosciences, 14(23): 1–19.

Kamal, A. S. M., Hossain, F. and Shahid, S. (2021). Spatiotemporal changes in rainfall and droughts of Bangladesh for1. 5 and 2°C temperature rise scenarios of CMIP6 models. Theoretical and Applied Climatology, 146(1): 527–542.

Katz, R. W. and Brown, B. G. (1992). Extreme events in a changing climate: variability is more important than averages. Climatic Change, 21(3): 289–302.

Khudri, M. M. and Sadia, F. (2013). Determination of the best fit probability distribution for annual extreme precipitation in Bangladesh. European Journal of Scientific Research, 103(3): 391–404.

Kroll, C. N. and Vogel, R. M. (2002). Probability distribution of low streamflow series in the United States. Journal of Hydrologic Engineering, 7(2): 137–146.

Kumar, A. (2000). Prediction of annual maximum daily rainfall of Ranichauri (Tehri Garhwal) based on probability analysis. Indian Journal of Soil Conservation, 28(2): 178–180.

Kwaku, X. S. and Duke, O. (2007). Characterization and frequency analysis of one day annual maximum and two to five consecutive days maximum rainfall of Accra, Ghana. ARPN J. Eng. Appl. Sci., 2(5): 27–31.

Lee, C. (2005). Application of rainfall frequency analysis on studying rainfall distribution characteristics of Chia-Nan plain area in Southern Taiwan. Journal of Crop, Environment & Bioinformatics, 2: 31–38.

Li, Z., Li, Z., Zhao, W. and Wang, Y. (2015). Probability modeling of precipitation extremes over two river basins in northwest of China. Advances in Meteorology.

Mahgoub, T. and Abd Allah, A. (2017). Fitting probability distributions of annual rainfall in Sudan. Journal of Engineering and Computer Science (JECS), 17(2): 35–39.

Mahmud, S., Islam, M. A. and Hossain, S. S. (2020). Analysis of rainfall occurrence in consecutive days using Markov models with covariate dependence in selected regions of Bangladesh. Theoretical and Applied Climatology, 140(3): 1419–1434.

Maity, R. (2018). Statistical methods in hydrology and hydro climatology. Springer.

McCuen, R. H. (2016). Modeling hydrologic change: statistical methods. CRC press.

Mohsenipour, M., Shahid, S., Ziarh, G. F. and Yaseen, Z. M. (2020). Changes in monsoon rainfall distribution of Bangladesh using quantile regression model. Theoretical and Applied Climatology, 142(3): 1329–1342.

Nadarajah, S. and Choi, D. (2007). Maximum daily rainfall in South Korea. Journal of Earth System Science, 116(4): 311–320.

Ogarekpe, N. M., Tenebe, I. T., Emenike, P. C., Udodi, O. A., Antigha, R. E. et al. (2020). Assessment of regional best-fit probability density function of annual maximum rainfall using CFSR precipitation data. Journal of Earth System Science, 129(1): 1–18.

Ogunlela, A. O. (2001). Stochastic analysis of rainfall events in Ilorin, Nigeria. Journal of Agricultural research and development, 1(1): 39–50.

Olofintoye, O. O., Sule, B. F. and Salami, A. W. (2009). Best–fit Probability distribution model for peak daily rainfall of selected Cities in Nigeria. New York Science Journal, 2(3): 1–12.

Rahman, M. R. and Lateh, H. (2016). Meteorological drought in Bangladesh: assessing, analysing and hazard mapping using SPI, GIS and monthly rainfall data. Environmental Earth Sciences, 75(12): 1–20.

Seier, E. (2002). Comparison of tests for univariate normality. InterStat Statistical Journal, 1: 1–17.

Sharma, M. A. and Singh, J. B. (2010). Use of probability distribution in rainfall analysis. New York Science Journal, 3(9): 40–49.

Singh, R. K. (2001). Probability analysis for prediction of annual maximum rainfall of Eastern Himalaya (Sikkim mid hills). Ind. J. Soil Conserv, 29: 263–265.

Tabish, M. O. H. D., Ahsan, M. J. and Mishra, A. (2015). Probability analysis for prediction of annual maximum rainfall of one to seven consecutive days for Ambedkar Nagar Uttar Pradesh. IJASR, 6: 47–54.

Index

N

O

P

Q

R

S

T

U

V

W